T0318390

WHISKY AND OTHER SPIRITS

WHISKY AND OTHER SPIRITS

Technology, Production and Marketing

THIRD EDITION

Edited by

INGE RUSSELL

International Centre for Brewing and Distilling (ICBD), Heriot-Watt University, Edinburgh, United Kingdom

GRAHAM G. STEWART

International Centre for Brewing and Distilling (ICBD), Heriot-Watt University, Edinburgh, United Kingdom

JULIE KELLERSHOHN

Ted Rogers School of Hospitality and Tourism Management, Ryerson University, Toronto, ON, Canada

ELSEVIER

ACADEMIC PRESS
An imprint of Elsevier

Academic Press is an imprint of Elsevier
125 London Wall, London EC2Y 5AS, United Kingdom
525 B Street, Suite 1650, San Diego, CA 92101, United States
50 Hampshire Street, 5th Floor, Cambridge, MA 02139, United States
The Boulevard, Langford Lane, Kidlington, Oxford OX5 1GB, United Kingdom

Notices
Knowledge and best practice in this field are constantly changing. As new research and experience broaden our understanding, changes in research methods, professional practices, or medical treatment may become necessary.

Practitioners and researchers must always rely on their own experience and knowledge in evaluating and using any information, methods, compounds, or experiments described herein. In using such information or methods they should be mindful of their own safety and the safety of others, including parties for whom they have a professional responsibility.

To the fullest extent of the law, neither the Publisher nor the authors, contributors, or editors, assume any liability for any injury and/or damage to persons or property as a matter of products liability, negligence or otherwise, or from any use or operation of any methods, products, instructions, or ideas contained in the material herein.

Library of Congress Cataloging-in-Publication Data
A catalog record for this book is available from the Library of Congress

British Library Cataloguing-in-Publication Data
A catalogue record for this book is available from the British Library

ISBN 978-0-12-822076-4

For information on all Academic Press publications
visit our website at https://www.elsevier.com/books-and-journals

Publisher: Nikki P. Levy
Acquisitions Editor: Nancy Maragioglio
Editorial Project Manager: Lena Sparks
Production Project Manager: Bharatwaj Varatharajan
Cover Designer: Mark Rogers

Typeset by STRAIVE, India

Working together
to grow libraries in
developing countries

www.elsevier.com • www.bookaid.org

Contents

13. Contamination: Bacteria and wild yeasts in whisky fermentation

Nicholas R. Wilson

14. Batch distillation

Denis A. Nicol

15. Grain whisky distillation

Douglas Murray

16. Maturation

John Conner

17. Blending

Stephanie J. Macleod

About the authors

Donna Abdelrazik (Razik), MSc (Packaging Value Chain), is a design leader, educator, and technical packaging specialist with 20 years of experience in the CPG alcohol beverage sector and graphic arts industry. She has been teaching in the School of Graphic Communications Management at Ryerson University, Toronto, Canada, since 2017. Previously, a senior leader in creative services at Molson Coors, she led the development and commercialisation of packaging, print innovations, and graphic communications for the organisation's brand portfolios. She is an active member of PAC, a previous chair of the global packaging leadership executive committee, and a current judge in the International and Canadian packaging innovation category.

Anne Anstruther is an expert on the history of Scotland, in particular Edinburgh, and the evolution of beer and whisky in this geographical region. Her degrees are in computing science and management studies. Before retiring from a number of positions in this area, she worked as an administrator of the International Centre for Brewing and Distilling (ICBD) at Heriot-Watt University.

Irene Aylott currently partners Ross Aylott in Aylott Scientific consultancy. Irene has a degree in chemistry and a PhD in pharmacology from Edinburgh University and worked previously in the biotechnology contract research sector. Her role as Director of Quality Systems included ensuring compliance with Good Laboratory Practice and Good Manufacturing Practice Regulations, quality improvement, management of the audit team, and client liaison internationally.

Ross Aylott is Chartered Chemist, Fellow of the Royal Society of Chemistry, and former Chairman of the Analytical Division, Scottish Region. He formed Aylott Scientific consultancy in 2009. Ross joined the predecessor of Diageo in 1981, staying for the next 28 years. He specialised in process, quality, regulatory, and analytical aspects of distilled spirits. Scientific highlights include ethyl carbamate investigations, using science to combat counterfeit product, and technical liaison with trade associations and government. Anticounterfeit contributions include authenticity analyses for distilled spirits and their packaging, training of laboratories and enforcement agencies internationally, and being an expert witness in legal proceedings.

Kirsty Black is the Distillery Manager at Arbikie Highland Estate Distillery in Inverkeilor, Scotland. She joined Arbikie in 2014 and oversaw all aspects of its conversion from disused cattle shed to a multiaward-winning farm to bottle distillery. In conjunction with working at Arbikie, she sits on the board of examiners and the awards committee for the Institute of Brewing and Distilling. She holds a BSc in plant science from the University of Edinburgh, an MSc in brewing and distilling from Heriot-Watt University, and is studying to obtain a PhD from Abertay University and the James Hutton Institute, researching both the potential to use legumes in intercropping systems and the conversion of the resulting crops into beverage-grade alcohol. In 2018, she was honoured with the title 'Young Scientist of the Year—Scotland Award', from the Institute of Food Science and Technology.

Tom A. Bringhurst is Fellow of the Institute of Brewing and Distilling. He worked in the Scotch whisky industry for more than 42 years, gaining essential hands-on experience in a wide range of Scotch whisky malt and grain distilleries. Before his retirement in 2014, he was the Senior Scientist leading the raw materials research projects of the Scotch Whisky Research Institute (SWRI) and has been heavily involved with the Institute's research into raw materials, cereals processing, fermentation, and distillation, together with various supply chain collaborative projects.

James Brosnan is the Director of the Edinburgh-based Scotch Whisky Research Institute (SWRI), the research and technology organisation for the UK distilling industry. He entered the distilling industry in 1991 as Management Trainee and Process Development Scientist with Diageo. James joined SWRI in 1994, as Cereals Research Scientist, where his work focused on engaging with the supply chain and academia to improve the quality of barley and wheat for Scotch whisky production. In 2007, he was made Research Manager with overall responsibility for the SWRI research programme 'from barley to bottle', which provided plenty of scope to enjoy both the scientific challenges and precompetitive camaraderie of Scotch whisky production. James was appointed Director of SWRI in 2015 and has continued his interest in collaboration with the wider world of science to benefit the Scotch whisky industry. He currently sits on several industry and science committees, including the International Barley Hub, the AHDB Barley and Wheat Committees, and the UK Malting Barley Committee. James is Fellow of the Institute of Brewing and Distilling and is Honorary Professor at Heriot-Watt University. Prior to employment in the Scotch whisky industry, James was trained as a plant biochemist with a PhD in beetroot from the University

of York—a subject he notes was harder to engage with socially than whisky!

Jeng-Ing Chen is currently the distillery manager of the Laizhou Distillery of the Shanghai Bacchus Group in China. He started as a metallurgy engineer, and in 2014, he obtained his MSc in brewing and distilling from Heriot-Watt University, thereby starting his distilling career. He is an enthusiast of the spirit beverage industry and spent 10 years to visit more than 200 distilleries, breweries, wineries, cooperages, and maltsters around the world. He published *Gaze at Scotland: Alchemist and Malt Story*, which records his Scotch whisky distillery journey in 2012. He is the Chinese translator of the second edition of *Whisky: Technology, Production and Marketing*.

John Conner has more than 25 years of research experience in distilled beverage flavours. He is the Senior Scientist of the Maturation Research Group at the Scotch Whisky Research Institute (SWRI). He has in-depth expertise in all aspects of maturation, ranging from the origins of the flavours created using different cask types, to the role environmental conditions play on the evaporative loss of spirit. A consistent aim of his research has been an understanding of mature spirit quality and the relationships between the chemical composition and sensory properties of distilled beverages in general and whisky in particular.

Shinji Fukuyo has been in charge of Suntory's whisky brand development as the Chief Blender for Suntory since 2009. He joined Suntory in 1984 and was assigned to their Hakushu Malt Distillery in Japan and moved to their Blending Department in 1992. Between 1996 and 2002, he worked in Scotland at both the Heriot-Watt University International Centre for Brewing and Distilling and Morrison Bowmore Distillers Limited, a subsidiary of Suntory. He has also taken the position of the Whisky Quality Advisor to Beam Suntory since 2018.

Grant E. Gordon is a liveryman and member of the Worshipful Company of Distillers. He chairs the Distillers' Charity, which promotes alcohol education in the trade and with young people. He is a member of the management committee of the William Grant Foundation, which focuses on creating greater opportunities for people in Scotland. In his early career in the drinks industry, he was the Commercial Director of William Grant & Sons. His main focus currently is working as a philanthropist and social entrepreneur with a portfolio of UK-based charities. In 2021, he was awarded an OBE (Order of the British Empire Award) for services to philanthropy, including during the COVID-19 response.

Nami Goto is the President of the National Research Institute of Brewing (NRIB), which conducts advanced analysis, evaluation, and fundamental and applied research on the alcoholic beverages of Japan and disseminates the relevant information.

Barry M. Harrison is Senior Scientist at the Scotch Whisky Research Institute, where he leads the Raw Materials and Processing Research Group. His group is involved in projects relating to raw materials, cereals processing, fermentation, and distillation. He has a personal interest in flavour chemistry and understanding how to control flavour in the production process. Barry studied Biochemistry at Edinburgh University before carrying out a PhD on whisky flavour chemistry at Heriot-Watt University. His PhD project was based at SWRI, and he joined the SWRI staff upon the completion of his degree in 2007.

Frances Jack is Senior Scientist at the Scotch Whisky Research Institute (SWRI), where she specialises in research into whisky flavour. In addition to her research, she provides support and training to the whisky industry in these areas and works closely with companies in the development of their sensory functions. She has a background in food science and a PhD in sensory method development from the University of Strathclyde in Glasgow. She has published widely on sensory and whisky-related topics.

Julie Kellershohn is an assistant professor at the Ted Rogers School of Hospitality and Tourism Management at Ryerson University, Toronto, Canada. She holds an honours BSc from Queen's University, an MBA from Harvard Business School, and a PhD from Harper Adams University. Her research focuses on food and beverage consumer behaviour, beverage marketing, beverage innovation, and beverage evolution.

Tae Wan Kim has more than 19 years of experience both within the industry and as a researcher in a research institute that focuses on the R&D of alcoholic drinks, especially distilled spirits. He was initially trained as a chemical engineer at Hanyang University in Seoul, Korea, and he completed his MSc and PhD in brewing and distilling at Heriot-Watt University, Edinburgh, Scotland. He is currently Principal Research Scientist at the Korea Food Research Institute (KFRI).

James W. Larson is a chemical process engineer who has worked for more than 40 years in the fermentation industry, including the G. Heileman Brewing Company, Siebel Institute of Technology, and Alltech. He has expertise in the areas of brewing and brewery engineering, distillation, and fuel ethanol production. He has lectured globally in the area of plant sanitation and consulted on plant design and construction with an emphasis on sanitary design. He holds BS and MS degrees in chemical engineering and an MSc in brewing and distilling from Heriot-Watt University.

Natalia Lumby is an educator and researcher with a passion for packaging, design, and technology. Currently, an associate professor in the School of Graphic Communications Management (Ryerson University, Toronto, Canada), she has been teaching since 2008. Her research interests are largely in consumer packaging. She looks at packaging systems from various stakeholder perspectives. She has investigated innovations in packaging technology and evaluated the performance of systems and processes in packaging. She is also a judge for the PAC Global Leadership Awards in the International Packaging Innovation category.

Stephanie J. Macleod is the Master Blender at Bacardi (John Dewar & Sons Ltd) and is responsible for the Blended Scotch and Single Malts portfolio for Bacardi's Scotch whisky brands. She assumed this position in 2006, following a 3-year training period. Past positions include Manager of the Dewar Spirit Quality Laboratory in Glasgow and previously as Sensory Analyst at the University of Strathclyde working on projects to attempt to unlock the maturation secrets of Scotch whisky. She has also studied rum, olive oil, wine, and cheese and has published a number of scientific papers on the results of her research.

Binod K. Maitin is an independent technical consultant for Beverage Industries and lives in Lucknow, India. He worked with United Spirits Limited for 25 years and headed their Corporate Technical Centre and has extensive knowledge and experience in analytical and sensory sciences to drive various technical functions, including quality management and consumer-driven new product development. He holds a PhD in analytical chemistry from Queen's University Belfast and has published and presented more than 50 papers on various aspects of analytical chemistry and beverage product development and quality. He is Fellow of the Royal Society of Chemistry, United Kingdom, and a member of the Institute of Brewing and Distilling

and the Sensometric Society. He enjoys serving as a member of the judging panels of International Wine and Spirit Competitions in the United Kingdom and Germany.

Mike Mitchell is a master brewer. Before retirement, he worked for Diageo in Scotland as Site Operations Manager in malt distilling before becoming the Operations Manager at their Kilmarnock bottling plant. Previous to working with Diageo, he worked for 10 years with Bass Brewers. He holds a diploma in management studies and a degree in brewing and distilling from Heriot-Watt University.

Douglas M. Murray has been involved in all aspects of the distilled spirits industry for more than 48 years, from selection of raw materials, to processing, distillation and maturation, and packaging. He has gained extensive knowledge of almost every type of distilled spirit, including both malt and grain distilling. He is currently a master distiller and blender for Diageo and manages the Global Technology Development and Innovation Team. He is the President and Fellow of the Institute of Brewing and Distilling, a member of its examination board, a member of the Scotch Whisky Association Scientific Committee, SpiritEurope Scientific Committee, Keeper of the Quaich, and serves as a judge for several prestigious spirits competitions.

Yoshio Myojo is in charge of Suntory's whisky inventory management and quality control for their whisky production. He joined Suntory in 1996 and was assigned to their process development in Japan. He was transferred to the Blending Department in 2010 and worked there as a blender. He has been in his current position since 2020.

Brian Nation is Master Distiller at the O'Shaughnessy Distilling Company in Minneapolis, MN, USA. Brian's responsibility is for the commissioning and start-up of this new distillery as well as developing new innovative distillate styles and whiskeys. These whiskeys will be released under the brand name Keeper's Heart. Brian has over 23 years of experience in the Irish Whiskey Industry. He has held a number of different positions including, Project Engineer, Process Engineer, Distiller, Engineering Manager and Master Distiller all at Irish Distillers Pernod Ricard, Midleton, Cork, Ireland. He holds a BEng in Chemical and Process Engineering along with the IBD Diploma in Distilling.

Denis Arthur Nicol has extensive experience in the alcohol industry, including holding the following positions over the years: Analytical Chemist at Hiram Walker & Sons (Scotland) Ltd.; Manager at the Tormore Distillery, Speyside; General Manager at the Laphroaig Distillery, Islay; Project Scientist at Long John International; Laboratory Manager at James Burrough Distillers Ltd; Laboratory Services Manager at Allied Distillers Ltd; and Technical Manager at Demerara Distillers Ltd., Guyana, South America. In addition, he is also a teacher of chemistry, physics, and science.

Jonghun (Jay) Park is an assistant professor in the School of Graphic Communications Management at Ryerson University, Toronto, Canada. His research focuses on sustainable packaging systems, packaging distribution, and human factors in packaging design. He is particularly interested in understanding design factors leading to sustainable packaging and developing data-driven packaging design methods. Before joining the faculty of Ryerson in 2017, he worked at Samsung Electronics as a senior packaging engineer. He is the Director of the International Safe Transit Association's North America Division Board and an active member of the International Association of Packaging Research Institutes.

Heather Pilkington is the Managing Director and CEO of First Key Consulting Inc., a brewing and beverage industry consultancy that delivers projects for craft to large-scale clients, private equity groups, associations, and academic institutions around the world. First Key counts well-known major and craft beverage companies among its clients, and its team of consultants includes experts from every facet of the industry. Earlier in her career, Heather held positions at Labatt and AB InBev, managing process development and technical support departments both at the national and international levels. Heather is a licensed professional engineer and received her PhD in biochemical engineering from Western University in Canada.

David Quinn is Technical Director with Irish Distillers Pernod Ricard and has more than 30 years of experience in the Irish whiskey industry. During this period, he has held various positions, including Biochemist and Process Manager at the Midleton Distillery and Head Distiller at the Old Bushmills Distillery. Currently, he has responsibility for group quality management, technical innovation, new product development, analytical services, and whiskey planning. He holds an MSc in both biotechnology and brewing science and technology.

Inge Russell is the Editor-in-Chief of the *Journal of the American Society of Brewing Chemists*, an honorary professor at Heriot-Watt University, Edinburgh, Scotland, and a Fellow of the Institute of Brewing. She has more than 40 years of research experience in the brewing and distilling industry. She served as Editor of the *Journal of the Institute of Brewing*, President of the American Society of Brewing Chemists (ASBC), and President of the Master Brewers Association of the Americas (MBAA). She holds a PhD and DSc from the University of Strathclyde in Scotland. She has authored more than 150 papers in the area of yeast biotechnology and is a cofounder and past coeditor of the journal *Critical Reviews in Biotechnology*.

Duncan McNab Stewart retired from his position as Technical Manager for Cameronbridge Distillery and Leven Spirit Supply in 2017, after 40 years of working within various roles across primary production of grain spirits. Throughout his career, he has always had an interest in energy management, which led to a global energy project in 2004. Through this work, he saw a need for rethinking the use of coproducts as an energy source as well as an animal feed. Since retiring, he is now a full-time art student but still retains a keen interest in distillation coproducts and environmental issues.

Graham G. Stewart has been Emeritus Professor of Brewing and Distilling at Heriot-Watt University, Edinburgh, Scotland, since he retired in 2007. From 1994 to 2007, he was Professor of Brewing and Distilling and Director of the International Centre for Brewing and Distilling (ICBD), Heriot-Watt University. For 25 years prior to this, he was employed by the Labatt Brewing Company in Canada, holding a number of scientific/technical positions and from 1986 to 1994 was its Technical Director. He holds a PhD and DSc from Bath University and is Fellow of the Institute of Brewing. He was President of the Institute of Brewing and Distilling in 1999 and 2000. He has more than 300 publications (books, patents, review papers, articles, and peer-reviewed papers) to his name and is a cofounder and coeditor of the journal *Critical Reviews in Biotechnology*.

Matt Strickland is the Head Distiller for Distillerie Côte des Saints in Mirabel, Quebec. As a farm distillery, Côte des Saints grows their own barley and produces high-end single malt whisky as well as gin and liqueurs. Matt is an active teacher in the distilled spirits industry, sitting on the faculty of the Distilled Spirits Epicenter and the Siebel Institute. He is also the only American to sit on the Board of Examiners for Distilling at the Institute of Brewing and Distilling. Matt is also a writer, producing numerous technical scripts for industry publications such as *Distiller Magazine*, *Artisan Spirit*, and *Brewer Distiller International*. His first book, *Cask Management for Distillers*, was published in the fall of 2020 by White Mule Press.

Hitoshi Utsunomiya is the Director of the Japan Sake and Shochu Makers Association and the UTSUNOMIYA Sake & Food Lab. His expertise is in the flavour chemistry of Sake, and his most famous work is 'The Flavor Terminology and Reference Standards for the Sensory Analysis of Sake'. He is also the editor of the book *A Comprehensive Guide to Japanese Sake*.

Nicholas R. Wilson currently works for Whyte & Mackay at the Invergordon Grain Distillery in Scotland as the company's Microbiologist/Process Scientist. He graduated from Heriot-Watt University with a BSc degree in biological sciences, followed by a PhD in microbiology. His research has focused on the effects of bacteria in malt whisky fermentations pertaining to congener composition and sensory characteristics. He also has teaching experience in chemistry and life sciences and was a teaching fellow at the International Centre for Brewing and Distilling (ICBD), Heriot-Watt University.

Steve Wright is the Founder and Proprietor of Spiritech Solutions Inc., a technical services consultancy and developer of new products for distilled spirits producers. He is also the cofounder and blender of a Canadian whisky brand: Kavi Reserve Coffee Blended Whisky. Steve has 40 years of distillery experience and previously held positions with McGuinness Distillers, Hiram Walker & Sons Ltd., and Allied Domecq PLC. He is a regular instructor at several prominent educational institutions, a member of the examination board of the Institute of Brewing and Distilling, and a team member of Thoroughbred Spirits Group and First Key Consulting Inc.

Foreword

The first two editions of this book, published in 2003 and 2014, met a very real need for a modern textbook covering the technology, production, and marketing of whiskies. I am delighted to be writing the foreword to this third edition, which now covers a much wider field of potable spirits—an innovative and exciting category that continues to expand worldwide.

Whilst the most obvious growth sectors have been gins and whiskies, other products, including vodka, rum, tequila, and shochu, have all seen innovation and growth in many diverse markets. The trend towards premiumisation is very evident across all sectors. This is especially welcome as consumers who may be drinking less are prepared to experiment and pay more for high-quality products with individual heritage and brand equity.

In whiskies, the number of styles, brands, and limited editions has grown exponentially, led by a huge interest in single malts, but the thirst for knowledge is also generating serious interest in innovative blends and renewed respect for the art and science of blending. Whilst Scotch continues to lead internationally, 'world whiskies' are growing at a great pace and are creating tremendous interest from a new generation of very discerning whisky enthusiasts. Specialist retailers have more brands than ever before. The growth of online channels and social media commentators has opened up new audiences, who are eager to learn and to experiment by sampling at home, at whisky events, and by increasing numbers visiting the distilleries. The industry has responded to the growing demand for insight and knowledge by developing distillery tours, lectures, websites, and tutored tastings and by encouraging informed debate.

Gins, flavoured gins, and gin liqueurs have enjoyed an even greater rate of growth than whiskies, albeit from a lower base. Having started in well-developed markets such as Europe and the United States, the popularity of gin continues to spread widely across the world, including Asia and Eastern Europe.

In the past decade, small craft distilleries producing gin, whisky, rum, vodka, and other spirits have been established by entrepreneurs on almost every continent, and many existing distilleries have been expanded in capacity. Large, new, and highly efficient distilleries have also been built by several of the long-established international spirits companies.

One very obvious consequence in these premium-driven markets has been an essential emphasis on product quality. There is no doubt that the intense scrutiny afforded to new premium spirits brands will ensure that only those companies that offer the highest consistent quality at the right price point will be successful.

The earlier editions of this book met the needs of both students and specialist practitioners within the whisky industry. Whilst this third edition will be welcomed by many of the same readers and their successors, it will also be of great interest as a reference book to a much wider audience, including writers, entrepreneurs, investors, and consumers—whether they are seasoned experts or new entrants to the spirits category. The fact that a third edition is warranted with an extended remit after a gap of only 6 years bears witness to the accelerating pace of change and the desire for knowledge from many diverse participants across the vibrant world of international potable spirits.

I must once again congratulate the editors on assembling a group of highly esteemed experts who have been willing to share their extensive knowledge and experience in the chapters of this book.

Dr. Alan G. Rutherford OBE, BSc, PhD,
CChem, CEng, FRIC, FInstE, FIBD
Chairman of Compass Box Whisky Co
Ltd and of The Lakes Distillery Co. plc.
(Formerly Scotch Whisky Production
Director at Diageo and Visiting Professor
at the International Centre for Brewing and
Distilling (ICBD), Heriot-Watt University)

Preface

The first edition of *Whisky: Technology, Production and Marketing* was published in 2003. It was a detailed exploration of Scotch whisky's characteristics and included aspects of the technology, production and marketing of whisk(*e*)y that had not been previously reviewed.

The second edition of this book was published in 2014. The types of whisk(*e*)y discussed were expanded beyond Scotch whisky, to include other forms of whisk(*e*)y: Irish, Japanese, Indian, and North American (bourbon, Tennessee, sour mash, and rye). In addition, the sensory spectra of these whiskies were reviewed. Important relevant raw materials, such as water, barley, and yeast, and key production elements, such as manufacturing requirements, processing, fermentation, and microbiology, were discussed. Whisky packaging developments were considered. Finally, the technical aspects of all matters pertaining to the manufacturing process were reviewed.

In the third (current) edition of the book, as well as updating the chapters on whisk(*e*)y,
chapters on gin, saké, shochu, soju, rum, and vodka have been added. The packaging aspect of spirits has been expanded, as has the marketing of spirits, including a focus on the implications of COVID-19 on the marketplace. With these additions, the book's title has been modified to read *Whisky and Other Spirits: Technology, Production and Marketing*.

We are very grateful to the authors of all the chapters, without whose efforts we would not have a book to publish! We are thankful for the many colleagues who have provided us with information and assistance during the 2-year duration of this project. Finally, Anne Anstruther deserves special mention and thanks for all her hard work, patience, and support from the inception to the completion of this project. Her support cannot be overstated!

Inge Russell
Graham G. Stewart
Julie Kellershohn

1

An introduction to whisk(e)y and the development of Scotch whisky

Anne Anstruther and Graham G. Stewart

GGStewart Associates, Cardiff, Wales, United Kingdom

Early days

Recent research on humankind's ability to metabolise alcohol, by an enzyme called *alcohol dehydrogenase*, suggests that this ability was present 10 million years ago in gorillas, chimpanzees, and humans when they first identified alcoholic fruit as a safe and consumable food (Dudley, 2004). Thus began our relationship with alcohol, which would in time become a safe (and social) drink and, in the correct proportions, an effective medicine and important for religious ceremonies. Most geographical regions provided raw materials suitable for brewing and oenology (beer from barley, wine from grapes, saké from rice, etc.). The earliest evidence of alcoholic beverages was found in China ca. 7000–6600 BC, where their consumption was probably common (McGovern et al., 2005). Eventually, through trial and error, the process, joys, and benefits of distillation were discovered. Early Egyptians used distillates to produce cosmetics and aromatics, and, reputedly, monks carried this knowledge from the Mediterranean to Ireland and then on to Scotland. It is believed that potable whisky was discovered in the search for the elixir of life (known by many as *usque beatha*). The monasteries retained exclusive whisky production until the 1500s. In 1545, Henry VIII dissolved the English monasteries and Scotland followed suit in the 1560s, allowing the knowledge of distillation to become wider spread. Each farm and large house owned its own still to preserve its cereal harvest excess. This distillate had to last until the next harvest, for all purposes (medicinal, rents, etc.), and was of course treasured as a rare and exhilarating beverage!

Effects of the agricultural and industrial revolutions

The next four centuries saw radical changes in the relationship between whisky and people. The agricultural and industrial revolutions steadily improved farming efficiency. Ease, speed, and volume of transport gradually increased as canals and roads were built, and steam-powered boats reduced travel time and costs. Grains could be imported quickly and less expensively. Redundant land workers sought work in urban areas as mechanisation increased production. The distillers' share of the potential market increased, and governments kept an eye on their profits.

Four main forces impacted the evolving distilling industry: (1) weather, (2) process efficiency, (3) taxation, and (4) food and drink regulations. Long-term weather was (and still is) unpredictable. Bans on distilling occurred in 1579, 1660, 1757–1760, and in the 1840s due to particularly bad harvests and the necessity of reserving the grain for food. These restrictions saw the demise of many good distilleries, but those that remained became stronger.

Taxes were raised to fund wars. The first excise duty raised by the Scottish Government in 1644 supported its army. All people resist taxes, through umbrage or necessity, and many distilleries went underground. The 1725 Malt Tax caused a brewers' strike in Edinburgh and rioting in Glasgow, and illicit distilling flourished. The 1757–1760 ban on distilling exempted private stills that were up to 10 gallons in capacity. In 1779, this maximum was reduced to 2 gallons. In 1781, when private distilling became illegal, smuggling increased. Meanwhile, famine and land management forced many Scots and Irish to migrate to Canada and to the United States, where their distilling skills were welcomed. Canada's first distillery opened in 1769 and numerous US distilleries opened shortly thereafter.

Crippling taxes, changes in regulations, and the need for survival led legal distillers in Scotland to produce quantity at the expense of quality. However, the Highland distillers could not expand to meet the high taxes, and illicit distilling increased yet further. Consequently, the best whisky was illegal! In 1784, in Scotland, the Wash Act defined the Highland Line and reduced taxes for the Highland distillers but stipulated severe restrictions. The Lowland distillers increased output at the expense of quality but were decimated by subsequent restrictions and taxes. By 1816, the government had begun a regime more suited to the larger producers (who were more easily controlled) with passage of the Small Still Act (no still under 40 gallons in capacity) and then the 1823 Excise Act, which required distillers to become licensed. In this way, the government finally gained some control, and illicit stills and smuggling faded.

In 1831, an Irish exciseman, Aeneas Coffey, perfected the continuous still, which was rapidly embraced by brandy, rum, and some Scotch whisky distillers. Preferring their smooth, flavoursome, and successful thrice-distilled whiskies, the Irish did not embrace the blandness and volume offered by the Coffey still. In less than 20 years, the first whiskies had been successfully blended to suit the London palates and, with its increased popularity, whisky's reputation spread. At this point, whisky was considered the 'poor man's strong alcoholic drink'. Adulteration of food and drink was rife, many with poisonous substances (Burns, 2012). Each change of hands from source to customer introduced a new watering down and more adulterants to mask the change. Content was less important than effect. The adulterated whisky often caused blindness, violence, and death, particularly amongst the poor, due to the additives and fusel oils, as the whisky was still sold immature. Temperance societies gathered strength from the 1830s, and by the 1860s, whisky consumption had decreased by a third (MacLean, 2003).

Controls, taxation, and amalgamation

In 1865, eight Lowland grain distilleries combined to become the Scotch Distillers Association (MacLean, 2003). In 1872, the UK Act for Regulating the Sale of Intoxicating Liquors made an early attempt at food and drink regulation (though not extended to Scotland) together with the Act to Amend the Law for the Prevention of Adulteration of Food and Drink and Drugs (Burns, 2012). By the 1880s, the steady advance of the beetle *Phylloxera vastatrix* and a succession of poor grape harvests had destroyed wine and cognac production in France. Brandy, the preference of the upper classes, became very rare, and whisky was ready to take its place. Demand worldwide rose and exports escalated, but, without proper controls, enthusiastic investments in new distilleries created a massive whisky surplus. This resulted in a catastrophic collapse of Scotch whisky sales in 1900, causing the public to lose faith in the industry; many companies and individuals were bankrupted!

In a world becoming increasingly exacting, potable whisky needed an identity, which could be an emphasis on single malt, the blend, or its production origin. In 1908, a Royal Commission was formed to investigate the whisky situation. A year later, they defined *whiskey* as 'a spirit obtained by distillation from a mash of cereal grains saccharified by the diastase of malt' and *Scotch whisky* as 'whisk(e)y so defined, distilled in Scotland' (MacLean, 2003). This was a momentous decision, which would allow whiskies the freedom to maintain their identities with pot and Coffey stills, whilst working together to extend the product's range and capacity to further worldwide appeal.

In 1912, the Wine and Spirit Brand Association was formed, which would become the Scotch Whisky Association in 1917 (http://www.scotch-whisky.org.uk/). The organisation's prime concerns are to represent the industry locally and worldwide, at both government and legal levels, to protect the industry against unfair legislation, and to ensure that Scotch whisky is not compromised by adulteration, misrepresentation, or fraud. By 1916, the Immature Spirits Act determined that Scotch whisky should be matured in casks for a minimum of three years, and by 1917 the Central Control Board (formed to control alcohol consumption) had established the strength at which whisky could be sold, which was 70° proof (40% ABV). Although distillers contributed much during the 1914–18 war, the 1909 tax was increased by a factor of five until 1939.

The Japanese determination to produce whisky as a Scotch-type spirit became evident when their first distillery opened in 1924. The first whisky marketed in 1929 was not popular, but later was successfully blended to suit the Japanese palate. Since 1938, the industry has flourished (see details in Chapter 2).

Irish whiskey suffered three catastrophic events. Between 1919 and 1921, Ireland's War of Independence prevented access to overseas markets, following which Ireland was denied access to England's market. The final blows came with prohibition in the United States from 1920 to 1933 (Ireland's second largest market was the United States) and with World War II. From an industry of 160 distilleries in 1880, the Irish industry was reduced to only three (see Chapter 9).

Due to the stringencies of two world wars and prohibition, the distilling industry endured catastrophic bans, although some companies produced distilled products such as glycerol and butanol–acetone for the war effort. Despite further cuts, the industry found ways to survive, strongly supported by appeals to the government for tax equality with wine and beer. The export of the existing stocks of mature whisky contributed extensively to reducing the UK's debt after World War II. As had been experienced in the 1700s, taxes became prohibitive, and bootlegging and the black market flourished.

The determination of whisky in 1909 led to ongoing legislation of Scotch (and other) whiskies. The Scotch Whisky Act 1988 specified alcoholic strengths, minimum maturation, country of production, etc. The Scotch Whisky Regulations 2009 encompassed the tenets of the 1988 Act and further extended them to cover labelling, packaging, and advertising. Meantime, Regulation (EC) No. 110/2008 of the European Parliament and the Council applied to all spirit drinks, categorising rum, whisky, etc., and their geographical locations, whether produced in the European Union (EU) or in another country (see Chapter 19; Scotch Whisky Act 1988, 1988; Statutory Instrument 2009, 2009; EC Regulation, 2008).

In 2019, there was an amendment to the Scotch Whisky Technical File (which incorporates the relevant UK law—The Scotch Whisky Regulations 2009) that expanded the range of casks potentially allowed for the maturation of Scotch whisky (SWA, 2019). Also, Scotch Whisky's status as an EU Geographical Indication (GI) was continued under Regulation (EU) No 2019/787. HMRC is tasked with verifying compliance with the Scotch Whisky Technical File and their 2021 detailed Scotch whisky technical guide is available for download from the UK Government website at https://www.gov.uk/guidance/producing-scotch-whisky.

Whilst developments in mechanisation, specialisation, and research have improved the efficiency and quality of the distillation process, similar advances in communication and transport have reduced dependence on local grains. Access to world grains widened the markets for whisky and opened up new possibilities. In the 1980s, competition from the wine market and further increases in whisky taxation contributed to more distillery closures and amalgamations, which eventually led to a controlled balance between production and demand. The merger of Guinness and Distillers Company Limited (DCL) created United Distillers (UD), which then merged with International Distillers and Vintners (IDV) to form Diageo. These mergers, combined with consolidation of Seagram and a number of smaller companies into Pernod Ricard, resulted in the industry having greater unification and being able to concentrate on the marketing of Scotch whisky, particularly aiming at specialty blends. More recently, a rise in interest in single-malt products has led to a burgeoning market. Nevertheless, blended whisky is still the 'bread and butter' of the industry.

Growth and sustainability

In 2012, there were 98 malt distilleries and seven grain distilleries in Scotland. Diageo owned 28 malt and 1.5 grain distilleries. Pernod Ricard owned 12 malt distilleries and one grain distillery. The remaining distilleries were owned by a large number of distilling companies.

In 2020, there were 134 operating Scotch whisky distilleries in Scotland, with more planned openings for 2021. More than 10,000 people are directly employed in the industry, many located in rural areas in the north of Scotland. Scotch whisky accounts for over one-fifth of all UK food and drink exports. The industry exports to over 200 markets with the top export markets by value being the United States, France, Singapore, Taiwan, Spain, and Germany, adding over £5 billion per annum to the UK economy (SWA, 2021).

In terms of sustainability and lowering their carbon footprint, the distilling industry has an ambitious vision of achieving net-zero emissions by 2040 (SWA, 2021).

There is a focus on sustainability with new farm management techniques to help local ecosystems thrive, water management, and the identification of barley varieties with improved agronomic characteristics, which are likely to be more resilient to the effects of climate change.

Environmentally friendly packaging is also a focus. For example, in 2020, Diageo announced the world's first-ever paper-based spirits bottle made entirely from sustainably sourced wood in accordance with their commitment towards Goal 12 of the United Nations Sustainable Developments Goals: 'Responsible Consumption and Production'. A variety of bottle shapes and sizes are planned for spirits and other beverages and will be available with the cooperation of suppliers for other consumable products (Diageo, 2020).

Brexit and COVID-19

Since the second edition of this book, the spirits industry, and whisky in particular, has been presented with unexpected obstacles in ways that mostly could not have been predicted.

In 2016, the UK voted to leave the European Union and the spirits industry prepared itself for the potential challenges and opportunities that this decision brought, specifically through the significant uncertainty regarding the nature of future trade arrangements for both the single market and the world in general.

In early 2017, the United States proposed a border tax, in retaliation against EU subsidies given to the aircraft maker Airbus. In October 2019, the United States imposed these measures against its EU imports, including a 25% tariff on Scotch whisky. With the advent of the new US President in January 2021, this tariff on single malt Scotch has been suspended for at least 4 months. At least one company considered shipping in bulk and bottling in the United States. Although some US-based companies stated their determination to absorb a percentage of the cost, to date the effect has been traumatic (SWA, 2021).

In early 2020, the virus COVID-19 was declared an international pandemic. Subsequent 'lockdowns' saw whole countries brought to a standstill! Social contact and travel were banned. Imports and exports were restricted to 'essential only requirements'. Despite financial help from governments, many companies have continued to fail. For many whisky companies (dependent on tourism), closed shops, public houses, and social areas also hit the industry hard. Many were able to continue by producing significant volumes of hand sanitisers and ethanol to support health and other services during the crisis. Global exports of Scotch whisky fell by more than £1.1bn during 2020 (SWA, 2021).

Home drinking has increased during the pandemic and, although whisky has been hit hard by the US tariff, sales of tequila and bourbon have increased, benefitting companies with sufficient scope. However, with the exception of China, generally, markets have decreased (SWA, 2021).

The future

Whisk(e)y is an alternative drink to discerning adults, and it has developed an epicurean status. The appreciation of whisk(e)y has spread globally. It has been found in locations as diverse as Antarctica and the Sahara and is currently being distilled on the International Space Station (ISS) (a no-gravity environment). Many countries have developed their own distinctive brands and this trend will continue with each country developing its own standards, regulations, and specific criteria in order to protect its exclusivity as well as addressing the issue of counterfeit distilled spirits. Expanding populations will also exert pressure on the industry when the increasing demand for food, water and energy conflicts with the development of potential new whisk(e)y markets and other beverages. In 2020, the UK industry suffered as exports decreased by more than £1.1bn due to COVID-19 and the 25% US tariff. Uncertainty caused by Brexit will need to be resolved and market competition with changing consumer tastes will mean an ever-fiercer battle for the global whisky market share. New challenges will continue to impact the industry, but its notable past history suggests that resilience is its strength!

Acknowledgements

The authors thank Alan Park and Alan Rutherford for their contribution regarding the Scotch Whisky Technical File.

References

Burns, E., 2012. Bad Whisky, third ed. Neil Wilson Publishing, Castle Douglas, Scotland, pp. 33–34 (Chapter 2).

Diageo, 13 July 2020. Diageo Announces Creation of World's First Ever 100% Plastic Free Paper-Based Spirits Bottle. Press Release https://www.diageo.com.

Dudley, R., 2004. Ethanol, fruit ripening, and the historical origins of human alcoholism in primate frugivory. Integr. Comp. Biol. 44, 315–323.

EC Regulation, 2008. EC Regulation No. 110/2008 of the European Parliament and of the Council of 15 January 2008 on the Definition, Description, Presentation, Labelling and the Protection of Geographical Indications of Spirit Drinks and Repealing Council Regulation (EEC) No. 1576/89.

MacLean, C., 2003. Scotch Whisky, A Liquid History. Cassell & Co, London (Chapter 7, p. 132; Chapter 9, p. 170).

McGovern, P.E., Underhill, A.P., Fang, H., Luan, F., Hall, G.R., Yu, H., Wang, C.-S., Cai, F., Zhao, Z., Feinman, G.M., 2005. Chemical identification and cultural implications of a mixed fermented beverage from late prehistoric China. Asian Perspect. 44, 249–275.

Anon., 1988. Scotch Whisky Act 1988. Her Majesty's Stationery Office, London (Chapter 22).

Statutory Instrument 2009, 2009. No. 2890, The Scotch Whisky Regulations 2009. Her Majesty's Stationery Office, London.

SWA (Scotch Whisky Association), 2019. Scotch Whisky Technical File Amended—Guidance on Allowable Casks for Maturation. https://www.scotch-whisky.org.uk/.

SWA (Scotch Whisky Association), 2021. Newsroom and Insights. https://www.scotch-whisky.org.uk/.

Japanese whisky

Shinji Fukuyo[a] and Yoshio Myojo[b]

[a]Whisky Blending & Planning Department, Suntory Spirits Limited, Osaka, Japan [b]Whisky
Development & Production Department, Suntory Spirits Limited, Tokyo, Japan

Introduction

Ian Fleming, a journalist of Scottish descent and author of the famous James Bond series, wrote about Japan in his *Thrilling Cities* travelogue, which was based on his visit to Japan in 1959 (Fleming, 1963). In that book, he made a complimentary note on whisky in Japan. The whisky market in Japan was more difficult to please than Fleming, however, and consequently, sales often struggled. During the downturn, Japanese whisky companies tried to attract consumers through improvements in their marketing efforts and product quality. In this chapter, we introduce the history, market, production processes, and, in part, research and development activities of Japanese whisky companies and related research institutes.

History

The dawn of Japanese whisky (1850–1950)

The first record of whisky being imported into Japan was in 1853, at the end of the Edo era. Commodore Matthew Perry of the US Navy visited Japan to persuade the *Bakufu Shogunate* (Japanese feudal government) to open the country to the world, and whisky was served to Japanese officers on his ship. In the following year, he brought a cask of whisky as a gift to the Japanese imperial families and their aides (Hawks, 1856). A whisky cask is shown in a painting that depicts the disembarkation of the gifts (Fig. 2.1; Dower, 2013).

The flavour of whisky fascinated some people working in the pharmaceutical industry in Japan, which led them to produce whisky-flavoured drinks made by adding artificial essence to white spirits (Sekine, 2004). In 1918, Kihei Abe of Settsu-Shuzo Co. in Osaka sent a young technical employee named Masataka Taketsuru (who later founded the Nikka Whisky Distilling Co., Ltd.) to Scotland to learn Scotch whisky production techniques (Buxrud, 2008). After receiving training at several Scotch whisky distilleries, Taketsuru returned to Japan in 1920 only to find that the company that sent him to Scotland had abandoned its plans to make whisky in Japan.

FIG. 2.1   Disembarkation of gifts from the United States at the Uraga port in 1854. *Courtesy of the Yokohama Archives of History, Yokohama, Kanagawa, Japan.*

At around the same time, though, Shinjiro Torii, the founder of Suntory Holdings, Ltd., was considering launching a whisky business in Japan. It was the time when Japanese people were beginning to accept the Western culture, and Shinjiro Torii predicted that whisky would soon become popular in his country. When he decided to construct a whisky distillery in Japan, he initially intended to invite a whisky technician from Scotland to come to Japan. But, when he learnt that Taketsuru had recently returned to Japan after already learning whisky production in Scotland, Torii hired him, and they selected Yamazaki as the place to build a distillery. The first Japanese distillery (Suntory Yamazaki Distillery) opened in November 1924 (Taketsuru, 1976).

Torii began selling the first made-in-Japan whisky, Shirofuda ('White Label'), in 1929. Unfortunately, it did not sell well because, generally speaking, Japanese people at that time did not like the strong smoky flavour of Shirofuda, which was typical of some Scotch whiskies. Following numerous blending and tasting trials, it was decided that what Japanese people wanted was a flavourful whisky with less smokiness. Torii created a new blend, Kakubin, which went on sale in 1937 and was well received in the Japanese market (Koshimizu, 2011).

After serving his 10-year contract with Suntory, Taketsuru moved to Hokkaido and opened the Yoichi Distillery. He believed that the climate on that island was much closer to that of Scotland. The first new-make spirit was produced from his still in 1936 (Taketsuru, 1976).

A period of growth for Japanese whisky (1950–83)

In 1945, after the end of World War II, Western culture and lifestyle grew in popularity in Japan, and whisky became a symbol of this trend. As the aspiration to Western culture continued to grow, sales of whisky increased. The Japanese whisky market maintained its growth until 1983, led by skilful marketing. All over Japan, whisky bars served highballs (whisky mixed with soda) in the summer and *oyu-wari* (whisky mixed with hot water) in the winter. Around 1965, *mizu-wari* (whisky mixed with ice and water) became accepted as an appropriate drink to accompany Japanese foods. At about the same time, the 'bottle-keeping' system gained popularity in bars and restaurants. Rather than paying by the glass, a customer could buy a bottle of whisky to share. The customer wrote his/her name on the bottle, and the bar kept the bottle for the patron's next visit. Pouring drinks for each other is a Japanese custom, and this bottle-keeping system is a good example of the introduction of Japanese culture to whisky bars. *Mizu-wari* and the bottle-keeping system contributed greatly to the increase in whisky consumption rates (Suntory Ltd., 1999).

The fall and rise of Japanese whisky (1983–2008)

After 1983, whisky production rates began to decrease (Fig. 2.2). This decline was attributed primarily to a tax hike on whisky and the increase in consumption of *shochu*, a Japanese white spirit. The consumption rates of *shochu* increased dramatically around 1980, supported by the popularity of *chu-hai*, a type of highball that used *shochu* as its base spirit. *Shochu* can be mixed with a variety of flavours, and its sweetness attracted younger consumers. Consecutive revisions of the liquor tax act in 1981 and 1984 resulted in more than a 25% price increase in the price of whisky (Suntory Ltd., 1999). In spite of redressing the liquor tax disparities between *shochu* and whisky in 1989 and 1998, the unpopularity of whisky amongst consumers persisted, and whisky consumption continued to fall until 2008. Despite the difficult environment, whisky companies continued their marketing and research and development (R&D) efforts, which resulted in a solid increase in the consumption of single-malt products beginning around 2005 and helped to gradually redirect consumers back towards the consumption of whisky.

FIG. 2.2   Changes in market volumes of whisky and *shochu* (National Tax Agency, 2019).

The Japanese whisky market (2008–2018)

The decline in whisky consumption, which commenced in 1983, finally ceased in 2008. A campaign promoting highballs was a great success, and the number of people who enjoyed the character of a single-malt whisky grew. The total whisky market in 2018 was approximately 180 million litres in Japan, and the share of Japanese whisky was more than 85% of that total (National Tax Agency, 2019). Single-malt products accounted for 2.2% (approximately 4.0 million litres) of the whisky market in Japan. Amongst that 2.2% share, Japanese single malts accounted for 1.5% (approximately 2.6 million litres) (Suntory, unpublished data).

Export of Japanese whisky

Most Japanese whisky is consumed domestically, with a relatively small volume being exported. In 1934, Suntory first exported whisky to the United States, shortly after the repeal of the National Prohibition Act (Koshimizu, 2011). The ratios of export vs domestic shipments fluctuated between 2% and 5% from 1990 to 2018. About 6.2 million litres of Japanese whisky were exported in 2018 (Ministry of Finance, 2019).

Production processes

The distilleries

Japan had eight whisky distilleries, both malt and grain, by the end of 1980s. Except for Yamazaki and Yoichi, six of these distilleries were built during the whisky boom in Japan; Chichibu was opened later, in 2008 (Ingvar, 2012). In the list below, which shows when these nine distilleries began operation, the current owners' names are in parentheses:

    1923: Yamazaki Distillery (Suntory Liquors, Ltd.)
    1936: Yoichi Distillery (Nikka Whisky Distilling Co., Ltd.)
    1969: Miyagikyo Distillery (Nikka Whisky Distilling Co., Ltd.)
    1973: Chita Distillery (Sungrain, Ltd.)
    1973: Hakushu Distillery (Suntory Liquors, Ltd.)
    1973: Fuji Gotemba Distillery (Kirin Distillery Co., Ltd.)
    1984: White Oak Distillery (Eigashima Shuzo)
    1985: Mars Distillery (Hombo Shuzo Co., Ltd.)
    2008: Chichibu Distillery (Venture Whisky, Ltd.)

Following the Chichibu Distillery, new whisky distillers have risen one after another throughout Japan.

Technology for the production of Japanese whisky was brought back from Scotland in 1920, and its influence on Japanese whisky production has been widespread and significant. Only a brief overview of each company's operations is presented here; for further information, refer to *Japanese Whisky—Facts, Figures and Taste* (Buxrud, 2008).

Malt whisky production

Malting

Historically, barley was malted in-house at Japanese distilleries. Today, most distilleries purchase malted barley from commercial maltsters. Each company has its own specifications for its malted barley and these are agreed with the maltsters. Nonpeated or very lightly peated malts are preferred, but small amounts of medium and heavily peated malts are also used.

Mashing

Water used for mashing is soft to moderately hard, with a pH in the neutral range (Buxrud, 2008). The mashing process is almost the same as for Scotch whisky. Malted barley is milled, mixed with hot water, and set to a temperature of about 64°C for efficient extraction. The liquid (wort) is filtered through the husk bed. As wort clarity can influence the development of flavour components such as esters and fatty acids, each company adopts different policies to produce wort to suit their needs. Kirin incorporates its knowledge of brewing beer to keep the wort clear, thereby producing whiskies rich in fragrant and fruity esters. Suntory also aims for good wort clarity to ensure the production of a fragrant and complex aroma in the distillate.

Fermentation

Stainless steel and wooden washbacks are used for fermentation. Research into whisky and beer fermentation by yeast and other microorganisms has been actively conducted with all three major distilling companies in Japan producing beer as well as whisky. Each company adopts its own policies for selecting their yeasts and fermentation conditions. Kirin chooses specific yeasts from their collection of hundreds of yeast cultures to produce particular fragrant ester flavours. Nikka utilises multiple yeasts to produce a variety of whiskies with different fragrances and tastes. Suntory uses both distilling and brewer's ale yeasts in order to create complexity of flavour.

Distillation

Similarities between Scotch and Japanese whisky production continue into the still house with the use of double distillation in copper pot stills. Each company has its own policy for selecting their stills. During the first distillation, wash (fermented wort) at 7% to 8% alcohol by volume (ABV) is heated to obtain a distillate with 20% to 25% ABV. This first distillate, the low wines, is then redistilled, and the distillate is divided into three fractions: heads, heart, and tails. The heads and tails are mixed with the low wines from the wash distillation and are distilled again. Each company has its own specifications regarding the optimal alcohol strength of the new-make spirit.

Kirin uses an indirect heating system, and the narrow ABV range of their heart fraction makes their new-make spirit rich in esters. Nikka employs different types of pot stills in different distilleries. The lyne arms of their stills in Yoichi are descending, and heating is carried out by direct firing of coal; at the Miyagikyo Distillery, the distillation pots are ball shaped, the lyne arms are ascending, and an indirect heating method is used. At Suntory, a variety of pot stills differing in shape, size, and heating methods are used in their two malt distilleries to produce different types of spirit.

Whisky produced by continuous distillation

The three companies that produce blended whiskies in Japan possess their own grain whisky distillation facilities. Various types of grain whiskies are produced to broaden the characteristics in the final blend. Production processes vary between companies as they pursue different flavours in their products.

Fuji Gotemba Distillery (Kirin Distillery Co., Ltd.)

Kirin produces grain and malt whiskies at their Fuji Gotemba Distillery. The production processes are based on both Scotch grain whisky and bourbon whisky. This makes their grain whiskies quite diverse. Maize (corn) is the main ingredient. It is heated with water, and malted two-row barley is added as the source of amylolytic enzymes and additional extract. Rye can be used at this point, depending on the flavour types to be developed. The whole mash is cooled and fermented without wort separation. This produces wash with a concentration of 8% to 10% ABV. Three types of distillates are produced, each with a different flavour. Heavy flavour types are developed using a beer column and a doubler; the medium type uses a kettle (a tank with steam coil) and a rectifying column, and the light type uses multiple columns. The strengths of the distillates range from approximately 70% to 94% ABV, which differ depending on the distillation methods; the light types tend to have a higher ABV content.

Miyagikyo Distillery (Nikka Whisky Distilling Co., Ltd.)

Nikka produces grain whisky in their Miyagikyo Distillery. Like Kirin, their production processes are based on those of Scotch grain whisky. The main ingredient is maize, and malted two-row barley is used. In mashing, maize is mixed with water, then heated. Heating is conducted in a pressure cooker. When cooking is complete, the temperature is lowered to around 65°C, and milled malted barley is added. The gelatinised starch and other mash products are separated into solids and liquids using a belt press. Only liquids are used for the fermentation process. The wash, obtained after the fermentation, is distilled using a continuous Coffey still. The strength of the distillate is approximately 94% ABV. It is noteworthy that a product known as Coffey malt is produced with only malted barley as its raw material (further details in Chapter 10).

Chita Distillery (Sungrain, Ltd.)

The grain whisky from the Chita Distillery is used for Suntory's blended whiskies. Although their production processes are based on the Scotch grain whisky, the combination of their distillation columns is unique as a result of Suntory's R&D activities. Maize is the raw material, and malted six-row barley, which has high diastatic power, is used for saccharification of the maize starch. During mashing, maize is mixed with water and heated. Heating is performed in a pressurised continuous cooking tube. After cooking, the temperature is brought to around 65°C. Water-slurried malted barley is continuously added for saccharification. Unfiltered wort is used for fermentation. The distillery has four column stills, and various flavours are obtained by changing the combination of columns—two, three, or four columns. In each case, the strength of the final distillate is greater than 94% ABV.

Maturation and blending

The wood used for the maturation casks is mainly American and Spanish oak; the four cask types are barrel, hogshead, puncheon, and butt. The majority of the casks are new charred, ex-bourbon, and ex-sherry. Recently, ex-wine (including Madeira and Port) French oak casks have begun to be used. Kirin uses small (barrel) casks and new-make spirit with a strength of 50% ABV at the start of maturation for better transfer of the cask-derived flavour to the whisky. In addition to American oak and Spanish oak, Suntory also uses Japanese oak to produce whisky with a unique flavour said to be reminiscent of shrines and temples.

Climatic differences in Japan affect the environmental conditions in the warehouse, with seasonal differences in Japan being greater than in Scotland (Fig. 2.3).

The location of the Yamazaki Distillery is characterised by a warm climate, with a year-round temperature higher than that of Scotland. In contrast, the Yoichi Distillery, which is located in northern Japan, is characterised by cold winter temperatures and temperatures between November and March that are lower than those in Scotland. Thus, the environmental conditions at different locations affect the maturation processes, resulting in the production of a variety of whiskies. Based on the data from Suntory, the average loss of spirit from the warehouses is 2.5% per year. Matured whiskies are then made ready for blending, which serves two main purposes: to develop new products and to manage the quality of existing products.

Research and development

Each distilling company has developed its own technological processes, which have contributed to quality improvement and shaping of the character of the Japanese whisky.

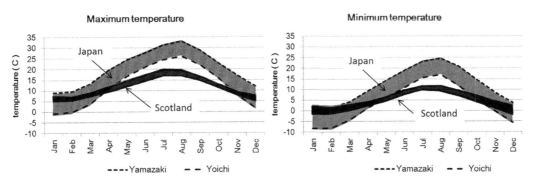

FIG. 2.3 Monthly changes in maximum and minimum temperatures (normal mean between 1981 and 2010) at meteorological stations near eight distilleries in Japan (Chichibu, Gotemba, Hakushu, Mars, Miyagi, Yamazaki, Yoichi, and White Oak distilleries) and seven areas in Scotland (Banff, Inverbervie, Inverness, Keith, Port Ellen, Strathallan, and Tain Range). *Data from Japan Meteorological Agency, 2014. Weather, Climate & Earthquake Information, http://www.data.jma.go.jp/obd/stats/etrn/index.php; Met Office, 2014. U.K. Climate—Synoptic and Climate Stations, http://www.metoffice.gov.uk/public/weather/climate-network/.*

Mashing

Esters are one of the many important flavour groups that contribute aroma to whisky. Unsaturated fatty acids on yeast cell membranes block the enzymatic activity of alcohol acetyltransferase, resulting in inhibition of the development of acetic esters (Yoshioka and Hashimoto, 1983). Clear wort can reduce the total amount of fatty acids including the unsaturated acids on yeast cell membranes. As stated earlier, some Japanese distilleries make great efforts to maintain good wort clarity in order to produce fragrant and fruity whiskies rich in esters.

Fermentation

Two types of microorganisms are mainly involved in whisky fermentation: yeast and lactic acid bacteria (LAB). In Scotland, both distillery yeasts and excess yeasts from the ale brewing industry have traditionally been used, but ale yeasts have become less popular recently. In contrast, some Japanese distilleries still attach the utmost importance to the role of ale yeasts in their whisky flavours. The use of both yeast types characterises the composition of the survival ratio profiles during the course of the fermentation and contributes complex flavours to the new-make spirit. It has also been reported that exposing yeasts to starvation conditions during the later stages of fermentation will enhance flavour complexity and add a fullness of body to new-make spirits (Yomo et al., 2005).

Lactic acid bacteria also enhance sweet and fatty flavours, which are produced by combining LAB with other microorganisms near the end of fermentation. The production of such flavours follows two stages: (1) ale yeasts become nonviable during the later stages of fermentation and (2) intracellular substances are excreted. These eluted substances are then hydroxylated by the LAB and converted into lactone by the distillery yeasts, which produce sweet and fatty flavours (Wanikawa et al., 2000).

Distillation

Pot stills of various shapes and sizes are used in Japanese malt whisky distilleries. It is generally recognised that such differences affect the flavour and quality of new-make spirit. Research into wash distillation has shown that the transfer of nonvolatile materials cannot be explained by the vapour–liquid equilibrium. Nonvolatile materials contained in the foaming wash are transferred into the low wines only by entrainment. It has been noted that the rate of entrainment is dependent on the distance between the top of the wash stills and the top level of the foams generated during distillation. Thus, the entrainment rate during the operation of the first distillation can be controlled by altering this distance (Ohtake et al., 1994).

Maturation

Recently, casks made of Japanese (mizunara) oak have attracted extensive attention. Sensory tests have indicated that whisky matured in mizunara casks has a stronger coconut flavour than that obtained in white oak casks. Chemical analysis of these matured whiskies has shown that the ratio of *trans*-lactone is higher for mizunara casks than for white or

European oak casks. Previously, spiking *trans*-lactone samples into whisky had been shown to contribute a similar coconut flavour (Noguchi et al., 2008).

Final thoughts

The production of Japanese whisky dates back to the transfer of knowledge and technologies from the Scotch whisky industry. Influences of the natural environment in Japan, efforts towards creating an identifiable Japanese style, maintaining a continuous *kaizen* of quality, and the gradual acceptance by the Japanese people of the taste of whisky have all contributed to the development of a particular Japanese whisky character. The blending processes for Japanese whisky are similar to those used in blending Scotch, and the recipes are equally confidential. It is noteworthy that the quality of Japanese whisky, as a whole, has improved significantly during the last 20 years, as evidenced by the wealth of awards won by many Japanese distillers at top-class competitions. For a long period of time, Japanese whisky was consumed almost entirely domestically, but recent efforts have increased its share of the international market. It is hoped that Japanese whiskies will be exported not merely as stand-alone products but also as integral parts of the Japanese food and drink culture that can be enjoyed all over the world. In order for this to be realised, it will be important to extend efforts not only to the continual improvement of whisky quality but also to developing comprehensive strategies to build global brands and increase international awareness of the Japanese whisky culture.

Acknowledgements

The authors are grateful to the Yokohama Archives of History for giving permission to use its collected items. For the process information, we are greatly indebted to Kirin Distillery Co., Ltd. and Nikka Whisky Distilling Co., Ltd.

References

Buxrud, U., 2008. Japanese Whisky—Facts, Figures and Taste. DataAnalys Scandinavia AB, Malmö, Sweden, pp. 41–117.

Dower, J.W., 2013. Black Ships & Samurai: Commodore Perry and the Opening of Japan (1853–1854). Massachusetts Institute of Technology, Boston. http://ocw.mit.edu/ans7870/21f/21f.027/black_ships_and_samurai/bss_essay07.html.

Fleming, I., 1963. Thrilling Cities. Vintage, London, p. 71.

Hawks, F.L., 1856. Narrative of the Expedition of an American Squadron to the China Seas and Japan, Performed in the years 1852, 1853, and 1854, Under the Command of Commodore M. C. Perry, United States Navy, by Order of the Government of the United States. D. Appleton & Co, New York.

Ingvar, R., 2012. Malt Whisky Yearbook 2013. Magdig Media, UK, pp. 220–227.

Koshimizu, S., 2011. Whisky Has Become Native to Japan. Shinchosha Publishing, Shinjuku, Tokyo (in Japanese).

Ministry of Finance, 2019. Trade Statistics of Japan. https://www.customs.go.jp/toukei/info/index_e.htm.

National Tax Agency, 2019. National Tax Agency Reports. http://www.nta.go.jp/kohyo/tokei/kokuzeicho/tokei.htm.

Noguchi, Y., Hughes, P.S., Priest, F.G., Conner, J.M., Jack, F., 2008. The influence of wood species of cask on matured whisky flavor—the identification of a unique character imparted by casks of Japanese oak. In: Walker, J.M., Hughes, P.S. (Eds.), Worldwide Distilled Spirits Conference. Nottingham University Press, Nottingham, UK, pp. 243–251.

Ohtake, K., Yamazaki, H., Kojima, K., 1994. Evaluation of mass-transfer by mist as a new parameter in the control of wash distillation. In: Proceedings of the Aviemore Conference on Malting, Brewing and Distilling. Institute of Brewing, London.

Sekine, A., 2004. A Page of Making Western Style Liquors (in Japanese). Barrel Publishing, Ltd., Osaka, Japan.

Suntory, Ltd., 1999. Let's Start Every Day Afresh—Commemorative Publication for 100 Years History of Suntory, Ltd (in Japanese). Suntory, Osaka, Japan.

Taketsuru, M., 1976. Whisky and I. Nikka Whisky Distilling Co., Ltd, Tokyo (in Japanese).

Wanikawa, A., Hosoi, K., Kato, T., 2000. Conversion of unsaturated fatty acids to precursors of gamma-lactones by lactic acid bacteria during the production of malt whisky. J. Am. Soc. Brew. Chem. 58, 51–56.

Yomo, H., Noguchi, Y., Yonezawa, T., 2005. Effect on new-make spirit character due to the performance of brewer's yeast. In: Walker, J.M., Hughes, P.S. (Eds.), Worldwide Distilled Spirits Conference. Nottingham University Press, Nottingham, UK, pp. 109–122.

Yoshioka, K., Hashimoto, N., 1983. Cellular fatty acid and ester formation by brewers' yeast. Agric. Biol. Chem. 47, 2287–2294.

Sake and shochu

Hitoshi Utsunomiya[a] *and Nami Goto*[b]

[a]Japan Sake and Shochu Makers Association, Tokyo, Japan [b]National Research Institute of Brewing, Hiroshima, Japan

Introduction

Japanese sake is a fermented alcoholic beverage made from rice that has a history dating back more than 2000 years; during which time, the Japanese have continuously improved the brewing technique. By contrast, shochu is a distilled alcoholic beverage with only five centuries of history that was created through an encounter between the sake brewing technology of Japan and the distillation technology of Asia. A characteristic common to both sake- and shochu-making is the use of koji, a culture of mould comprised of *Aspergillus* spp. as the enzyme source instead of malt.

In 2018, 1433 sake makers and 271 shochu makers were members of the Japan Sake and Shochu Association. The five prefectures with the most sake makers were Niigata (n=89), Nagano (n=80), Hyogo (n=71), Fukushima (n=63), and Fukuoka (n=60). By contrast, most shochu makers were based in the Kyushu region and Okinawa, including Kagoshima (n=110), Okinawa (n=46), Miyazaki (n=37), Kumamoto (n=31), and Nagasaki (n=10) prefectures (Fig. 3.1).

Koji is central to Japanese food culture

Koji is a substance produced by growing koji fungi (*Aspergillus* mould) on steamed grain. It contains a wide variety of enzymes and vitamins that are a source of nutrition for other microorganisms, such as yeast. Since ancient times, koji fungi have been used in various foods and drinks, such as sake, shochu, mirin, miso, and soy sauce, and they contribute widely to Japanese food culture.

In addition to the food industry, Jokichi Takamine extracted and produced a digestive medicine, Taka-Diastase, from this fungus about 110 years ago. Indeed, koji is likely to be useful for the production of valuable materials in a wide range of fields, including pharmaceutical science. To commemorate this, The Scientific Conference of the Brewing Society of Japan (2006) declared koji fungi as 'The National Fungi'.

17

FIG. 3.1 Map of Japan.

There are three main types of koji fungi:

1. *Aspergillus oryzae* is a yellow koji mould used for sake making.
2. *Aspergillus sojae* is used for soy sauce-making.
3. *Aspergillus luchuensis* is a black koji mould; its albino mutant, *Aspergillus luchuensis* mut. *Kawachii* (*Aspergillus kawachii*), is a white koji mould used for shochu-making (Hong et al., 2013).

Aspergillus oryzae. Yellow koji mould

The first written record of sake brewing using koji dates back to the early 8th century. According to the Chronicles of the Land of the Harima (now Hyogo prefecture), the steamed and dried rice offered to God got wet and mouldy, so people used it to make sake and held a banquet dedicated to God. Since that era, the use of koji in the brewing of sake has been popular in Japan.

Fully grown koji mould with conidia (spores), termed 'seed koji' (*tane-koji* in Japanese), is used as a seed in koji-making. Because the colour of the seed koji is yellow–green, it is called yellow koji mould. Seed koji has been made since the 18th century. An early koji producer developed the special technique of adding wood ash to steamed rice, which reduces the growth of harmful bacteria and improves the durability of seed koji. The providers of seed koji might be considered to be amongst the world's first biotech companies.

Aspergillus luchuensis. *Black koji mould and white koji mould*

The fungus *Aspergillus luchuensis* has been traditionally used in the production of Okinawa's spirit, awamori, and so was formerly called *Aspergillus awamori*. It is referred to as black koji mould because of its black conidia. The black koji mould produces a large amount of citric acid, which is essential to suppress the propagation of harmful bacteria, even in warm climates.

In 1918, Genichiro Kawachi discovered a white mutant within the black koji mould and isolated it as white koji mould. White koji mould is now used by many shochu producers because it does not soil workers' clothing, amongst other reasons.

Enzyme activities in koji and malt

Koji produces various enzymes including glycolytic enzymes (e.g. amylase and glucoamylase), proteolytic enzymes (e.g. acid protease and acid carboxypeptidase), and lipolytic enzymes. The main enzyme activities in koji and malt are summarised in Table 3.1 (Nishiya, 1993). Koji has higher glucoamylase and protease activities as compared with malt. Owing to the activities of these enzymes in koji, glucose, rather than maltose, is mainly produced from starch. A comparison of shochu koji and sake koji shows that the amylase activity of shochu koji is relatively low, but enzyme stability is high under acidic conditions (Iwano et al., 1986, Table 3.2).

Sake yeast and shochu yeast

The essential characteristics of sake yeast are the production of a pleasant aroma, high fermentation ability at low temperatures, and low production of acid. The practice of isolating

TABLE 3.1 Major enzyme activities of koji and malt (U/g dry weight).[a]

Koji or malt	α-Amylase	Glucoamylase	Acid protease	Acid carboxy-peptidase
Sake rice koji (*A. oryzae*)	1270	223	2285	7750
Shochu rice koji (*A. kawachii*)	159	282	29,101	9224
Shochu barley koji (*A. kawachii*)	88	189	13,933	3574
Awamori rice koji (*A. luchuensis*)	102	143	17,529	4447
Japanese pale malt	1156	80	943	1431
European pale malt	1452	94	375	3465

[a] *According to standard analytical methods of NRIB (Nishiya, 1993).*

TABLE 3.2 pH range where each koji enzyme is stable.

Koji	α-Amylase	Glucoamylase	Acid protease	Acid carboxy-peptidase
Sake rice koji (*A. oryzae*)	5.0–8.0	3.5–8.0	3.0–6.0	3.0–6.5
Shochu rice koji (*A. kawachii*)	2.5–6.5	2.5–8.0	3.2–6.0	2.5–6.0
Awamori rice koji (*A. luchuensis*)	3.0–6.0	2.5–8.0	3.0–5.3	2.5–5.0

pH stability: treated for 60 min at 30°C (Iwano et al., 1986).

and selecting sake yeast from the mash of a brewery that produces good sake has a long history. Since 1906, yeast selected in this manner has been distributed by the Brewing Society (*Jozo Kyokai* in Japanese) of Japan as 'Kyokai yeast'. The Kyokai yeast strains are numbered: currently, the most widely used strains are #6, #7, #9, and #10. Each strain produces unique aroma and taste characteristics; thus, the choice of strain depends on the desired sake quality. More recently, sake brewers have been using newly bred yeasts designed to increase the production of esters delivering a fruity aroma.

By contrast, shochu yeast must have the following characteristics: (1) tolerance of low pH; (2) good growth at temperatures higher than used for sake mashes; and (3) vigorous fermentation of alcohol and production of excellent flavour. At present, strains of shochu yeast are provided not only by the Brewing Society of Japan but also by prefectural laboratories and cooperatives of shochu-producing regions.

Analysis of yeast genomes demonstrates that shochu yeast is more closely related to sake yeast than to wine or whisky yeast (Fig. 3.2). However, shochu yeast differs from sake yeast in its ability to ferment at around 30°C and in its flavour productivity. It is thought that these small genetic differences characterise shochu yeast, but this has not yet been accurately defined. We can expect further development of this research in the future.

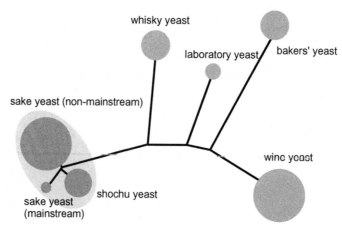

FIG. 3.2 Outline of phylogenetic relationship amongst industrial yeast. Reproduced with permission from Dr. Akao, National Research Institute of Brewing (NRIB), 2020. https://www.nrib.go.jp/English/index.htm.

A unique fermentation method

There are a number of characteristics common to the production methods of both sake and shochu:

– Use of koji as a saccharification enzyme source.
– The amount of water added to the raw materials is lower than in beer and whisky production; fermentation proceeds under very concentrated conditions.
– Parallel fermentation: saccharification by koji enzymes and fermentation by yeast work simultaneously in the fermentation mash.

The parallel fermentation method is effective for avoiding high sugar concentrations. As a result, a moderate osmotic pressure does not repress fermentation speed and does not induce the production of acetic acid. In addition to these fermentation methods, the characteristics of sake and shochu yeast are why the alcohol content of the fermentation mash reaches 16%–20% (v/v) at the end of fermentation. This is the reason that the alcohol concentration of sake is higher than that of wine and beer, and why shochu is made by a single distillation process.

History of sake

If we consider the history of sake as the history of rice-based liquor, its origins go back as far as 2500 years when rice growing became prevalent in Japan. Historical records compiled by the imperial court in the eighth century show that many types of sake were produced at that time. In the 12th to 15th centuries, sake came to be brewed at Shinto shrines and Buddhist temples, and many types of brewing techniques used today, such as the use of polished rice, the three-step preparation of fermentation mash, and pasteurisation, were developed during this period. In the 17th century, during the early Edo period, sake came to be made in large volumes in the winter. From the late 19th century onwards, the processes involved in sake brewing have been scientifically elucidated. Kikuji Yabe first isolated sake yeast from sake mash in 1895. In 1904, a national institute (now the National Research Institute of Brewing) was established and has contributed to the development of sake brewing technology (Akiyama, 2010).

Raw materials for sake

Rice

Roughly 270 varieties of Japonica-type rice are grown in Japan, including specific varieties known as 'sake rice' that are suitable for use in sake making. Sake rice grains are large and typically have a white core (*shinpaku*, the white opaque section at the centre of the rice kernel formed by a matrix of starch granules pocked with voids) and low protein content. Sake rice should absorb water well and be resilient when steamed. It should also be readily digestible by koji enzymes at low temperatures in the fermentation mash. The price of sake rice is, on average, more than 15%–60% higher than that of table rice.

In 2018, 227 thousand tons of brown rice were cropped for sake making in Japan. Sake rice accounts for only 88 thousand tons (38.8%) of the total rice used because certain varieties that are grown mainly as table rice are also used for brewing sake (Ministry of Agriculture, 2020a).

Water

Most water in Japan is soft—the overall hardness expressed in calcium carbonate equivalent is less than 60mg/L—but in some areas, the water is much harder. For example, in the Nada district in Hyogo prefecture, there is an area of hard water with calcium carbonate equivalent of 150mg/L. Calcium stimulates the production and extraction of enzymes. Other minerals in hard water, such as potassium, magnesium, and phosphates, assist the fermentation process by promoting the proliferation of koji fungi and yeast. For this reason, the sake that is produced in areas where the water is hard tends to have plenty of body and dry taste with an excellent finish. In addition to hardness, the most important property of water is a low iron content. Too much iron results in a brown colour and an unpleasant flavour of sake.

Sake production processes

Outline of sake making

The polished rice is washed and steamed. Twenty percent of the steamed rice is used to make koji. A small part (ca. 7%) of koji, steamed rice, and water are mixed as yeast seed culture. The rest of the koji, steamed rice, and water are then added to the seed culture for the main fermentation. After fermentation, the mash is pressed to separate sake from the residue or 'sake cake'. After storage and bottling, the sake is shipped (Brewing Society of Japan, 2016).

Rice polishing

Although the main component of the rice grain is starch, the outer layers and germ of brown rice contain many nutrients, such as protein, fats, minerals, and vitamins. These nutrients are important for the proliferation of koji mould and yeast; however, an overabundance of such substances tends to cause an unpleasant taste, inhibit the production of aroma, and overaccelerate fermentation. For this reason, not only the germ but also the outer layers of brown rice are removed by polishing.

The 'rice-polishing rate' is used to indicate how much the grain has been polished and is calculated as follows:

$$\text{Rice polishing rate} = \text{wt. of polished rice} / \text{wt. of brown rice}(\%).$$

The rice polishing rate of table rice is 92%, whereas that of rice used in sake making is usually 70%–40%.

A type of roller mill is used to polish the rice for sake brewing. Rice grains fed from a hopper are polished by a roller in the milling room. The roller, which is coated with a surface layer of carborundum (silicon carbide), rotates around a vertical axis and scrapes the surface of the grains. The scraped grains drop onto a sieve, which removes the rice bran, and then a bucket elevator carries the grains back to the hopper (Fig. 3.3). The process is repeated until the grains are polished to the desired extent. The lower the rice polishing rate, the higher the

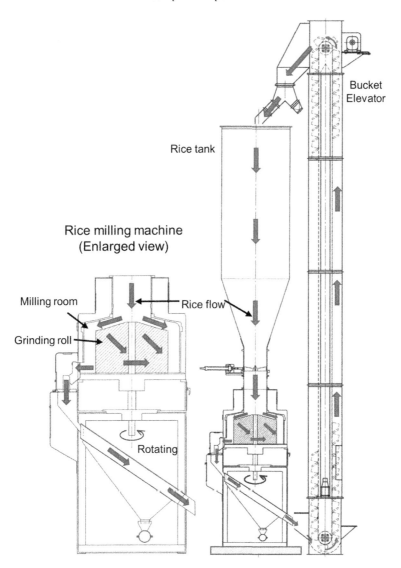

FIG. 3.3   Rice milling machine. From Shinnakano Industry Co., Ltd.

cost of producing the sake, but the result is a well-balanced sake with a pronounced aroma, smooth mouthfeel, and pleasant aftertaste. By contrast, sake made from rice with a high rice polishing rate generally has a rich taste with high acidity and a moderate aroma. Thus, the optimal rice polishing rate depends on the desired style of sake.

Rice steaming

The polished rice is washed in water and then left to steep in water. When it has absorbed water equivalent to 25%–30% of its original weight, the rice is removed from the water and steamed. During steaming, the starch is gelatinised, and the protein is denatured. Moreover, the grains are sterilised and absorb more water equivalent to about 10% of the weight before steaming. It takes 30–60 min to steam the rice completely. Traditionally, steaming was performed by a batch method using a large vessel. In the 1960s, steel belt-type equipment for continuous rice steaming was developed. Similarly, steel belt-type continuous cooling equipment was also designed for the cooling stage.

Koji making for sake

Koji making is said to be the most important process in the production of sake. Unlike barley malt, which is produced in external factories exclusively designed for the purpose, koji is prepared individually by each brewery.

The first step of koji making is to inoculate steamed rice with seed koji. The spores are allowed to germinate and start to spread their hyphae. After about 2 days, the steamed rice is entirely covered with hyphae. As the koji mould grows, it produces enzymes, which accumulate within the koji. Koji mould is most active at a temperature of around 36°C and ceases all activity above 45°C. In addition to temperature, enzyme production is affected by the moisture content of the rice grains. For this reason, the process is carefully controlled in a room where the temperature is kept at approximately 30°C and the relative humidity is maintained in the range of 50%–80%.

In small-scale breweries, koji is produced by hand using wooden trays in a dedicated room. During the manufacturing process, the koji is stirred by hand and its growth is homogenised with heat dissipation. Nowadays, semiautomatic or automatic koji making equipment that has mechanised the processes of heaping, turning, mixing, and discharging under optimal conditions is widely used amongst the larger sake breweries (Fig. 3.4).

Starter culture

The starter culture (also called seed mash or *shubo* in Japanese) is the preculture of the yeast that is used in the main fermentation. It should contain sufficient numbers of yeast cells, as well as lactic acid, which keeps the pH low. Essentially, the starter culture can be divided into two types: *kimoto* or *yamahaimoto*, in which lactic acid bacteria are used to produce lactic acid; and *sokujo* (meaning quickly made), in which brewing-grade lactic acid (90% solution) is added to the starter.

In *kimoto* and *yamahaimoto*, steamed rice, koji, and water are mixed at approximately 8°C. The temperature is gradually raised, which increases the number of lactic acid bacteria. About 2 weeks later, once enough acid has formed, the yeast is added. As the temperature is further raised slowly to around 22°C, the formation of alcohol and the increased acidity kill the lactic acid bacteria and any other bacteria, and only the yeast proliferates. It takes a month to make this type of starter. Therefore, to reduce the length and complexity of the *yamahaimoto* and *kimoto* process, the *sokujo* method was developed. The *sokujo* process is now the most widely used. However, sake made with a *yamahaimoto* or *kimoto* starter culture is said to have a more complex flavour than sake made with a *sokujo* starter culture, because *yamahaimoto* and *kimoto* involve the use of complex microbial interactions. The resulting sake is found to be rich in peptides.

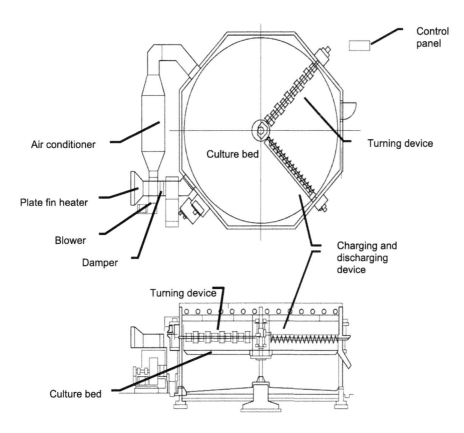

FIG. 3.4 Rotary-type automatic koji making equipment. From Fujiwara Techno-Art Co., Ltd.

Main fermentation

For the main fermentation (also called main mash, fermentation mash or *moromi* in Japanese), a standard ratio of steamed rice, koji (expressed as the weight of polished rice), and water of 80:20:130 is placed in the fermentation tank. The total amount of rice used for the main fermentation ranges from less than 1 to more than 10 metric tons.

The ingredients are not all added at once, but in three steps over 4 days. On the first day, the amount of steamed rice and koji placed in the tank is one-sixth of the total. The starter culture is also added on this first day. Nothing is added on the second day, allowing the yeast time to multiply. Two-sixths of the total amount is added on the third day and the remaining three-sixths on the fourth day. The temperature of the mix is 12°C in the first step but is gradually lowered to 10°C in the second step, and 8°C in the third step. This three-step preparation method is effective for maintaining the yeast cell density and avoiding contamination.

During the main fermentation, the enzymes in koji hydrolyse the components of steamed rice and the yeast simultaneously ferments the resulting sugars. The fermentation temperature is usually maintained in the range of 8–18°C. The process generally takes 3–4 weeks, yielding an alcohol content of about 16%–20% (v/v). Using a lower fermentation temperature of 12°C or less prolongs the fermentation period to around 4–5 weeks. Under this condition, the action of the yeast and the process of enzymatic hydrolysation are retarded, reducing the acidity and resulting in sake with a highly fruity aroma and clear taste. To take the advantage of natural cool conditions, sake has traditionally been made during the winter. Now, however, temperature-controlled tanks are widely used.

Pressing

When the fermentation is complete, the main mash is filtered through a cloth, and the residual rice, koji, and yeast are removed, leaving the newly prepared sake. This process is carried out either by placing the mash in a cloth bag and using equipment to apply pressure from above or by using a machine similar to a beer mash filter press. The cake leftover from the process is called sake cake (*sakekasu* in Japanese). In addition to residual rice and yeast, it contains about 8% alcohol by weight.

The sake obtained after pressing has some turbidity. Thus, after sedimentation at a low temperature, the sake is filtered. To avoid protein haze, fining is carried out with persimmon tannin or colloidal silica in some cases. The use of activated carbon to remove substances that cause colouring and flavour changes is also authorised for decolouring, flavour adjustment, and control of the ageing process.

Pasteurisation, ageing, and bottling

Even though at this stage, the sake contains approximately 16%–20% (v/v) alcohol, some species of *Lactobacillus* that are highly alcohol tolerant can spoil it. Thus, most sake undergoes pasteurisation at a temperature of 60–65°C before storage. In addition to killing microorganisms, pasteurisation will also inactivate the residual enzymes. If enzyme activity is allowed to continue, glycolytic enzymes will increase the sweetness and oxidases will alter the aroma. Many sake products are pasteurised again during bottling.

Some breweries store high-flavour sake called *ginjo-shu* below 10°C, but usually, the sake is stored at room temperature. Sake brewed in the winter is stored over the summer before shipping starts in the autumn, so it is sold about 1 year after production. Just before bottling, water is added to sake to lower its alcohol concentration to 14%–16% (v/v), which is still higher than the alcohol content of wine but suitable to enjoy with a meal.

Although most sake is consumed within one to 2 years after production, aged sakes are also produced. Sake kept in long-term storage undergoes colour changes due to the Maillard reaction between amino acids and sugars. There is also a decline in the fruity aroma derived from esters, and the aroma takes on a sweet, burnt quality. Sake aged for several years or even decades turns amber or dark amber in colour, and the flavour becomes more complex, resembling that of soy sauce, dried fruits, and nuts. In some cases, the sake may develop a sulphury aroma of dimethyl trisulfide, amongst other compounds (Isogai et al., 2005). The taste loses its astringency and sharpness and becomes more complex and bitter. Temperature and oxygen accelerate these reactions.

Honkaku shochu and awamori

Under Japanese liquor tax laws, shochu is classified based on distillation methods: 'pot-distilled shochu' and 'column-distilled shochu'. This section outlines the history and production of the former, a liquor up to 45% (v/v) alcohol distilled in a pot still, which is traditionally made in Japan. These beverages are called *honkaku shochu* and *awamori* in Japanese, where *honkaku* means authentic or genuine.

History of shochu

The distillation technique in Asia is said to have begun in Siam (now Thailand) around the 15th century. The technique was passed through Southeast Asia and China and eventually introduced to the Ryukyu Kingdom (now Okinawa Prefecture) in Japan, where awamori, a specialty shochu in Okinawa, was created. The oldest western record concerning shochu appeared in a 'Report on Japanese Matters' written by a Portuguese, Jorge Álvarez, in 1546. According to this record, an *'Orraqua* (a kind of spirit) made from rice' was drunk in the Yamagawa area (Ibusuki City, Kagoshima Prefecture). Around the same time, the first Japanese–Portuguese dictionary published in 1603 also explained shochu as Xŏchŭ: *Vinho que sesaz do fogo como vrraqua*, a liquor that is made with fire-like orraqua. Around that time, it seems that shochu was being made across Kagoshima Prefecture, and subsequently, it was introduced to other south Kyushu regions, Miyazaki Prefecture, and the Kuma region of Kumamoto Prefecture.

Historically, *A. luchuensis* was used to produce the koji for awamori making in the Okinawa region, whilst *A. oryzae* was used for shochu as well as sake making in Kyushu. Subsequently, it is thought that microbial contamination of shochu often occurred in the warm south Kyushu region. In the early 20th century, *A. luchuensis* was introduced to shochu making in Kyushu and was highly effective for avoiding microbial contamination. At present, shochu is made all over Japan; however, Kyushu and Okinawa remain the centres of production (Pellegrini, 2014).

Raw materials for shochu

In shochu making, the main materials and koji materials are different in many cases. The main materials are usually starchy materials such as rice, barley, sweet potato, or buckwheat. In addition, rice is often used to make koji, although barley and sweet potatoes may also be used. When rice is used to make koji and sweet potato is used as the main material, the final product is called 'sweet potato shochu'. In many cases, the main materials depend on the specialty of the individual shochu making regions. The shipment volume of different shochu types by raw material is shown in Table 3.3. Although shochu is made from many kinds of raw materials, below the three most-used raw materials (rice, barley, and sweet potatoes) are described.

TABLE 3.3 Shipment volumes of shochu by raw material (2018).

Raw materials	Volume (million litres)
Sweet potato	189.9
Barley	176.6
Awamori	20.0
Rice	16.5
Buckwheat	8.7
Brown sugar	7.5
Sake cake	0.5
Others	5.6
Total	425.3

Data from Japan Sake and Shochu Makers Association (JSS), 2020. https://www.japansake.or.jp/.

Rice

Rice shochu is made from rice with a polishing rate of about 90%, the same as that of table rice, whereas sake is generally made from rice with a polishing rate of 70%–40%, as described above. Rice is also used to make koji for sweet potatoes and some barley shochu because its physical properties are highly suitable for koji making.

Barley

Barley was introduced to Japan about 1800 years ago. Two-rowed barley is used to make barley shochu. In 2018, 76 thousand tons of two-rowed barley was cropped in Japan for shochu making. The remainder of the 164 thousand tons of barley needed for production is mainly imported from Australia (Ministry of Agriculture, 2020b).

Barley grains contain more lipids in the outer layer as compared with rice; therefore, about 30%–35% of the outer layer is milled away (barley polishing ratio 70%–65%) to make shochu (Iwami et al., 2005). Barley also absorbs water more rapidly and to a greater extent than rice, and the grains may easily stick to one another. Thus, control of water absorption is a key factor in the successful use of barley for koji making (Shimoda et al., 1998); as a result, rice is used to make koji for barley shochu in some regions.

There are a number of differences between barley shochu and whisky in terms of the production process, sulphur compounds, and maturation:

- Polishing process: In barley shochu making, the polishing process removes lipids and sulphur-containing amino acids to some extent before fermentation.
- Use of koji, instead of malt: S-methylmethionine increases during the germination of malt; by contrast, it is thought that the shochu mash made using koji is relatively low in sulphur components before distillation.
- Distillation: Barley shochu is often made by a single distillation under reduced pressure (see below). The temperature of reduced pressure distilling varies from 45°C to 55°C, and the distillation time is shorter than in whisky making. There is little change in

sulphur-containing amino acids and vitamins to volatile sulphur compounds at these lower temperatures as compared with whisky distillation.
– Maturation: the majority of barley shochu is shipped after a shorter maturation period relative to whisky; in addition, maturation of shochu in oak casks is not so common.

Sweet potatoes

Sweet potatoes arrived in Kagoshima from China via Okinawa in the 18th century and were first grown mainly in the Kyushu region. In 2018, 797 thousand tons of sweet potatoes were harvested in Japan; of which 278 thousand tons were harvested in Kagoshima Prefecture. Of these, 213 thousand tons were used for shochu making (Ministry of Agriculture, 2020c).

Sweet potatoes rot easily, and the resulting off-flavours and bitterness can reduce the quality of the shochu; thus, the ends of the potatoes and any damaged or scarred sections are removed soon after the potatoes are harvested and washed. Sometimes, the skin is also peeled. The cleaned potatoes are then steamed in a large vessel.

Sweet potato shochu contains monoterpene alcohols, which have the aroma of Muscat grapes, lychee, citrus fruits, and flowers and are unique to this type of shochu. Monoterpene alcohols in sweet potatoes exist in the form of glycosides and are broken down into sugar and monoterpene alcohols by koji enzymes (Ohta et al., 1990). Yeast alters the composition of the monoterpene alcohols and distillation under acidic conditions changes their structures even further. β-Damascenone, a norisoprenoid with a sweet aroma similar to apple compote, is characteristic of the aroma of sweet potato shochu (Yoshizaki et al., 2011).

Shochu made from purple-coloured sweet potatoes also contains diacetyl, which imparts an aroma similar to dairy products, and shochu made from orange-coloured sweet potatoes contains β-ionone derived from carotenoids, which has an aroma associated with boiled carrots or pumpkin and violets (Kamiwatari et al., 2006). Relative to the influence of cereal crop varieties, different types of sweet potatoes have a more significant effect on the flavour of sweet potato shochu (Photo 3.1).

PHOTO 3.1 Sweet potato. From Japan Sake and Shochu Makers Association (JSS), 2020. https://www.japansake.or.jp/.

Other raw materials

In some types of shochu, buckwheat and chestnut are used as starchy raw materials. Sake cake, a by-product of sake, has also been used for shochu since the 17th century. The Amami Islands in Kagoshima Prefecture produce a unique shochu, called brown sugar shochu, which uses koji for the starter culture and brown sugar for the main fermentation.

Shochu production processes

Outline of shochu making

First, koji, *Aspergillus luchuensis* cultured on rice or other grains, is made and then the starter culture of yeast is prepared by using all of the koji, water, and a small amount of yeast (first step). After the yeast has fully grown, the main ingredient, namely rice, barley, or sweet potato, is steamed and added to the starter culture with water for the main fermentation (second step). When the fermentation has finished, it is transferred into a pot still for single distillation, which produces the shochu.

In ancient times in Kyushu, the fermentation mash was prepared by mixing all of the materials at once or by a two- or three-step preparation with a starter culture similar to sake making. The two-step method mentioned above was developed for sweet potato shochu in Kagoshima in the early 20th century and has become widely used for barley and rice shochu in other regions (Fig. 3.5).

Koji making for shochu

To prepare rice koji made with either black or white koji mould, freshly steamed rice is cooled, koji mould spores are sprinkled all over it, and the koji is left to develop over a period of about 40h. The temperature is initially set high (40–42°C) in the first step to encourage the growth of koji mould and the production of enzymes. The temperature is then lowered (33–35°C) in the second step to accelerate the production of citric acid. Shochu producers use the same equipment for koji production that is used for sake. Rotary drum-type koji making equipment is also used for washing and steaming the raw materials and making koji.

Starter culture (seed mash, first *moromi*)

The starter culture (also called seed mash or first *moromi*) is a preculture used to obtain a sufficient number of yeast cells so that the main fermentation can proceed without contamination. Koji, water, and a small amount of yeast cells are placed in a suitable container (stainless steel tank or earthenware vessel). Enzymes from the koji convert the starch in the koji into sugar, and the yeast uses the sugar to multiply. The main difference between the starter culture of shochu and that of sake is that all of the koji is used for shochu, whilst only part of the koji and steamed rice is used for sake. In addition, the citric acid of shochu koji lowers the pH to around pH3.0 to avoid contamination, whilst lactic acid is used for this purpose in sake making.

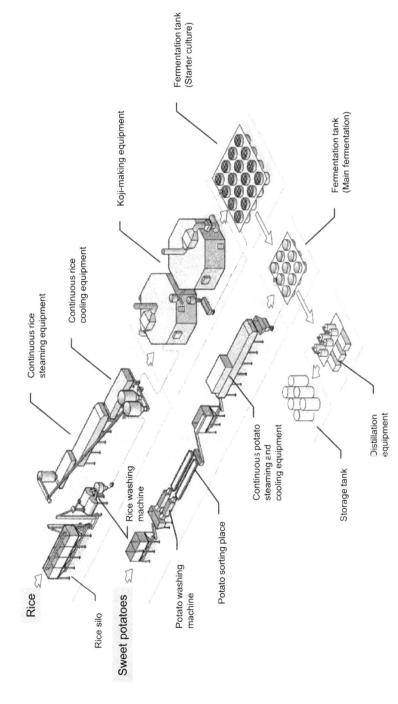

FIG. 3.5 The outline of sweet potato shochu making. From Fujiwara Techno-Art Co., Ltd.

After 3–8 days of fermentation at a temperature of 30°C, the starter culture is complete. Sake mash ferments at temperatures below 15°C, but the temperature of the shochu mash may be higher than 30°C at times. Thus, the shochu yeast strains must be able to tolerate low pH and high temperature conditions.

Main fermentation (fermenting mash, second *moromi*)

The main fermentation (also called the fermenting mash or second *moromi*) is the step in which the steamed main ingredient is added to the starter culture and fermented. The starch in the main material is hydrolysed into sugars by the koji enzymes present in the starter culture. The resulting sugars are simultaneously converted into alcohol by the yeast. The maximum temperature of the main fermentation reaches 28–32°C. To avoid bacterial contamination, a low pH and high yeast density are essential. Fermentation is complete in about 2 weeks, producing a final alcohol content of approximately 16%–19% (v/v). For sweet potatoes, the final alcohol content of the mash is a bit lower, around 14%–15% (v/v), because the potatoes have a higher water content than grains.

Distillation

At the end of fermentation, the mash is moved to a pot still for distillation. Two heating methods are used for distillation: a direct method, in which steam is blown into the fermented mash; and an indirect method, in which steam is blown into a steam jacket or other device. The selected method depends on the physical properties of the mash. For example, for mashes with high viscosity, such as for sweet potato shochu, directly blown steam is used so that the mash does not burn and stick to the still.

Most systems for distilling shochu are made from stainless steel. Historically, copper stills have not been used to make shochu, although there are a few examples in which copper has been used in a small part of the still, namely the rising neck or swan neck. A reduced pressure still, also called a vacuum still, was developed in Fukuoka Prefecture in 1973 and introduced into the production of rice and barley shochu. Distillation under reduced pressure, which means decreasing the internal pressure of the distillation equipment, lowers the boiling point. Usually, this type of shochu is distilled at 45–55°C. Thus, the distilled shochu contains a lower concentration of compounds with higher boiling points and has a more flowery aroma and a lighter taste (Photo 3.2).

These lighter types of shochu have contributed to an increase in the consumption of shochu. Although the main production and consumption region of shochu remains south Kyushu, consumption in other regions such as Tokyo has increased significantly. To produce shochu that appeals to the tastes of diverse consumers, both atmospheric pressure stills and reduced pressure stills are used as the situation demands. In some cases, producers may blend the two types of shochu to benefit from the advantages of each method or may lower the degree of decompression to distill at a pressure somewhere between atmospheric and reduced pressure (i.e. slightly reduced pressure).

In general, shochu is made by single distillation, resulting in an alcohol content of approximately 37%–43% (v/v). Some shochu distilleries separate out the foreshots, whilst others do not. However, feints are carefully separated to avoid a burnt aroma.

PHOTO 3.2   Shochu still, dual-use still of normal pressure and reduced pressure. From Sanwa Shurui Co., Ltd., Hita distillery.

Maturation and packaging

After coarse filtering, the distillate (undiluted shochu with an alcohol content of 37%–45% (v/v)) is stored and matured in a container, often a stainless steel tank or an earthenware vessel or cask, to adjust the harmony of flavours. Freshly distilled shochu has a rather piquant smell and a rough pungent taste; however, with maturation, it develops a smooth and mellow taste.

The distillate obtained under atmospheric pressure will contain plenty of fatty acid derivatives originating from the main materials and the koji. These are primary factors in the sweetness and smoothness of the shochu but, when the alcohol content or the temperature is lowered, they become insoluble and make the shochu cloudy. Furthermore, when they rise to the surface and come into contact with air, the lipids undergo oxidative decomposition and change into components with an unpleasant or 'oily' smell. Since the development of a method to remove the oil by cooling and filtering, the problem of an oily smell has been resolved.

Just before the shochu is bottled and shipped, water is added to lower the alcohol content to 20% or 25% (v/v). In the 2018 shipment of shochu, the ratio of products with a 20%, 25%, and 35% or higher alcohol content was 12.6%, 86.4%, and 1.0%, respectively.

One of the features of shochu is the flavour derived from its raw materials. As compared with other kinds of spirits such as whisky, many shochu products are not aged for a very long time. Usually, shochu is shipped within 1 year of being distilled. Occasionally, it is stored long term in earthenware vessels or wooden casks. When it is stored in earthenware vessels, the contact with oxygen and the catalytic effect of some minerals in the container create a unique flavour.

Varieties and regions

Four main varieties of shochu are currently produced: awamori, rice, sweet potato, and barley shochu (Table 3.4). In addition to the main raw material, each type of shochu may differ in the production method as outlined below.

TABLE 3.4   Major honkaku shochu and awamori.

Shochu	Awamori	Rice shochu	Sweet potato shochu	Barley shochu
Main production area	Okinawa	Kumamoto	Kagoshima Miyazaki	Oita
Koji material	Indica rice	Rice	Rice	Barley
Polishing rate (%)	86–90	86–90	86–90	65
Main material	Rice koji	Rice	Sweet potato	Barley
Polishing rate (%)	–	86–90	–	65–70
Starter culture	Rice koji	Rice koji	Rice koji	Barley koji
Main fermentation	–	Rice	Sweet potato	Barley
Alcohol content of fermented mash (%)	16–18	18–19	14–15	17–19
Alcohol content of distillate (%)	44–46	42–45	37–40	43–45
Spirit yield (L/t)	420–440	450–470	200–220	420–440

Data from Kumamoto Taxation Bureau. Shochu Report BY2017 (in Japanese); Okinawa Regional Taxation Office. Awamori Report BY2017 (in Japanese).

Awamori, GI Ryukyu

Geographical area: Okinawa Prefecture
Raw materials:

– Rice koji produced by black koji (*Aspergillus luchuensis*)
– Water collected in Okinawa Prefecture

Rules for GI Ryukyu:

– Must be fermented, distilled, stored, and bottled in Okinawa Prefecture
– Must be distilled from fermented mash made using rice koji and water in a pot still

Number of distilleries (2018): 45

The name 'awamori' appears in lists of items presented to Shogun in the Edo period (1671). In the production of awamori, all of the ingredient rice is used to make koji. In other words, awamori uses an all-koji preparation method. Furthermore, long, narrow, and Indica-type rice varieties from Thailand have been used for koji making since the early 20th century. The characteristic aroma compound of awamori, 1-octen-3-ol, is derived from black koji (Osafune et al., 2019).

During production, the fermenting mash reaches a maximum temperature of 27–31°C and, after 14–18 days, the alcohol content reaches 17%–18% (v/v). Distillation under atmospheric pressure is often used, but some products are distilled under reduced pressure. Awamori that has been matured for 3 years or more after production is called *kusu* (long-aged spirit). *Kusu* has an alcohol content of 40% (v/v) or higher, and storage and maturation impart a smooth taste with less alcoholic sharpness and the sweet aroma of vanillin. The vanillin in *kusu* is derived from ferulic acid released from the cell walls of rice by the koji enzymes. Ferulic acid is converted into 4-vinylguaiacol mainly by the koji enzyme ferulic acid decarboxylase and also by the heat of distillation in a small part. Some yeast strains also have this enzymatic activity. The 4-vinylguaiacol goes through further chemical changes during storage to become vanillin.

Sweet potato shochu, GI Satsuma

Geographical area: Kagoshima Prefecture*
Raw materials:

- Sweet potatoes produced in Kagoshima Prefecture*
- Rice koji or sweet potato koji made from sweet potatoes
- Water collected in Kagoshima Prefecture*

Rules for GI Satsuma:

- Must be produced, stored, and bottled in Kagoshima Prefecture*
- Must be distilled from fermented mash made using koji, sweet potatoes, and water in a pot still
  *Except for Amami City and Oshima District.

Number of distilleries (2018): 89

Satsuma was the name of a fiefdom in the Edo period. This area, which is covered with volcanic ash and sand, is not suited to rice cultivation. Thus, the Satsuma clan encouraged the cultivation of sweet potatoes. Satsuma (Kagoshima) is the largest region producing sweet potatoes, and sweet potato is also called satsuma *imo* (potato). Sweet potato shochu has been made in this region since about the early 18th century. In addition to Kagoshima, sweet potato shochu is made in the southern part of Miyazaki Prefecture.

Barley shochu, GI Iki

Geographical area: Iki City, Nagasaki Prefecture
Raw materials:

- Barley
- Rice koji
- The weight ratio of rice koji to barley must be approximately 1:2
- Water collected in Iki City in Nagasaki Prefecture

Rules for GI Iki:

- Must be produced, stored, and bottled in Iki City in Nagasaki Prefecture
- The starter culture must be prepared with rice koji and water; the main fermentation is then prepared by adding steamed barley and water; the shochu must be distilled in a pot still

Number of distilleries (2018): 7

Iki Island is located between Kyushu and the Korean Peninsula and was an important transportation route for culture and goods in ancient times. On Iki Island, which is said to be the birthplace of barley shochu, it is thought that shochu was first made with barley for private use in the mid-18th century. Iki shochu is made by a traditional method that uses rice koji and barley.

Barley shochu, Regional collective trademark: Oita *mugi* (barley) shochu

Area: Oita Prefecture
Raw materials:

– Barley koji and barley

Number of distilleries (2018): 23

In 1973, a shochu distillery in Oita Prefecture developed a method to make barley koji, which facilitated the production of all-barley shochu. At the same time, the innovation of vacuum distillation enabled the production of Oita barley shochu with a soft and refreshing flavour, which created a boom in shochu consumption throughout Japan.

Rice shochu, GI Kuma

Geographical area: Kuma District and Hitoyoshi City in Kumamoto Prefecture
Raw materials:

– Rice and rice koji made in Japan
– Water collected in Kuma District and Hitoyoshi City in Kumamoto Prefecture

Rules for GI Kuma:

– Must be produced, stored, and bottled in Kuma District and Hitoyoshi City in Kumamoto Prefecture
– Must be distilled from fermented mash made with rice, rice koji, and water, or with rice koji and water in a pot still.
   Note: Even when the fermented mash is made with rice koji and water, it should be prepared with the two-step method with a starter culture and a main fermentation.
   Number of distilleries (2018): 27

Kuma shochu is thought to have originated around the middle Edo period (late 16th to early 17th century). In the Kuma region, boiled brown rice and yellow koji were used in the mash preparation until the 1920s. Boiled rice, koji, and water were put into a small bucket or earthenware vessel, allowed to ferment naturally for about 30 days, and then distilled. The lactic acid bacteria would have produced lactic acid at low temperatures, and then the yeast would have grown slowly and fermented the substrate to produce alcohol, in a fermentation style similar to the *kimoto* starter culture of sake. Lastly, wood ash was added just before distillation to neutralise the acid. Today, Kuma distillers use polished rice and white koji and have increased the use of vacuum distillation.

Japanese market

Due to the decrease in population and changes in the demographic composition of Japan, the total domestic shipment of alcoholic beverages has declined from a peak of 10.17 million KL in FY1999 to 8.68 million KL in FY2018 (National Tax Agency, 2020a,b).

The domestic shipment of sake dropped to 0.49 million KL in 2018 from its peak of 1.77 million KL in FY1973. Despite the decline in shipment volumes, the share of premium sake in the specially designated sake category has increased, and the average shipment value (yen/L) has been on an upward trend since 2012.

In 1970, the shipment volume of shochu was only 53 thousand KL, and it was a local spirit consumed in the southern Kyushu area. Around 1975, sweet potato shochu 'Satsuma Shiranami' (Satsuma Shuzo Co., Ltd., Kagoshima) became popular in northern Kyushu. Then, barley shochu 'Iichiko' (Sanwa Shurui Co., Ltd., Oita), which debuted in 1979 with a light flavour generated by vacuum distillation, expanded the market to Tokyo. In the 2000s, sweet potato shochu 'Kuro Kirishima' (Kirishima Shuzo Co., Ltd., Miyazaki) dramatically increased the shipment volume. Since 2009, the shipment of sweet potato shochu has replaced barley shochu as the number one preferred shochu.

In 1987, the GATT panel mentioned the liquor tax rate on shochu as being a violation of the principle of national treatment. Subsequently, the tax rate was increased five times—1989, 1994, 1997, 1998, and 2000—resulting in a quadrupling of the original rate. At present, the tax rate on all spirits, including whisky and shochu, is the same. Despite the increase in retail prices due to the tax hike, shochu shipment has continued to increase and is now close to 0.44 million KL, which is similar to sake (2018). Shochu has already overtaken sake in terms of 100% alcohol equivalent shipments (Fig. 3.6).

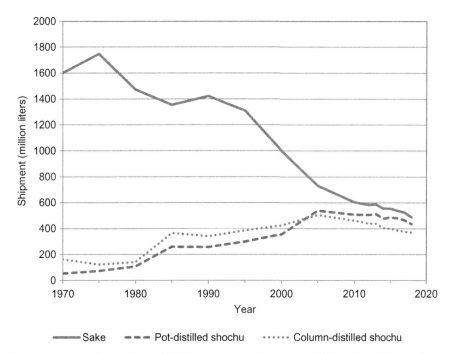

FIG. 3.6 Changes in market volumes of sake and shochu. Data from National Tax Agency, 2020b. Alcohol Beverages Report. https://www.nta.go.jp/taxes/sake/shiori-gaikyo/shiori/01.htm (in Japanese).

Export of sake and shochu

The value of sake exports has grown for 10 consecutive years, tripling from 7.2 billion yen to 23.4 billion yen between 2009 and 2019. In 2019, the United States was the top exporting country worth 6.76 billion yen followed by China with 5.0 billion yen and Hong Kong with 3.94 billion yen. Export volume is estimated to be 5% of domestic shipments. Sake has long been produced overseas in the order of South Korea, the United States, China, Taiwan, and Brazil, and more recently, the establishment of craft sake breweries is also expanding globally. The increase in exports of sake is related to the growth in the number of Japanese restaurants. The future of sake will depend on whether it can be enjoyed not only with Japanese cuisine but also with international cuisine.

The export value of shochu was 1.56 billion yen, with China 0.53 billion yen, the United States 0.38 billion yen, and Thailand 0.1 billion yen. Exports in Asia, totalling 1.07 billion yen, are 68.8% of the total. The export volume is estimated to be 0.5% of the domestic shipment volume. Shochu is rarely produced in other countries, except in a few craft distilleries.

Shochu is usually diluted from 40% or more of the original alcohol to 25% before bottling, and then it is also diluted with hot or cold water to about 12%–15% when being consumed and enjoyed during meals, in the same way as wine or sake. These unique characteristics of these spirits, which are advantageous in Japanese markets, have not been communicated to the rest of the world. Some mixologists perceive shochu in the bar scene as a new spirit using koji, which may be of considerable significance for future exports.

References

Akiyama, H., 2010. SAKE: The Essence of 2000 Years of Japanese Wisdom Gained From Brewing Alcoholic Beverages From Rice. Brewing Society of Japan, Tokyo.

Brewing Society of Japan, 2016. Textbook of Sake Brewing. Brewing Society of Japan, Tokyo.

Hong, S., Lee, M., Kim, D., Varga, J., Frisvad, J., Perrone, G., Gomi, K., Yamada, O., Machida, M., Houbraken, J., Samson, R., 2013. *Aspergillus luchuensis*, an industrially important black *Aspergillus* in East Asia. PLoS One 8, e63769. https://doi.org/10.1371/journal.pone.0063769.

Isogai, A., Utsunomiya, H., Kanda, R., Iwata, H., 2005. Changes in the aroma compounds of sake during aging. J. Agric. Food Chem. 53, 4118–4123. https://doi.org/10.1021/jf047933p.

Iwami, A., Kajiwara, Y., Takashita, H., Omori, T., 2005. Effect of the variety of barley and pearling rate on the quality of shochu koji. J. Inst. Brew. 111, 309–315. https://doi.org/10.1002/j.2050-0416.2005.tb00689.x.

Iwano, K., Mikami, S., Fukuda, K., Shiinoki, S., Shimada, T., 1986. The properties of various enzymes of shochu koji (in Japanese). J. Brew. Soc. Jpn 81, 490–494. https://doi.org/10.6013/jbrewsocjapan1915.81.490.

Kamiwatari, T., Setoguchi, S., Kanda, J., Setoguchi, T., Ogata, S., 2006. Effects of a sweet potato cultivar on the quality of imo-shochu with references to the characteristic flavor (in Japanese). J. Brew. Soc. Jpn 101, 437–445. https://doi.org/10.6013/jbrewsocjapan1988.101.437.

Ministry of Agriculture, 2020a. Rice for Sake Brewing (in Japanese) https://www.maff.go.jp/j/seisaku_tokatu/kikaku/sake.html.

Ministry of Agriculture, 2020b. Supply, Demand and Price of Wheat and Barley (in Japanese) https://www.maff.go.jp/j/seisan/boueki/mugi_zyukyuu/.

Ministry of Agriculture, 2020c. Data on Potatoes and Starches (in Japanese) https://www.maff.go.jp/j/seisan/tokusan/imo/siryou.html.

National Tax Agency, 2020a. Information on GIs Protected in Japan. https://www.nta.go.jp/english/taxes/liquor_administration/geographical/02.htm.

National Tax Agency, 2020b. Alcohol Beverages Report (in Japanese) https://www.nta.go.jp/taxes/sake/shiori-gaikyo/shiori/01.htm.

Nishiya, T., 1993. Technological relationship between the brewing of sake and beer. In: Proceedings of the 24th EBC Congress, Oslo. IRL Press, Oxford, pp. 619–634.

Ohta, T., Ikuta, R., Nakashima, M., Morimitsu, Y., Samuta, T., Saiki, H., 1990. Characteristic flavor of kansho-shochu (sweet potato spirit). Agric. Biol. Chem. 54, 1353–1357. https://doi.org/10.1271/bbb1961.54.1353.

Osafune, Y., Toshida, K., Han, J., Isogai, A., Mukai, N., 2019. Characterization and threshold measurement of aroma compounds contributing to the quality of Honkaku shochu and Awamori. J. Inst. Brew. 126, 131–135. https://doi.org/10.1002/jib.589.

Pellegrini, C., 2014. The Shochu Handbook—An Introduction to Japan's Indigenous Distilled Drink. Telemachus Press, LLC, Dublin, Ohio.

Shimoda, M., Ogawa. K., Takashita, H., Omori, T., 1998. Characteristics of water uptake of Australian polished barley in shochu-making. J. Inst. Brew. 104, 33–35. https://doi.org/10.1002/j.2050-0416.1998.tb00971.x.

The Scientific Conference of Brewing Society Japan, 2006. Koji fungi (Kōji-kin) as the National Fungi. https://www.jozo.or.jp/gakkai/wp-content/uploads/sites/4/2020/01/gakkai_koujikinnituite2.pdf.

Yoshizaki, Y., Takamine, K., Shimada, S., Uchihori, K., Okutsu, K., Tamaki, H., Ito, K., Sameshima, Y., 2011. The formation of β-damascenone in sweet potato shochu. J. Inst. Brew. 117, 217–223. https://doi.org/10.1002/j.2050-0416.2011.tb00464.x.

4

Spirit beverage development in the Asia-China region

Jeng-Ing Chen

Laizhou Distillery, Qionglai, China

Introduction

The relationship between alcoholic fruit/beverage and humans dates back at least to 7000 BCE when solid evidence can be found for the production of alcoholic drinks. Clay pots excavated in Jiahu, Henan Province, suggest that alcoholic beverages were made from mixed substrates, such as grape juice, cereal grains, and honey (McGovern et al., 2004; McGovern et al., 2005). The taste of the fermented drink might have been sweet, sour, and 'funky' because of the lack of modern brewing techniques and the concept of pasteurisation. The flavour may have been more like Belgium Lambic beers. The Chinese character '酒' (alcohol or wine) illustrates a very vivid concept. The radical (left part) is 'water' and the suffix (right part) is the shape of a 'clay pot'. The meaning of '酒' is water flowing out from a clay pot, similar to Mesopotamian brewed ale/beer or Roman grape wines also made in clay pots.

Alcoholic beverage brewing activity may have started during the Xia Dynasty in China. One sentence stating 'Yidi makes wine' was found in an ancient Chinese history record the 'Lüshi Chunqiu' (a history record of Chunqiu-era written by Lüshi). Also, a story recorded in the 'Zhan Guo Ce' (a history record of Zhanguo-era) is described as follows:

Once upon a time, the daughter of the emperor Yu asked Yidi to make wine. Yu drank the wine and said 'wine is beautiful and I believe there will be emperors losing their kingdoms by binge drinking.' After that, Yu stopped drinking wine and stayed away from Yidi.

Before baijiu was created, the major fermented alcoholic beverage enjoyed in China was 'rice wine', which has a long history. The earliest category of rice wine is described in 'Rites of Zhou'. It marked three different types of wine: event, season, and clear wine. Event wine had the shortest brewing time and was used for ceremonies and banquets. Seasonal wine was brewed in the winter and drunk in the spring. Clear wine was the best wine and was conditioned for longer durations.

In the past, brewing techniques for rice wine were held by royal families, but after wars between clans and countries, brewers were forced to relocate, thereby passing the techniques to other regions. This differs from the case in the western world, where the most brewing techniques were held by monasteries. The Chinese religion, Taoism, had no interest in brewing. In addition, drinking is forbidden in the Buddhist discipline. Therefore, the brewing techniques were documented and taught by the Imperial Court or by the local brewers. The *'Qi Min Yao Shu'* (techniques of how common people make a livelihood), written by Sixie Jia (circa 544 CE), was the first official agricultural document that recorded rice wine brewing methods. The first brewing technical document, the *'Bei Shan Jiu Jing'*, was composed by Yizhong Zhu in the Song Dynasty (no specific time). This book collected various methods of making 'Qu' and fermented substrates and how different styles of rice wine (also called yellow wine ever since) were produced.

Interestingly, 'boiled wine' was invented in the Tang Dynasty (618–907 CE). Many scholars thought it was a kind of distilled spirit but based on the description that boiling the wine could turn the colour into amber-yellow and extend its storage time, the boiling step was more likely to be a means of pasteurisation. People at that time already knew how to preserve the wine and keep it from turning sour by using heat.

It was thought by many researchers that the first distilled spirit appeared in the Han Dynasty (202 BCE to 220 CE) after a Han bronze still was unearthed. However, if the discovered bronze apparatus was used for spirit distillation, more descriptions of it should have been found in written documents. The Emperors of the Han Dynasty fancied the concept of reaching immortality, so those stills could also have been an apparatus for extracting a floral essence or to make an elixir stone.

The first distilled spirit was recorded by Derun Zhu (1343 CE) (Yuan Dynasty, 1271–1368 CE). According to his article *'Zha Lai Ji Jiu Fu'* (a compilation of liquor production), he described the shape of the still and how it was utilised. Additionally, he also documented when rice wine should be poured into the still, how the fire was controlled, an imagination of how distillation occurred in the still, and about the spirit that flowed from the spout. The technique of spirit distillation originated prior to this record, as a bronze still dated to the Song-Jin Dynasty (1115–1234 CE) was discovered in Qinglong County, Hebei Province, and its shape fitted exactly the description in the record found in the Yuan Dynasty.

One of the most important and interesting facts was found in the *'Yin Shan Zheng Yao'*, a book of food and nutrition written by a Mongolian, Sihui Hu, in 1330. A spirit drink named and pronounced 'a-la-ji' in Mongolian, which one can speculate, might have meant 'alembic'. During the Yuan Dynasty, Mongolia expanded their territory into eastern Europe and northern Iraq. The alembic still and distillation techniques were also obtained by them at that time. This hypothesis is supported by two statements. *'Ju Jia Bi Yong Shi Lei Qun Ji'*, a book of common households and livelihoods, published in the Yuan Dynasty, noted that a strong beverage named 'Southerners' spirit drink' was produced by an alembic-shape-like still with a horizontal straight pipe. The vapour could move freely in the pipeline and be discharged from the nozzle. Another statement was written by Mu Zhang in *'Tiao Ji Yin Shi Bian'*, in 1823. This was a book of Chinese meals with healing and medicinal functions (during the Qing Dynasty and what a coincidence!). Zhang's finding was that burnt wine, fired wine in other words, which was also known as 'a-la-ji' in *'Yin Shan Zheng Yao'*, was a loanword. This alcoholic beverage was not made using traditional Chinese methods. Instead, its production was learnt from Siam (nowadays Thailand) and Holland. It can therefore be said that the Chinese and Europeans drank aqua vitae almost since the same era.

Baijiu consumption started to grow during the Yuan Dynasty, and the industry began to prosper. The process of distillation was employed to create various spirit drinks, such as north-western grape spirits, Mongolian horse milk spirits, northern sorghum spirits, south-western grain spirits, south-eastern rice spirits, and distilled spirits from yellow wine (also called huangjiu) spent grain (similar to a rice-wine-grappa). The variety of spirit drinks depends on the grain bill, the type of Qu, the microorganisms, solid/liquid fermentation, distillation process, distillation times, spent grain reprocessing methods, ageing time… etc. More information can be found on the Chinese Alcoholic Drinks Association website (https://www.cada.cc/).

Chinese baijiu is the representative of Chinese spirit drinks, on a par with Scotch Whisky, American Bourbon, French Cognac, and Mexican Tequila. It has a long history, and the flavour types are diverse. The combination of terroir and blender's effort can create an astonishing baijiu!

The flavours of Chinese baijiu

There are many flavour types of baijiu but the first and second most popular flavour types are strong-flavoured (68.5%) and mild-flavoured (13%). Table 4.1 lists the major types of baijiu. The well-known Jiang flavour baijiu, Maotai, occupies only 2% of the types (Yu, 2018).

The flavour type is complex. The relationship and similarity are demonstrated in Fig. 4.1. Most Chinese consumers cannot distinguish the types. The taste notes of baijiu are always

TABLE 4.1 Flavour types of baijiu and famous brands.

Flavour	Brand
Strong flavour	Wuliangye, Jian Nan Chun, Luzhou Laojiao, Yanghe, Wenjun, Shui Jing Fang
Mild flavour	Fenjiu, Red Star Erguotou
Rice flavour	Guilin Sanhua
Feng flavour	Xifeng
Chi flavour	Yuping shao
Zhima flavour	Meilanchun
Te flavour	Site
Nongjiang flavour	Baiyunbian
Laobaigan flavour	Hengshui Laobaigan
Jiang flavour	Maotai, Langjiu
Yao/Herbal flavour	Dongjiu
Others	–

Adapted from Yu, C.W., 2018. Traditional Baijiu Production Techniques, second ed. China Light Industry Press. Ltd, Beijing. and GB/T 17204-2008, Chinese Standard. 2008. Whisky. Available in English from https://www.chinesestandard.net/PDF/English.aspx/GBT11857-2008: Classification of Alcoholic Beverage.

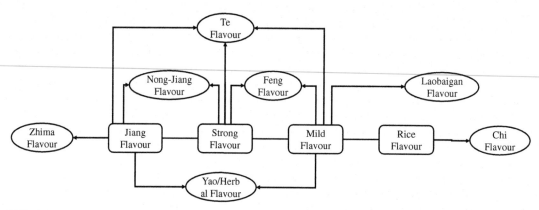

FIG. 4.1   The relationship and similarity of flavour types. *Adapted from Yu, C.W., 2018. Traditional Baijiu Production Techniques, second ed. China Light Industry Press. Ltd, Beijing and GB/T 17204-2008, Chinese Standard. 2008. Whisky. Available in English from https://www.chinesestandard.net/PDF/English.aspx/GBT11857-2008.*

obscure and abstract and so it is quite difficult for consumers to describe the flavours. Usually, only well-trained panels and blenders can distinguish differences in flavours. Therefore, consumers tend to drink local or well-known types.

A baijiu distillery would not produce only one kind of baijiu. Many blenders and distillers modify the production and blending to produce different flavour types. The boundary of flavour types is therefore blurred. Baijiu experts reached an agreement that the classification of baijiu should be upgraded to 'less flavour types but more styles' (Wang et al., 2009) and that a specific flavour wheel for baijiu should be developed, as this would lead to more customer appreciation of styles.

In addition to the flavour types, there is another method for distinguishing product types of baijiu. The types are now listed in the new Terminology and Classification of Alcoholic Beverages (GB/T 17204-2021) and the extraction is as follows:

Traditional baijiu: The spirit is produced using grain cereals in a solid or semisolid state which are saccharified and fermented by daqu, xiaoqu, or fuqu, and distilled, matured and blended. This spirit product cannot directly or indirectly have added additional edible alcohol or any nonbaijiu-fermented flavouring and colour additive. This baijiu should have its own flavour characteristics.

Liquid fermentation baijiu: The spirit is produced using grain cereals in a liquid state fermenting process (note that this includes saccharified and distilled but this is not mentioned). Edible grain spirits can be added but nonbaijiu-fermented flavouring and colour additives are not permitted.

Traditional and liquid fermentation baijiu: This spirit could be produced by (1) mixing liquid fermentation baijiu or edible grain spirits with solid fermentation substrate or specially prepared substrate to steam-distill or normal distill (liquid distillation) or (2) blending liquid fermentation baijiu or edible grain spirits with traditional baijiu.

These spirit products cannot directly or indirectly have any added nonbaijiu-fermented flavouring and colour additive. This baijiu should have its own flavour characteristics.

The baijiu production process is varied and complex. This chapter will introduce the basic production process of mild flavour baijiu and strong flavour baijiu. The process of baijiu production is explained in conventional brewing or distilling descriptions, which is more comprehensive for readers who are unfamiliar with baijiu production.

The basic production process

The process of making traditional solid fermentation baijiu is very complicated, but it can be summarised in these five steps:

1. Qu production
2. Substrate preparation
3. Distillation
4. Maturation
5. Blending and bottling

Qu production

Qu is the starter for solid fermentation. It is an extremely important step for the characteristics of the end product. The raw materials for Qu are sorghum, rice, wheat, barley, maize, and/or pea. Certain types of Qu will have Chinese herbs added for the suppression of bacterial growth or for special flavours. Microorganisms are allowed to grow naturally in a Qu room (Fig. 4.2) and to fall on the Qu chunk (Fig. 4.3).

FIG. 4.2 A typical Qu room.

FIG. 4.3   A Qu chunk.

FIG. 4.4   The fundamental process of Qu production.

The cultivation duration depends on the specific distillery. Some will grow for only 1 month, while others will last for a year. Some distilleries control the temperature and moisture of the Qu room. The Qu chunk/cake is turned or flipped to aid growth during cultivation.

The main microorganisms in the Qu chunk are *Aspergillus, Rhizopus, Mucor, Saccharomyces, Lactobacillus,* and *Acetobacter.* These microorganisms will saccharify and ferment at the same time, producing unique aroma congeners and ethanol. The process of Qu production is illustrated in Fig. 4.4.

Substrate preparation

A substrate is the nutrition for the Qu and the environmental microorganisms. Every distillery has its own grain bill, of which the ingredients affect the flavour and taste significantly. Generally, a sorghum-based aroma is considered as elegant, maize-based as sweet, rice-based as clean, glutinous-rice-based as soft, and wheat-based as astringent.

After milling, the grist is mixed with water. Strong flavour baijiu distilleries use a large proportion of the distilled spent substrate to blend with the new grist but mild flavour baijiu distilleries do not carry out this process. Steamed rice husks are added to the substrate. This creates a void/space for the steam to penetrate the substrate during the cooking process and allows oxygen to enter the substrate during fermentation.

The cooking process is conducted after the blending of the substrate. The purpose of cooking is to both gelatinise the starch and disinfect the substrate. Strong flavour baijiu distilleries cook the substrate and distil the fermented substrate at the same time, while mild flavour baijiu distilleries separate the cooking and the distillation process. The previous extracts are more grainy flavour congeners.

A cooked substrate is cooled rapidly and mixed with Qu at the correct temperature and then the mixture is transferred to pits for the fermentation stage (Figs. 4.5 and 4.6).

FIG. 4.5   Substrate being mixed with Qu.

FIG. 4.6   Substrate in a fermentation pit.

The duration of fermentation again depends on the baijiu flavour types and the specific distillery; typically, 40–60 days, 21–28 days, and 10–15 days for strong flavour distilleries, mild flavour distilleries, and rice flavour distilleries, respectively. A Jiang flavour baijiu distillery proceeds with a special eight-times technique and this technique is detailed in reference Xu (2012).

Distillation

When the substrate fermentation is complete, the spent liquor (called 'yellow water' in Chinese) generated during fermentation is drained dry. The pH value of the spent liquor is quite low (between pH 2 and 4) and is full of bacteria, dead yeast, unfermented starch, and protein. Spent liquor may contain 1%–3% vol. ethanol. Experienced distillers can diagnose the status of the fermentation from the spent liquor and predict the aroma of the new-make spirit. Large distilleries sell the spent liquor to chemical or essence factories. A few inferior distilleries will recycle the ethanol from the spent liquor, which increases the yield but sabotages the quality of the baijiu because the off-note congeners are collected in the recycling process.

Solid distillation has been used as a method to extract essence/perfume with steam since ancient times. Figs. 4.7 and 4.8 illustrate a typical baijiu stainless steel pot still. There is a sieve in the pot, where the fermented substrate is laid. A lid is used to cover the top of the pot, and then steam is introduced. Steam heats up the substrate, forcing the ethanol and many congeners out of it. When the temperature in the still is raised to a certain degree, the preforeshot will come out. Usually, 0.5 kg of preforeshot is discarded, and then the foreshot, heart, and

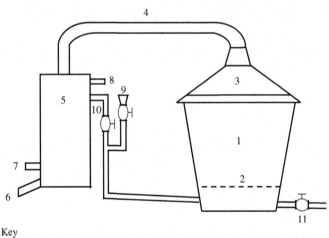

Key

1. Pot
2. Sieve
3. Pot lid
4. Lyne arm
5. Condenser
6. Distillate outlet spout
7. Cooling water inlet
8. Hot water outlet
9. Feints inlet
10. Hot water outlet to pot bottom
11. Steam inlet

FIG. 4.7   Schematic of a typical baijiu stainless steel pot still.

FIG. 4.8   A typical baijiu stainless steel pot still.

feint flow out from the spout. It should be noted that in Chinese baijiu distillation, there are 4 sections, unlike whisky production, which lacks the preforeshot stage. The Chinese explanation of foreshot confuses many readers because in whisky production, foreshot may or may not be collected with feints and distilled again; but in a baijiu distillery, the foreshot is collected and matured separately as a kind of new-make spirit.

Laying the substrate in the still is a professional skill. The layout affects the quality and yield of the final distillate. If the substrate is spread too loose, the amount of substrate distilled will be low, which wastes energy. Laying it too dense will stop the steam from penetrating the substrate completely, causing a severe drop in alcohol yield. Therefore, laying the substrate with a low amount of steam can help in judging the optimum substrate density. An accomplished distiller can determine how the substrate is placed by looking at how the vapour flows out from below the substrate. The eight fundamental principles of substrate laying are 'thin layer by every shovel', 'every layer should be smooth', 'every layer should be equally thick and compact', 'cover the steam pinhole', 'separate the substrate clearly', 'moderately press the substrate', 'the lining should be moderately loose', and 'lay the substrate patiently'.

In a mild flavour baijiu distillery, all of the fermented substrates is distilled. The foreshot, heart (new-make), and feint are collected separately. The new-make spirit is around 65%–67% vol. The utilisation of the foreshots and feint are explained later. Several mild flavour baijiu distilleries mix new Qu into the first-time distilled substrate to ferment it again. After the second distillation, this substrate is recycled as DDGS or biofuel.

The distillation in a strong flavour baijiu distillery is different. The substrate is dug out by layers of the pit. A certain depth will be a layer. The different layers of the substrate are distilled separately, except for the top layer, which is discarded after distillation, and the rest of the substrate is blended with new grist and then distilled. The new-make spirit

FIG. 4.9 Distillates stored in traditional clay pots.

is graded before being charged into maturation vessels. The new-make spirit is approximately 65%–70% vol.

The utilisation of foreshots and feints differs from a whisky distillation. The main substances in the foreshot are high volatile congeners. The aroma is strong and there is an off-note. The foreshot is collected and matured in different vessels and is blended into end products as flavouring. The feint has higher alcohol, acids, and long-chain ester concentrations and is redistilled with the next substrate.

Maturation

The foreshot and new-make are graded and stored in separate vessels according to the final products. Distillates are stored in traditional 50–1000 L clay pots (Fig. 4.9) or in 5000–8000 large pots (jiuhai), or in giant modern stainless steel tanks.

Jiuhai is a special and traditional maturation vessel for baijiu. It is built with wood and chaste tree twigs. The inner surface of the vessel is lined with hundreds of layers of cotton cloth and glued together with pig blood and limestone powder. After the vessel is shaped, the inner surface is coated with a mixture of egg white, bee wax, and rapeseed oil to prevent leakage.

Compared to oak maturation (whisky, brandy, rum, and tequila), baijiu maturation shares similar mechanisms, such as evaporation, oxidation, and esterification. One noticeable difference in baijiu ageing is the lack of migration of oak lactones, vanillin, and tannins from the wood to the distillate. Also, baijiu distillate does not acquire colour from the maturation vessel as occurs with oak maturation; however, sometimes the colour may turn to a very pale yellow during a long maturation.

With various sinter temperatures during clay pot production, free ions such as sodium, potassium, calcium, magnesium, and iron are released and will dissolve into the baijiu, enriching its flavour. The jiuhai also releases a number of chemicals into the baijiu and changes its flavour and colour. Normally, the angels' share in jiuhai is around 2%–3% (similar to Scotch), while in the clay pot it is 6%–9%.

During maturation, baijiu is regularly inspected by sensory panels. Some distilleries blend baijiu of different grades to average the quality after a certain amount of time (half-year or longer) and then divide the baijiu again into pots or tanks for a longer maturation time until the bottling of final products.

Blending and bottling

Blending of baijiu is similar to whisky or brandy blending. Sensory panels evaluate the base liquors, and the master blender confirms the results. A final baijiu product is usually a blend of basic liquor and flavouring liquor. It is quite similar to blending grain and malt whiskies. The basic liquor is the bone and the flavouring liquor is the flesh. They are mixed depending on recipes, styles, aroma types, and marketing demands. The final products are conditioned for a short duration. Readers can consider it similar to the 'marrying' process in whisky production, enhancing fusion and making the flavour more stable. Before bottling, cold filtration is conducted to prevent cloudiness. Some distilleries also employ carbon filtration to absorb any excessive off-notes.

Figs. 4.10 and 4.11 illustrate substrate production, distillation, maturation, and the blending process of mild flavour and strong flavour baijiu to help readers gain the fundamental concepts of baijiu production.

The Chinese baijiu industry is similar to the Scotch industry in that many companies purchase different styles and grades of baijiu from various distilleries in order to blend their products. Also, many large distilleries buy liquors from small local distilleries if they cannot produce enough themselves. These baijiu companies are more like blending houses in the Cognac industry. Several baijiu brands have no distillery, rather they contract with distilleries to produce for them, and some brands exist as independent bottlers. Like many spirit drinks in other parts of the world, Chinese baijiu is abundant in history, flavour, and creativity

The future challenges and opportunities of Chinese baijiu

Baijiu occupies an enormous spirit market in China. The market share approximates 63 billion GBP (2015–2019) and the volume is around 10.7 million kL 65% vol. product (2015–2019) (China National Bureau of Statistics, 2020). The Chinese government's discouragement of over-luxury activities hit the baijiu market severely in 2013, but the market share and volume are still large compared to other foreign spirit drinks. The trend of drinking different styles of alcoholic beverages is currently on the rise because the younger generation wishes to try new and fashionable choices. Many traditional baijiu companies are facing this issue and are trying to address this problem.

The major occasion for drinking baijiu is dining. Many Chinese cuisines are enjoyed with baijiu. Chinese 'bottom-up' and 'drink-for-respect' drinking cultures tend to be disliked by

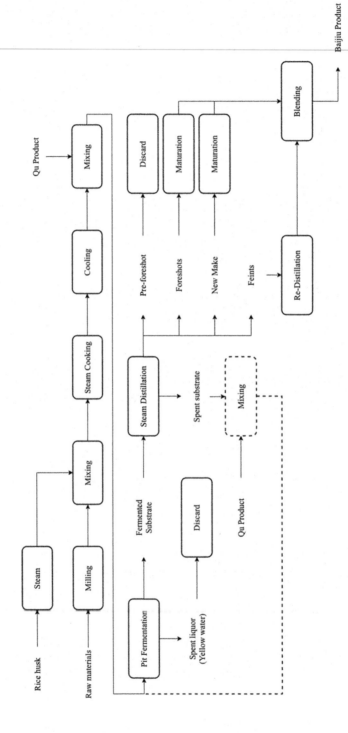

FIG. 4.10 Mild flavour baijiu production.

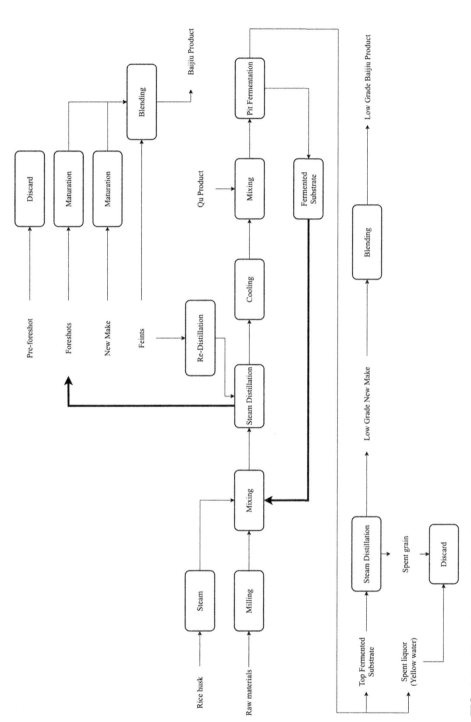

FIG. 4.11 Strong flavour baijiu production.

the younger generation, and baijiu is seen as 'elder's drink'. Some hygiene and food safety issues were also exposed several years ago, especially adulteration, mislabelling, and fictitiousness, which provoked customers. In addition, consumers are hindered by the complexity of baijiu's history, many brands, and production methods.

The observation is that currently, younger Chinese consumers are immensely proud of China-made products. They are eager to approach the legacy of Chinese spirits in different ways. The baijiu industry could be more open and friendly to attract new customers, but the essential concept is to address the needs of young consumers. Several elements could be employed, such as entry-level products, new flavours, and promoting the fashion and styling of the products. Furthermore, there are many baijiu distilleries scattered across China. These distilleries could consider converting into tourist-oriented distilleries to help people understand the process of making baijiu, thereby fostering interest in baijiu's history and culture.

A history of whisky development in China

As a part of the world whisky promotion and development, consumption and production of whisky are now becoming a trend in China. After joining the World Trade Organisation (WTO) in 2001, the whisky import volume started to grow. In China, the direct import value of Scotch whisky has grown from under 10 million GBP in the 2000s to 89 million GBP in 2019 (Scotch Whisky Association, 2020). Data from the General Administration of Customs, P.R. China stated that imported whisky volumes were 18.9 (2018), 21.5 (2019), and 21 (2020, Dec. excluded) million litres and the values were 139 (2018), 175 (2019) and 203 (2020) million GBP (Data from China Administration of Customs, 2021). It is anticipated that the whisky market will grow further during the next decade.

Although a type of *uisge beatha* was produced in China centuries ago, the first 'whisky' by contemporary definition was produced by the German-built Tsingtao Winery (called Tsingtao Melco Weinkellerei) in 1914. This winery made sparkling wine, brandy, and occasionally whisky. These alcoholic beverages were consumed mostly by foreigners and sailors. After several decades, the Tsingtao Winery became state-owned and was merged into the Tsingtao Beer Brewing Company. Fig. 4.12 shows a historical Chinese whisky distillery and its workers in 1949. Several whisky brands were created in the Tsingtao Winery. Labels can be seen in Fig. 4.13. Very small amounts of whisky were produced and selling was problematic. In 2004, the company filed for bankruptcy. The old distillery building is shown in Fig. 4.14. The remaining equipment inside of the building is currently mothballed and is shown in Fig. 4.15. The Tsingtao City Council is planning to convert this winery (distillery) into a historical museum.

Many Chinese wineries, breweries, and baijiu distilleries have tried to produce whisky. Many eye-catching products have been released, such as sorghum-baijiu-infused whisky and strong-flavour-baijiu-mixed herbal whisky (made from 5 grains: sorghum, rice, maize, waxy rice, and wheat). Even the famous Moutai Baijiu Company launched a Moutai Whisky in the 1990s.

The whisky industry was not prosperous in China before the 21st century. The Chinese consumers preferred baijiu and the taxes and price for luxurious imported goods were high;

FIG. 4.12 The workers at the Tsingtao Winery (Tsingtao Melco Weinkellerei) in 1949. Photograph compliments of Yenming Li and Dali Wu (Editors of Tsingtao Winery History).

FIG. 4.13 Labels from whisky brands produced in the Tsingtao Winery. Photograph compliments of Yenming Li and Dali Wu (Editors of Tsingtao Winery History).

FIG. 4.14   The front gate of old Tsingtao Winery. Photographs compliments of Funxun Sun.

therefore, whisky was not generally affordable at that time. After the growth of the middle class in China in the 2000s, tastes in the country have shifted and many of the younger Chinese consumers now look for different choices. Various interesting products have appeared on the market, for example, oak-matured baijiu, clay-pot-matured whisky, and baijiu-infused whisky (Chen et al., 2020).

Whisky regulations in China

The law regulating whisky in China is GB/T 17204-2008, Chinese Standard, 2008. Because the Chinese whisky industry is just at its dawn, the Chinese Alcoholic Drinks Association has organised the committee to modify the whisky regulations in 2021. The new Chinese whisky regulations may not be issued yet for several years. Therefore, the current regulation is introduced in this chapter. This regulation refers to the European Parliament and of the Council Regulation (EC) No 110/2008. If distillers want to produce

FIG. 4.15   Equipment present in the historical site. (A) The mash tun and lauter tun; (B) The pot stills (side view); (C) The pot stills (front view); and (D) The blending room. Photographs compliments of Funxun Sun.

whisky in China, they should at a minimum comply with this recommended standard. The terminology, definitions, category, requirements, and criteria are included. The following is an abstract of GB/T 11857–2008:

Some Terminologies and Definitions (Adapted from Chen et al., 2020).

Whisky—A spirit drink produced from a mash made from malt and cereals that has been through mashing, fermentation, distillation, maturation, and blending.

Malt whisky—A spirit drink produced from a mash made from all malted barley, which has been through mashing, fermentation, distillation, maturation in oak cask for 2 years, and blending.

Grain whisky—A spirit drink produced from a mash made from cereals (rye, wheat, maize, Highland barley and/or oat), which has been through mashing, fermentation, distillation, maturation in oak cask for 2 years, and blending.

Blended Whisky is a whisky produced from blends of certain portions of malt whisky and grain whisky.

The sensory characteristics are listed in Table 4.2 and the chemical characteristics are listed in Table 4.3.

The substance limitation in whisky is listed in GB 2757–2012 National Safety Standard for Food: Distilled Spirits and Related Mixing Products. The methanol limitation is less than 0.6 g/LA and cyanide is less than 8.0 mg/LA. If distillers want to produce whisky in China, these standards must be followed.

TABLE 4.2 The sensory characteristics of whisky in China.

Grade	Premium	Good
Observation	Clear with no suspension and sediment	
Colour	Light golden to golden yellow	
Aroma	Should have malty and/or grainy flavour, the harmony flavour given from oak, strong aroma, and/or the aroma acquired from peat	Should have malty and/or grainy flavour, the inferior harmony flavour given from oak and/or the aroma acquired from peat
Taste	Rich, rounded, and sweet. Should have malty and/or grain flavour, the taste/flavour given from oak. No apparent off-note	Less rich, rounded, and sweet. Should have malty and/or grain flavour, the taste/flavour given from oak
Style	Should have its unique style	Should have its obvious style

Adapted from GB/T 17204-2008, Chinese Standard. 2008. Whisky. Available in English from https://www.chinesestandard.net/PDF/English.aspx/GBT11857-2008.

TABLE 4.3 The chemical characteristics of whisky in China.

Grade	Premium	Good
Alcohol (%vol)[a]	≥40.0	
Total acid (as acetic acid)/[g/L (pure ethanol)]	≤0.8	≤1.5
Total ester (as ethyl acetate)/[g/L (pure ethanol)]	≤0.8	≤2.5
Total aldehyde (as acetaldehyde)/[g/L (pure ethanol)]	≤0.2	≤0.4

[a] *The error for alcohol content and the label is 1.0% vol.*
Adapted from GB/T 17204-2008, Chinese Standard. 2008. Whisky. Available in English from https://www.chinesestandard.net/PDF/English.aspx/GBT11857-2008.

The future of Chinese whisky

In the next decade, the future development of Chinese whisky will probably be more focussed on middle-sized (above 1 million litre pure alcohol, MLPA) and large-sized distilleries (above 7 MLPA) and small plants will have less opportunity to grow. Craft distilling will be an interesting area to monitor. The above forecast is based on the observations of the Chinese whisky market, drinking habits, and the attitude of the government and related regulation in the last 5 years.

Whisky, compared to brandy or cognac, is less famous in China. Consumers tend to choose brandy, baijiu, and red wine at banquets. Off-line trading is dominated by wine, fruit wine, and ready-to-drink beverages. Consumers of whisky are located mainly in large coastal cities such as Shanghai, Shenzhen, Guangdong, Qingdao, and Xiamen, with a smaller proportion of customers in Beijing, Xi-an, Chengdu, and Changsha. Many whisky festivals are held in these cities, especially from May to October. Whisky has not been a top Chinese choice for a spirit beverage, although many key opinion leaders promote whisky history, culture, and enjoyment via different social media. The forecasted figures of trade value and volume are both growing, reflecting well the effort of many large alcoholic beverage importers and marketing companies.

Japanese whisky is now well-known all over the world and is very popular in China. Many non-whisky drinkers have turned into whisky lovers because of the unique and light flavour and are starting to appreciate different types of whisky, such as Scotch whisky and American Bourbon. This phenomenon has drawn the attention of venture capitalists. Several have chosen to first invest in independent bottling companies, and one has acquired a Scotch whisky group. Only a small number of domestic whisky distilleries are currently under construction, with one malt distillery owned by a famous international alcoholic beverage corporation. Whisky is a long-term business, and it is only at an early stage in China, although many believe that despite its immaturity, there are ample growth opportunities in the near future.

A history of whisky development in Taiwan

Taiwanese whisky is the whisky produced in Taiwan and Taiwan is now recognised as one of the international whisky regions. The whisky market in Taiwan is sophisticated and it is the third largest region for single malt whisky after the United States and France. The volume ratio of blended and malt whisky sold in the rest of the world is around nine to one. Taiwan is the only whisky market where more single malt is drunk than blended whisky.

In the Asia-Chinese region, Taiwan has the earliest and most developed whisky market. Many Scotch whisky enthusiasts are often surprised by the Taiwan whisky market, as it rates in the top rank of Scotch whisky exports. Astonishingly, over 90% of the whisky market share on the island is Scotch whisky (Taiwan Ministry of Finance, 2020).

One key factor for such acceptance is its colonial history. Taiwan was colonised by the Netherlands (Southern Taiwan, 1624–1662), Spain (Northern Taiwan, 1626–1642), and Japan (1895–1945). Between 1950 and 1979, the US Military was also present in Taiwan. As a result, the Taiwanese have a stronger cultural link to the west and accept whisky and different types of western alcoholic beverages more easily than other East Asian areas. Before whisky, sorghum baijiu was the top-selling spirit consumed in Taiwan. Today, the younger consumers do not choose baijiu, except when drinking with elders or on certain cultural occasions. This is similar to mainland China's situation; therefore, many international alcoholic beverage companies conduct their pilot tests of products or marketing strategies in Taiwan first.

Taiwan entered the W.T.O. in 2002 as a separate customs territory. Before that, the Taiwan Tobacco and Liquor Corporation (TTL) (the former Taiwan Tobacco and Wine Monopoly Bureau) had monopolised the alcoholic beverage market since the 1900s. After 1987, the government started to permit the importation of alcoholic beverages. Many brandy brands came to Taiwan, but at that time very few whiskies were imported. The whisky importing business was completely granted in 1991. Some whisky importers utilised a number of misleading 'medical reports', which implied that the sugar in brandy might result in diabetes, to convert customers into drinking more whisky (Zuo, 1999).

The brands Macallan, Famous Grouse, Johnnie Walker, and Chivas Regal were the pioneers in the Taiwanese market. Branding strategies were often highly creative and fashionable. Whisky was promoted in private or commercial guest houses, hotel bars, karaoke bars, Taiwanese-style-stir-fry restaurants, and even at wedding banquets. 'Brand ambassadors' were appointed and experienced bartenders were hired to host tasting events and to invent new cocktails, such as green tea and whisky, to increase whisky consumption.

The price of whisky is low in Taiwan compared to Europe and North America. The import tax for whisky is zero and the alcohol tax is 250 TWD (approx. 6 GBP/LPA) (Taiwan Ministry of Finance, 2020). This has played an important role in accelerating market development (You, 2019) and has encouraged the Taiwanese to adopt whisky as their favourite spirit drink while fostering the development of the island's own whisky culture.

The TTL company already attempted to produce whisky in 1948, but the procedure of whisky production was not fully understood at that time. The 'whisky' was manufactured by mixing edible alcohol, flavouring, and caramel. Jin-Dan Lin, considered to be the Father of Taiwan whisky, was sent by the TTL to the International Centre of Brewing and Distilling, Heriot-Watt University in 1979. On his return, he designed a whisky production process for TTL. The wash was produced in the Wuri brewery and distilled and matured in the Nantou distillery (the Nantou distillery was a brandy distillery at that time). After blending with Scotch and Indian whisky, the final product was bottled and released. The Nantou distillery was fully converted into a whisky distillery in 2009 and the first Nantou whisky 'Omar' was launched 4 years later (according to Jin-Dan Lin and TTL unpublished data).

Another world-famous whisky distillery in Taiwan is the Kavalan distillery. Before the enactment of 'Tobacco and Alcoholic Beverage Regulation' in 1999, privately owned wineries, breweries, or distilleries were not permitted in Taiwan. The creation of the Kavalan distillery was helped by Dr. Jim Swan and this operation was started in 2006. The first Kavalan whisky was launched in 2008 and exhibits a tropical fruit style. Whisky products from Nantou and Kavalan have been awarded prizes in many world-class competitions and Taiwan has now been recognised as an international whisky-producing region (Liu, 2017).

The whisky regulations in Taiwan

The Taiwan whisky regulation is listed in the Enforcement Rules of the Tobacco and Alcohol Administration Act 2014. In Article 3;

"Alcohol" set forth in Article 4, Paragraph 1 of the Act is classified into the following categories:

5. Distilled spirits: The following spirit drinks made from fruit, grain, or starch- or sugar-containing plants by saccharification or not, fermentation, and then distillation:

(2)Whisky: Distilled spirits made from grain by saccharification, fermentation, distillation, and maturation in wooden casks for at least 2 years, with an alcohol content of not less than 40%.

This is the definition of whisky in Taiwan. And there is no specific geographical indication of Taiwanese whisky. The limitation of substances is set in 'The Standard of Food Hygiene and Safety for Alcoholic Beverage' (Table 4.4; Taiwan Ministry of Finance, 2016).

TABLE 4.4   Substance limitations in whisky in Taiwan.

Substance	Limitation
Methanol	$\leq 1000 \, mg/L$ (100% pure alcohol)
Pb (Lead)	$\leq 0.3 \, mg/L$ (product)
SO_2	$\leq 0.03 \, g/L$ (product)

Adapted from Taiwan Ministry of Finance, 2016. The Hygiene Standards for Alcohol Products: 2016. https://www.nta.gov.tw/Eng/singlehtml/310?cntId=nta_1186_310.

The craft whisky movement in Taiwan

After winning many international awards, Taiwanese whisky enthusiasts are now eager to produce or blend more fine whiskies. Several craft distillery projects are currently ongoing and many craft breweries have plans to be upgraded to distilleries. The craft whisky movement is growing, and innovation is strong in this area, and one can expect to see new concepts of use of local grains, solid fermentations, and single-farm whisky in the near future.

Acknowledgements

The author would like to thank Jie Xu, former manager of the department of quality assurance and control of the Wenjun Baijiu Distillery, who provided valuable advice, abundant materials, and photos of baijiu production. Also, thanks to Jin-Dan Lin, who supplied information on the Nantou distillery and on the history of Taiwanese whisky development. Finally, thanks to Dr. Thomas Tan, a researcher at the University of Edinburgh, who proofread and provided valuable ideas for this chapter.

References

Chen, J.I., Osborne, D., Singh, P., Davies, S., 2020. The China syndrome. Brewer Distiller Int. 16 (6), 34–37. Institute of Brewing and Distilling, London, UK.

China Administration of Customs, 2021. website http://www.customs.gov.cn/.

China National Bureau of Statistics, 2020. website http://www.stats.gov.cn/.

GB/T 17204-2008, Chinese Standard, 2008. Whisky. Available in English from https://www.chinesestandard.net/PDF/English.aspx/GBT11857-2008.

Liu, M., 2017. How a Taiwanese whisky became a global favorite. Available from: https://www.cnn.com/travel/article/taiwan-whisky-kavalan/index.html.

McGovern, P.E., Zhang, J.H., Tang, J.G., Zhang, Z.Q., Hall, G.R., Moreau, R.A., Nunez, A., Butrym, E.D., Richards, M.P., Wang, C.S., Cheng, G.S., Zhao, Z.J., Wang, C.S., 2004. Fermented beverages of pre- and proto-historic China. Proc. Natl. Acad. Sci. U. S. A. 101, 17593–17598.

McGovern, P.E., Underhill, A.P., Fang, H., Luan, F., Hall, G.R., Yu, H., Wang, C.-S., Cai, F., Zhao, Z., Feinman, G.M., 2005. Chemical identification and cultural implications of a mixed fermented beverage from late prehistoric China. Asian Perspect. 44, 249–275.

Scotch Whisky Association, 2020. International Trade https://www.scotch-whisky.org.uk/insights/international-trade/.

Taiwan Ministry of Finance, 2014. Enforcement Rules of the Tobacco and Alcohol Administration Act: 2014.

Taiwan Ministry of Finance, 2016. The Hygiene Standards for Alcohol Products: 2016. https://www.nta.gov.tw/Eng/singlehtml/310?cntId=nta_1186_310.

Taiwan Ministry of Finance, 2020. Statistic Database Query of Customs. Port and Trade of Customs Administration.

Wang, Y.T., Wang, G.Z., Song, Y.S., 2009. Advanced Baijiu Distiller Training Textbook (Translated Title). China Alcoholic Drinks Association, Beijing.

Xu, Y., 2012. Moutai (Maotai): production and sensory properties. In: Piggot, J. (Ed.), Alcoholic Beverages: Sensory Evaluation and Consumer Research. Woodhead Publishing, Cambridge, pp. 315–330. Chapter 15.

You, S., 2019. Innovative Localized Marketing Strategy of Whiskey in Taiwan-Case Study: The Macallan Single Malt from 2002 to 2012. Master thesis of Department of Business Management, College of Commerce, National Chengchi University, Taipei. Available from https://ah.nccu.edu.tw/item?item_id=140699. https://doi.org/10.6814/THE.NCCU.AMBA.001.2019.F08.

Yu, C.W., 2018. Traditional Baijiu Production Techniques, second ed. China Light Industry Press. Ltd, Beijing.

Zuo, Z.Y., 1999. The Effect of Alcoholic Beverage Importer Advertisement Strategy by Taiwanese Culture. Master thesis of Department of Business Management, College of Management, National Sun Yat-sen University, Kaoshiung.

5

Korean soju

Tae Wan Kim

Korea Food Research Institute, Wanju-gun, Republic of Korea

Introduction

The history of soju can be traced back to the 13th century, the Koryo Dynasty in the Korean Peninsula (Kim, 2014). During the expansion phase of the Mongol Empire, they had acquired the technique of distilling *Araq (Arak)* from the Arabic peninsula, developing it into soju in Koryo (Lee et al., 2016) (Fig. 5.1.).

Literally, soju means 'burnt alcoholic drink', referring to the heat of distillation. From the Korean Liquor Tax Act, soju is defined as '*Alcohol content more than 1% (v/v) distilled spirit from the fermentation of starch-containing substrates, Guk (fermentation starter, also called Gok, Gokja, the other name is Nuruk), and water as a source material (excluding germinated cereals)'.* The Act details the additives specified by law (sugars, acids, and herbs) that can be used during fermentation and blending. Wood maturation can also be employed. The ash content of the final product must be less than 2% (w/v).

Soju is classified into two types: pot-distilled and diluted. They differ based on their manufacturing procedures, mainly the distillation methods employed.

For the pot-distilled soju, new-make spirit, by a single-batch pot distillation, contains alcohol at 40%–60% (v/v) from a fermented wash with an alcohol strength of 15%–20% (v/v). Traditionally, a ceramic pot still (*sojut-gori*) has been used for distillation. Currently, stainless steel and copper stills are generally employed. The final products of pot-distilled soju vary in alcohol strength, from 20% to 50% (v/v). Normally, 40% (v/v) is the main alcohol strength of pot-distilled soju.

In the case of diluted soju, continuous distillation methods developed in the 20th century are used, with four (or more) distillation columns (stripping, extracting, rectifying, and demethylating), and the new-make spirit's alcohol strength is over 95% (v/v) and contains a hint of congeners. For the production of diluted soju, dilution and additional purification processes are applied. Generally, the final products of diluted soju have around 20% (v/v) of alcohol strength. More recently, the average of the diluted soju category has been 16%–18% (v/v) alcohol (due to the trend to produce lower alcohol soju in Korea).

63

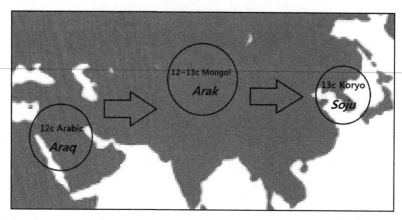

FIG. 5.1   Origin of soju.

Interestingly, there are statistics relating to alcohol consumption by Koreans. They drink 13.7 shots of spirits per week and drink the hardest liquor in the world, more than twice as much as the 6.3 shots consumed in Russia (Ferdman, 2014). Per-person-alcohol-consumption is 8.7L per year, on a pure alcohol basis, in which soju is 6.5L and most alcohol is consumed from hard liquor, soju (Statistics Korea, 2020).

Distilled spirit industry in Korea

In 2019, the alcoholic drink market in Korea was 3843 million litres, worth 8.9 billion USD (ex-factory price, Statistics Korea, 2020). In the spirit industry as a whole, 1250 million litres, worth 4.5 billion USD, was consumed in Korea. Among the spirits, diluted soju is the most consumed alcoholic beverage (Table 5.1).

Despite its unique characteristics, pot-distilled soju consumption is very low, and most pot-distilled soju makers are very small-scale companies. Their average annual production is less than 100 kL, worth 1 million USD. On the other hand, diluted soju makers are mostly large-scale producers and can manage to develop their products by themselves. The situation in the soju market is disproportionate.

Currently, traditional pot-distilled spirit soju is gradually becoming recognised as being of high quality and an important traditional part of the culture in Korea. In several areas, significant efforts have been made to revive traditional pot-distilled soju.

Every year, the number of soju distilleries in Korea has steadily increased, and a total of 133 distilleries were producing soju in 2019 (Statistics Korea, 2020). Among them, pot-distilled soju distilleries are 112 in number, five times more than diluted soju distilleries (21 in number), but the production scale of a pot-distilled soju distillery is less than one-hundredth of that of the diluted soju distilleries. Although pot-distilled soju production is growing gradually to reflect the requirements of consumers, the market is still small, the production process is long, and the production per batch is not large; therefore, more efforts are still required in order for it to become a stable industry (Table 5.2).

TABLE 5.1 The volume and price of spirits sold in Korea in 2010–2019 (Statistics Korea 2020).

Category	2010		2015		2019			CAGR (2010–2019)	
	Volume (kl)	Price (000USD)	Volume (kl)	Price (000USD)	Volume (kl)	Price (000USD)	MS (by price)	Volume	Price
Total	1,229,167	3,557,485	1,301,693	4,260,677	1250,296	4,547,744	100.0%	0.2%	2.8%
Pot-distilled soju	717	10,596	954	19,458	1714	38,382	0.8%	10.2%	15.4%
Diluted soju	930,605	2,867,317	955,507	3,466,624	915,596	3,738,247	82.2%	−0.2%	3.0%
Whisky (bottled in Korea)	3417	192,940	439	29,506	72	5098	0.1%	−34.9%	−33.2%
Spirits (others)	3774	15,012	5520	22,679	1681	12,629	0.3%	−8.6%	−1.9%
Liqueur	426	4963	29,856	194,783	2536	19,218	0.4%	21.9%	16.2%
Others (including raw alcohol)	290,228	466,657	309,417	527,627	328,697	734,170	16.1%	1.4%	5.2%

TABLE 5.2 Production capacity and unit price of soju in Korea in 2019 (Statistics Korea 2020).

Category	Distilleries (number)	Production capacity (kL/year)			Unit price (USD/L)
		Max	Min	Average	
Soju total	133	–	–	–	4.12
Pot-distilled soju	112	3000	10	15	22.39
Diluted soju	21	500,000	50,000	43,600	4.08

Soju production process

As previously discussed, soju is divided into two types according to the distillation methods employed. There are also some differences in the use of raw materials, starch degradation enzyme sources, fermentation microorganisms, alcohol fermentation process, postdistillation treatment, and product packaging. As a result, there can be large differences in the flavour profiles between the products, similar to differences observed between malt whiskies and vodkas (Kim, 2017b) (Table 5.3).

Pot-distilled soju

The 700 years of soju history has included a variety of raw materials and manufacturing methods, and it is manufactured in various scales, ranging from several tens of litres to tens of kilolitres. Pot-distilled soju uses rice, barley, and sweet potatoes as the main raw materials, but is not limited to these materials. There is also soju produced with unique flavours by the use of corn, buckwheat, and sorghum.

In addition, as a traditional fermenting agent (starting culture), Guk (Nuruk) adds to the flavour of soju due to the complex microbial community of yeast and fungi that produces the starch degrading enzymes (Carroll et al., 2017).

Distillation is performed in a single-stage (batch pot) system, which is usually stainless steel and operated in selectively reduced and atmospheric pressure. In the case of some small-scale distilleries (less than a thousand litres per batch), the pot-distilled soju is produced using a sojut-gori (an earthenware-based traditional Korean distillation pot), which contributes to the diversity of the products and the inherent traditional distillation techniques used (Fig. 5.2.). Sojut-gori may be one of the few earthenware materials in the world that uses the traditional pot distillation still (Kim, 2019).

A treasure trove of microbial diversity, Nuruk (Guk)

One of the large differences between Eastern and Western liquor brewing is the enzyme source for starch degradation in the saccharification and fermentation processes. In the Western case, the endogenous enzymes produced by germination of grains (malt) are used to produce soluble sugars prior to alcoholic fermentation. In the Eastern case (Korea, China, and Japan), fungi are grown on starch sources (naturally or artificially) to produce enzymes and, using this, alcoholic fermentation is simultaneously carried out during starch degradation.

TABLE 5.3 Overall production process of two soju types.

Process	Pot-distilled soju	Diluted soju
Raw materials	Rice, barley, (sweet potato)	Starch cereals (Rice, barley, sweet potato, tapioca, wheat, corn, etc.)
Enzyme sources	Guk (Nuruk)	Modified Nuruk, purified enzyme
Yeast	Dry, slurry	Slurry
Fermentation	Batch system. Less than 20 kL/batch 12–20 days	Continuous/batch system 100–250kL/batch 3–5 days
Water grist ratio	160–170 (70–80)	320–340
Final wash	17–19 (14–16)% (v/v)	9%–11% (v/v)
Distillation	Batch pot distillation	Continuous column distillation
New-make spirit	45–55 (35–45)% (v/v)	95%–96% (v/v)
Alcohol yield	420–450 (200−220)L/ton	460–480L/ton
Post-treatment	Chill filtration	Activated carbon filtration
Additives	None	Approved sweeteners, sugars, sugar alcohols, (amino) acids
Packaging	One-way bottle (Nonreturnable) 300–700mL	Returnable standard bottle (shared bottle system) 360mL
Major Products	20%–45% (v/v)	16%–22% (v/v)

Also, in the Eastern case, the processes of producing enzyme sources differ somewhat in the formulation of the methods and the types of microorganisms involved, but they are similar in their traditional methods to produce spirits in each country: Korea's Nuruk (= Guk) use soju, China's Qu use Baijiu (Liu and Sun, 2018), and Japan's Koji use Shochu (Yoshizaki et al., 2010).

Traditional Korean Nuruks vary in raw materials and shapes (Lee et al., 2017) (Fig. 5.3.) and are naturally inoculated with various microorganisms from the manufacturing site. Recently, the functional value of Nuruk has been re-examined (Jeong et al., 2018), and useful microorganisms, such as *Saccharomyces* spp., *Pichia* spp., *Aspergillus* spp., *Rhizopus* spp., *Mucor* spp., *Lactobacillus* spp., *and Leuconostoc* spp. have been separated from Nuruk and used as bioresources (Bal et al., 2017). In order to selectively utilise the brewing properties of screened microorganisms, Nuruk is modified to artificially inoculate single microbes (Choi et al., 2014). The *Aspergillus* genus (species *oryzae* and *luchuensis*) is generally used in the alcoholic beverage industry.

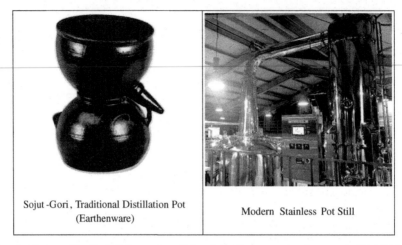

FIG. 5.2   Pot distillation stills. (Left) Sojut-gori, traditional distillation pot (earthenware). (Right) Modern stainless pot still. Panel Left: Credit: Gwangju folk museum, Hwayo.

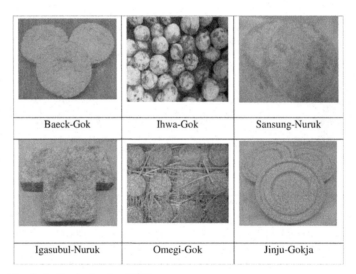

FIG. 5.3   Various Nuruks in Korea. Credit—KFRI.

Traditional pot distillation still, sojut-gori

The exact time of the sojut-gori invention is not certain. It is thought that the distillation technology introduced from Mongol in the 13th century met the ceramic manufacturing technology that flourished at the time when it entered the Korean peninsula, and as a result, the pot distillation still had changed. From the Koryo Dynasty's (918–1392) celadon, manufacturing technology to the Joseon Dynasty's (1392–1897) white porcelain manufacturing technology would have given the shape and functional characteristics of sojut-gori.

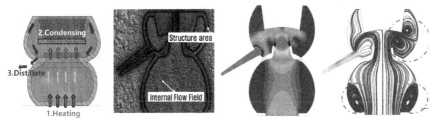

FIG. 5.4   Sojut-gori distillation CFD (computational fluid dynamics). Credit: KFRI.

Unlike modern metal stills, the heating part, the cooling part, and the distillate outlet part are composed of one body, and the distillation vapour is liquefied through a static cooling system (Fig. 5.4.). Due to its internal structure, heated fermented wash and vapour do not mix during distillation. Due to the properties of the material, thermodynamically, the distillation energy consumption of sojut-gori is less than that of a metal still under the same conditions (Kim, 2017a).

In terms of distillation energy, the heat loss that occurs during operating a still is measured and calculated as convection heat, conduction heat, and radiation heat (Fig. 5.5). Due to the characteristics of ceramic materials, the heat loss is smaller than that of a metal. The use of sojut-gori is one of the characteristics of traditional pot-distilled soju made by a small-scale craft, and it contributes to various consumption preferences, along with soju, made with modern metal stills.

Diluted soju

In the 1910s, continuous distillation stills were introduced on the Korean Peninsula, and diluted soju appeared, for the first time. Since then, the market has gradually expanded, and now, over 100years later, it has become the most consumed spirit in Korea.

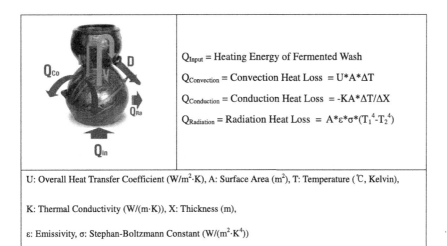

FIG. 5.5   Distillation heat loss calculation. Credit: KFRI.

Diluted soju is a distilled spirit that does not have an aroma like vodka. It is characterised by a clean and smooth taste, from mouth to throat, and it is usually drunk cold (7–14°C).

The odourless and light taste of dilute soju is achieved by the production of its high strength [over 95% (v/v)], new-make spirit through continuous distillation and the removal of trace congener compounds as a result of activated carbon filtration. The name of 'diluted soju' is distinguished from 'pot-distilled soju' as it means that the produced high alcohol strength new-make spirit is reduced in its final product alcohol content by dilution.

Large-scale rapid fermentation of new-make spirit for diluted soju

With more than 2.5 billion bottles of diluted soju being sold in the domestic market per year, the fermentation process to meet the demand is also large, and the fermentation time of distilled soju is shorter. This is achieved by fine milling the starch material and increasing the water supply ratio (320%) and the fermentation temperature (32–34°C). The fermentation scale ranges from 100 to 250kL, and continuous fermentation is carried out to minimise the space (some plants are batch systems). As an enzyme source, modified Nuruk (single culture) and purified enzyme are used together to increase the alcohol yield.

Continuous distillation of diluted soju

The two-column distillation system (i.e. the 1830 Coffey still) continued to be developed and several columns with different functions were added to further improve the alcohol concentration and component separation:

(1) The first column (mash column and stripping column) is used to separate the solid content of the fermented wash around 10% (v/v) of concentration. The 72%–73% (v/v) alcohol strength of upstream is generated and exits on the top of the still.
(2) Then, the mainstream enters the middle of the second column (pre-distillation column and extracting column) and is diluted to 10%–12% (v/v) with hot water supplied from the top (diluted to increase the degree of separation of components). This flows out to the bottom of the column. High volatile compounds, such as aldehydes, are removed from the top of the second column.
(3) The diluted stream of 10% (v/v) that enters the middle of the third column (rectifying column) is concentrated to 95%–96% (v/v) and flows out to the top, and low volatile compounds including fusels are removed from the middle and lower decanting points of the column.
(4) The concentrated alcohol (from the top of the third column) enters the upper part of the fourth column (product column and demethylating column), and methanol is removed (on the top of the column), and high purity new-make spirit is finally produced at the lower part of the column. The main flow is carried out through four columns, and the congener compounds are removed from each column flow to an extra line and improve material separation and yield through the operation of auxiliary columns (feint column, side-stripping column, and stillage water column) (Kim, 2019) (Fig. 5.6.).

Activated carbon filtration for purifying new-make spirit

New-make spirit produced through continuous distillation contains trace amounts of congener compounds, such as fusels (higher alcohols) and aldehydes, which are removed,

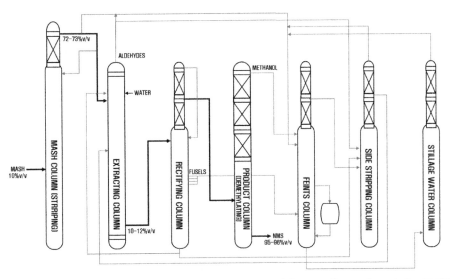

FIG. 5.6   Continuous Distillation Column Operation (four main and three auxiliary columns). Credit: KFRI.

to some extent, through activated carbon filtration (Siřišťová et al., 2012). At the same time, this process can reduce the astringency of newly produced spirit (Cai et al., 2016). The activated carbon filtration process is also referred to as the 'alcohol deodorisation' process. Two methods are used. In most cases, powdered activated carbon is applied for use in a batch process, while some diluted soju manufacturers use a column process with granular activated carbon.

First, the high alcohol strength of the new-make spirit is diluted to 45%–55% (v/v) (an efficient removal concentration) and activated carbon is added at a concentration of 300–1000mg/L. An operation of periodic mixing and settling is repeated for 6–12h. The contact time and concentration of the activated carbon are determined through pre-experimental work. The activated carbon used is mainly wood-based or peat-based, but the type is not limited and can be used as one or mixed together. Powdered activated carbon is not reused. For the column-type process, granulated activated carbon is filled into several columns (connected in series), and the reaction time with the activated carbon is controlled through the flow rate (space velocity [SV]) control to produce purified new-make spirit.

Soju flavour compounds

As for the differences in the production processes, the two types of soju also differ in the content of sensory-active flavour compounds. For pot-distilled soju, flavour compounds such as higher alcohols, esters, acids, aldehydes, and other carbonyl compounds have a role in the sensory parameters. The flavour compounds of pot-distilled soju originate from material usage to the applied processes (brewing and distilling). On the other hand, diluted soju has fewer flavour compounds, except ethyl alcohol, and they are hardly detectable in the new-make spirit of diluted soju (Table 5.4).

TABLE 5.4 Flavour compounds in two types of soju.

Concentration	Pot-distilled soju	Diluted soju	
Product alcohol (%(v/v))	40	20	40 (Recalculated)
Ash content (%(w/v))	0.02	0.05	0.1
Total (mg/L)	**1856.6**	**3.8**	**7.5**
Higher alcohols	1488.9	1.6	3.1
Esters	267.4	Not detected	Not detected
Aldehydes	51.8	1.2	2.4
Acids	22.3	Not detected	Not detected
Methanol	26.2	1.0	2.0

DATA supported by LOTTE Company, Seoul, Korea

New product development of soju in Korea

The alcoholic beverage market is increasingly seeking diversity and premium values of products, and an awareness of the environmental and sustainable industrial ecosystem is becoming more important. In this regard, various studies are being conducted.

Development of characteristic strains and formulation technologies on demand

It is possible to produce various products due to the many types of microorganisms involved in these alcoholic beverages. Research on the discovery and development of microorganisms is underway with the goal of strengthening specific flavours, obtaining high alcohol levels, and utilising by-products. Along with this, stability and ease of use are important for small-scale manufacturers that have difficulty managing these microorganisms. For them, it will be necessary to commercialise strains in the form of powders or slurries and to adopt the technology that maintains the activity of the formulation strain over time (KFRI, 2020).

Energy efficiency of distillation stills

Most of the energy in the distillery goes into the distillation process. Therefore, it is important to increase the efficiency of resource utilisation and to reduce the environmental impact by decreasing the amount of energy consumed during distillation. Based on the principles of manufacturing spirits, it is necessary to develop improved distillation stills through research on the efficiency of materials, structures, heating regimes, and cooling systems (KICET, 2020).

Development of maturation materials using soil and forest resources

The final major step in the technical aspects of alcoholic beverage production is maturation. Maturation highlights the technical elements of the alcoholic beverage while, at the same time, connecting the industrial factors in the new fields. The use of soil and forest resources as maturation materials can help build a sustainable industrial ecosystem and increase the value of resource use. Since ancient times, various container (earthenware) manufacturing techniques produced from soil have existed, and maturation studies of foods, including alcoholic beverages, are being conducted (KICET, 2020). In the case of forests, over 70% of the land area in Korea is mountainous (the forestation rate is world-class) and the amount of resources is abundant (NIFoS, 2020). Research on the value utilisation of native oak trees (*Quercus aliena, Quercus variabilis, Quercus dentata, Quercus acutissima, Quercus mongolica, and Quercus serrata*) is being conducted in the food and beverage industry (Cha et al., 2020).

Development of matured pot-distilled soju

A variety of competitive products have been developed through the fusion of empirical knowledge from tradition and brewing technologies and modern science (KFRI, 2020). For vitalisation of the alcoholic beverage industry, continuous feedback should be based on market responses. Innovative products based on technology can become a tradition in the future as time accumulates in relation to consumers. Matured pot-distilled soju could be such an item.

Acknowledgements

This chapter was supported by the Korea Food Research Institute (KFRI Project No. E0201100) and the National Research Foundation of Korea (NRF) (Project No. 2018M3C1B505214813). There was also assistance from other organisations: Korea Ceramic Engineering and Technology (KICET), National Institute of Forest Science (NIFoS), Lotte Chilsung, Lotte R&D Center, Hitejinro, Hwayo, Kooksoondang, Andong soju, Muhak, Daesun, Chungbuk, Bohae, Hallasan, and Boogle.

References

Bal, J., Yun, S.-H., Yeo, S.-H., Kim, J.-M., Kim, B.-T., Kim, D.-H., 2017. Effects of initial moisture content of Korean traditional wheat-based fermentation starter Nuruk on microbial abundance and diversity. Appl. Microbiol. Biotechnol. 101, 2093–2106.

Cai, L., Rice, S., Koziel, J.A., Jenks, W.S., Van Leeuwen, J., 2016. Further purification of food-grade alcohol to make a congener-free product. J. Inst. Brew. 122, 84–92.

Carroll, E., Trinh, T.N., Son, H., Lee, Y.-W., Seo, J.-A., 2017. Comprehensive analysis of fungal diversity and enzyme activity in Nuruk, a Korean fermenting starter, for acquiring useful fungi. J. Microbiol. 55, 357–365.

Cha, J., Chin, Y.-W., Lee, J.-Y., Kim, T.-W., Jang, H.W., 2020. Analysis of volatile compounds in soju, a Korean distilled spirit, by SPME-Arrow-GC/MS. Foods 9, 1422. https://doi.org/10.3390/foods9101422.

Choi, H.-S., Kim, E.-G., Kang, J.-E., Choi, J.-H., Yeo, S.-H., Jeong, S.-T., 2014. Effect of varying the amount of water added on the characteristics of mash fermented using modified Nuruk for distilled-soju production. Korean J. Food Preserv. 21, 908–916.

Ferdman, A. R. 2014. South Koreans drink twice as much liquor as Russians and more than four times as much as Americans. Quartz. <https://qz.com/>, 2014 (last accessed December 2020).

Jeong, S.-T., Choi, H.-S., Kang, J.-E., 2018. Korean alcoholic beverages: Makgeolli/Yakju. Chapter 18. In: Park, K.Y., Kwon, D.Y., Lee, K.W., Park, S. (Eds.), Korean Functional Foods: Composition, Processing and Health Benefits. CRC press, Boca Raton, FL, pp. 441–462.

KFRI (Korea Food Research Institute), 2020. Food research information. <https://www.kfri.re.kr//>, last accessed December 2020).

KICET (Korea Ceramic Engineering and Technology) 2020. Research information and service. <https://www.kicet.re.kr//>, (last accessed December 2020).

Kim, B.S., 2014. Exploring the possibility of becoming a textbook for the history of food culture exchange between Mongolia and Koryo. Soc. Stud. Educ. 53, 117–130.

Kim, T.W., 2017a. Globalisation of traditional pot distilled spirits through fusion research. Food Ind. Nutr. 22, 49–53.

Kim, T.W., 2017b. New Product Development of Korean Distilled Spirits, Soju. Heriot-Watt University. Doctoral dissertation.

Kim, T.-W., 2019. Distillation technology and history of Korean distilled spirit, soju. Food Sci. Ind. 52, 410–417.

Lee, J., Moon, S., Bai, G., Kim, J., Choi, H., Kim, T.W., Jung, C., 2016. Distilled Spirits. Ministry of Agriculture, Food and Rural Affairs, Korea Agro-Fisheries & Food Trade Corporation, Gwangmungak, Gyeonggi.

Lee, J.-E., Lee, A.R., Kim, H., Lee, E., Kim, T.W., Shin, W.C., Kim, J.H., 2017. Restoration of traditional Korean Nuruk and analysis of the brewing characteristics. J. Microbiol. Biotechnol. 27, 896–908.

Liu, H., Sun, B., 2018. Effect of fermentation processing on the flavor of baijiu. J. Agric. Food Chem. 66, 5425–5432.

NIFoS (National Institute of Forest Science) 2020. Research project information and publication. <https://nifos.forest.go.kr/>, (last accessed December 2020).

Siřišťová, L., Přinosilová, Š., Riddellová, K., HajŠlová, J., Melzoch, K., 2012. Changes in quality parameters of vodka filtered through activated charcoal. Czech J. Food Sci. 30, 474–482.

Statistics Korea. 2020.Liquor shipment status in Korea. <http://kostat.go.kr/>, (last accessed December 2020).

Yoshizaki, Y., Yamato, H., Takamine, K., Tamaki, H., Ito, K., Sameshima, Y., 2010. Analysis of volatile compounds in shochu koji, sake koji, and steamed rice by gas chromatography-mass spectrometry. J. Inst. Brew. 116, 49–55.

Further reading

The National Law Information Center, Liquor tax law. <https://www.law.go.kr/>, 2020 (last accessed December 2020).

Korea Customs Service, Trade statistics. <https://unipass.customs.go.kr/>, 2020 (last accessed December 2020).

National Research Foundation of Korea (NRF), Food research information. <https://www.nrf.re.kr/>, 2020 (last accessed December 2020).

Andong soju, Liquor products information. <http://www.andongsoju.com/>, 2020 (last accessed December 2020).

Bohae, Liquor products information. <http://www.bohae.co.kr/>, 2020 (last accessed December 2020).

Daesun, Liquor products information <http://c1.co.kr/>, 2020 (last accessed December 2020).

Hallasan, Liquor products information. <http://www.hallasan.co.kr/>, 2020 (last accessed December 2020).

Hitejinro, Liquor products information. <https://www.hitejinro.com/>, 2020 (last accessed December 2020).

Hwayo, Liquor products information. <http://hwayo.com/>, 2020 (last accessed December 2020).

Lottechilsung, Liquor products information. <https://company.lottechilsung.co.kr/>, 2020 (last accessed December 2020).

Kooksoondang, Liquor products information. <http://www.ksdb.co.kr/>, 2020 (last accessed December 2020).

Muhak, Liquor products information. <https://www.muhak.co.kr/>, 2020 (last accessed December 2020).

Indian whiskies

Binod K. Maitin

Independent Technical Consultant, Lucknow, India

History of alcoholic beverages in India

Alcohol consumption in India dates back to the pre-Vedic era when it was called *somras* or *sura*. The pre-Vedic Harappan civilisation mentioned the production of toddy from palm trees. According to Hindu mythology, in the Vedic era (1500–700 BCE), alcohol was believed to be liberally imbibed by both gods and humans. *Somras*, or *soma*, was the drink of gods, and *sura*, a form of beer, was popular among the general population. Alcohol was produced from flowers, grains, and fruits. It was consumed for its invigorating effects and as an integral part of the Ayurvedic system of medicine. Some of the traditional alcoholic drinks are still popular in modern India. Examples include *toddy*, which is made from palm, and *fenny*, which is made from cashews and coconuts and is popular in tourist spots such as Goa and Kerala. Another local drink known as *mahua* is made from mahua flowers (*Madhuca latifolia*) and comes from the state of Madhya Pradesh in Central India.

A more detailed account of the history of alcohol and drinking habits around the world has been presented by several authors (Bennett et al., 1998; Hocking and Dunbar, 2019; Fernandes and Desai, 2013; Hanson, 1995; Maitin and Stephen, 2004; SIRC, 1998).

The Indian alcoholic beverage market

After China, India is the second largest liquor market in the world based on population. Whisky is a key drink for Indian consumers, and brands cover all affordable price points. In 2019, whisky volume accounted for 17% of the global spirits market, and almost half of that consisted of Indian whisky brands (Drink International, 2020).

Although India has traditionally lacked a domestic drinking culture, liquor has gained popularity in the last few decades. The key drivers for this growth are increasing urbanisation, favourable demographics, enhanced social acceptance for consumption of alcohol, rising per capita income, exposure to and availability of a wider brand variety, and a shift towards branded spirits by country liquor consumers. Innovative marketing campaigns and the propagation of ideas such as potential health benefits of alcohol have also influenced the increase in consumption. In recent years, the Indian alcoholic drinks market has shown strong growth in terms of both value and volume and healthy signs of premiumisation. The industry is highly complex and subject to local and national taxes, and there are significant regional variations in consumer preferences. Despite this, India is emerging as a key market for the global spirits industry, fuelled by the growth in consumption.

India is one of the largest producers of distilled spirits in the world, and alcoholic beverages in India can be broadly divided into distilled spirits, country liquor, beer, and wine (Fig. 6.1).

IMFL and country liquor account for the bulk of alcohol consumption. Currently, the total IMFL consumption in India is 3.5 billion litres. Country liquor is another mass-market product, amounting to an additional 3.0 billion litres (~ 30% of the alcohol beverage industry in India). Beer consumption is also around 3.3 billion litres. Wine has had a relatively low consumption rate of only about 1%, but this segment is projected to increase, especially in urban centres, with the increasing young population and with female consumers.

The major share of the distilled spirits segment is composed of brown spirits (whisky, brandy, rum) and white spirits (gin, vodka), which are collectively known **as Indian-made foreign liquor** (IMFL).

IMFL is the official government and business term for liquor manufactured in India and also **includes** spirits produced in foreign countries and imported and **bottled** in India (BII). Excluded are indigenous alcoholic beverages such as arrack, feni, and toddy. Beer, wine, and imported foreign liquor (IFL) are also not included in IMFL. The term IMFL has been explained amply in a monograph by Fernandes and Desai (2013).

Country liquor (also referred to as Indian Made Indian Liquor—IMIL) is an alcoholic beverage sold with an alcohol content of ~ 30%–36%. It is usually produced locally with flavours

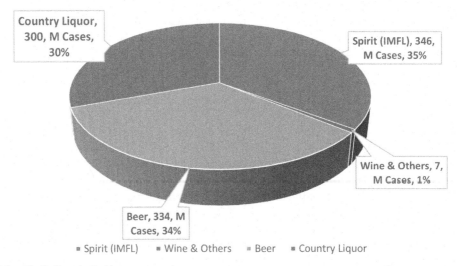

INDIAN ALCOHOL BEVERAGE MARKET

Country Liquor, 300, M Cases, 30%

Spirit (IMFL), 346, M Cases, 35%

Wine & Others, 7, M Cases, 1%

Beer, 334, M Cases, 34%

▪ Spirit (IMFL)   ▪ Wine & Others   ▪ Beer   ▪ Country Liquor

FIG. 6.1   The Indian alcohol beverage market.

influenced by regional taste preferences. Both legal, licensed distilleries, and a segment of illicit brew in the unorganised sector are included in this designation. Several country liquor producers now use higher quality extra neutral alcohol (ENA) to subdue the odour and promote the drink as one where no water or soda needs to be added to make it palatable. This also offers the opportunity of moving this product to IMFL consumers.

Spirits are far more popular than beer and wine and account for the largest segment of the market. Within the IMFL segment (Fig. 6.2), whisky predominates and is by far the most popular category, unlike any major market in the Western world. Whisky accounts for over 63% of annual sales, followed by brandy (21%), rum (12%), and gin and vodka combined (3%) (IWSR, 2020).

India consumes 48% of the world's whisky, and Scotch whisky is very popular. In 2020, for Scotch whisky imports, it was the third largest by volume (SWA, 2020). However, high government tariffs (150%) and a market devolved to individual states have created separate markets, each with different regulations, taxes, and restrictions.

Indian whisky

Whisky, the most popular distilled alcoholic beverage in India, was introduced to India in the 19th century during the time of the British Raj. It has gained popularity among affluent Indians, and India has become one of the largest markets for whisky in the world. As in other countries, Indian whiskies also have their traditions. Whisky making in India is somewhat different than in other countries, as Indian whiskies are traditionally blended with neutral alcohol, commonly known as extra neutral alcohol (ENA). ENA is produced from sugarcane molasses due to its abundant availability in the country and has been the preferred base for alcoholic beverages in India for many decades. This process contributes to the distinct identity of Indian whiskies, similar to Scotch or local whiskies in other countries.

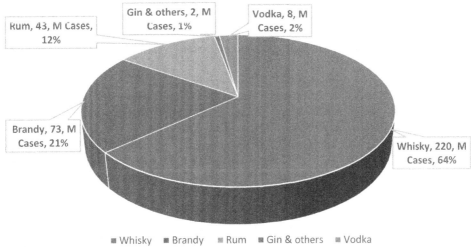

DISTRIBUTION OF INDIAN MADE FOREIGN LIQUOR

Gin & others, 2, M Cases, 1%
Vodka, 8, M Cases, 2%
Rum, 43, M Cases, 12%
Brandy, 73, M Cases, 21%
Whisky, 220, M Cases, 64%

■ Whisky   ■ Brandy   ▨ Rum   ■ Gin & others   ▨ Vodka

FIG. 6.2   Relative distribution of Indian-made foreign liquor (IMFL) (IWSR, 2020).

Indian whiskies are available in a broad range of price segments to cater to consumers of diverse socio-economic status. Economy whiskies are produced using only ENA from sugarcane molasses, with flavourings added. At the other end of the spectrum are single-malt whiskies produced using fermentation processes and pot-still distillation and matured in oak wood casks, similar to the practices of Scottish distilleries. An intermediate segment is comprised of premium blended whiskies containing varying proportions of Indian malt whiskies and Scotch malt whiskies, with neutral alcohol being used as the base and with or without the use of flavourings.

Indian whiskies are broadly classified into premium, prestige, regular, and economy segments. The prestige and premium segments have grown continuously with India's thirst for higher-end spirits, with their consumption a marker of 'the good life'. This is described as a key driver in Euromonitor's forecast of a 25% growth in the Indian spirit market by 2022 (Euromonitor, 2020). Demand for premium liquor has also increased due to overseas travel and to high-net-worth Indians' driving sales in the premium segment.

Raw materials

References to whisky as the 'water of life' do not specify its raw materials. The raw materials, composition, definitions, and laws concerning whiskies in various countries are diverse. They are dependent on the availability of raw materials, convenience of use, the local economy, regulations, and environmental issues. Scottish, Irish, American, and Canadian whisk(e)y processes use malt and several other cereal grains, such as barley, corn, rye, and wheat, to produce their whiskies. Although they all use cereal grains, they digress to varying degrees from the concept of only using malt, originally practised in Scotch whisky production. Raw materials other than barley malt are generally adopted to facilitate the production of whiskies using locally grown materials. For example, some highly rated global whiskies include rice as a key ingredient. Indian whiskies, using various raw materials (molasses and various grains), have acquired their own enthusiasts.

Traditionally, IMFL products, including whiskies, used to be based solely on ENA distilled from fermented molasses. However, many premium Indian whiskies are currently produced from malt and other grains and include some popular single-malt varieties by Amrut, Radico, and Paul John.

The use of alcohol derived entirely from grain was not considered a critical requirement for blending in Indian whiskies, as there is no perceivable difference between the neutral alcohol produced by column distillation from cane molasses and that from grain in terms of congener profiles. Both are rectified in order to make them pure and congener free. Indian whiskies from the two substrates cannot be differentiated organoleptically or chemically. Thus, a grain-based whisky does not offer any benefits in terms of consumer preferences for aroma and palate.

In view of food shortages, the scarcity of grain for potable alcohol has been a major reason to restrict grain use for whisky in India. The abundant availability of molasses in India (currently 11.1 million tonnes) has consequently been extensively utilised for alcohol production (~ 2500 million litres). Of this, only about 1000 million litres is used as potable alcohol. The remaining is used for industrial, medical, and other purposes. Even though grains such as barley, wheat, corn/maize, jowar, bajra (pearl millet), and sorghum are produced in large quantities ($\sim$ 300 million tonnes), only a negligible fraction is utilised for spirit and beer production, and the rest is used as food for the country's large population.

In 2020, India produced 26 million tons of sugar, and from that, 11.1 million tons of molasses gave a total alcohol production of 2500 million litres and from nonfood stocks (grains) 2400 million litres (AIDA, 2021). It is projected that by the end of 2026, more than 16 billion litres of alcohol will be sold across India, while in 2016, that number was over 8 billion litres.

Volatility in the price and availability of molasses and the cyclical nature of the sugarcane crop, combined with allied users' needs, necessitated the adoption of grain-based alcohol by liquor companies to meet the growing demand for potable alcohol products. The increasing sufficiency in India's grain production encouraged the IMFL industry to produce and use grain alcohol in Indian whiskies, solely or interchangeably with molasses alcohol.

IMFL manufacturers adopted grain alcohol for use in premium whiskies and white spirits and used molasses-based alcohol in less expensive brands and country liquors, in addition to its use for industrial purposes and mandatory fuel blending.

Using grain alcohol allowed the IMFL industry to launch cereal-derived Indian whiskies for both domestic and export markets. Although grain whiskies, such as Scotch grain whiskies are distilled at higher alcohol strength, as occurs for ENA in India, they are not neutral and therefore possess a different congener profile, similar to that of molasses-derived rectified spirit, and contain esters, acetaldehydes, methanol, and fusel alcohols.

No significant improvement in terms of depletion of these congeners could be envisaged during the maturation of such spirits. Instead, additional wood-derived tannins and aromatic aldehydes are leached from casks (barrels) into the whisky during maturation. Excessive extraction of tannins from oak in India's tropical climate suppresses the desired aroma and taste profile. Perhaps, for similar reasons, unlike Scotch, American whiskies also use unaged neutral grain in their blended products.

For Indian whiskies, pure ENA derived from sugarcane, with negligible or no congeners (as in vodka), is the preferred base compared to malt and grain whiskies loaded with congeners, such as American and Canadian whiskies, with their high levels of congeners and a woody character.

Indian whisky production

Neutral alcohol is produced from molasses using column distillation, similar to the processes used worldwide. As discussed earlier, malt whisky is also produced from 100% barley malt using the processes of fermentation, pot-still distillation, and maturation in oak wood casks, similar to those employed by Scottish distilleries. Indian whiskies use molasses or grain ENA as the base. This is blended with Indian and/or imported Scotch malt in varying proportions depending on the whisky segment. Premium whiskies have the highest malt content and are generally produced without added flavours.

When ageing whisky in India, the whisky derives benefit from the warm, tropical, humid climate, which allows for a shorter maturing process and a more rapid extraction from the wood, which can leave the whisky less susceptible to effects from the barrel. However, the warmer temperature results in a higher loss to evaporation (the angel's share) and an abv rise during maturation. A smoother, softer whisky, at a younger age, allows for more diverse options on flavours and aromas. Many distilleries are nearing a 10-year cycle of maturation, and these products at ten years can be compared to a 20-year matured product in Scotland.

On the contrary, economy whiskies do not contain malt, and the whisky characteristics are accomplished through the use of minuscule quantities of food-grade flavours. Colour is adjusted using a spirit-soluble caramel. The strength of the whisky and other IMFL products is adjusted to 42.8% abv with demineralised water. All of the ingredients used and the final products are assessed for sensory and chemical quality.

Flavour profiling of whiskies

The flavour profiles of three types of Indian whiskies (economy, premium, and single malt) have been discussed by Maitin and Stephen (2004). A similar evaluation was conducted recently for the taste and aroma attributes of three whiskies: economy Indian (A), blended Scotch (B), and premium blended Indian (C). The observations were similar to those reported earlier, and the profiles are shown in Fig. 6.3.

The blended Scotch and the premium blended Indian whisky both demonstrated a rich aroma and taste profile, with comparably high levels of malty and woody aroma and tastes, robust body, smoothness, pleasant mouthfeel, and lengthy finish. The premium blended Indian whisky also had a substantial peaty aroma. The economy Indian whisky displayed

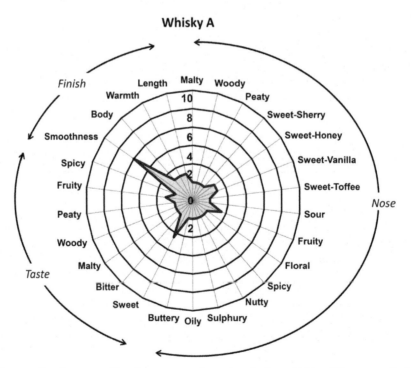

FIG. 6.3   Comparative flavour profiles of economy and premium Indian whiskies: (A) economy Indian whisky,

Whisky B

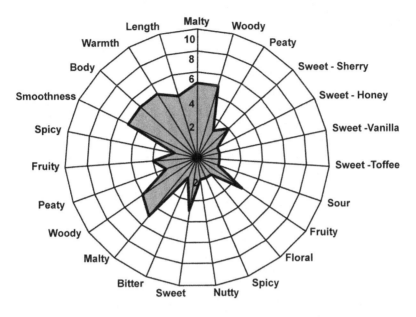

Whisky C

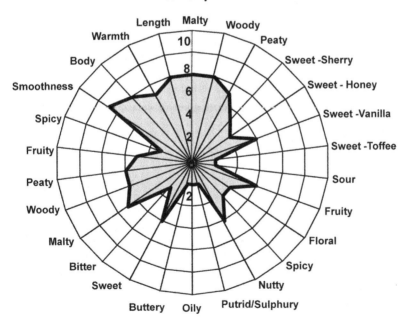

FIG. 6.3—cont'd   (B) blended Scotch whisky, and (C) premium blended Indian whisky. Nose, taste, and finish as labelled for whisky A also apply to whisky B and whisky C.

a profile quite different from the other two, with a lean body, as it did not contain any malt and had no malty or woody attributes. Only fruity, sweet, and vanilla attributes derived from added flavours were observed, but the whisky was perceived as being smooth.

Gas chromatographic profiling of whiskies

Gas chromatographic (GC) analysis data for the three types of whiskies discussed earlier (economy Indian whisky, blended Scotch whisky, and premium blended Indian whisky) are illustrated in Fig. 6.4. As observed earlier, the economy whisky had negligible levels of congeners and did not contain any malt. The levels of congeners were highest in Scotch and moderate in the premium blended malt whiskies, as a function of their substantial malt content.

Consumer research: Home use test

A home use test (HUT) on superpremium and economy whiskies was conducted for the three whiskies described earlier. Economy Indian whisky, blended Scotch whisky, and premium blended Indian whisky were subjected to sensory evaluation by 62 consumers. The results (not shown) indicated that most consumers could not discriminate between the premium and nonpremium whiskies. This could be attributed to the typical Indian style of drinking, which often inhibits the accurate perception of aroma and taste required for appreciating whisky. Nevertheless, the data suggested that the less expensive whisky was preferred over both the premium blended Indian whisky and the Scotch whisky by most non-Scotch consumers, although they also liked the premium brands. This observation reflects an evolution

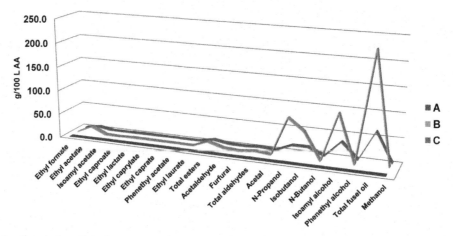

FIG. 6.4 Chemical profiles of economy and premium Indian whiskies analysed by gas chromatography: (A) economy Indian whisky, (B) blended Scotch whisky, and (C) premium blended Indian whisky.

in consumer taste perception compared to previous data (Maitin and Stephen, 2004), wherein the evaluators disliked the premium whiskies. Younger consumers also liked both types of whiskies, but the premium whisky preference was more pronounced. This observation is in line with the increasing trend of premium whiskies in India. Scotch consumers showed mixed preferences for premium and Scotch whiskies. Interestingly, as the observations reported earlier, several experienced Scotch consumers did not categorically dislike the economy whisky. Hence, the premium attributes of the whiskies, as judged by the level of malt, did not have any impact on consumer preference. The aforementioned observations are in line with the considerable consumption of economy Indian whisky in the domestic market.

Benefits of neutral alcohol-based whisky

The use of neutral spirit as a base is economical and offers many advantages. It gives better uniformity to the final product and serves as a common inventory for several products ranging from whiskies to vodka. Its usage facilitates the production of preferred light-bodied whiskies and helps to retain the characteristic aroma and taste of malt whiskies in the premium blended Indian whiskies. Also, it has lower congener levels, leading to lesser chances of a hangover. With negligible amounts of added flavours ($\sim 0.002\%$–0.005%), desired flavour attributes can be easily achieved. Use of neutral alcohol as a base is less problematic for East Asian consumers, as almost 50% of people from this region are known to be deficient in the ALDH2 isozyme required to detoxify acetaldehyde (Maitin and Stephen, 2004; Matsushita and Higuch, 2017).

By Indian law, molasses-based ENA can be used in domestic whisky production. Due to regulations in many countries, the product cannot be exported as a whisky. It may require to be labelled 'Spirit Drink' rather than 'Whisky' for those locations. Most Indian whisky producers, therefore, do not export to Western markets as their consumer base in India is in the tens of millions, so there is little need to export with such high domestic sales. The product is designed for the Indian palate in terms of sweetness, mouthfeel, and character.

Drinking patterns in India

Drinking practices vary substantially among various countries and different populations. In India, the drinking style varies with socio-economic class and consumer segment. Most people in India normally drink alcohol in the evenings, before dinner and with spicy snacks that tend to mask the real taste of the drink. Consumers generally mix whisky with a generous quantity of soda, water, or both and drink it with or without ice. This significantly dilutes the actual taste of the whisky, diminishing the sensory qualities of a premium product. The drinking time is often compressed, sometimes less than 30 min. Fast drinking or gulping is very common. For some new consumers, such as those who may have just turned legal drinking age, the process of drinking and becoming inebriated is more important than actually drinking for enjoyment.

Whiskies that are blended and lack intense aroma have become popular as being 'family friendly' with a less noticeable aroma on the breath. While such consumers are continually looking for superior blended whiskies with a lesser aroma of malt-heavy whiskies, the

TABLE 6.1   Top ten Indian whisky brands by sales volume in 2019.

Whisky brand	2019 Sales volume
1. McDowell's No.1 (United Spirits[a])	30.7 million 9-L cases
2. Officer's Choice (Allied Blenders and Distillers)	30.6 million 9-L cases
3. Imperial Blue (Pernod Ricard)	26.3 million 9-L cases
4. Royal Stag (Pernod Ricard)	22.0 million 9-L cases
5. Original Choice (John Distilleries)	12.7 million 9-L cases
6. Haywards Fine (United Spirits)	9.6 million 9-L cases
7. 8PM (Radico Khaitan)	8.5 million 9-L cases
8. Blenders Pride (Pernod Ricard)	7.7 million 9-L cases
9. Bagpiper (United Spirits)	6.1 million 9-L cases
10. Royal Challenge (United Spirits)	5.5 million 9-L cases

[a] *United Spirits is the world's second-largest spirits company by volume and is a subsidiary of Diageo (who own 55.9% of its shares in 2020).*
Source: The Spirits Business, 2020. Indian Whisky Brand Champions. (https://www.thespiritsbusiness.com/2020/06/indian-whisky-brand-champion-2020-bangalore-malt-whisky/).

palates of new-age consumers are changing, and as they become more evolved, they tend to move towards whisky with a more malty character as well as single and pure malts.

Worldwide ranking of Indian whiskies

India is the largest whisky market in the world. The Indian whiskies listed in Table 6.1 account for six of the top 20 whiskies ranked by global volume of sales in 2019. The top 4 highest selling whiskies are Indian Whiskies, totalling 111 million cases and superseding the highest selling Scotch and American Whiskies at fifth and sixth position. Also, of the top 12 growth brands within the world's top 100 brands of spirits, three are Indian whiskies, with ranks of 1st, 11th, and 12th. Interestingly, 60% of the total volume (321 million cases) of 52 whiskies come from only 17 Indian whiskies (Drinks International, 2020; The Spirits Business, 2020). India's major whisky producers are as follows:

- United Spirits Limited, A Diageo plc UK Company (Bangalore, Karnataka) https://www.diageoindia.com/
- Pernod Ricard India (Gurugram) https://www.pernod-ricard.com/en-in/
- Radico Khaitan (Rampur, Uttar Pradesh) www.radicokhaitan.com
- Allied Blenders and Distillers Pvt. Ltd. (Mumbai) https://www.abdindia.com

India's major single-malt whisky producers are as follows:

- Amrut Distilleries Private Ltd. (Bengaluru) www.amrutdistilleries.com
- John Distilleries (Bangalore, Karnataka) https://jdl.in/
- Radico Khaitan (Rampur, Uttar Pradesh) www.radicokhaitan.com
- Piccadily Agro Industries Ltd., Gurugram (Haryana) www.piccadily.com

World-class Indian single-malt whiskies

Whiskies from outside the typical production areas (e.g. Japan and Taiwan) have had outstanding products on the market in recent years. India is clearly also now firmly on the world whisky map with its award-winning single malts and whisky blends. The next generation of high-end Indian-made single malts are becoming much sought-after and valued products in Western markets.

Currently, India has four main malt distilleries—Amrut, Rampur, Paul John, and Piccadily. Piccadily in the north is able to produce rich, fruity whiskies despite wide temperature variations (from 5°C in winter to 48°C in the summer), while the malt distilleries in the south produce more mellow spirits.

The following are the examples of Indian-made single-malt whiskies that have won global accolades, and each features very different styles and characteristics.

Amrut Fusion is an Indian single-malt whisky that uses **Indian barley and peated barley from Scotland** (hence the name Fusion). In addition to other awards, this whisky won the silver medal at the World Whisky Awards in 2019. Amrut has been the leader in the Indian single-malt expression and today supplies over 60 countries.

Rampur Select won a double gold medal winner at the San Francisco Wine and Spirits competition in 2017 for its **single-malt whisky aged in select casks** in the Himalayas' foothills.

Kanya by John Paul is from Goa (**unpeated single malt**) and won the accolade as the best Asian whisky in 2018. It was aged for 7 years in first-fill bourbon white oak casks and is a single-malt product.

Woodburns is an Indian malt whisky made by the family-owned Fullarton Distilleries in Goa, **with all-Indian ingredients** (no foreign malts or grains). It was awarded a silver medal in the world blended section in 2019 at the International Spirits Competition, and at the Singapore World Spirits Competition, and a silver medal at the 2020 New York International Competition.

Indian whiskies—Future perspectives

India is one of the largest spirits markets based on volume. Though it is driven primarily by whisky growth, other subsegments such as brandy, rum, vodka, and gin have also been posting steady numbers. The country's massive population (1.3 billion strong), rising disposable income, and changing lifestyle are expected to keep the spirits demand high for the foreseeable future.

Alcohol consumption has become more socially acceptable, especially with women, at social events, and in family environments. In the cities, there is wider exposure to a variety of alcoholic products. Both gin, once considered an older man's drink, and vodka, as only for women, are now being embraced by the younger generation.

The growing demand in India for spirits can be attributed to its huge young population base and a younger generation that is starting to consume alcohol. The majority of the alcohol is consumed by people between the age of 18 and 40. Demographically, India has one of the youngest populations, with 50% of the population under the age of 25 and 65% below the age of 35.

With 20 million new legal drinking age consumers each year in the country, fascination for new and exotic whiskies from around the world is continually increasing. There is a new focus being put on Indian whisky, as well as other New World distilling nations, to make products that are not only high quality but also highly marketable.

An increasing number of bars in urban areas have opened opportunities for youths to socialise. Multisensory experience of pairing food with drinks in restaurants is a top trend. Whisky cocktails and the cocktail culture are becoming very popular among young consumers. The rise in the independence of women has also led to an increase in whisky consumption by women.

The drive towards the premiumisation of the Indian spirits industry is another important feature of the Indian spirits market. In fact, future growth in the liquor industry in India will be mirrored by continued consumer trends towards premiumisation, motivated by rising affluence, globalised outlooks, urbanisation, and progressive lifestyles. With more Indians travelling abroad, rising aspirations, favourable environments for imported liquor, and higher disposable income, consumers are upgrading towards premium segments in the country. Whisky packaging is being improved to better appeal to consumers and to gain their confidence (e.g., Guala Closures, Tetra Paks, and other innovative packaging). Anticounterfeit/antitamper features help to ensure that the liquor is not adulterated (see Chapter 20 for additional details).

While Indian single malts from various producers have already gained immense popularity, blended whiskies with premium Indian grain and malt should also create demand among consumers in India and other countries.

Acknowledgements

The author wishes to acknowledge United Spirits Limited (USL) for the permission to write this chapter and the USL Technical Centre team for their contributions.

References

AIDA, 2021. Personal communication—all India Distillers Association. (www.aidaindia.org).

Bennett, L.A., Campillo, C., Chandrashekar, C.R., Gureje, O., 1998. Alcoholic beverage consumption in India, Mexico, and Nigeria: a cross-cultural comparison. Alcohol Health Res. World 22, 243–252.

Drinks International, 2020. Drinks Report. (https://drinksint.com/).

Euromonitor, 2020. Spirits in India Report. (https://www.euromonitor.com/spirits-in-india/report).

Fernandes, V., Desai, T., 2013. Cheers: The Indian Alcobev Industry Era. Special Audience Publications, Mumbai, pp. 13–51.

Hanson, D.J., 1995. History of Alcohol and Drinking around the World. State University of new York. Potsdam, New York. http://www2.potsdam.edu/hansondj/Controversies/1114796842.html.

Hockings, K., Dunbar, R. (Eds.), 2019. Alcohol and Humans: A Long and Social Affair. Oxford University Press, UK.

IWSR, 2020. Drinks Market Analysis. https://www.theiwsr.com/.

Maitin, B.K., Stephen, S.N., 2004. Indian whisky: an overview. In: Bryce, J.H., Stewart, G.G. (Eds.), Distilled Spirits Traditions and Innovation. Nottingham University Press, Nottingham, U.K, pp. 149–154.

Matsushita, S., Higuchi, S., 2017. Use of Asian samples in genetic research of alcohol use disorders: genetic variation of alcohol metabolising enzymes and the effects of acetaldehyde. Am. J. Addict. 26 (5), 469–476.

SIRC, 1998. Social and Cultural Aspects of Drinking: A Report to the European Commission. Social Issues Research Centre, Oxford, UK. http://www.sirc.org/publik/social_drinking. pdf.

SWA, 2020. International Trade. https://www.scotch-whisky.org.uk/insights/international-trade/.

The Spirits Business, 2020. Indian Whisky Brand Champions. https://www.thespiritsbusiness.com/2020/06/indian-whisky-brand-champion-2020-bangalore-malt-whisky/.

Whiskies of Canada and the United States

Steve Wright[a] *and Heather Pilkington*[b]

[a]Spiritech Solutions Inc., Kingsville, ON, Canada [b]First Key Consulting Inc., London, ON, Canada

Although whisky's origins predominantly lie overseas, North American whiskies have evolved into their own creations with distinct characteristics that have both shaped and been impacted by historical events. Clearly differentiated not only by different spellings, American *whiskey* and Canadian *whisky* (we will stick with *whisky* for continuity) are also unique in style and taste and are products of their own individual cultures, climates, and laws and regulations. In this chapter, we will look back at the history of whisky in North America and then move forward to view the present-day whisky markets and the key North American producers. Finally, we will discuss what sets these whiskies apart by describing the methods used in producing the beloved bourbon, rye, and Tennessee whiskies of the United States and the equally desirable blended and straight whiskies from Canada.

History

Alcoholic beverage consumption came to North America with the first settlers. The production and consumption of alcoholic beverages provided some practical benefits during those early years. Spirits were often healthier to drink than the available water and, from an economic standpoint, farmers with too much grain on hand would operate their small distilleries rather than risk the crop being spoiled. The spent grains could then be used to feed livestock. Early colonists were distilling whatever grains they had on hand with their stills being considered pieces of farm equipment. The distilled products also had the advantage of commanding a higher value than grains within a smaller volume in the market.

In the early 1600s, the alcoholic beverage styles consumed in America were beer, mead, and ciders. Rum was also brought in from the West Indies. In the 1700s, the tides shifted with the American revolution driving a decrease in consumption of rum from the West Indies where the British were involved in trading of raw materials. Americans still wanted alcoholic beverages and consumption of locally produced whiskies increased as a result.

Later in the 1700s, rye predominated as the grain of choice for whisky production because rye was harder and easier to grow than barley in New York, Pennsylvania, Virginia, and Maryland which were the most populated states at that time. There were thousands of small farm-based distilleries operating with diverse production methods. Further south where corn was natively grown, it was also distilled, but to a much lesser extent since most of the population was located in the north.

In colonial times, the two styles of rye whisky that arose were the Maryland style, which was a blend of rye and corn with a sweet and mellow taste and the Pennsylvania or Monongahela style, which was almost all rye, more full-bodied, and spicy. Given the growing conditions, it is no surprise that most of the rye further north produced the Monongahela rye, where corn was not native, and that the southern parts of the United States made the Maryland-style rye, where corn was more readily available.

In the late 1790s, George Washington started a distillery at Mount Vernon. His distillery initially produced rum but then turned to rye whisky production (predominantly Maryland style). By the time he died, he had made a sizeable profit from the distillery, and whisky sales made up a significant portion of his revenues (Regan and Regan, 2009).

Whisky tax was introduced by Washington in 1791. This was the first federal excise tax and it was put in place to pay down the debt from the Revolutionary War. This new tax resulted in a series of rebellions between 1791 and 1794 that centered around Pennsylvania, Virginia, New Jersey, and Maryland, which were eventually quelled by Washington and his soldiers. However, by the end of the rebellion, many rye whisky distillers decided to move farther afield and headed west towards Kentucky, where there were plenty of farms that were mainly growing corn. The distillers also planted their rye, and by 1810, there were thousand distilleries in Kentucky. With the growing climate favouring corn over rye, the distilleries in Kentucky produced Maryland-style rye whisky with the characteristic of rye-corn mix, and when using a higher corn content, they produced a unique spirit that would soon be known as bourbon (Pickerell, 2018).

Rye whisky predominated over bourbon from the civil war until Prohibition because the US population was centered around the rye-producing eastern and northern states and shipping routes were favourable. Prohibition, which was in effect between 1919 and 1933, caused American whisky stocks to become severely depleted. When Prohibition abruptly ended, well-aged American whisky was in short supply, and it was very difficult for American whisky producers to compete with fully matured imported spirits from Canada, the Caribbean, and overseas. This left American producers having to compete on price rather than quality as they replenished their stocks over time. Meanwhile, the distilleries in Kentucky had more land to build larger and more efficient facilities, and it was also easier to produce bourbon than rye. On top of that, farmers growing corn were supported by the farm subsidy a few years earlier than those growing rye. All of this resulted in very few rye distilleries being able to come back after the Prohibition ended, resulting in bourbon growth and rye whisky falling into decline.

Around the time Washington's distillery was being built at Mount Vernon, immigrants from the European mainland and Britain were also coming to Canada and bringing their distilling practices with them as outlined in Davin de Kergommeaux's engaging Canadian Whisky textbook (de Kergommeaux, 2017). By 1767 in Quebec City, a distillery was operated that produced commercial-scale quantities of rum. Throughout the 1700s, numerous small-home distilleries were operating throughout Lower Canada (Quebec) and Upper Canada (Ontario), making spirits from surplus grain. During the American Revolution (1775–1783), United Empire Loyalists also made their way to Canada, bringing their farm-based distilling practices with them, and by the first half of the 1840s, there were over 200 distilleries operated in Canada. It is commonly thought that the United Empire Loyalists first introduced whisky making to Canada. However, most of the Loyalists' small operations disappeared by the late 1800s and the key players, who successfully built the commercial whisky industry in Canada, hailed from England and continental Europe (Germany and the Netherlands).

Interestingly, the first major Canadian whisky distillery was founded in Montreal in 1801 by John Molson, the founder of Molson Brewery, when he purchased a copper pot still that was once used to make rum. With his son Thomas Molson and business partner James Morton, Molson's distillery had operations in Montreal and Kingston, until Molson's withdrawal from the whisky business in 1867. In 1831, James Worts and William Gooderham, brothers-in-law from England, started a milling company in Toronto and, by 1837, they were also brewing and distilling. Their distilling business grew rapidly, becoming a major competitor to Molson. By the 1850s, Gooderham and Worts Company had surpassed Molson and built a large-scale production distillery in Toronto. With continued expansion, the distillery was the largest producer of whisky in the world during the 1860s. The former Gooderham and Worts production site in Toronto has recently undergone commercial and residential redevelopment to become a Canadian national historic site known as the Distillery District.

The mid-1800s saw the start-up of many iconic Canadian whisky producers including Corby's (Belleville, Ontario), J.P. Wiser (Prescott, Ontario), Hespeler and Randall's Waterloo Distillery (later to become Seagram's), and Hiram Walker (Windsor, Ontario, now part of Pernod Ricard), which is the largest alcoholic beverage distiller in North America to this date.

In the 1880s, the Canadian Government passed a 'Bottled-in-Bond' law that certified the time a whisky spent in ageing and allowed distillers to defer taxes for that time period. This law supported the distillers ageing their whisky and giving their consumers' confidence in Canadian producers. In 1890, the government further passed a law that Canadian whisky must be aged a minimum of two years before being sold. This was the first law of its kind in the world and it reinforced Canada's reputation around the world as producers of high-quality whisky. The United States followed suit with legislation in 1897. Today, Canadian whisky must be aged no less than three years in oak barrels.

With the temperance movement and Prohibition reducing the size of the market and the challenges of the new whisky ageing law, the Canadian whisky business was ripe for consolidation in the early- to mid-1900s. Despite the Canadian whisky industry going through a period of instability and risk during American Prohibition, some Canadian distillers profited considerably from supplying the United States with whisky during this period. Larger distilleries such as Seagram and Hiram Walker also put into place a successful strategy of developing whiskies with longer ageing periods in anticipation of the Prohibition reaching an end (de Kergommeaux, 2017). When the US Prohibition ended, the love of Canadian whisky was strongly etched into the mindset of whisky consumers in America. Canadian whiskies prevailed and continued to outsell bourbon in the United States until 2010.

Present-day markets and producers

American and Canadian whiskies today are experiencing strong growth, and history suggests that there is still considerable room for continued growth (per capita consumption of whisky is still lower than it was in the 1960s). Whisky expansion in North America is being aided by the explosive growth of craft distilleries in the past 15 years, which produced volumes of whiskies and other spirits that represented 5.8% of the total US spirit value in 2018, according to the Craft Spirits Data Project (Park Street Imports, 2019). Insights into the craft distilling movement in North America can be found in Chapter 8.

The growth of North American whiskies has been assisted by export sales. According to data from the Distilled Spirits Council of the United States (DISCUS), American whisky exports grew about 60% between 2009 and 2015. However, the imposition of retaliatory tariffs by the EU on the import of American Whisky in 2018 resulted in a 27% decrease in whisky exports to the EU the following year and a total of 16% decline across all export markets. The US and Canada are strong whisky-trading partners. Canadian whisky represents 32% of whisky volumes in the US. Data from DISCUS and Spirits Canada show that the US exported $52 M (USD) whisky to Canada in 2019, while Canadian whisky exports to the US reached $252 M (USD). The global Canadian whisky export value that year was $276.7 M (USD), indicating that the US received 91% of Canada's exported whisky.

Bourbon and Tennessee whisky (a close cousin of bourbon) have shown remarkable growth at home in recent years. In the 1980s and 1990s, domestic whiskies in both the US and Canada had fallen out of favour with many consumers, and whisky distillers were reducing, and in some cases even halting new-make spirit production. Distillers in both countries posted substantial declines in whisky sales in the final decades of the 20th century. The new millennium, however, marked a return to growth in North American whisky as consumers came back to the spirits that their parents and grandparents grew up on!

The re-emergence of a cocktail culture in larger urban US cities has sparked renewed interest in American rye whiskies. As a key ingredient in many classic cocktails, rye whiskies have been in very strong demand in recent years and have grown more than 13-fold since 2009 (Swonger, 2020). A modern-day trend towards lower alcohol consumption and a preference for higher-quality spirits along with the growing demand for more whisky choices and an interest in the bold spicy characters of ryes have also fuelled the increase in demand for rye whisky. The growing demand for premium whiskies is particularly evident in the US where, for more than 10 years, growth in premium-priced whiskies has greatly exceeded the growth in value-priced whiskies.

The major players in the North American whisky market are diverse and range from multinational corporations to family-owned independents. The biggest companies include Brown-Forman, Beam Suntory, Diageo, Heaven Hill, Sazerac, and Gruppo Campari, with a longer list of smaller players.

The family-owned Brown-Forman Company is based in Louisville, Kentucky and is the maker of Jack Daniel's Tennessee Whiskey, the largest selling whisky brand in the US and the largest American whisky worldwide by volume. The company's leading bourbon brand is Woodford Reserve, which was recognised in 2019 as the fastest-growing American whisky. Brown-Forman exited the Canadian whisky category in 2020 when it sold its Canadian Mist

Distillery and the brand to Sazerac. Canadian Mist is the third-largest-selling Canadian whisky.

Beam Suntory produces a variety of whiskies, including Jim Beam, the world's largest-selling bourbon (second to Jack Daniel's in global American whisky sales) and the fast-growing Maker's Mark brand. Beam Suntory also owns Canadian Club Canadian whisky, which has been produced at the same site, the Windsor Ontario-based Hiram Walker distillery, since 1858, even as the distillery is now owned by the French company Pernod Ricard.

Diageo PLC is a leading global spirit producer with a large presence in North America. Its Crown Royal Canadian whisky is the third-largest-selling North American whisky brand and the largest Canadian whisky brand in the US. With two distilleries in Canada, Diageo is Canada's largest whisky producer. Diageo has two notable American whiskies—George Dickel Tennessee Whiskey and Bulleit Bourbon. Bulleit was launched by Seagram in the mid-1990s and later acquired by Diageo and manufactured by Four Roses until 2017, when Diageo completed construction of a namesake Bulleit distillery in Shelbyville, Kentucky. Today, Bulleit is a popular and successful brand that sells at a premium price.

Heaven Hill is a family-owned American distiller with interests in both American and Canadian whiskies. A large volume producer with many notable bourbon brands, Heaven Hill manages one of the world's largest volume inventories of maturing bourbon. In 2019, Heaven Hill acquired Black Velvet Canadian whisky, the #2 selling Canadian whisky and a high-volume-exported brand, along with the Black Velvet Distillery in Lethbridge, Alberta, and a variety of smaller Canadian whisky brands, making it an important player in Canadian whisky.

Sazerac is another American family-owned company that owns several distilleries and a wide range of bourbons and Canadian whiskies. Bourbons include brands from Buffalo Trace, Barton, and A. Smith Bowman distilleries and include the venerable line of Van Winkle bourbons. Sazerac's 2020 purchase of Canadian Mist gave them a second production site in Canada and greatly increased their capacity to supply their growing legion of Canadian whisky brands.

Italian company Gruppo Campari has interests in North American whiskies through its acquisition of the Kentucky-based Wild Turkey Bourbon, a midsize brand with broad export distribution, and the Canadian whisky producer Forty Creek. Other notable players in the North American whisky business include Kirin Co. with its ownership of Four Roses bourbon and Luxco, which owns a variety of smaller bourbon and Canadian whisky brands and two boutique bourbon distilleries in Kentucky. Others include multinational players—Pernod Ricard, William Grant & Sons, and Bacardi—which have established a presence through investment in craft whisky distilleries. All said, the whisky landscape in North America is dynamic and diverse in styles, brands, and players and is continually growing and changing to meet the increasing demand.

Regulations

Distillers produce their whiskies on both sides of the US/Canada border according to federal regulations, which help protect the characteristics and identities of the unique and distinctive whiskies of North America. In the United States, standards of the identity for American whiskies are reported in the Code of Federal Regulations (27 CFR 5.22). Regulations state the requirements for distillation strength (not exceeding 160° proof [80% abv]), barrelling conditions (stored at no more than 125° proof [62.5% abv] in charred new oak containers) and bottling strength (not less than 80° proof [40% abv]). Requirements for mash bills are also stipulated. In order to be labelled 'bourbon', 'rye', 'wheat', 'corn', or 'malt' whisky, the whisky mash bill must contain a minimum of 51% of the named grain. Note that corn is the named grain in bourbon and the malt type required for malt whisky is barley malt. The US regulations do not establish a minimum required maturation period, except for whiskies designated as *Straight Whisky*, which must be aged a minimum of two years in charred new oak containers. However, if the whisky is under four years of age, the actual age of the spirit must be listed on the bottle.

Bourbon can be produced anywhere within the United States, whereas Tennessee whisky, predictably can only be produced in the state of Tennessee. While CFR standards do not distinguish between Tennessee whisky and Bourbon, it is Tennessee state law that enacts the requirements for Tennessee whisky. State law, along with some international trade agreements, identify Tennessee whisky as a straight bourbon whisky produced in Tennessee. State law also mandates whisky makers to implement the 'Lincoln County process', a practice of filtering whisky spirits through sugar maple charcoal prior to barrel-ageing. This process is believed to smooth the whisky and, in the minds of state regulators, producers, and some consumers, to aptly distinguish it from bourbon. Interestingly, Prichard's Distillery in Kelso, Tennessee gained special exemption from using the Lincoln County process in the production of their Tennessee whiskies.

In Canada, a more modest approach to regulating whisky provides a great opportunity for diversity in manufacturing methods. Canadian whisky is to be fermented, distilled, and matured in Canada (bottling in other countries is permitted) and bottled at no less than 40% abv. It can be produced from any cereal grain, distilled to any strength no greater than 94.8% of alcohol, and matured at any preferred strength in 'small' (<700 L) wooden barrels for a minimum of three years. The caveat is that the final matured product must 'possess the aroma, taste, and character generally attributed to the Canadian whisky' (Government of Canada C.R.C., c. 870), which establishes broad boundaries for the distillers but offers great prospects for creativity. The common approach to whisky making in Canada is to ferment, distil, and mature the individual elements of the final whisky separately and then bring them together to a final blended product. Most Canadian whiskies are blends of a light *base whisky* (similar to the *grain whisky* used as a foundation for blended Scotch whiskies) and lesser amounts of *flavouring whiskies* that tend to be ferments of rye, malts, and other grains that are distilled in manners in order to retain an abundance of character.

Crafting methods

Whisky making practices in North America follow the same sequence of operations as those in other whisky making countries, including milling, mashing, fermentation, distillation, maturation, and blending, but history, tradition, and agronomic and market factors have resulted in processes and, of course, products, that are distinctive and unique to the United States and Canada. In this section, we will highlight the steps used in the making of whiskies that may be exclusive to the United States and Canada and that contrast with many of the practices used in the making of other major whisky types covered in various chapters throughout this publication.

Grains

The choice and proportions of grains used in American whiskies are regulated to some degree. The named grain must comprise at least 51% of the mash bill. Bourbon producers commonly use more than 51% corn in their mash, with 68%–78% of corn most used. The balance is made up of malted barley and rye or wheat. The use of malted barley is not mandatory, but all major bourbon and Tennessee whisky producers use some amount of malt, ranging from as low as 5% to seldom more than 12%. Rye or wheat make up the remainder of the grain, with levels of 10%–20% common. Rye is more commonly used than wheat in bourbon, but Maker's Mark bourbon has popularised the use of wheat and wheat bourbons have grown in popularity in the past 25 years. Malted barley is widely used in whisky making in North America, but it is not mandated for any whiskies other than American malt whisky. Some malted grains are produced specifically for distillers, which require high diastatic power (DP) and high alpha-amylase (AA) levels in order to provide suitable starch conversion at relatively low-use levels.

Yellow dent corn is the most widely used grain for North American whiskies, as it is the predominant grain in bourbon mash bills and is the grain of choice for the base whisky produced by most distillers in Eastern Canada. Western Canadian distillers make more use of local rye and wheat, which are more favoured by growing conditions in the West. Canadian whiskies are often distinguished by their rye character, and while there is no regulatory requirement for rye in Canadian whisky, virtually, all brands use one or more rye-flavouring whiskies in their blends.

Rye grain is widely used in both Canada and the United States. American rye whisky must be made from at least 51% of rye and, while many of the large whisky producers use 51% of rye, with corn and malted barley making up the balance, a wide range of rye contents are used in the different styles of rye whiskies produced in different parts of the country with traditions dating back to pre-Prohibition times when mash bills were often dictated by the grain types available in the region. High rye levels can be problematic to mashing and fermentation due to the high content of arabinoxylans in rye, which causes high viscosities and can lead to foaming in mashing and fermentation in the high rye content of mash bills. A few whisky brands from Canada claim to be '100% of rye', in which case the mash is all rye. Some distillers will include malted rye in the mash to take advantage of its enzyme activities, while others will rely on commercial enzyme sources. Some rye distillers, especially those using 100% of rye, will choose to add a commercial xylanase enzyme during mashing to assist in the control of mash viscosity and foaming issues.

Milling and mashing

In North America, cereal grains are milled mostly by a hammer mill and are mashed by various methods to hydrolyse the starches into fermentable sugars through cooking and enzymatic activities. Most whiskies in the United States and Canada are produced from a whole grain mash, known also as *grain in* mashes, where the grain particles are not strained or filtered out prior to fermentation, and hammer milling provides a suitable grist composition for whole-grain mash. Exceptions include the producers of American single malt whisky, who are growing in number and are mostly producing clear wort and using roller mills on their malts. Roller mills can also be found in at least one major bourbon distillery. Maker's Mark proudly claims that their choice of a roller mill for grinding their corn, wheat, and malted barley helps protect the grains from scorching from the heat generated by a hammer mill. Each distiller has their own preference and their own justification, but it is evident that a roller mill can be effective for the milling of grains other than malted barley.

Mashing methods vary considerably in North American whisky, where both *batch* and *continuous* mashing methods are used. Bourbon, Tennessee and American rye whisky makers use batch mashing systems, which have slurrying, cooking, and conversion of the grains conducted in a single vessel in a batch-wise manner. Cooling of the finished mash may also take place in the cooker using a cooling coil or external heat exchanger, but it is common to pass the completed mash through a cooler during transfer to the fermenter. Some distillers use a two-tank system, where the mash is cooked in a batch cooker and then transferred to a holding tank for enzymatic conversion before cooling during transfer into the fermenter.

Bourbon is cooked either at atmospheric temperature (100°C) or in a pressure vessel at 100–120°C, where the higher temperature of pressure cooking achieves more complete gelatinisation of the corn starch. Distillers understand, however, that there is a trade-off when cooking to a high temperature between achieving complete gelatinisation and hydrolysis of starches and loss of sugars due to higher caramelisation by the Maillard reaction at a higher temperature and they will establish a cooking time/temperature that balances the two competing factors to optimise performance. Energy cost weighs into this decision as well, and there is a trend towards accepting small yield losses by cooking at lower temperatures to reduce energy consumption.

When mashing rye, wheat, and malted grains, cooking temperatures ranging between 62°C and 85°C are common, understanding that these grains do not need the same high cooking temperature as corn. Fig. 7.1 illustrates the broad nature of time/temperature conditions used in batch mashing of various cereal grains. The figure depicts two mashing scenarios for rye grain, which are distinguished by a low (64°C) peak temperature that enables the malt enzymes to survive throughout the mashing sequence. Notice that both rye sequences employ a rest time on the temperature rise. This is to encourage protease and perhaps beta-glucanase activities and such rests can be built into any mashing sequence using malt when wishing to better express specific activities.

The mashing sequence shown for wheat in Fig. 7.1 has a cooking temperature of 90°C, although some distillers cook to as low as 80°C, which is sufficient for full gelatinisation of the wheat starch. In this example, malt is used for conversion, so rest at 58–60°C on the cool-down is added to enable suitable conversion of liquefied starch to maltose and other fermentable sugars. Examples of two batch corn mashing regimes are also displayed. The lower temperature cooking sequence depicts a bourbon mash, which is using malt enzymes for conversion and which shows a rest period at just under 60°C on cool-down; at which point, the majority of the malt is added for starch conversion. A small amount of malt, typically one-quarter to one-third of the total malt content, is added to the mash at the start of the cooking to allow for some thinning (liquefaction) of the mash as the increasing temperature

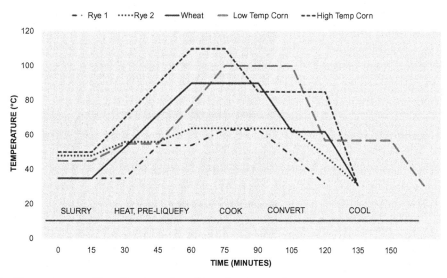

FIG. 7.1   Temperature profiles of batch mash cooking of various grains.

gelatinises the starches and before the heat inactivates the enzyme activities. This practice of partially liquefying the mash on the heating side of the cooking is called *pre-liquefaction*, and when this is done using malt, it is known as *pre-malting*.

A simplified bourbon mashing sequence is also depicted in Fig. 7.2, which shows the two malt addition points. Malt is slurried with cool water prior to addition to the mash to help activate the enzymes and to minimise the dust that is associated with hammer-milled malt. Note that all major bourbon/Tennessee whisky producers add *backset* to the mash, which is either mixed in with the mash water at the start of the cooking or in a more traditional manner added to the fermenter at the end of filling to adjust the final liquid/grain ratio and achieve the preferred pH. The addition of *backset* or the practice of *backsetting* denotes the return of thin stillage (centrifuged or strained still bottoms from a previous fermentation) to a subsequent mashing operation and represents the foundation of the original *sour mash* process that was developed in 1823 by Dr. James C. Crow at the original distillery site that is now the Woodford Reserve Distillery in Versailles, Kentucky (Cowdery, 2004). The addition of thin stillage to a grain mash is nearly universally practised in grain distilleries in North America, especially in whisky making, and yet the use of the term *sour mash* and the marketing of whiskies as *Sour Mash Whisky* is much more limited.

The addition of stillage helps balance the mash pH, which is especially useful in *continuous* mashing systems (described later in this chapter) that commonly add alkaline ammonium salts or aqueous ammonia to the mash as a source of supplemental nitrogen in order to benefit yeast nutrition. The stillage adds buffering capacity to the mash and provides some nutritional benefit to the mash as a minor source of vitamins and utilisable nitrogen, but one of the strongest justifications for backsetting is that it reduces mash water use and ultimately reduces hydraulic loading on the co-product drying operations that most North American grain distilleries operate. A review of co-product operations is found in Chapter 21. It is not uncommon for the backset to represent up to 35% of the total mash volume.

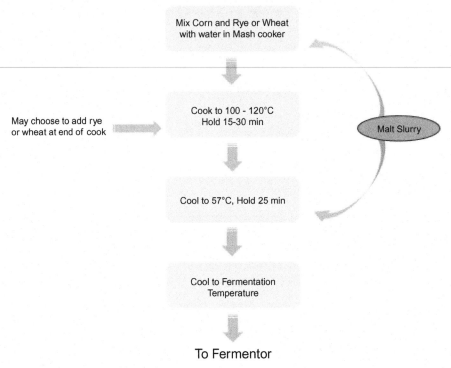

FIG. 7.2   Steps in bourbon mash preparation.

Fig. 7.1 also shows a high-temperature corn batch mashing sequence. In this scenario, the corn is cooked under pressure and is liquefied using commercial thermotolerant alpha-amylase. A portion of the alpha-amylase is added to the mash at the start of the temperature rise to initiate liquefaction and the balance is added post-cook, where a rest period at 85°C is often added to ensure optimal liquefaction. When using commercial enzymes for liquefaction (alpha-amylase) and saccharification (glucoamylase), it has become common practice to add the saccharifying enzyme to the mash once it has been cooled to fermentation temperature and is being transferred to the fermenter. This is a departure from what was the common practice 25 years ago (and still practised today by some distillers) of holding the mash at ~56°C—the ideal temperature for the activity of the enzyme—for up to 2h to enable near-complete saccharification of the liquefied mash. This practice gives rise to a mash with a very high glucose level, and as the industry developed a better understanding of the negative impacts of stresses on yeast performance in fermentation, it was realised that the osmotic pressure is associated with a high glucose content mash that contributed to a long lag time or a slower start to fermentation than what was seen in a high maltose mash, which does not have the same high osmotic pressure. The practice of adding glucoamylase at fermentation temperature is known as simultaneous saccharification and fermentation or 'SSF'. In this case, the distiller establishes a dose rate of the glucoamylase into the mash that will liberate glucose from the liquefied mash at roughly the same rate as it is taken up by the growing yeast. This avoids the glucose level accumulating in the mash, which keeps the osmotic pressure from being inhibitory and helps keep contaminant bacteria in check by reducing the availability of their food source.

It should be evident that a large variety of batch cooking scenarios are used by whisky distillers in North America and an equally large number of methods are used by those distillers operating 'continuous' mashing systems. Continuous systems are used by most distillers for producing the light base whiskies used in most Canadian whiskies, and most of the neutral grain spirits produced in North America also use continuous mashing. Continuous mashing systems provide a constant flow of mash through a chain of vessels held at specific temperatures and with residence times in each vessel determined by mash flow rate, vessel volume, and fill level. They are designed for steady-state operation with a fixed sequence of three mashing steps: *slurrying*, *cooking*, and enzymatic *conversion*. Continuous systems are used in high-volume plants for long uninterrupted production runs with minimal labour input. They are designed to maximise process efficiency and minimise energy consumption. Continuous systems are typically operated at higher gravity than batch systems and often employ methods for heat recovery and reuse. It is not uncommon for a continuous corn mash system to operate at greater than 33% w/w total dry matter content and produce a mash that ferments to higher 14% abv.

Fig. 7.3 depicts a continuous corn mashing system as would be used for the production of neutral spirits and Canadian base whisky. Many variations of this design are used, but they all follow similar steps, including: (1) *slurrying* of the milled grains with water and backstillage to wet and disperse the grist; (2) cooking of the grain mixture to disrupt the starch granules and gelatinise the starches; and (3) enzymatic conversion of the starches to dextrins or in the case of malt addition, conversion to maltose and other fermentable sugars. Malt is not commonly used in continuous mashing systems, which more typically make use of commercial alpha-amylase and glucoamylase enzyme concentrates.

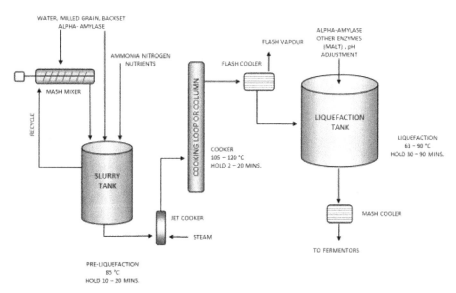

FIG. 7.3   Continuous corn mashing system.

In the slurry step, the milled grain is mixed with hot mashing water and thin stillage, usually with the aid of a mash mingler or mixer to provide aggressive mixing of the grain with high shear to achieve a homogeneous mixture and with added alpha-amylase to provide some early liquefaction of the mixture. The slurried grain then enters the slurry tank, which provides some retention time for further mixing as it feeds the cooker, and a certain portion of the slurry is recycled back to the mash mingler to aid in mixing and encourage flow through this device.

Ammonia salts or aqueous or anhydrous ammonia is often fed into the slurry tank to help serve the nutritional requirements of the yeast during fermentation, and this is especially important in mash bills with low or no use of malt and with high levels of corn, which includes most Canadian base whisky mash bills. As indicated in Chapter 12, the nitrogen needs of yeast are satisfied in malt whisky wort by the amino acids, peptides, and ammonium ions produced by the proteolytic activities of malt during mashing, but the lower or no use of malt and the predominant use of less protein-rich grains than malted barley in North American whisky mash will benefit from the addition of supplemental nitrogen sources. A growing number of distillers are discovering the benefits in adding commercial nitrogenase enzymes during mashing to hydrolyse grain proteins and are finding that they can reduce or eliminate the addition of inorganic nitrogen and still achieve complete and efficient fermentation.

The slurried mash is then cooked to as high as 140°C in order to fully gelatinise the grain starch. The cooker typically consists of a steam jet cooker (hydroheater) attached to a stretch of narrow pipe configured into a 'U' shape or a series 'U' bends of an adequate length to give the desired amount of residence time in the conversion tank at the high temperature to provide sufficient cooking. Mash exiting the cooking tube is flash-cooled to about 85°C into a low-pressure tank (flash tank) and then transferred into a liquefaction tank to provide residence time for complete dextrinisation and limited saccharification of the gelatinised starch. Additional alpha-amylase, or in some cases malt, is added at the entry to the liquefaction tank. Residence time in the liquefaction tank can range from 30 min to 4 h. The converted mash exits this tank and is pumped to the fermenters, passing through in-line heat exchangers to cool it to fermentation temperature (30–31°C).

The reader can find more detail on batch and continuous mashing methods in Chapter 10 of this textbook and publications by Collicutt (2009) and Wright (2017).

Fermentation

Grains are fermented in vessels constructed of various materials. Early American distillers used fermenters made of cypress wood, which resisted rotting and did not impart resinous flavours to the mash. Today, ancient open-top cypress fermenters (Fig. 7.4) continue to be used by a small number of bourbon distillers, where they add interest and value to the visitor experience, and may also contribute key flavours to the fermenting mash by stubbornly resisting sanitation efforts and effectively harbouring a microflora of flavour-producing bacteria and wild yeast. Modern fermenters are usually of stainless construction with a closed top and conical or sloped bottom. They are commonly equipped with mixing capability and temperature control and are usually amenable to cleaning-in-place (CIP). The distilleries still operating fermenters with no active temperature control will set the

FIG. 7.4   Open Cypress fermenters at Maker's Mark distillery in Loretto, KY.

starting mash temperature to as cool as they can or at an established temperature that will avoid overheating by the heat of the fermentation based on the expected ambient temperature during the fermenting procedure. Non-temperature-controlled fermentations are commonly set at 18–21°C and those with temperature control may be set at 27–29°C and controlled at 30–31°C (Ralph, 1999). A 3- to 5-day fermentation period is typical for whisky and this may be dictated somewhat by the chosen fermentation temperature and the wort gravity.

Many bourbon producers, as well as Jack Daniel's, the pre-eminent Tennessee whisky brand, propagate their own proprietary yeast strain at the distillery, while others purchase a commercial yeast strain of their choice in either dry or liquid format. This yeast is either pitched directly into the fermenter, sometimes involving an initial hydration step in warm water, or maybe held in a dilute mash for a number of hours to allow the yeast to propagate for one or more generations before transferring to the fermenter. In Canada, it is common practice for producers of blended Canadian whiskies to use their own proprietary yeast strain for producing their flavouring whiskies and to source a commercial strain for their higher volume base whisky fermentations. Details on the methods used in propagating yeast at the distillery can be found in Chapter 12. Some of the whisky distillers still cultivating their own yeast also maintain the traditional practice of souring their yeast mash with lactic acid bacteria. In the making of its Tennessee Sour Mash Whisky, Jack Daniel's conducts two souring practices. Firstly, they add backset to their mashing water, as is the convention for most grain distillers, which drops the mash pH to about 5.2, and they also sour their yeast mash using a culture of *Lactobacillus delbrueckii*, which they maintain using rigorous microbiological practices and use for inoculating their yeast tank to sour the mash ahead of propagating their daily yeast crop. The lactic acid bacteria thrive in the yeast tank at an elevated temperature and grow suitably for over a 3- to 7-h incubation period to produce about 1% of lactic acid concentration and drop the yeast mash pH to 4.0 (Smith, 2017). Souring of the yeast mash helps protect the yeast against the outgrowth of contaminating bacteria and, when pitched into the fermenter, this soured yeast mash lowers the starting mash pH to about 5.0, which is desirable for the activities of both the yeast and the malt enzymes that remain active in the prepared mash. It is also believed that metabolites from the souring bacteria contribute to the sensory profile of the final whisky.

Distillation

Details on the theory and practice of continuous distillation are covered in Chapter 15 of this volume; therefore, this section will only describe the distillation methods used for American and Canadian whiskies that may be unique to these whisky styles. The majority of bourbon and Tennessee whisky volume is distilled using a two-still traditional design that is characterized by a continuous single beer column which feeds a continuous pot still. The beer column is relatively simple in design. Steam is injected at the base of the column and travels upward, while the beer, containing 8%–11% abv, enters the column close to the top and flows downward. The still is composed of two sections, each containing perforated trays designed to encourage liquid–vapour contact to permit the recovery of alcohol and volatile congeners from the beer. The lower portion of the still is the stripping section, which contains a minimum of 15 sieve trays, constructed of copper or stainless steel, that serve to strip the beer of its volatile substances and carry them up the still. The rectifying section sits above the stripping section and is where the vapours are concentrated by further reaction with the steam on a series of 3 to 8 rectifying trays. The rectifying trays are made of copper and are usually of bubble cap or valve tray design. Common configurations of the American whisky column-pot still design have the beer still feeding the pot with either liquid condensate, as shown in Fig. 7.5,

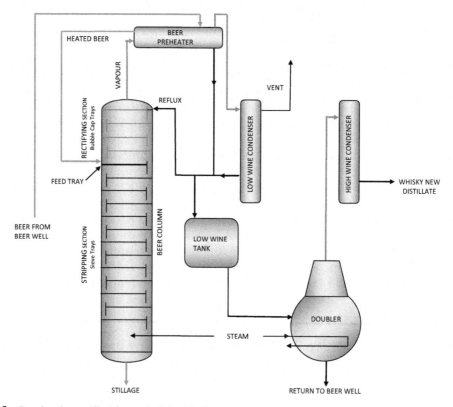

FIG. 7.5 Bourbon beer still with attached doubler.

or noncondensed vapour directly off the still. In the former design, the pot is referred to as a doubler, and the latter is known as a thumper; a term no doubt referencing the sound made when the steam collapses into the liquid volume in the pot.

As shown in Fig. 7.5, fermented beer is pumped from the beer well to the beer still, passing first through a beer preheater which heats the beer to the same temperature as the tray onto which the beer will be pumped when it enters the still. The beer enters onto the feed tray, the top tray of the stripping section of the still, and exits the base of the still as stillage after having been stripped of alcohol and most volatile congeners. Alcohol vapours rise to the top of the still and are concentrated in the rectifying section. The concentrated vapours exit the top of the still and enter the beer preheater, where they give off some of their heat to the incoming beer and are partially condensed. The vapours from the preheater flow to the low wine condenser, where they are fully condensed into low wines of about 60% alcohol content. The low wines then take two different paths. A portion is mixed with the preheater condensate and is returned to the top tray of the still as reflux at a rate established to balance the alcohol strength at the top of the still , while the balance is transferred to a low wine tank, which then provides a continuous feed of low wines to the doubler. The doubler typically raises the alcohol content of the spirit by 5%–8% abv and the end product, or high wine, is typically recovered at 65%–70% abv (130–140 proof). The second distillation not only raises the alcohol content of the whisky spirits, but also provides additional exposure of the spirits to the reactive surfaces of the copper still. Copper exposure allows for the additional catalysis of flavour compounds and serves an important role in reducing the concentration of ethyl carbamate in the spirits. More information on ethyl carbamate can be found in Chapter 19.

The relatively simple design of the conventional American whisky distillation system gives the distiller only limited ability to modify the profile of the distillate through operation of the stills alone. The number of rectifying trays in the beer column establishes the lower limit to the concentration of alcohol coming off the still, and the upper limit is controlled by the volume of reflux that is able to be sent back to the still. There may be some ability to vent small volumes of Heads off the low wine condenser, but beyond that, there is little ability to vary the taste profile of the spirits by control of the stills alone. Efforts to better modify the profile of an American whisky spirit, perhaps to better differentiate one whisky from another when expanding a whisky portfolio, have some distillers bringing a conventional pot still into play. Redistilling either the whisky low wines or high wines on a pot still on a batch basis enables the distiller to adjust the congener levels by varying the Heads and Tails cuts and allows for a wide range of profiles to be created from the same base spirits. We may see greater use of batch pot distillation in American whisky production in the future as consumer interest in new products continues to drive innovation.

In Canada, continuous distillation systems are used for the production of most base whiskies that make up a large portion of whisky blends. A three-column *Extractive Distillation* system is the most used to produce the base whisky, which must be distilled below 95% abv. The reader is encouraged to refer to Chapter 15 for a detailed description of extractive distillation. The flavouring whiskies used in the Canadian blends are distilled to a lower strength to deliver a bold character and higher flavour intensity. Many still types and distillation methods are used to distil various flavouring whisky mash recipes. These include the following: continuous single column, batch hybrid column, Coffey still, column plus pot, single pot, and double pot, with each imparting its own distinctive flavour characteristics to the spirits.

Maturation and blending

American whiskies are mostly products of a single spirit type from a single distillery, where the term *blending* has traditionally been used sparingly and mostly in reference to the practice of combining various production batches or bonds of the same whisky from different warehouse locations or from various warehouses owned by the same distiller. Times, however, are changing. The legal definitions for American whiskies leave considerable room for creativity and originality, allowing for unique mash bills and fermentation conditions, together with a wide range of distillation techniques.

The art of blending today is contributing a bigger role in creating new and original American whiskies. Producers are more likely today to blend whiskies from different barrel finishes, different char levels and various toast regimes, or from specific locations or environments within the maturation warehouses and to create unique final whisky blends. The current growing interest in whiskies with special *finishes* is further expanding the role of the blender. *Finishing* is the practice of adding fully matured whiskies to casks formerly used in maturing other beverage types such as sherry, port, Spanish wine, and cognac and resting the whisky in these casks for short-time periods to allow the whisky to take on some of the characteristics of the original cask contents. American whisky makers are responding to the demand for special *finished* bourbons and the market is supporting many fine examples of bourbons finished in sherry and port casks or finished in barrels with different char or toast levels. Opportunities for unique finishes are almost limitless, as blenders seek new flavour opportunities by utilising cooperage from rum, tequila, and other whisky types, Cognac and American wine.

Canadian whiskies are also answering the call of consumers wanting unique finishes, and Canadian blenders are pleased to add this to their already long list of capabilities. The Canadian approach to whisky production is to develop different flavour elements separately through separate fermentation distillation, and maturation protocols for the various whisky components and then blend them together in order to create a final unique product. As previously stated, most Canadian whiskies are blends of light grain whiskies with bold Canadian rye, malt, and other *flavouring* whiskies that add a unique character to the blend in a manner similar to how malt whiskies contribute to a blended Scotch whisky. Contrary to the situation in Scotland, Canadian distillers tend to produce most components of their blends under their own roof and there is little of the trading of whiskies between distillers that is the foundation of blended Scotch whiskies. Canadian regulations permit blenders to further differentiate their whiskies by allowing the use of other blending ingredients, including other matured spirit types, and wine, sherry, and other fortified wines to a maximum level of just over 9% on a total alcohol basis. This gives the blender even a greater opportunity for creating new distinctive whisky blends to answer the call for new whisky experiences.

American and Canadian regulations establish the framework for methods used in maturing whiskies. American whiskies are to be matured at no greater than 125 proof (62.5% abv) in charred new oak containers, with *straight* whiskies requiring a minimum of 2 years of maturation time. The requirement for new oak means that whisky barrels can only be used once for bourbon and other such American whiskies, and this produces a steady source of once-used barrels for export to all other whisky-producing countries, particularly in Scotland.

Canadian whisky must be matured for at least 3 years and regulations permit maturation in any type of wooden barrel or *cooperage*, so long as it is no greater than 700 L capacity given that the matured product possesses the taste and aroma generally attributed to Canadian whisky. This gives whisky makers in Canada plenty of latitude for diversifying the maturation elements of their whiskies. They tend to be matured using a variety of barrels of different pedigrees, including mostly *refill* barrels and ex-bourbon and charred new American oak barrels. Distillates for the base whisky component of Canadian blends are more highly rectified and lighter in character than the flavouring whisky distillates and tend to favour maturation mostly in refill (multi-used) barrels, with new oak and ex-Bourbon (once-used) barrels used in the mix in varying but lesser proportions to balance the lighter distillates. On the other hand, most of the Canadian flavouring whiskies, with their bolder and more complex compositions, are matured in charred new oak barrels.

Barrels are filled at alcohol strengths that have been established to give the desired effects on the specific whisky style. Bourbons and other popular American whisky styles are mostly matured at or slightly below the allowable upper limit of 125 proof (62.5% abv), with some matured at lower strengths of 115 proof or less. In Canada, a range of 68%–74% abv is common for maturing the base whisky portion of Canadian whisky, while the flavouring whisky barrels are commonly filled at 55%–60% abv. Fill strength has a considerable impact on congener extraction from the barrel and on in-barrel flavour formation, with higher alcohol strengths favouring extraction of alcohol-soluble components from the wood and lower strengths favouring water-soluble flavour recovery, as discussed more thoroughly in Chapter 16 of this textbook. Overall, higher fill strength promotes a slower maturation rate and lower recovery of total extractable solids from the barrel, and producers use this understanding to fill barrels at a strength that best matches the composition and complexity of the spirits entering the barrel and the style and expected profile of the final matured whisky. Common practice is to mature lighter whisky distillates at higher fill strengths and bolder distillates at lower strengths to best balance the flavour contributions of the distillate and the barrel to the final whisky product.

Outlook

Whisky in North America has a long and varied past. Production methods are steeped in history and tradition while producers continue to innovate to meet market needs and continue to advance technologically in terms of production efficiencies and sustainability. They are constantly mindful of upholding tradition and honouring the past. The future for North American whiskies is attractive. Domestic sales continue to grow as existing producers diversify their product ranges to appeal to consumer demands as new whisky distillers come into being through the explosive growth of craft distilling in North America. Export sales are anticipated to continue to escalate and new consumers are expected to come into the whisky category as interests in all types of whisky worldwide continue to flourish.

References

Collicutt, H., 2009. Whisky: Grain mashing and fermentation. In: Ingledew, W. (Ed.), The Alcohol Textbook, Fifth Edition. Nottingham University Press, Nottingham, pp. 413–430.
Cowdery, C.K., 2004. Bourbon, Straight, the Uncut and Unfiltered Story of American. Whiskey, Chicago. ISBN: 0975890100.

de Kergommeaux, D., 2017. Canadian Whisky, Second Edition: The New Portable Expert. Appetite by Random House, Toronto.

Park Street Imports, 2019. Annual Craft Spirits Economic Briefing. American Craft Spirits Association. https://americancraftspirits.org/news/craft-spirits-data-project/.

Pickerell, D., 2018. The Rise and Fall and Rise of Rye. Retrieved from DISCUS: https://www.distilledspirits.org/wp-content/uploads/2018/09/The-Rise-and-Fall-of-Craft-Whiskey.pdf.

Ralph, R., 1999. Production of American whiskies: Bourbon, corn, rye and Tennessee. In: Jacques, K. (Ed.), The Alcohol Textbook, Third Edition. Nottingham University Press, Nottingham, pp. 211–224.

Regan, G., Regan, M.H., 2009. The Book of Bourbon. Retrieved from: https://americanwhiskeytrail.distilledspirits.org/american-whiskey-history.

Smith, K., 2017. Yeast practices in the production of American whiskies. In: Walker, G. (Ed.), The Alcohol Textbook, Sixth Edition. Hobbs The Printers, Hampshire, pp. 335–362.

Swonger, C.R., 2020. 2019 Economic Briefing. Retrieved from DISCUS http://www.distilledspirits.org/wp-content/uploads/2020/02.

Wright, S.A., 2017. Grain mashing for beverage alcohol. In: Walker, C.A.G.M. (Ed.), The Alcohol Textbook, Sixth Edition. Hobbs The Printers, Hampshire, pp. 193–207.

Craft distilling in North America

Matt Strickland[a], Steve Wright[b], and Heather Pilkington[c]

[a]Distillerie Cote des Saints, Mirabel, QC, Canada [b]Spiritech Solutions Inc., Kingsville, ON, Canada [c]First Key Consulting Inc., London, ON, Canada

Introduction

Craft distilling has become a prominent force in the alcoholic beverage market within North America during the past decade. Throughout the United States and Canada, small distilleries have been opening their doors to thirsty and curious imbibers who are eager to sample locally made distillates produced in small facilities by a small number of people.

Considerable marketing has taken place to promote the craft distilling movement and it is not uncommon to hear terms such as 'handmade', 'traditional', and 'artisanal' attached to every label from vodka to whiskey. How much value these terms hold in the eyes of consumers is debatable, but the amount of expansive growth in what was barely a market presence two decades ago is now extensive! In this chapter, we hope to explore the craft distilling movement in North America, how it was born, how these distillers define and identify themselves as well as some of the techniques they employ to produce their products differently from the larger distilling facilities. Finally, we examine how the craft distilling landscape is evolving with an eye towards future commercial prospects.

History

The current market climate in North America for distilled beverages indicates that whiskey and vodka both sit comfortably on the throne of the distilled spirits world. Interestingly, the history of craft distilling in North America is more indebted to brandy and rum producers rather than their grain-based cousins. So, before we examine the current craft spirits revolution, we will briefly discuss the history of North American distillation in general to avoid too much of a myopic point of view.

North America is expansive. Impressively massive in scale, one can only imagine the thoughts running through the typical colonists in the 17th and 18th centuries as they first set their sights on their new home. During this time, colonists arrived by sea to the mid- and northern Atlantic stretches of coast on the eastern seaboard of the United States and Canada. The majority of colonists coming to the nascent United States (then a colony of the British Empire) were from Great Britain, and many were of Scottish and Irish descent. These people had their own distilling traditions and brought their skills with them to the new world.

Distilling was very much a seasonal affair with farmers using their small stills to produce another farm commodity that would help keep food on the table and a stable roof above the families' heads. These farm distillery operations were incredibly common. Concrete production numbers are hard to come by given the nature of these types of distilleries, but it has been estimated that between 1810 and 1840, there were 14,000–20,000 distilleries operating in the United States (Pickerell, 2018).

However, by 1920, because of the Prohibition Act, the 18th amendment—the production, sale, and consumption of alcoholic beverages in the United States were effectively made illegal and the distilling industry was brought to its knees. By the end of this decade when the great economic depression had set in, most people had soured on prohibition, and in 1933, it was ended with the ratification of the 21st amendment to the US Constitution. Unfortunately, by this point the damage to the distilling industry had been done. Where once thousands of small distilleries had dotted the North American landscape, only a handful remained and even these were quickly bundled in a series of astute acquisitions by the larger firms.

Today, we are seeing a rise in small distilleries once again. It has taken several decades to arrive at this point and many an antiquated law had to be changed but here we are with nearly 3000 registered distilleries in the United States alone, Canada's own distilling fortunes are quick to follow suit.

This renaissance in the craft-distilling sector began in the early 1980s with a couple of brandy distillers in California. One was Hubert Germain-Robin, a French immigrant with a deep familial tie to classical cognac production, whilst the other, Jörg Rupf, was a German immigrant who fell in love with the fruits growing around the San Francisco Bay area. Both started distilleries in the Northern California region around 1982–83 producing French- and German-style brandies with California fruits. The successes of each company were hard-won, and in the early years, it was often not apparent to the average consumer exactly what these distilleries were producing.

During the 1980s in North America, vodka was experiencing a meteoric rise as the spirit to keep in one's liquor cabinet. Whiskey was what the older generation drank, and few people understood brandy or 'eau de vie'. The fact that these distilleries survived and continue to thrive to this day is quite remarkable, and perhaps, the founders realised that the time was appropriate for something new. After all, not too far from these distilleries and only a few years prior several well-known craft breweries had also started. This included the Anchor Brewing Company in San Francisco, and Sierra Nevada Brewing in Chico, California. Along with the budding Northern California's wine industry (which had just made waves from two local wineries coming out on top of prestigious French wines in the famed Judgement of Paris), the regional population was becoming increasingly more accustomed to these localised alcoholic beverage offerings. A sense of regional pride set in and these companies were able to thrive.

Indeed, the craft distilling movement's fortunes have mirrored that of the craft beer resurgence following a 10- to 20-year delay. During the 1980s, there was a growing number of small breweries opening with 'the floodgates' truly opening in the 1990s and early 2000s. A small handful of distilleries opened during these years, some proving to be massively influential such as the Clear Creek Distillery in Portland, Oregon, known for its brandies and single malt whiskey. However, it was not until the mid-to-late 2000s that small distillers began to increase in significant numbers. During the early 2000s, there were less than 100 small- to mid-sized distilleries in operation in the United States. As of August 2019, there were 2046 such operations, which was an 11.5% increase over the previous year (Park Street Imports, 2019).

In Canada, similar trends in the growth of craft distilling are seen. Due to a myriad of factors including legislative restrictions, the Canadian craft distilling movement is a few years behind the trendlines forged by its southern neighbour. However, that is rapidly changing. With roughly a tenth of the population of the United States, Canada arguably has more distilleries per capita. At the time of writing, there are roughly 350 distilleries operating in Canada, with dozens more in the planning stages. The indisputable majority of those distilleries are considered craft entities (Artisan Distillers Canada, 2020).

The definition of craft distilling

A definition of 'craft distilling' is fraught with difficulties. It could be argued that defining 'craft' in modern terms is somewhat of a pointless task, with efforts yielding few results. The trade organisations involved with the craft spirits industry (Fig. 8.1) have all tried to develop their own definitions on the subject with varying degrees of success and almost constant consternation from multiple parties. No definition appears to be good enough to encompass all those who want to be included without subsequently excluding some others. Having the word 'craft' as part of a company's identity is worth quite a bit in marketing cache. Even some larger distillers producing millions of litres of spirit a year have co-opted the term for their own operations, which some argue dilutes the word's ever-evolving definition even more.

It is tempting to define 'craft' based on the company size, but this route carries with it quite a bit of baggage. Some companies inevitably begin small enough to meet the definition and will grow beyond the seemingly arbitrary production limit of 'craft', but still feel that they meet the qualitative feeling and intent of the definition. Confusing matters even further is that different trade organisations have offered different production capacities. The American Distilling Institute established a sales cap of 100,000 US proof gallons per year as part of its definition of a Certified Craft Spirit (American Distilling Institute, 2020). Meanwhile, the American Craft Spirits Association partly defines a craft distillery as producing less than 750,000 US proof gallons per year (American Craft Spirits Association, 2020).

Other qualifiers for the craft definition relate to the amount of independent ownership of the company, which has become increasingly important in recent years. The American Distilling Institute requires that 'less than 25% of the distillery (distilled spirits plant or DSP) is owned or controlled (or equivalent economic interest) by alcoholic beverage industry members who are not themselves craft distillers' (American Distilling Institute, 2020). The second half of the 2010s saw a large number of acquisitions by large distilling and winemaking companies of smaller craft brands. This is not the place to discuss the merits or problems with this changing of the craft landscape but what is certainly apparent is that the larger distilleries have taken notice of the effects that craft distilleries have had on the market. Certainly, they have noticed that consumers are paying more attention to these brands. It is only natural for

FIG. 8.1 Several trade organisations that have formed to lobby and advocate on the behalf of craft distillers in North America.

the larger companies to want to take back some of that market share by growing their portfolios with smaller brands.

How these acquisitions have been generally perceived by the distilling community is complex. There are some that have complained about their peers selling a majority stake to companies such as Pernod Ricard, Constellation Brands, and William Grant & Sons. Others regard these acquisitions as a necessary step for small brands to obtain capital, distribution channels, and additional resources to further grow their brands in a hypercompetitive environment.

Beyond the size and ownership criteria, craft spirits tend to be defined in somewhat nebulous terms surrounding 'handmade' and 'traditional' processes. These are highly interpretable ideals and not readily enforceable. One distiller's 'traditional' process may be completely anathema to another. Bourbon has traditionally been produced at least partially in large continuous column stills for more than a century, but some small distillers feel that these behemoths of chemical engineering have no place in 'proper' bourbon production. Others take an opposite position. It is an interesting and passionate debate that will likely never end especially as new technologies come into operation and older ones are phased out.

Important craft whiskey styles

It can be difficult to talk specifically about craft spirits, particularly whiskey, without generalising for the sake of brevity. There are too many distillers with different visions and personalities injected into thousands of products to take anything but a broad approach here. Of course, this makes things all the more exciting for the craft spirit consumer who is currently spoiled for choice.

American bourbon and rye

In the United States, bourbon whiskey is undisputedly the champion of the whiskey world. Consumers want it and their thirst requires that a lot of liquid be produced. Following bourbon, there is a renewed interest in American rye whiskey.

The interest in these whiskeys amongst craft distillers stems not just from consumer demand but also their historical importance. Rye was arguably the first truly successful American whiskey style before bourbon supplanted its dominance, during the 1800s, on into today. Due to their historical value, many distillers have spent considerable amounts of time researching old recipes, techniques, and grains to reproduce modern facsimiles of the original whiskies. Bourbon and rye production are discussed at greater length in Chapter 7 of this book.

Canadian whisky (Canadian rye)

Once relegated to production by only the largest distilling firms, the practice of fermenting, distilling, and maturing individual whiskies separately, then combining them to produce the final blend is being practised by a growing number of Canadian craft distillers. The whiskies produced for blending generally include a light-styled base whisky produced at high alcoholic strength, with a very subtle and nuanced character, distilled by some using a continuous column still, and one or more 'flavouring' whiskies, which are used in smaller proportions and often (but not always) contain some amount of rye in the mash bill. This concept of producing separately the various whisky components and blending them as a final step is not unlike the practices used for blended Scotch whisky. This ultimately provides the distiller with an inventory of various whiskies that can be blended in different combinations and proportions to yield a variety of blended whisky products. The art of blending takes a prominent role in this style of whisky.

Other craft Canadian whiskies are produced more in the style of bourbon, using single grains or mixed-grain mash bills and distilling a single spirit of a complex character. These spirits, however, are most often matured in a manner more in keeping with the Canadian whisky style, mostly using once-used and reused bourbon barrels to deliver a softer maturation character that may require longer rest times in the barrel to develop that character.

It is worth noting that Canadian whisky is often colloquially referred to as 'rye' whisky. This is largely a reference to historical Canadian whiskies that contained large amounts of rye grain in their recipes. However, today despite often being called 'rye', there is no legal requirement for any Canadian whisky to contain rye and some in fact contain no rye grain whatsoever.

Regardless, Canadian whisky whether from large or small distillers must be matured in cask for a minimum of 3 years in order to affix the term 'whisky' on the bottle. This has proven to be a huge financial and temporal barrier to entry for many small firms that might not be able to afford to wait for such a long time to begin selling their products. In contrast, the United States has no minimum maturation times for most whiskey styles and consequently many craft distillers release their products at a relatively young age.

Single malt

If there is one whiskey type to truly capture the imaginations of craft distillers in North America, it would certainly be single malt whiskey. Distillers in both the United States and Canada have taken to this time-honoured whiskey style and are currently crafting their own unique expressions that pay homage to the traditional malt whisky produced in Scotland.

These whiskies should not be mistaken as being mere clones of their Scottish inspirations. There has been a drive in both Canada and the United States to define what it means to be a single malt in each country. For instance, in the United States, the American Single Malt Whiskey Commission was founded by a handful of malt whiskey distillers aiming to define the category as the number of distilleries rapidly grew. Whilst they have not obtained an official legal definition as recognised by the US Tax and Trade Bureau, they have established some self-governing criteria for the American single malt style. In order to be considered to be an American single malt whiskey, the whiskey must be

- Made from 100% malted barley
- Distilled entirely at one distillery
- Mashed, distilled, and matured entirely in the United States
- Matured in oak casks not exceeding a capacity of 700 L
- Distilled to no more than 80% alcohol by volume
- Bottled at 40% alcohol by volume or more (American Single Malt Whiskey Commission, 2016)

Looking beyond the efforts of the American Single Malt Whiskey Commission, many distilleries are attempting to produce regionalised spins on the single malt style, using different varieties of malted barley, maturing in used casks of various provenances, and even kilning the grain with unique woods and peat sources.

Experimental whiskey

Moving away from traditional whiskey styles, many small distilleries have opted for experimental production approaches and flavours to further differentiate themselves from the rest. The Corsair Distillery in Nashville, Tennessee, was founded on this very principle (Fig. 8.2). To date, they have launched an array of whiskeys produced using nontraditional grains such as quinoa and spelt, to whiskeys that have been distilled through a bed of pelletised hops for an intensely floral and herbaceous aromatic effect. Not to be outdone, Balcones in Waco, Texas, produces a bourbon made entirely from blue corn, as well as a whiskey smoked with Texas scrub oak. Perhaps one of the most seemingly bizarre creations has been from Tamworth Distilling in Tamworth, New Hampshire. Their 'Eau de Musc' is produced using oil extract from the castor gland of the North American beaver (which is located next to the animal's anal glands).

FIG. 8.2   Preprohibition era pot still used by Corsair Distillery in Nashville, Tennessee, to make some of their experimental whiskies.

Major craft whiskey distillers in North America

With thousands of distilleries currently operating in North America, it can be hard to select even a few to discuss without feeling that you have left too many other important examples out in the literary cold. Nevertheless, we have discussed a handful of the more influential distillers to begin operating in the past several years.

The first whiskey distiller to discuss would be the St. George Distillery in Alameda, California. Whilst it is true that St. George initially earned its reputation from eau de vies and gins, it has quietly and confidently been releasing its own single malt whiskey since 2000. A yearly release in relatively small batches and this is one of the more difficult American single malts to acquire!

In some ways, the Canadian counterpart to St. George is the Glenora Distillery on Cape Breton in Nova Scotia, which claims to be North America's first single malt whisky distillery. Following production methods very much akin to those employed in Scotland for Scotch malt whisky, production started soon after the facilities were constructed in 1990, with the first release of this whisky in 2000. Glenora has developed a large selection of matured stock and is able to release several aged products onto the market ranging from 10 to 25 years old.

The Province of Ontario has the second highest number of craft distilleries. A relative Ontario newcomer is the Spring Mill Distillery in Guelph, Ontario, founded by John Sleeman of Sleeman Breweries. This distillery is in a historic building, which is a former mill and distillery, which operated on the site in 1835. Spring Mill is currently laying down whisky stocks that have been fermented in Douglas Fir washbacks (Fig. 8.3) and have been variously distilled using a custom three-column continuous still and two pot stills, all manufactured by Forsyths of Scotland. Spring Mill is a few years away from launching its line-up of whiskies, which will include a Canadian rye whisky, a single malt, a bourbon style, and an Irish style barley whisky.

In the United States, there are several prominent craft distillers residing in Colorado and Utah. These include the noted High West Distillery in Park City, Utah, which has made a name for itself by sourcing and blending high-quality whiskies from larger distilleries. They also distil their own whiskies and have steadily been blending more of their own distillates into their products as they mature. The Leopold Bros. distillery is in Denver, Colorado. There the brothers produce traditional American rye whiskey (amongst many other products) and also commissioned the production of a 1800s-era three-chamber still so that they can distil rye whiskey in the fashion of many distilleries 150 years ago.

The Koval Distillery in Chicago, Illinois, is one of the largest craft distillers in North America (Fig. 8.6). It produces an impressive array of products including several whiskies, all using organic grains. In addition to bourbon and rye, they also produce oat whiskey and millet whiskey. Seattle, Washington, is where Westland Distillery is located, and Matt Hoffman directs this operation and he has been an advocate for American single malt whiskey for many years. The distillery produces several single malt whiskeys, with an emphasis on showcasing the regional terroir through its barley and maturation programme. Rémy-Cointreau purchased control of this brand in 2016 (Barton, 2016).

FIG. 8.3 Douglas Fir washbacks at the Spring Mill Distillery, Guelph, Ontario. Photograph compliments of the Spring Mill Distillery.

In Kentucky, there is the Alltech family-owned Lexington Brewing and Distilling company, which produces the award-winning Pearse Lyons Reserve whisky, the first malt whisky produced in Kentucky since 1919.

In Canada, the Province of British Columbia has the highest number of craft distilleries. There are a number of craft distilleries on Vancouver Island, which is the home of the Shelter Point Distillery. They distil distinctive *single-grain* whiskies, including a barley whisky from barley grown on their own farm, a single malt and a rye, along with a few blended products. The whiskies are matured a few hundred yards from the Salish Sea, which separates the island from the Canadian mainland.

Craft distilling in Mexico

It is worth mentioning the craft distilling efforts in Mexico. Craft and 'boutique' distilling in Mexico has existed for a long time. Certainly, there are megalithic tequila manufacturers producing an immense amount of agave distillate, but most distilleries throughout Mexico are incredibly small. Many of them produce products such as bacanora, mezcal, raicilla, sotol, and aguardiente. However, recently there have been a few intrepid distillers that have appreciated the liquid possibilities of native corn and grain varieties to make truly Mexican whiskey. The Sierra Norte Distillery in Oaxaca currently produces three corn whiskeys, all made with a different strain of local corn, and the resulting distillate is then matured for around 10 months in French oak before being bottled.

Not to be outdone, Destileria y Bodega Abasolo produces Abasolo whiskey, which is made from 100% corn. The corn is put through a process called nixtamalisation, which is a similar process used to make traditional tortillas in Mexico, where the corn is soaked in an alkaline solution for several hours. This process has several nutritional benefits to the tortilla consumer, and in whiskey production, it allows for the partial gelatinisation of starch, the saponification of lipids, and the solubilisation of some protein materials surrounding the starch granules (Arendt and Zannini, 2013). After nixtamalisation, the grain is mashed, fermented, and distilled. The resulting new-make whiskey is then matured in ex-bourbon casks for 24 months (Miller, 2020).

Raw materials and processes

There are several things that distinguish 'craft' distilleries from their larger corporate counterparts. Whilst many of these differences may be attributed to size and ethos, the small production capacities and subsequent budgets of these companies often mean that raw materials and processing techniques must be amended and approached differently. A small company knows that they cannot compete with larger distillers based on volume or price, and the strategy often switches to competing on philosophy, ingredients, and perceived product quality. Whilst raw materials, processing, and distillation are discussed more deeply in other chapters (e.g. Chapter 10), here we will focus our efforts on the aspects of these practices that are sometimes treated differently by the craft distillers in North America.

Raw materials

Craft distillers want to differentiate themselves and there is often an incentive or a desire to use heritage grain varieties or grains not commonly used in whiskey production.

The Jeptha Creed Distillery of Shelbyville, Kentucky, produces its straight bourbon whiskey using 70% bloody butcher corn, a red heritage corn variety grown on their own farm (Fig. 8.7). Similarly, in Charleston, South Carolina, the High Wire Distilling Company has been working with a nearly extinct variety of red corn called Jimmy red. Both companies are using these corn varieties in some of their bourbon production to introduce new flavour characters to the bourbon market.

Perhaps no one has worked with as many whiskey grains as the Corsair Distillery in Nashville, Tennessee, where Andrew Webber and Derek Bell began their move into the world of 'alt-grain whiskey' in the late 2000s. Since then, they have produced whiskeys from quinoa, spelt, amaranth, buckwheat, oats, and many more. Some of their standard whiskeys utilise roasted forms of traditional brewer's grains such as barley and rye. They have also worked intensely in the realm of smoked whiskeys, burning different types of wood and herbs to add unique smoked characters to their base malted barley. They have released whiskeys smoked with hickory wood, and cherrywood.

Santa Fe Spirits in Santa Fe, New Mexico, has also begun producing smoked whiskeys. Their single malt named 'Colkegan' is purportedly produced with malt kilned over mesquite wood.

Some of the craft distillers have even introduced their own malting into the blend. They argue that this gives them maximum control over the ingredients and flavour profiles. Leopold Bros. in Denver began their malting operations using locally grown grain in 2010 and are currently in the process of expanding their operations. The company says that completion of the newly expanded facilities will make them one of the largest floor malting operations in the world!

Raw materials processing

With the space and capital restraints that often go with these small distilleries, there seems to be no end to ingenuity in solving basic production problems when the need arises. For example, in larger distilleries, if a fermentation vessel needs repair, it would be taken out of the production system until the problem was fixed and production would be reduced to a lower level. Smaller distilleries are often not afforded such luxuries. If a fermenter were to go out of operation and orders needed to be filled, then a workaround would need to be found. In the case of a downed fermenter, some distillers might opt to temporarily ferment their mashes in 1000-L plastic IBC totes. Whilst this may be far from ideal, at least it keeps the product flow for the smaller distillers.

Some distillers alter their production processes to produce something wholly unique or to restore more traditional methods. Some years ago, several American distillers set about helping to recreate George Washington's distillery at his Virginia residence of Mount Vernon. The original distillery operated in the latter years of the 1700s for only a brief time but was at one point considered one of the largest whiskey manufacturers in the United States. Mount Vernon and its Board have made it their mission to share the life of the United States' first president in a relatively accurate fashion and that includes the rebuilding of the distillery and mill (Fig. 8.4). The resulting distillery uses traditional mashing techniques that require the addition of boiling water to large wooden tubs in order to heat and cook the grains (Fig. 8.5). It is an immensely laborious process that is fascinating from a historical perspective.

Fermentation

Fermentation practices in small distilleries are generally similar to those of many of the larger whiskey facilities. Dry yeast is typically the preferred format as opposed to pressed cake or liquid formats. It is easier to handle and manage for smaller facilities with less throughput than their larger peers.

Fermentation vessels are usually simple in design with distillers utilising everything from 1000-L plastic IBC totes, to large-scale cypress, or Oregon pine open-top fermenters. A common design is square or rectangular cubic steel IBC totes, which are readily available on the market at a low price and sometimes can even be rented. Whilst some distillers do instal cooling systems to mitigate large temperature swings during fermentation, this is not considered common. Most distillers ferment their grain washes at moderately warm temperatures for 3–4 days, similar to the larger whiskey houses.

FIG. 8.4 One of the five stills used at the Mount Vernon Distillery in Northern Virginia.

Distilling

Distilling operations in craft distilleries have traditionally been almost entire batch processes. Though this is beginning to change with a growing number of brands studying the potential in continuous still designs, most companies continue to operate the humble pot still.

Of course, one of the aspects of craft distilling that makes these companies so unique is their seemingly improbable ability to often produce many different types of products using the same equipment. For these companies, versatility is the key, and many have opted to purchase distillation equipment that affords them the most amount of flexibility when it comes to product creation. This is commonly achieved by using custom-designed 'hybrid'

FIG. 8.5 Cask used for the mashing of cereals at the Mount Vernon Distillery (note the wooden mash rake).

distillation systems. These systems include a distillation pot that is attached to a small column with some form of distillation trays installed. The number of trays may vary from only a few (four with copper bubble caps is common) to more than 40 for a single system. Often, the trays have a toggle design that allows the distiller to control their function as on or off.

The use (or nonuse) of trays on a hybrid system allows small distillers the option to produce multiple spirit types and styles on one still. It is not uncommon to see small distillers in both Canada and the United States produce an array of products including whiskies, gins, rums, vodkas, and more on the same still. These products all require a variety of distillation techniques and different degrees of rectification. A well-designed hybrid still allows the distiller to approach each of these spirits on a somewhat level playing field. Whilst it is true that the rectification capabilities of even the most highly engineered hybrid will still fall short of the efficiencies inherent in a well-designed continuous system. These hybrid systems will still allow many distillers to produce their own vodka from grain, often reaching spirit concentrations of 95%–96% alcohol by volume with patience and proper techniques. Many of the hybrid still designs allow for individual trays to be disabled, which gives maximum flexibility to the distiller. The trays are often built with bubble caps though this heavily depends on the manufacturer and other designs are also commonly used.

Another increasingly common addition to batch stills in craft distilling is a vapour basket. Traditionally used in the production of gin, it allows hot alcohol-enriched vapours to pass through an isolated bed of botanicals. These devices are now finding their way into

FIG. 8.6 Koval Distillery (Chicago, Illinois) uses a 5000-L hybrid pot still to produce a wide array of spirits from whiskies to gins to neutral spirit.

some experimental whiskey distillations. One distillery has cleverly used their vapour basket to produce 'hopped' whiskies by passing alcohol vapours through a bed of hops inside the basket.

A growing number of distilleries are benefitting from the use of continuous column stills. Some bourbon producers operate a small beer column with attached doubler to more closely emulate the distillation practices used by the larger commercial bourbon producers. Others are using a column still for the first stripping distillation run in vodka production, or in Canada for producing rye flavouring whiskies. These small, typically 12-in.-diameter column stills enable rapid and consistent distillation, with minimal operator input and add substantial distillation capacity to a distillery.

FIG. 8.7   'Doubler' attached to a small-diameter column still used for bourbon production. Jeptha Creed Distillery, Shelbyville, Kentucky.

Maturation

It would be an understatement to say that small distillers feel financial pressures differently than the large distilling houses. Many companies are founded by small groups of individuals pooling their resources and taking large financial risks. The difficulties are compounded when factoring in the long turnout times for new products. Vodka, gin, and other unaged spirits can be produced, packaged, and sold within a few weeks, but traditionally aged spirits require long periods of time before they are ready to be bottled and consumed.

It is a common practice for bourbon to be distilled and diluted to a cask fill strength of 62.5% abv, where it will enter a new American oak heavily charred whiskey barrel of approximately 200 L in size. Once in cask, it will spend anywhere from 2 to 6 years or more being matured before it is decanted, diluted, and bottled. A timescale of years is difficult for some small distillers to overcome when waiting for product sales.

To circumvent these longer production cycles for whiskey, many North American distillers, particularly in the United States, make use of smaller casks ranging from 5 to 30 gal in size (Fig. 8.8). Some will even use cask sizes of a mere 3 gal! The aim of using such small casks is to increase the maturation speed by increasing the surface area of cask wood that comes into contact with the spirit. This in turn increases the extraction rate of various cask components into the spirit, producing a 'mature' spirit in a fraction of time. Some producers can produce a spirit in a matter of 2–3 months as opposed to the considerably longer maturation period often required when using larger casks.

However, the increase in maturation speed does not come without its own costs and special considerations: first is the cask price. A new 200-L American oak whiskey barrel costs around $200 (USD), whereas a new 39-L (10-gal) barrel costs around $175 (USD) (Barrel Mill,

FIG. 8.8   Many small distilleries in North America use small casks for their whisk(e)y maturation. Here a set of 15-gal casks lay atop a row of 30-gal casks (Corsair Distillery, Nashville, Tennessee).

2020). A cask that is only ~20% of the capacity of a standard-sized cask costs almost as much, making the maturation programme considerably more expensive in the short term. The hope is that the extra incurred cost would be recuperated at a faster rate with the shorter maturation time.

Another issue with small barrels is that they take up considerably more space per unit volume of liquid stored. For small distilleries, which may not have the financial outlay to afford larger warehousing, space is an important consideration. Once again, the hope is that the turnover of maturing spirit is faster and therefore space constraints can be mitigated.

Perhaps the most important issue regarding the use of small casks is the different ways that the spirit interacts with the wood. Having a greater surface area-to-volume ratio of cask wood to spirit does allow for faster extraction of oak compounds, but other reactions (ester formation, oxidative reactions, and diminishment of sulphur and other grain notes) are not sped up at the same rate. The resulting cask character experienced by tasters of these small cask spirits is subsequently different than that of spirits matured in standard casks. One study with Scotch whisky showed that smaller casks did not enhance the sensory character of the immature spirit nor were they useful for modelling faster maturation of Scotch whisky (Withers et al., 1995).

Similarly, the Buffalo Trace Distillery (Frankfort, Kentucky) in 2012 released conclusions from an internal study comparing the sensory differences between bourbon matured in standard casks versus smaller cask formats. The distillery took their standard high rye bourbon distillate and matured it in 53-gal (200L), 5-gal (20L), 10-gal (39 L), and 15-gal (57 L) barrels. The barrels were sampled and assessed periodically. The hope was that the small barrels would produce a similar-quality distillate to the larger casks in a shorter amount of time, but the distillery found that the quality of the small casks never reached the same level of their standard barrel-matured distillate. Buffalo Trace deemed the small casks a 'failure' and moved on (Buffalo Trace, 2012).

The reality of small cask usage in craft distilleries is a bit more complicated than these experiments would suggest. It would be appropriate to say that the small cask character is *different*. Consumers have their own opinions on whether this difference is *good* or *bad*. Many craft distillers have discovered techniques to produce distillates that are a better match during the maturation conditions provided by smaller casks and are producing high-quality distillates as a result. Many small distillers have argued that putting distillates with higher levels of low and high boiling point congeners in small casks is a mistake because oxidation reactions in these casks do not keep pace with the extraction rate, thus ameliorating the immature spirit character caused by higher levels of distillation congeners. Therefore, some distillers purposely distil lighter whiskey styles with highly conservative 'cuts' during the distillation process to better handle the special conditions provided by the smaller cask size. Even so, many craft distillers start out using smaller casks but eventually abandon them completely for larger sizes as their stocks become older, and they are able to introduce increasing numbers of standard barrels into their maturation programmes.

Blending and packaging

For many distillers in the United States, there is an immense amount of pride to be had in distilling every single drop of distillate themselves. Recently, however, there has been increased acceptance of companies that practise the art of blending. High West Distillery in Park City, Utah, may be the most famous of the new craft whiskey blending movement. Whilst they have been patiently distilling and maturing their own stocks for many years, to produce immediate cash flow, the company has long made a policy of purchasing good casks of bourbon, rye, and other whiskeys from larger distilleries and blending them together in unique ways. The High West 'Bourye' whiskey is a blend of rye and bourbon whiskeys. Their 'Campfire' Whiskey is a blend of bourbon, rye, and Scottish blended malt whisky.

Canadian distillers, on the contrary, have been practising the art of whisky blending since the beginnings of their industry. The concept and practice of blending whiskies from different distilleries or different marks from inside the same distillery have proven to be a viable and exciting avenue for distiller expression in Canada. Whilst many Canadian craft distillers are focusing more on single expressions, quite a few focus heavily on blending as a way to diversify their product portfolios.

As distillery stocks continue to grow in volume and age, these smaller brands are finding that blending internal stocks from different lots, barrel types, and mash bills can yield interesting results and provide easy portfolio line extensions to an eagerly accepting audience. The increasing interest in single malt and Scotch-style whisky production has also fuelled some of these North American blending endeavours. There is a growing list of distilleries that not only blend their whiskeys in unique ways, but also finish their products in casks that fall well outside the traditional heavily charred bourbon casks. Wine casks, especially oloroso sherry barrels, are seeing more use within North America than ever before and new flavours are constantly being marketed.

Packaging operations are often similar to larger facilities, but usually with less technology and more hand labour involved. Most craft distillers cannot afford large automated packaging lines and must often rely on the physical help of friends, family, and loyal customers to be available to help bottle the product. For some distilleries, this has been an incredibly beneficial method of packaging. A call for volunteers may be put out through social media channels for a given date. People show up and assist in bottling and labelling and may be rewarded for their afternoon efforts with food and/or a complimentary bottle. Not only does this give the distillery a source of readily available inexpensive labour, but it helps 'sell the brand' by giving these consumer volunteers a sense of 'brand ownership' and pride in the products. New Columbia Distillers in Washington, District of Columbia, has achieved great success with this type of system. Locals come into the distillery to help with packaging and then leave to tell friends and acquaintances about their experience and the brand.

The future of craft distilling in North America

Currently, future prospects for the craft distilling market in North America remain strong. The US craft spirit industry currently employs 25,500 people and is expected to grow substantially in the coming years. In 2018, the US craft market reached nine million cases in sales, which was a volume growth of 25.5% and a value growth of 27% over the previous year. The market share of total US spirit volume for the craft sector was at 3.9%, with a value share of 5.8% (Park Street Imports, 2019).

If the trends of craft beer throughout North America and the rest of the world are any indications, these numbers should hardly be surprising. Consumers have taken notice of craft spirits. Large distilling firms have also taken notice and are seemingly determined to not repeat the mistakes of their big brewer counterparts, by 'coming late to the party'. Big Beer's reticence to acknowledge the growing thirst and excitement for craft brands cost them a substantial market share throughout the mid-2000s into early 2010. Large distilling companies have decided to take notice in their own industry!

Interestingly, the methods and goals behind the efforts of these firms have largely not been hostile towards craft, but instead have been warm and welcoming. It seems that many of these companies feel that the best way to handle the rise of craft spirits is to invest in them rather than limit their access to customers. Many of the major spirit firms have their own unique ways of handling this. Some have opted to buy into many small distillers in minority stakes, whilst others prefer to purchase them outright. Other companies such as Diageo have opted to form entire intercompany departments that serve as craft brand incubators. In return, the craft distillers gain access to more capital, support, expertise, and distribution channels in order to continue growing their brands.

Conclusion

Craft distilling has become a serious market force to be reckoned with in North America. Consumers and larger distilling companies have certainly taken notice. The small players are proving to be nimble and innovative in a fast-moving market with interesting takes on traditional products, whilst also offering completely new ideas and flavours to consumers. The techniques used to make these spirits are often almost as fascinating and innovative as the traditional spirits themselves. Indeed, within an age where the consumer is more knowledgeable about distilling than ever before, these small craft companies have been able to tap into the growing locavore movement and provide unique tastes and experiences for their local markets.

The growth of this part of the industry does not seem to be slowing down. Of course, infinite growth is nigh impossible so the next few years will prove interesting as larger craft firms find ways to further separate themselves from an increasingly crowded market. We are also likely to continue to see larger companies find innovative new ways to tap into the potential of the burgeoning craft market through strategic investments and acquisitions.

References

American Craft Spirits Association, 2020. Craft. American Craft Spirits Association. Retrieved from: https://americancraftspirits.org/about-acsa/craft/.

American Distilling Institute, 2020. Craft Certification. American Distilling Institute. Retrieved from: https://distilling.com/resources/craft-certification/.

American Single Malt Whiskey Commission, 2016. American Single Malt Whiskey Commission. Retrieved from: http://www.americansinglemaltwhiskey.org/.

Arendt, E.Z., Zannini, E. (Eds.), 2013. Maizes. In: Cereal Grains for the Food and Beverage Industries. Woodhead Publishing, UK, pp. 67–113.

Artisan Distillers Canada, 2020. Artisan Distilleries. Artisan Distillers Canada. Retrieved from: https://artisandistillers.ca/artisan-distilleries.

Barrel Mill, 2020. Barrels. The Barrel Mill. Retrieved from: https://www.thebarrelmill.com/barrels.

Barton, S.S., 2016, December 1. Westland Distillery Sold to Remy-Cointreau. Whiskey Advocate. Retrieved from: https://www.whiskyadvocate.com/westland-distillery-sold-to-remy cointreau/.

Buffalo Trace, 2012. Buffalo Trace Distillery Announces Small Barrel Experiments Are Failures. Buffalo Trace Distillery, Frankfort.

Miller, K., 2020, May 11. Review: *Abasolo Aims to Put Mexican Whisky on the Map*. Inside Hook. Retrieved from: https://www.insidehook.com/article/booze/abasolo-mexican-whisky.

Park Street Imports, 2019. Annual Craft Spirits Economic Briefing. American Craft Spirits Association. Retrieved from: https://americancraftspirits.org/news/craft-spirits-data-project/.

Pickerell, D., 2018. The Rise and Fall and Rise of Rye. DISCUS. Retrieved from: https://www.distilledspirits.org/wp-content/uploads/2018/09/The-Rise-and-Fall-of-Craft-Whiskey.pdf.

Withers, S.J., Piggott, J.R., Conner, J.M., Paterson, A., 1995. Comparison of scotch malt whisky maturation in oak miniature casks and American standard barrels. J. Inst. Brew. 101 (5), 359–364.

9

Irish whiskey

David Quinn[a] *and Brian Nation*[b]

[a]Irish Distillers Pernod Ricard, Midleton, Ireland [b]O'Shaughnessy Distilling Company, Minneapolis, MN, United States

History and commercial development

When distilling began in Ireland is impossible to say, but it was probably in the 6th century and principally involved monks who brought the techniques of alcoholic distillation from outside Ireland. The history of whiskey distillation and its commercial development, particularly in Ireland and Scotland, have been well and truly covered in many books and articles that give much better detail and informed analysis than would be suitable for this chapter (e.g. McGuire, 1973; Mulryan, 2002; Nicol, 1997). However, some details of the history are warranted, particularly because the whiskey industry in Ireland has for centuries been at the heart of Irish social and political life. The *Red Book of Ossory*, dating from the late 15th and early 16th centuries, records *uisce beatha* being produced for consumption, but the art was still the preserve of the religious orders. In fact, it was not until dissolution of the monasteries in the Tudor period that whiskey ceased to be the drink of the elite. Queen Elizabeth I was known to be fond of the beverage. Peter the Great, Tsar of Russia, mentioned that

> Of all the wines, the Irish spirit is the best!

Much of the recorded history of more recent times relates to taxation and control of the manufacture and consumption of whiskey. With the Tudor settlement of Ireland, English law began to replace native Irish, or Brehon, law. Until 1607, home distillation was legal. However, the Crown needed funds and was anxious to begin extracting revenues. In 1605, licenses were granted in many areas of Galway, Munster, Leinster, and Ulster. At that time, it was a common practice for the Crown to lease the rights, or *patent*, to a particular activity, such as beer or whiskey production. For an agreed fee and over a specified period (usually 7 years), the patentee was authorised to realise whatever they could from the area of their licence

By the middle of the 17th century, the corrupt patent system was close to collapse and more cash strapped than ever. The English Crown had to devise a new way of raising revenue, and the modern concept of excise was born. Consequently, in 1661, the British Government introduced an excise duty on whiskey, and two new drinks were created: tax-paid *Parliament whiskey* and whiskey on which tax was not paid, which became known as *poteen*.

Irish whiskey became popular in Britain and its colonies during the 18th and 19th centuries. Records in the Irish House of Commons show that in 1779, there were over 1152 distilleries in Ireland. However, the significant presence of illicit distilleries soon forced the government to act. This led to a period of tax increases and greater government vigilance and control over distilling. In 1823, a comprehensive excise system was introduced.

The early 19th century saw unprecedented growth in Irish whiskey production, from 40 distillers in 1823 to 86 in 1840. Demand grew rapidly with rising incomes, and the availability of steam power led to more distilleries and larger pot stills. A wide variety of production processes (e.g. one, two, or three stills) and product types (e.g. malt, peated malt, products using varying percentages of malted and unmalted cereals) were in evidence. In 1823, the biggest pot still recorded could hold just 750 gal; however, by 1867, Midleton Distillery had the world's largest still with a capacity of 31,500 gal.

The whisk(e)y industries in Ireland and Scotland were closely interrelated in the 18th and early part of the 19th centuries, with much movement of people between the industries in the two countries; some owners had distilleries in both. During the mid-19th century, significant changes took place in the process as the Scots continued to make whisky using a malted barley mash, often peated, distilled in pot stills, but they also adopted on a large scale the new Coffey continuous still (patented by Irishman Aeneas Coffey in 1830), which produced a lighter flavoured whisky from a mash of malted and unmalted cereals.

The Irish did not embrace this new distilling technique until the close of the 19th century, when large-scale distilling facilities in Belfast, Dundalk, and Derry were established. Traditional pot still distilling also continued, usually in larger stills, with the majority of whiskey being produced using a combination of malted and unmalted barley and, in many cases, with some other unmalted cereals such as oats. Towards the end of this century, a majority of the pot still distillers adopted and modified the art of triple distilling, whilst the balance continued the practice of double distilling.

After the fallout from the 'what is whiskey?' debate and the resultant Royal Commission of 1909, there was further disruption to the Irish whiskey industry from 1916 until Ireland gained its independence in 1923. In the early 1930s, a trade war with Britain developed, denying Irish whiskey distillers access to the traditional Irish export markets of Britain and its Empire (Irish whiskey's largest market). Prohibition was instituted in the United States (Irish whiskey's second largest market) in 1920 and lasted until its repeal in 1933. When the market reopened, Irish distillers could not produce the volume required by the reawakened US demand, as they had not laid down sufficient maturing stock.

After World War II, there was a severe shortage of mature whiskey in the United Kingdom and the United States. At the same time, the Irish State was very short on revenue and was concerned by the potential loss of excise duty if available Irish whiskey stocks were sold in export markets rather than consumed domestically. As a result, the government imposed a quota system that allowed only a very small amount of Irish whiskey to be sold abroad.

The cumulative effect of all of these developments was decimation of the Irish whiskey distilling industry. By 1966, the number of distilleries operating in Ireland had dropped to four. Under pressure from the Irish State, this number became two, as the Jameson, Powers, and Cork distilleries merged to form Irish Distillers, then known as United Distillers of

Ireland. Finally, in 1977, the Bushmills distillery became part of the group. However, the tide had begun to turn when in 1975 a new distillery was commissioned by Irish Distillers in Midleton, Cork, with both pot and column still operations. In 1987, the Cooley Distillery was established, and in 1989, Irish Distillers was acquired by the French company Pernod Ricard. The industry, in recent times, has undergone further changes with the Tullamore Dew brand being bought by C&C in 1994 and subsequently by William Grant & Sons in 2010. In 2005, Diageo entered the Irish industry with their purchase of the Bushmills distillery but in 2015 was sold on to Jose Cuervo. In 2011, the US spirits company, Beam, Inc., purchased the Cooley Distillery and its brands.

The following technical descriptions of Irish whiskey production are primarily drawn from operations at the Midleton Distillery in County Cork.

Pot still whiskey production

Brewing

Traditional Irish pot still whiskey is made from a combination of malted and unmalted barley. Different combinations can be used, but an equal measure of both would be typical. The presence of barley in the mash can give rise to brewing difficulties and can vary from harvest to harvest. The use of spring barley is preferred, although trials with winter barley have been performed in the past.

Until recently, the brewing process at Midleton Distillery used lauter tuns for wort separation. These lauter tuns were coupled with mash conversion vessels, where the mash was put through a time–temperature brewing programme in order to optimise the range of malt enzymes and ensure sufficient conversion of the barley starch and also to facilitate some breakdown of the barley endosperm cell walls and protein. This brewing process typically delivered 92% extract efficiency compared to laboratory-derived extract. This equated to a recovered extract of 286 L°/kg (dry) compared to a laboratory extract of 310 L°/kg (dry).

The brewing programme involves conversion stands at 55°C, 65°C, and 72°C and a final heating to 76°C, just before transfer to the lauter tun. Full starch conversion, as indicated by an iodine check, is not usually complete at the end of the 65°C stand and usually requires the 72°C stand for completion. This is probably due to additional gelatinisation of small starch granules and their subsequent conversion at the higher temperature. This additional conversion is necessary to ensure good drainage in the lauter tun.

The total conversion programme takes 2½h to complete, whilst subsequent drainage in the lauter tun typically takes 4h from mash into draff out. The mash size is 12 tonnes of total grist (as is), yielding approximately 420 hL of strong wort with an original specific gravity of 1.074. A three-sparge programme gives approximately 200 hL of weak wort with a specific gravity of 1.013 for a subsequent mashing-in.

This brewing process served the Midleton Distillery well from its construction in 1974 until early 2013, when a new brewhouse was constructed as part of the distillery expansion programme. Currently, three Meura mash filters are utilised as the method of wort separation instead of lauter tuns. This decision was made following pilot-scale trials at the Meura pilot facility in Belgium, which showed no significant impact on wort composition. This wort was subsequently fermented and distilled at the pilot-scale distillery in Midleton, yielding normal pot still spirit with the expected sensory character and analytical profile.

The major benefit of using mash filters is the increase in recovered extract due to the ability to hammer mill the barley prior to mashing. The same brewing programme is used before the mash is transferred to the filter. Currently, this brewing process is yielding an extract of 307 L°/kg (dry), with an efficiency of 99%.

With a turnaround of approximately 120 min per filter, the brewhouse is processing 28 brews per day and is capable of supporting a pot still capacity of 22 million litres of pure alcohol (LAs) per year. Each brew contains 8.5 tonnes of a mix of malt and barley and yields 315 hL of strong wort at 1.074 SG.

Fermentation

The wort from the brewhouse is cooled to 26°C en route to fermentation. A typical fermenter (or washback) will take six brews (mash filter), with each brew being 315 hL, thus giving a working washback volume of 1890 hL. A wort portion (50 hL) is diverted to a yeast bub tank, where the yeast is grown for about 10 to 12 h before inoculation into a washback. Bubbing will normally increase yeast cell numbers sixfold and give an initial cell count of 2×10^7 cells/mL in the washback. Fermentation will typically take 60 h and give a wash of 10% alcohol by volume (abv) from an initial specific gravity of 1.074. External cooling coils on the outside of the washback are used with water from an underground cavern to control the fermentation temperature and prevent it from exceeding 32°C.

Pot still distillation

The fermented wash at 10% abv is transferred from the cold wash charger in the fermenter building to the hot wash charger in distillation, where it is preheated prior to charging the wash still. Each of the three wash pots takes a charge of 285 hL and can be distilled into a range of different types of low wines, depending on the final pot distillate style being produced. For traditional pot still spirit, all of the alcohol is effectively distilled, giving low wines of approximately 22% abv. For other styles, stronger low wines are produced, in some cases closer to 45% abv. In this case, residual alcohol left in the spent wash is recovered by using a two-column unit to produce a lightly flavoured feints stream.

The second distillation combines low wines with recycled weak feints, which charges the feints still with a combined volume in the range of 225 to 400 hL, depending on the style of pot distillate. A heads fraction is taken based on time (20 min), before cutting to strong feints. This distillation will last about 6 h, producing strong feints at typically 72% abv. The distillation then cuts to weak feints and continues to completion, giving a quantity of weak feints for a subsequent second distillation.

The strong feints continue forward into the spirit still for the third and final distillation. Again, a heads fraction is taken based on time, before cutting to spirit. The spirit distillation will typically require 13 h before cutting to strong feints. Distillation will continue on strong feints until preset distillate strength is achieved, before the final cut to weak feints.

In some cases, again depending on the style of pot distillate being produced, a rider of weak feints, or low wines plus weak feints, is added to the still to ensure that sufficient quantities of strong feints are produced. The final distillate will have an alcoholic strength of 82% to 85% abv, depending on the style of pot distillate.

Charge volumes and combinations of low wines, weak feints, and pot feints within the charge will also vary depending on what type of pot distillate is being produced. All of this is necessary in order to produce a range of different distillates in one distillery. This approach has been driven by the small number of distilleries in Ireland and therefore a lack of opportunity for reciprocal trading arrangements between distilleries. The pot distillation set-up is illustrated in Fig. 9.1

Due to the range of different pot distillate types produced and their associated low wines and feints streams, the Midleton Distillery Still House is characterised by a substantial array of spirit vats and receivers so specific types of low wines and feints can be kept separate. As a consequence of this varied distillation activity and the multitude of in-process distillation streams, the notion of a completely balanced still house (as is normal in a double-distillation operation) is not a feasible objective.

The current pot stills in Midleton are quite large by industry standards. There are now two stillhouses at Midleton Distillery. The original (The Barry Crockett Stillhouse) houses four pot stills with two wash pots, one feint still, and one spirit still, giving a capacity of 11 million LAs. The introduction of the new Garden Still House has added a further six stills, two wash pots, two feint stills, and two spirit stills. This has increased the total pot still capacity to 22 million LAs. All pot stills are heated by external heat exchangers using steam from gas-fired boilers.

Malt whiskey production

Although the Midleton Distillery does produce malt whiskey from time to time, its primary product is traditional Irish pot still whiskey. The main production of malt whiskey in Ireland is carried out at the Old Bushmills Distillery (Bushmills malt is a triple-distilled spirit) in County Antrim, the Cooley Distillery in County Louth (Cooley malt is double distilled), the Great Northern Distillery in County Louth, and West Cork Distillers in County Cork. Many other new distilleries are also making single malt whiskey. Mash conversion is isothermal (64–66°C) and is carried out in the mash tun or lauter tun before drainage. Extract and alcohol yields are close to 100% of laboratory yields, as would be expected from all-malt mashes. Original specific gravity would be in the range of 1.058 to 1.065, resulting in a wash with 8.0 to 9.0% alcohol by volume.

Grain whiskey production

Grain brewing

Maize (corn) is milled through a cage mill. The flour is dropped into a corn slurry tank and mixed with water at a temperature of 55°C. The liquor-to-grist ratio at this stage is 2:1. This slurry is pumped forwards through a steam injector, and the mash is heated to 150°C in a jet cooker. The jet cooker takes the form of a loop with a residence time of 5 min. The cooked

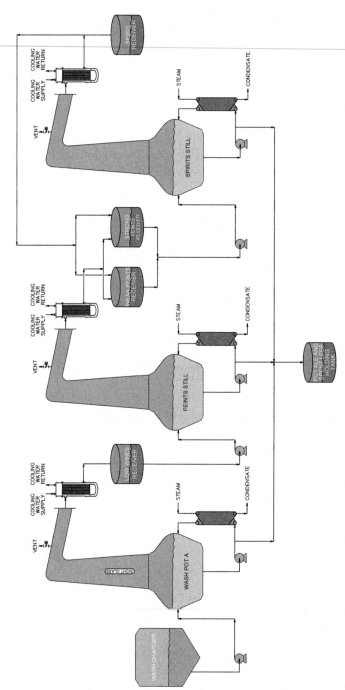

FIG. 9.1 Pot still configuration for the distillation of pot still whiskey.

mash next makes a tangential entry into a flash tank, where, under vacuum, steam is flashed off by dropping the mash temperature to 65°C. At this stage, the malt slurry is pumped into the mash as it enters the flash tank. A residence time here of about 20 min allows the malt to begin liquefaction of the maize starch, basically enough to allow easy pumping of the mash to the next vessel, the converter. This vessel is essentially a plug flow reactor, where the mash entering at the top takes about 30 min to reach the bottom. During this time, most of the starch is liquefied and a degree of saccharification takes place. On leaving the converter, the mash is pumped via a shell-and-tube cooler directly into the fermenter. This grain brewing system is illustrated in Fig. 9.2.

As is normal with grain whiskey production, no liquid–solid separation takes place, and the grains-in mash enters the fermenter at a temperature of 26°C and a specific gravity of 1.093.

Completion of the oligosaccharide and dextrin saccharification takes place during the course of the fermentation.

In 2014, the grain brewing process was upgraded in order to become more energy efficient. This involved moving from a high-temperature jet cooker at 150°C to a low-temperature cooker operating at 85–90°C and designing a more efficient energy recovery system for subsequent mash cooling.

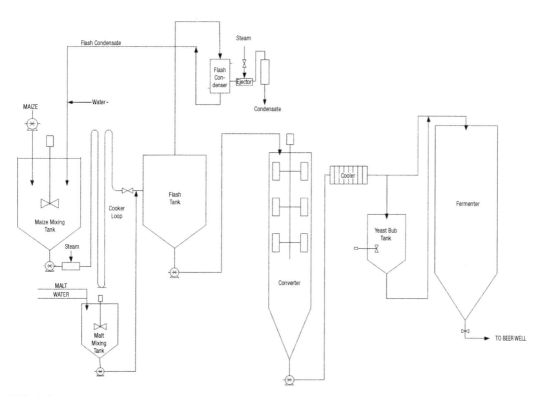

FIG. 9.2   Brewing system for grain whiskey production utilising a high-temperature cooker.

In this new brewing system, the mash is typically held at 87°C for 120 min to cook and partially liquefy the starch. On leaving the cooker, the mash is flash cooled from 87°C to 63°C via a three-stage flash cooler operating under vacuum. The flash vapour is condensed and recovered and is used as brewing water in the corn slurry tank. The mash is now pumped forwards into the conversion tank where the malt slurry is added. More complete liquefaction takes place at this stage combined with a degree of saccharification. On leaving the converter, the mash is flash cooled from 63°C to 26°C again via a three-stage flash cooler operating under vacuum. This flash vapour is condensed and recovered as brewing water as before. The mash continues on its way to the fermenter where a yeast bub is added on entry into the fermenter. This fermentation, at an O.G. of 1.093, will take approximately 85 h and yield a beer of 13% abv.

Grain distillation

Distillation of grain whiskey spirit at the Midleton Distillery uses a three-column system, combining a beer column with a two-column extractive distillation unit. This system is illustrated in Fig. 9.3. Fermented beer at 13% abv enters the beer column just above halfway (tray 22 of 37) onto the top tray of the stripping section. Steam introduced at the base of the column (via a reboiler) strips both alcohol and congeners as it travels up the column, bringing the alcohol-enriched vapour to the top of the column, where it is condensed and drawn off to produce beer-column high wines with a typical strength of 72% abv.

The extractive distillation column operates on the basis that dilution water added at the top of the column changes the volatility of the higher alcohols, aldehydes, and esters, which will now travel to the top of the column as the diluted alcohol stream flows down the column. As the diluted alcohol stream flows down, it reaches an area of the column known as the pinch, where the optimum congener and alcohol concentrations occur. At this point, the alcohol stream, now at approximately 20% to 22% abv, is transferred directly to the rectifying column.

At the top of the extractive column, the congener-rich vapour is condensed to produce a heads stream that flows through a decanter to allow for the separation of the fusel oils. The fusel oil stream is decanted and the remaining liquid is transferred back to the top of the extractive column as reflux. The diluted alcohol stream, also containing congeners not removed in the extractive column, enters the rectifier at tray 22 and is concentrated back up the column using steam entering at the base of the column. As the ethanol and congener stream travels up the column, a side stream is taken off between trays 56 to 62, where a majority of the remaining higher alcohols, particularly isoamyl alcohol, concentrate. This side stream is returned to the top section of the extractive column.

Near the top of the column, at tray 74, the product stream is removed, giving a spirit with a typical strength of 94.4% abv. The overheads from the top of the column are condensed to provide reflux, with the remainder being recycled back to the top of the extractive column. Bottoms from both the rectifier column and the extractive column are used to maintain a level in the dilution water tank with the excess going to the drain (Fig. 9.3).

In early 2014, a new three-column unit was installed at the Midleton distillery as part of the distillery expansion. This three-column unit is capable of producing up to 50 million LAs

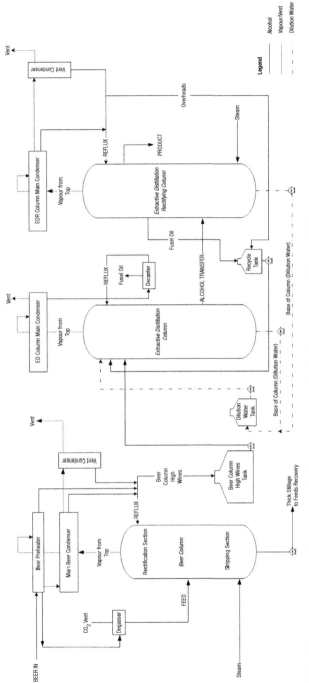

FIG. 9.3 Beer column and extractive distillation unit for the distillation of grain whiskey spirit (atmospheric).

per year. This unit operates on the same principle as described previously; however, from an energy point of view, it is quite different.

In the previous unit, all three columns operated at atmospheric pressure, which required steam to be applied indirectly to each column. For the new design, the beer column and the rectifier (EDR) are energy coupled, i.e., the beer column operates under vacuum and the EDR operates under pressure. Steam is used to heat the EDR and the overhead vapour produced from the EDR is condensed in the reboiler of the beer column, which ultimately heats the beer column.

The extractive distillation (ED) column operates at atmospheric pressure but gets most of its energy from a 3-stage mechanical vapour recompressor (MVR), which recovers the heat from the base liquid (dilution water) of the ED and EDR columns. This heat is compressed through the three fans of the MVR, which provides the energy to run the ED column. There is also a small amount of steam on the ED column. This type of energy coupling and energy recovery design has enabled for a reduction in the energy required to produce 1 L of pure alcohol (kw/LA) by over 50%.

This multipressure design has delivered very significant energy savings without compromising on the quality of the distillate. This was achieved through substantial trials being carried on the pilot distillation plant of the technical centre at the distillery. The trials were carried out at different vacuum levels for the beer column and different pressure levels for the EDR, to determine the optimal vacuum and pressure level without impacting the flavour of the final distillate (Fig. 9.4).

Maturation

Whiskey maturation at the Midleton Distillery follows a strict wood management policy, for both pot still whiskey and grain whiskey. As with whisky distilleries in Scotland, the United States is the main supplier of oak casks, which come via two cooperages in Kentucky. Selected distillery run barrels from a number of bourbon distilleries are checked, repaired if necessary, and then shipped to Midleton on a regular basis throughout the year. Current requirements are in the region of 160,000 first fill barrels (B1s) per year.

Midleton's maturation policy requires 40% of pot still spirit to be matured in B1s, whilst grain whiskey typically utilises 60% B1s in its maturation profile. This approach for grain whiskey maturation would not be typical for a Scotch grain whisky, where barrels will normally have performed a series of malt fills before being used for grain whisky maturation. All barrels from the United States are shipped as standing barrels, with no shooks being imported.

The use of sherry-seasoned casks also plays a significant role in the maturation of whiskey at Midleton Distillery. In this case, sherry butts (500 L) are commissioned from a specific cooperage in Jerez de la Frontera, Spain, which uses only European oak (*Quercus robur*). They are then seasoned with oloroso sherry for 2 years in three bodegas within the Jerez appellation. The company commits to the wood with the cooperage 2 years before manufacture, to allow for an 18-month air-drying period. All of the wood is sourced from the Galicia region of Northern Spain. In addition to sherry-seasoned oak butts, smaller quantities of fortified wine casks are sourced from Portugal (Port), Sicily (Marsala), Malaga (Malaga wine), and Madeira

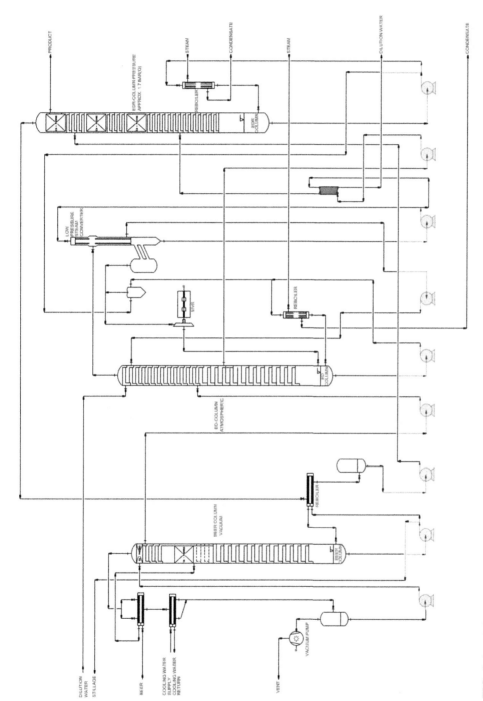

FIG. 9.4   Multipressure distillation unit for the distillation of grain whiskey spirit.

(Madeira wine). All casks (barrels and butts) are held in palletised warehouses during the maturation period. There are six barrels per pallet, and a stack is seven pallets high; there are four butts per pallet, going four high. All barrels remain on the pallet at all times, and they are both filled and emptied through the head. This significantly reduces the handling and rolling of barrels, which only come off the pallet for repair or when they are being culled from the population. Typical use of a barrel would be three maturation cycles before being culled. Currently, no cask rejuvenation is practised at the Midleton Distillery or in any other Irish distillery.

The future

The future for Irish whiskey looks exceptionally bright. During the past 5 years, the number of operating Irish whiskey distilleries has increased from 4 to 30. Some of these new distilleries are of a significant size, whilst many would be included in the craft category. Whilst most are dedicated malt/pot still distilleries, a few have installed column stills for grain whiskey production.

Acknowledgements

We would like to thank our colleagues Dagmara Dabrowska and Edel Scanlon for the help and assistance given to us in the preparation of this chapter.

References

McGuire, E.B., 1973. Irish Whiskey. Gill & Macmillan, Dublin.
Mulryan, P., 2002. The Whiskeys of Ireland. O'Brien Press, Dublin.
Nicol, D.A., 1997. Distilling—past, present and future. Ferment 10 (6), 382–391.

10

Scotch whisky: Raw material selection and processing

Tom A. Bringhurst, Barry M. Harrison, and James Brosnan

The Scotch Whisky Research Institute, Edinburgh, United Kingdom

Introduction

The production of Scotch whisky has long been the subject of an impressive volume of publications, both as a layman's guide for consumers and as a technical account for people directly involved in the distilling industry. Recent editions of *The Alcohol Textbook* (Walker et al., 2017; Ingledew et al., 2009; Jacques et al., 2003) provide useful backgrounds for all aspects of alcohol production. Hume and Moss (2000) have provided a concise, well-illustrated, historical perspective on both malt and grain whisky production. Accessible technical details about the production of malt whisky can be found in Piggott et al. (1989), Bathgate (1989, 1998, 2003), and Dolan (2003). Buglass (2011) summarises the production of a wide range of potable spirits, including malt and grain whisky. There are also some informative accounts of grain distilling, such as those of Pyke (1965), Rankin (1977), Bathgate (1989), Wilkin (1989), Piggott and Conner (1995), Bringhurst et al. (2003), Collicut (2009), and Kelsall and Piggott (2009).

Although some of the earlier references are now quite old, they all still contain relevant information. Briggs et al. (2004), although focused on brewing, provided a detailed account of the main stages of raw materials processing that is an essential reference for a more detailed technical understanding of the technology underlying grain intake, milling and mashing, and beyond and can be applied to both malt and grain distilling. More recently, the proceedings of the Worldwide Distilled Spirits Conferences (Bryce and Stewart, 2004; Bryce et al., 2008; Walker and Hughes, 2010; Walker et al., 2012; Goodall et al., 2015; Jack et al., 2018) are an essential source for important background information regarding recent developments in the production of Scotch whisky and other globally important distilled spirit products. Kelsall and Piggott (2009) have provided information on cereals, starch, milling, and a range of cooking processes, including low-temperature and no-cooking systems that are also relevant to Scotch whisky production.

Whisky and Other Spirits
https://doi.org/10.1016/B978-0-12-822076-4.00018-8

137

In recent times, the Scotch whisky industry has placed an increasing emphasis on reducing its environmental impact as well as meeting the challenges of climate change. Consequently, this is an appropriate time to review the cereal processing aspects of the production of Scotch malt and grain whisky.

Sustainability is a major consideration for distillers, for both grain and malt distilling. The main drivers for this are a strong interest in securing a ready supply of raw materials that are well suited to traditional production methods, together with ensuring that the industry's high-profile, globally branded products continue to be associated with a pristine environment (Hesketh-Laird et al., 2012). As a result, a significant investment by distilling companies must factor in their impact on long-term sustainability, not just for individual companies but also in terms of their impact on the entire industry and the various supply chains that support it, as well as the effects on the environment. This includes reducing the overall environmental impact of the industry, adapting to climate change, and taking active steps to reduce the carbon footprint.

In recent years, success in the export market has led to a sharp rise in distillery numbers. Between 2012 and 2019, 34 malt distilleries opened, increasing the total number of distilleries currently in operation to over 130 (Gray, 2019).

As a result, there is now a wide diversity of distillery processes, ranging from small traditional, cottage-type operations with a production capacity of around 500 L of pure alcohol (LPA) to large and complex modern technologically advanced processes with production capacities exceeding 100 million LPA (Gray, 2019). What they all have in common is that they produce spirit according to the legal definition of Scotch whisky (Statutory Instrument 2009, 2009), which specifies that it must be produced in Scotland from water and cereals and processed into a mash, which is converted to fermentable substrates only by endogenous (malt) enzymes fermented by yeast. No externally added enzymes or other process aids (aside from spirit caramel) are permitted. This requirement limits the process options that are available to Scotch whisky distillers, but is essential to maintaining the traditional aspects of Scotch whisky production.

In Scotland, the two types of whisky distilleries produce malt or grain spirit. When new-make spirit is produced in a distillery, it is matured for a minimum of 3 years, but typically for 8 or more years, before it is bottled as single-malt whisky or, rarely, as a single-grain whisky. Commonly, mature malt and grain spirits are blended together to produce a very wide range of blended Scotch whisky brands, which are the major products of the industry. Globally, the Scotch whisky market is changing and, while blended whiskies are still very important, demand for malt whisky continues to grow, as consumers show increasing awareness of the perceived quality of these very individual premium products.

At the time of writing, there are around 130 malt distilleries in production in Scotland, each producing a distinctive malt spirit character. Not only is this an essential attribute of an individual distillery, but it is also an important component of the blended product for which it is destined. A premium-brand blended whisky can contain as many as 50 different malt whiskies, together with grain whisky, which is the basic matrix for the blend. All of these must be at least as old as stated on the label. In contrast to malt whisky, the flavour characteristics of grain whisky are relatively neutral, as a result of the continuous fractionating (or Coffey) stills that are used to collect a very pure and clean spirit. However, Scotch whisky grain spirits are not entirely without flavour or aroma, and this has important implications for the raw materials and processing options.

The scale of grain whisky production is much greater than for malt distilleries. For grain distilleries, the production capacity ranges from 18 million LPA to 110 million LPA, averaging at 59 million LPA. On average, the capacity of a typical malt distillery is around 3 million LPA. In 2019, the overall capacity of the seven working grain distilleries (410 million LPA) was still greater than the combined total production of all of the malt distilleries (395 million LPA) (Gray, 2019). This emphasises that the production of both types of spirits complements each other and that both are essential facets of Scotch whisky production.

This chapter is focussed on the raw materials and processing aspects of the Scotch whisky process for both malt and grain distilling, and the aim is to provide a concise and accessible summary of the main features of distillery production. Malt and grain distilleries have some features in common. They both involve the processing of cereals (malted barley for malt distilling, and a mixture of malted barley and unmalted cereals for grain distilling). They use a mixture of milling, cooking/mashing, fermentation, and distillation processes to produce potable spirits according to the official legal definition of Scotch whisky. However, malt and grain distilleries differ in several important ways, both in the technology and in the scale of distillery production. In general terms, although there is a very large number of individual malt whisky distilleries, the basic production processes are effectively the same, differing only in relatively small details. In contrast, although there is only a small number of grain distilleries, they are able to use a much wider range of technology and process options (within the confines of the Scotch whisky definition), and as a result, there is much greater diversity in their production techniques.

It is not the aim of this chapter to describe each individual process in detail but rather to explain the fundamental principles behind distillery processing with reference to the various aspects of the technology that are used. However, before looking closely at the process itself, we must consider the main raw materials that are used in both malt and grain distilleries and summarise the basic biochemistry underlying the conversion of cereal starch into fermentable sugars, which are in turn converted by yeast into alcohol.

Supply chain sustainability

Brosnan et al. (2010) highlighted the importance of supporting the sustainability of the cereals supply chain, emphasising the differing economic, environmental, and social factors underlying sustainability. Each component of the supply chain has different perspectives and priorities influencing availability, price, and quality, and a careful balance must be achieved to make the entire chain sustainable. The key driver of this process comes from plant breeders, who are tasked with developing new cereal varieties that will support all stakeholders and offer a genuine technical route to innovation and sustainability (Brosnan et al., 2010).

Sustainability is a major target for Scotch whisky distillers, and the industry is reliant on securing the continuity of quality cereal supplies to maintain and underpin the development and growth of its products. It is essential for stakeholders in the supply chain to work closely together so that growers and grain merchants receive a clear message regarding the strategic needs of the industry (Brosnan et al., 2010; Rae, 2008).

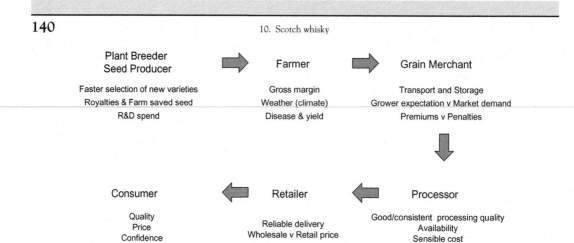

FIG. 10.1   Factors influencing the supply chain for cereals. Adapted from Brosnan, J.M., Bringhurst, T.A., Agu, R.C., 2010. Growing sustainability in the wheat supply chain. In: Walker, G.M., Hughes, P.S. (Eds.), Distilled Spirits: New Horizons: Energy, Environment and Enlightenment. Nottingham University Press, Nottingham, UK, pp. 27–32.

Fig. 10.1 highlights the production of wheat for distilling as a model for sustainability, where the major interlocking components of sustainability, environment, economics, and social development all have a major impact on the supply chain for wheat. Brosnan et al. (2010) identified the three main areas of concern: (1) distillers need readily available raw materials; (2) they would prefer to purchase their cereals at a reasonable price that avoids the uncertainties of the fluctuating market; and (3) the raw materials must perform consistently in the distillery. Therefore, in order to achieve sustainability, it is essential to balance these requirements with the complementary desires of the other supply chain stakeholders. One important feature of this has been the successful efforts of the Scotch whisky distilling industry to develop and maintain contacts with plant breeders, both on an individual company basis and as an industry, through the British Society of Plant Breeders, as well as through links with the UK agronomy sector. This is represented by bodies such as the Agricultural and Horticulture and Development Board (AHDB), and the Maltsters Association of Great Britain (MAGB), each of whom has major input into the decisions regarding new cereal varieties. It is through these links that distillers are able to influence the development and selection of new wheat and barley varieties that are suitable for distilling, and will also give a good return for growers, merchants, and other supply chain stakeholders.

Raw materials

The main cereal raw materials used by Scotch whisky producers are wheat and barley. Malted barley is used by malt distillers, and unmalted cereals such as wheat and maize are used by grain distillers. Over the years, although maize is still used by a small number of distillers, its importance has declined substantially, and distillers have developed clear preferences regarding the types of barley and wheat that they require for their processes.

In terms of barley, distillers are looking for modern varieties that can be malted by maltsters to provide a high level of alcohol yield and that are relatively easy to process. Currently, this means that they are looking for low-nitrogen spring barley varieties that will produce malt with a high level of fermentable extract and that are sufficiently well modified to give good distillery processing characteristics.

Grain distillers require low-nitrogen, soft winter wheat varieties, which are easy to process and give good alcohol yields. These are normally classed with feed and biscuit wheat and are at the opposite end of the quality spectrum from premium bread-making wheat, which is considered to be unsuitable for grain distillery production. Because of the nature of the market for feed wheat, distillers have traditionally traded wheat on a commodity basis and will generally not pay additional premiums for such a basic raw material. The involvement of distillers in the national UK selection process for new wheat has resulted in the increased availability of winter wheat varieties that have been identified as being suitable for grain distilling.

In order to secure sustainable supplies of their cereal raw materials, the Scotch whisky industry is closely involved in the UK selection process for both wheat and barley, and distillers maintain contacts covering the entire supply chain. Discussions with stakeholders including plant breeders, growers, agronomists, maltsters, and other end users are essential to ensure that sufficient supplies of both barley and wheat of suitable distilling quality are available, in relation to the differing (and sometimes opposing) requirements of other end users.

Communications between distillers and plant breeders have been essential to providing new barley and wheat varieties that are better suited to their processes in terms of increased alcohol production and good processability. As distillers have developed a better understanding of these qualities, they continue to encourage the development of new distilling varieties. Expansion in our knowledge of the genetics underlying important quality traits is now allowing plant breeders greater precision to identify suitable parent varieties in order to provide new barley (and potentially wheat) varieties more efficiently. These can offer significant advantages to distillers but will also provide better support for the cereal supply chain in general.

Cereal breeding

In the 19th century, Gregor Mendel established the basic principles of plant genetics. He discovered inherited traits, which are transferred from one generation to the next. The plant breeder's aim is to reassemble these units of inheritance, known as genes, to produce crops with improved characteristics. In practice, this is a complex, time-consuming, and expensive process, as each plant contains many thousands of genes, and the plant breeder seeks to combine a range of desirable traits in one plant to produce a successful new plant variety.

Conventional breeding involves crossing selected parent plants (genotypes), chosen because they have desirable characteristics such as high agronomic yield or disease resistance. The breeder's skill lies in selecting the best plants from the many and varied offspring. These are grown on trial plots and tested over a number of years. Typically, this involves examining large numbers of individual plants for different characteristics (phenotype traits) ranging from agronomic performance to end-use quality. Current estimates suggest that the breeder has to grow and assess up to 1 million individual plants in order to develop a single commercially successful wheat (or barley) variety, a process that can take up to 15 years before the product enters the marketplace (Brosnan et al., 2010).

Advances in modern (nongenetically modified [GM]) genetics have identified some of the genes associated with certain phenotype traits that are important to end users of both wheat and barley, and now this information is used actively by plant breeders to accelerate this process, using classical non-GM breeding techniques (Sylvester-Bradley et al., 2010). Next-generation plant breeding techniques, such as gene editing, do not involve the insertion or removal of whole genes, offering a more precise way to substantially improve crops. However, this approach still requires the establishment of acceptability amongst legislators and consumers (Harrison and Brosnan, 2017). One of the aims of modern research is to develop new varieties that will require fewer nitrogen and agronomic treatments, thus reducing the economic and agronomic costs of growing and supporting cereal varieties that are better suited for the UK market and providing support for the long-term sustainability of the cereal supply chain (Sylvester-Bradley et al., 2010). These initiatives are now resulting in the emergence of larger numbers of new high-quality barley and wheat varieties with strong potential for use by both malt and grain distillers (Bringhurst et al., 2012a) and which are also attractive to growers as a result of better agronomic performance and improved disease resistance. These developments make a strong positive contribution to sustainability, both for the distilling industry and within the cereal supply chain in general, and will help to stimulate plant breeders to provide continuing improvement of modern barley and wheat varieties.

Barley (and wheat) selection, in one form or another, is generally considered to have originated at a very early period in our history, at least 10,000 years ago (Badr et al., 2000). The modern two-row barley varieties now grown in the United Kingdom and Europe have gradually developed from early semiwild landraces adapted to the colder and wetter climate in areas of northern Europe. These aboriginal landraces provide much of the genetic diversity that modern barley breeders can use to develop new varieties. Descendants of these landraces can still be found, and a small amount of *bere* barley is still occasionally used in small-scale malt distillery production. The diversity found in the landraces will become increasingly important as we look for traits that will allow for robust new varieties able to cope with a changing climate.

Raw materials for malt distilling

Barley selection

There are now well-established systems for the breeding and selection of modern malting barley varieties. Barley breeding is carried out by commercial plant breeding companies using classical crossing techniques to produce a range of new varieties each year. The process of selection and testing of new varieties is both time-consuming and expensive, and each new variety requires up to 15 years development. The development time includes the initial selection and testing of each variety by the plant breeder (9–11 years), followed in the United Kingdom by several years of external national listing (NL) trials. Each new variety entering the UK assessment system represents an investment by the breeder of around £1 million.

Epiheterodendrin (EPH), a glycosidic nitrile (GN) present in some barley varieties, is the main precursor of ethyl carbamate (Cook, 1990), a contaminant in certain distilled spirits and regulated in certain markets. One of the complications in breeding new barley varieties for Scotch whisky production is the requirement for non-GN-producing varieties. However, not all barley varieties produce EPH, and breeders can select those that do not contain this precursor using a recently developed genetic marker to screen potential new distilling barley varieties (JHI, 2018). A well-established marker was already in place (Swanston et al., 1999), but the new marker system improves upon that approach as it is easier to execute, more robust, and can also be used on a wider range of germplasm. The current policy of the industry is to insist that all new barley varieties aimed at the distilling market should be identified as GN-non producers. Genetic markers have been developed for other important end-user traits, which will help the 'classical' selection of potential new varieties be faster and more efficient.

When a new variety is put into the market, it must first undergo distinction, uniformity, and stability (DUS) testing, prior to being entered on the AHDB national list, which is a compulsory requirement for all new commercial barley varieties entering the UK growing market. This ensures that the variety is actually a new variety with distinctive properties that will be sustained throughout the lifetime of the variety. This also helps to identify the market at which the new variety is targeted, whether it is to be malted for brewing or distilling or employed simply unmalted for brewing and feed.

The initial 2-year period of testing (NL1 and NL2) is currently managed by the AHDB and is designed to assess whether the new variety is suitable for growing in the UK, whether it offers any agronomic advantage over existing varieties, and whether it is vulnerable to diseases and pathogens. Each year the AHDB Barley Committee considers the data generated in the trials and decides whether varieties should progress to a further year of Recommended List (RL) trials prior to final approval. These trials are primarily designed to assess the suitability of new barley (and wheat) varieties for growing commercially in the UK. Successful varieties are added to the annually updated AHDB Recommended Lists.

In Scotland, similar decisions are made by the Scottish Cereals Consultative Committee regarding additions to the Recommended Lists of the Scottish Agricultural College, now known as Scotland's Rural College (SRUC) (SRUC, 2020). The Recommended Lists are important to distillers because, although there is no requirement to use exclusively Scottish barley and wheat, grain merchants, maltsters, and distiller/maltsters supplying the Scotch whisky industry prefer to seek local supplies of barley (and wheat) if possible. This is also in line with current thinking regarding supply chain sustainability and reducing the industry's carbon footprint.

For malting barley, a closely related set of parallel trials is managed by the Maltsters Association of Great Britain (MAGB), under the banner of the Malting Barley Committee. This committee is jointly funded by the MAGB, AHDB, British Beer and Pub Association (BBPA), and Scotch Whisky Association, and is focussed on malting and end-user quality. Decisions are made on analytical data from barley micromalts provided by stakeholders of the Micromalting Group, who report to the Malting Barley Committee.

Although there are dual-purpose malting barley varieties, which serve both brewing and distilling markets, it is important to emphasise that distilling barley has specific attributes that make it distinct from material destined for brewing. There is normally a division between spring and winter barley varieties, as distillers currently require spring barley and winter barley is primarily destined for brewing. The system for spring barley is further complicated with the additional requirement that all new distilling varieties must be nonproducers of epiheterodendrin. There is an interest in improving winter barley quality for distilling. This would provide a more consistent supply in the face of challenges to the spring barley crop attributed to climate change (Looseley et al., 2020).

Out of around 40 spring varieties entering the system at NL1 each year, generally only one or two will reach the final approval stages. Prior to final approval, a new variety will be given provisional approval and will generally undergo a further period of commercial macroscale distillery trials (normally 1–2 years) before achieving full MBC approval status. A list of currently approved varieties for distilling and brewing is displayed on the MAGB website (www.ukmalt.co.uk) (MAGB, 2020).

The MBC-approved list is quite fluid and changes each year, as new varieties come to the fore and others become outclassed. The Concerto variety has been the major malting barley for distilling for the past 10 years, but it is now becoming agronomically outclassed by more recent varieties, such as Laureate, LG-Diabolo, and KWS Sassy.

Barley procurement and malting

At current production levels, distillers in Scotland use in the region of 800,000 tonnes of barley malt each year, primarily for malt distilling, with about 10% of this being destined for high-enzyme grain distilling malt. Distillers normally obtain their malted barley supplies under contract from commercial sales maltsters, and there are strong industry communications between distillers and maltsters to ensure that they are each aware of availability, supplies, and end-user requirements. In the UK, most barley malt is traded under a quality assurance scheme, such as the Assured UK Standard (MAGB, 2014) operated by the MAGB, which helps to support food safety, product legality, good operational practice, and product quality to ensure that good quality supplies of suitable malt are available from traceable barley sources that are free from problems with contaminants, such as mycotoxins, pesticide residues, heavy metals, and other potential food safety issues. Commercial maltsters normally obtain their barley supplies under contract from farms within a relatively large regional catchment area, but most have malting plants in close proximity to Scotch whisky distilling areas or close to transport links serving them. In some situations, particularly if there are local shortages or problems with supply or price, maltsters and distillers may also obtain barley malt from suppliers from further afield, outside the UK, but normally within the European Union.

Only a few floor maltings still survive, and these tend to be small undertakings supplying a relatively small proportion of the distillery's overall malt requirements. These may be based on local barley supplies, but barley may also be obtained from other sources through grain merchants or malting companies. Although one of the main reasons for the survival of this part of the old malt distillery process is to maintain links with the traditional heritage of the industry, some floor maltings still provide a real addition to the production capacity of the distillery and help to impart a more distinctive spirit character, such as those associated with Islay and Orkney and other Island whiskies.

Distiller-maltsters are usually larger distilling companies producing a significant proportion of their own barley malt supplies. These facilities generally supply their own process requirements but can also provide supplies to other distilling companies, particularly within a small localised region. These companies tend to obtain barley from a local catchment area, but they can also obtain supplies more widely through agreements with grain merchants.

Malt distillers normally require low-nitrogen spring barley which can be malted easily to provide high-quality distilling malt and will give a high alcohol yield with good processing properties. Typically, barley destined for malt distilling will have a total nitrogen level between 1.40% and 1.60% (dry). Because the amount of nitrogen is inversely proportional to the starch content, lower nitrogen barley will have more starch and hence higher alcohol yield potential. For grain distilling barley, nitrogen levels will be somewhat higher, closer to 1.80%–1.85%, to provide more potential for higher levels of enzyme and amino acid (free amino nitrogen, or FAN) development, which are essential for the barley to perform well in a grain distillery.

Maltsters will generally check the barley nitrogen level, the level of screenings, and specific weight (or thousand corn weight) of each batch of barley as it is delivered to the maltings and will also routinely carry out simple malting tests to check for dormancy, water sensitivity, and pregermination, all of which have a significant impact on the malting performance of each new batch of barley.

During malting, barley for malt distilling (pot still malt) is generally steeped in water (and air rested) for about 2 days and germinated conventionally for three to 4 days, depending on the quality of barley from which it is to be produced. The degree of malt modification resulting from germination has important implications for the processability of barley malt in the distillery, and it is important that the malting process be carefully optimised to give good performance in the distillery (Agu et al., 2012; Bryce et al., 2010). To minimise damage to the amylolytic enzymes, distilling malt is kilned at a relatively low temperature compared with brewing malt. Typically, the maximum kilning temperature for pot still malt is around 72°C (Dolan, 2003) compared to 75–80°C for brewer's malt. For grain distilling malt, to maximise the enzyme content, the germination time may be longer and the final kilning temperature much lower (50°C) (Bathgate and Cook, 1989).

Peating

One important aspect of malting for Scotch whisky production is that floor maltings can provide a more specific peaty or phenolic spirit character during the kilning of malt, which is an essential quality for certain spirits. This derives from the distinctive characteristics of the peat that has been cut from the local area (Harrison et al., 2006). Although commercial maltsters can also supply peated malt, this generally tends to be more generic and will not necessarily reflect the specific character required for an individual distillery. The exception is in specific areas, such as on Islay, where the commercial malting facility is in close proximity to established sources of peat.

Burning peat is a traditional way of imparting peaty or phenolic, smoky, burnt, or medicinal characteristics to malted barley during the kilning. The location from which the peat is extracted is important in defining the flavour characteristics of the peated malt that passes into the spirit to give highly distinctive products.

Peat is 'organic sediment formed under waterlogged conditions from the partial decomposition of mosses and other bryophytes, grasses, shrubs, or trees' (Harrison et al., 2006).

In Scotland, peatlands cover more than 20% of the land surface. Extraction for the whisky industry is currently carried out on a small number of sites, notably on Islay, Orkney, and at St Fergus in the North East of the mainland. Peatlands are a valuable carbon store—there are around 1600 million tonnes of carbon stored in Scotland's peatlands. However, it is estimated that through various land uses, over 80% of our peatlands are degraded and are emitting carbon. Though only extracting small volumes of peat—less than 1% of the UK current extraction figures—the industry is focussed on minimising its impact and helping to reduce carbon emissions from peatlands.

The structure and degree of decomposition of lignin and other polyphenol components from local plant species, such as grass, heather, moss, and woody plants and trees, are fundamentally important in influencing the flavour composition of the peat smoke (Harrison, 2012). The chemical composition of peat smoke is very complex. The main components are phenols and methoxyphenols (guaiacols), some of which have very low aroma thresholds and are hence highly flavour active. The degree of thermal degradation during burning also has an important impact on the composition of the peat smoke, and the kilning conditions are important in providing distinctive aroma characteristics that can pass into the spirit. Although there is now greater understanding of the chemistry underlying peaty character, this is still one feature of modern Scotch whisky production that is still to some extent an art rather than a science.

One of the long-term problems facing the Scotch whisky industry is that some sources of peat are becoming less accessible. In these areas, distillers and maltsters are facing challenges to find new sources as existing seams begin to run short and alternative sources that can match their required flavour characteristics may become more difficult to access.

Malt specifications

Malt specifications are agreed upon by distillers and maltsters on an individual basis and are subject to change as a result of commercial trading issues, availability, and harvest quality. Thus, it is not possible to be absolutely specific regarding all the parameters that are included. Another point worth noting is that when setting out specifications with suppliers, it is important not to set values for different parameters that might be contradictory or result in the supplier having problems meeting the specification. An informed dialogue between individual distilling companies and their suppliers is essential to ensure that commercially binding specifications are realistic and achievable and reflect the availability of suitable supplies of barley malt. Table 10.1 shows a typical malt specification compiled from a range of sources.

The numbers quoted in the table are guidance values and may change on a seasonal basis or depend on availability and supplier. Not all parameters are specified in every case, but those that are mentioned in the table will generally fall into four main categories: (1) production efficiency, (2) processability, (3) spirit quality, and (4) product protection/due diligence, each of which provides highly important information about whether a particular batch of malt is suitable for the purpose, will perform well in a distillery, and will provide assurance in minimising the probability of any problems affecting spirit quality or product protection.

These parameters are generally checked using official standard analysis methods, which were originally published as Recommended Methods of the Institute of Brewing (IOB, 1997), but which are now collected in *Analytica-EBC* (European Brewery Convention, 2010) under

TABLE 10.1 Typical malt specifications.

Attribute	Guidance value	Notes
Moisture (%)	3.5–5.0	>6 results in storage/milling problems. 4.0%–4.5% is typical received
Soluble extract (0.2 mm) (fine) (%)	>79 (83 dwb)	Best available. Varies by malt crop but 78–80 is a typical range
Soluble extract (0.7 mm) (coarse) (%)	>78 (82 dwb)	Best available. Varies by malt crop. Some distillers use 1.0 mm to more closely mimic the process
Fine/coarse difference (%)	<1.0	Properly modified
Fermentability (%)	87–88	Best available
Fermentable extract (%)	>68	Not always specified
Predicted spirit yield (Psy) (LA/tonne)	≥410 (430 dwb)	Best available. Varies by crop year. ≥408–412 range is typical for many malt distilleries. Can be as low as 405 or up to 420 in extreme years. Several distillers use their own PSY method, others use the EBC/IoB method
Diastatic power (DP) (α-, β-amylase)	Not typically specified for malt distilling	≥150–160 (dwb) is typical for grain distilling with 180 - 200 deemed as very good
Dextrinising units (DU) (α-amylase)	Not typically specified for malt distilling	≥60 for grain distilling. Mid- to high 60s are typical and can reach 80 in extreme years
Total nitrogen (% dry)	<1.45–1.6	Varies by crop year but a TN range of 1.3–1.7 is possible including extreme years
Soluble nitrogen ratio (SNR)	38–42	Soluble N/total N. Indicates the degree of modification. Lower values being an indication of undermodification and higher values of over-modification
Free alpha amino nitrogen (FAN) (ppm)	150–180	High-gravity operation will require more FAN
Friability (%)	>90	Grain well modified
Homogeneity (%)	≥98	Endosperm evenly modified
Phenols content (ppm)	0–50	Depends on customer requirements. Unpeated malt tends to be set at <1–2 ppm.
SO_2 content (ppm)	<15	Depends on fuel source/level of peating/ NDMA risk. Some distillers may go as low as 5 ppm
Nitrosamines content (ppb NDMA)	<1	–
Glycosidic nitrile (GN) (g/tonne)	<1.2	This is a typical value of GN for varieties not on the MBC recommended list. Preferably use a GN nonproducer
GN nonproducer (g/tonne)	~0.5	Typical background level
β-Glucan (mg/L)	<100	High β-glucan can slow wort run off
Screenings (% at 2.2 mm)	≤1.5	Not always quoted. Some distillers may set a max value for screenings
Processing	–	Regulatory requirement. No enzymes, hormones, or other additives permitted to be added to the malting process

Note: Values are subject to seasonal change and are for guidance only.
Adapted from Dolan, T.C.S., 2003. Malt whiskies: Raw materials and processing. In: Russell, I. (Ed.), Whisky: Technology, Production and Marketing. Academic Press, London, pp. 27–74, with additional comment from Mulholland (2020 personal communication, 28/07).

the auspices of the European Brewery Convention (EBC). These methods are designed for use in the laboratory to confirm the quality of batches of barley and are not suitable for intake testing. If a batch of barley malt does not match the agreed specification under which it was ordered, the distiller is entitled to some form of compensation.

The most important parameters in terms of distillery performance are those relating to production efficiency, which are moisture, soluble extract, fermentability, fermentable extract, and alcohol yield (predicted spirit yield, or PSY), all of which have a strong influence on the potential amount of alcohol that can be produced from each tonne of malt. In some cases, levels of enzymes (diastatic power [DP]/dextrinising units [DU]) (α-, β-amylase) may also be quoted, but these are more important for grain distilling malt, where enzyme activity is more critical.

Malt modification parameters, such as friability (homogeneity), fine/coarse difference, soluble nitrogen ratio, and free α-amino nitrogen (FAN) come under the heading of processability, as they help to indicate how well the malt will progress through the distillery and help to identify batches that are likely to give processing problems, either with poor wort separation or with fermentation efficiency. Malt that is undermodified will give reduced extract and alcohol yield, particularly in a traditional or semilauter tun, as the starch will be more difficult to access without very fine grinding. If it contains a significant proportion of unmodified grains, there could be additional problems with reduced enzyme development and poor protein modification, resulting in poor wort separation due to a lack of β-glucan breakdown and partial solubilisation of incompletely degraded proteins. Undermodified malt will also contain lower than normal levels of soluble nitrogen and FAN, which can affect the fermentation efficiency.

Overmodified malt tends to be highly friable and will break up more easily into a finer flour, which can block the mash tun plates and pass as solids into the process, potentially giving downstream problems by depositing on warm and hot surfaces, such as heat exchanger plates and steam coils in the stills. Hence, the required degree of malt modification is a balance between these two extremes that may vary for different distilleries.

Peating, an important characteristic of some malt whiskies, is normally measured in terms of the collective concentration of a range of phenolic compounds, which are normally analysed by high-performance liquid chromatography (HPLC), although some colorimetric methods are also used. It is worth noting that colorimetric methods are less specific and do not necessarily agree with more precise and detailed HPLC analysis, although they are useful in assigning high, medium, or low levels of peating. Table 10.2 shows typical levels of phenols associated with different degrees of peating. Although very high levels of peating are now less common in modern production, there is still a requirement for this for certain products.

Nitrosamines and sulphur

Levels of nitrosamines, such as *N*-nitrosodimethylamine (NDMA), deriving from the reaction of barley hordenine and ambient levels of nitrogen oxides (NO_x) produced during burning, are now well controlled in modern maltings that use indirect kilning. However, the peating of malt requires the direct application of peat smoke to the drying malt, either as part of the kilning process itself or as an additional input to the kiln air flow, and it is still

TABLE 10.2 Typical levels of phenols associated with low, medium, and high peating.

Peating	Total phenols (ppm)
Light	1–5
Medium	5–15
Heavy	>15
Very high	40–50

Source: Dolan, T.C.S., 2003. Malt whiskies: Raw materials and processing. In: Russell, I. (Ed.), Whisky: Technology, Production and Marketing. Academic Press, London, pp. 27–74.

necessary to prevent the formation of nitrosamines in the malt by burning sulphur, normally in the form of briquettes, to produce sulphur dioxide (SO_2), which is added to the airflow passing through the drying malt. This blocks the reaction that results in the formation of nitrosamines from the hordenine in malted barley. The degree of sulphuring can vary for different maltings but should be sufficient to ensure that the formation of nitrosamines is kept to a minimum. The acidity produced by excessively high levels of SO_2 can result in corrosion damage to the equipment and to the plant. In the past, very high levels of sulphur, up to 20–25 ppm, have been used but more recent specifications for highly peated malt give a maximum of 15 ppm. As high levels of sulphur (> 15 ppm) can also affect the fermentability of the malt by reducing the pH of the mash, heavily peated, sulphured malt is allowed to recover by maturing for several weeks before being shipped to the distillery. In some cases, unpeated malt may be treated with a low level of sulphur (up to 5 ppm) to minimise any potential background levels of NDMA.

Glycosidic nitrile and ethyl carbamate

Ethyl carbamate (EC) is a trace contaminant in distilled spirits and is regulated in several important international markets. The pathways to the formation of this material have been well documented. The major precursor of EC has been identified as epiheterodendrin (EPH), a glycosidic nitrile (GN) present in some barley varieties (Cook, 1990). This contaminant is now well controlled, partly by careful attention to various aspects of the distillery process (e.g. malting, distillation, copper placement), but primarily by selecting distilling barley varieties that do not produce EPH.

It is now the policy of the Scotch whisky industry to specify that all new barley varieties entering trials for distilling barley must be EPH (or GN) nonproducers (more details in Chapter 19).

Other contaminants

Distillers are required to have a regime of diligence testing in place for cereal contaminants such as mycotoxins, pesticides, and heavy metals on their raw materials, but in practice this will be often carried out by an external subcontracted laboratory specialising in these particular analyses. To help minimise the risk to the supply chain, organisations such as the MAGB,

AHDB, and other cereal quality assurance authorities will also commission regular surveys of cereals for these contaminants.

Raw materials for grain distilling

Grain distilling barley

Although it can be relatively inexpensive compared to maize and wheat, unmalted barley has rarely been used in grain distilleries because of the processing problems associated with high levels of gums, such as β-glucans (Walker, 1986). In grain distilleries, barley is generally used in the form of malt, and its prime function is a source of enzymes to convert cereal starch from unmalted cereals, such as wheat or maize, into fermentable sugars. According to the legal definition of Scotch whisky, all of the enzymes in a mash must come from the malt, and no other externally added enzymes are permitted.

Enzyme levels in barley malt are normally defined in terms of dextrinising units (DU), which is effectively a measure of α-amylase and diastatic power (DP), which is essentially the total enzyme (α- and β-amylase) activity. These enzyme activities ares measured using standard methodology, per *Analytica-EBC* methods 4.12 and 4.13 (European Brewery Convention, 2010). It is essential that barley malt for grain distilling, which is often described as high-diastase or high-DP malt, contains high levels of starch-degrading enzymes.

Barley for grain distilling must contain high levels of β-amylase and have the potential to produce high levels of α-amylase, limit dextrinase, and α-glucosidase. Barley malt with a DP of 180–200 units and a DU above 50 units is still considered to be a standard requirement for grain distilling malt (Bathgate, 1989). Because barley malt is a relatively expensive component of the production process in grain distilleries, there is a continuing drive to reduce the amount of malt used, and this has resulted in malt inclusion rates falling below 10 % in many cases. Rapid moisture and enzyme complement techniques allow the optimisation of malt use. It is, therefore, more important now than ever for grain distilling malt to meet high enzyme specifications. As a result, there is a drive to increase the enzyme concentration in the high enzyme malt to allow this to happen more effectively while choosing low or non-GN varieties.

Compared with pot-still (malt distilling) malt, there are some differences both in the type of barley that is used and in the process for producing grain distilling malt. Barley for grain distilling generally has a higher nitrogen content (1.8%–2.0% N) than pot-still barley (Bathgate and Cook, 1989). This is because grain distillers are less interested in the amount of starch that is present (although this does contribute a small but significant proportion of alcohol yield) than in developing the highest possible enzyme potential. In the production of grain distilling malt, barley is allowed to germinate for a longer period, typically 5–6 days, and a gentler kilning regime (50–60°C) is used (if at all) to develop and preserve enzyme activity to convert unmalted cereal starch into fermentable sugars and provide sufficient amino acids to support fermentation.

Originally, unkilned green malt was commonly used for grain distilling (Bathgate, 1989, 1998), and malting was often carried out at the distillery using local or domestic barley. Green malt was generally cheaper to produce due to lower energy costs and gave higher enzyme levels than kilned malt, so a lower dosage rate was required to achieve an efficient conversion of starch. The use of unkilned green malt preserves around 35%–50% of the enzyme activity that would otherwise be lost on kilning (Bathgate, 1989); however, these advantages are offset by higher transport costs and a shorter shelf life (Walker, 1986). Although at least one grain distillery still continues to use green malt, this is now rarely used in modern grain distilling and has gradually declined in favour of commercially produced kilned (white) malt. In 2002, the last onsite production facility was replaced by commercially produced green malt (Robson, 2002), which is delivered fresh to the distillery from commercial maltings. In the modern production environment, the vast majority of Scotch whisky grain distilleries use kilned malt supplied by sales maltsters. The kilned malt is more suited to modern commercial production and is more convenient to transport and store.

In the past, high-diastase (high-DP) grain distilling malt has been imported from places such as North America and Canada (Bathgate, 1989) and latterly from Scandinavia (Sweden, Finland, and Denmark). Barley malt deriving from Scandinavian-grown barley has had a reputation for high levels of enzymes; at one time, these were in great demand in the Scotch whisky grain distilling sector. However, the use of these barleys has declined in favour of domestic supplies amid concerns over unacceptably high levels of GN in some barley varieties. As part of their overall efforts to minimise ethyl carbamate levels, distillers now favour the production of new barley varieties that do not produce this precursor (Cook, 1990).

Established domestic barley varieties suitable for grain distilling, such as Fairing and RGT Asteroid, emerged through the MAGB/AHDB assessment process for malting barley. However, it has been generally acknowledged by distillers and agronomists that there is a shortage of new high-nitrogen grain distilling varieties reaching the market. One of the reasons for this is that the demand for grain distilling barley is very small compared to the overall barley market, and it is difficult to motivate barley breeders and growers to focus on such a small market segment. Modern grain distilling varieties tend to be normal malting barley varieties (i.e. pot-still varieties) that have been adapted to high-nitrogen growing conditions, rather than new types specifically designed for grain distilling (although Fairing was specially bred for this purpose). On occasion, a lack of suitable UK grown high DP malt leads to imports of high enzyme malt from Scandinavia. However, future development of new UK barley varieties for Scotch whisky distilling based on recent research will hopefully address the lack of availability of suitable barley for the production of high-DP grain distilling malt (Hoad et al., 2017).

Wheat

Wheat is one of the most important cereals grown throughout the world, with a global production of around 650 million tonnes annually (FAOSTAT, 2012), providing a large proportion of the world's nutrition (Uthayakumaran and Wrigley, 2010). Wheat is important not just for food production but also for a wide range of other uses, ranging from animal feed to a plethora of industrial applications.

Wheat is a member of the genus *Triticum*, which encompasses a very broad diversity of species, of which only two, *Triticum aestivum* and *Triticum durum*, are grown commercially in significant amounts. *T. aestivum*, which is genetically hexaploid (i.e. contains six sets of chromosomes), is a common bread wheat and is used for a very wide range of applications including bread, cakes, pastries, biscuits, and puddings, as well as other industrial applications, such as starch production, biodegradable plastics, and ethanol production (Uthayakumaran and Wrigley, 2010). *T. durum* is genetically tetraploid (i.e. contains four sets of chromosomes) and is primarily used for pasta production. Common wheats are classified as red or white wheats, based on the intensity of the red pigmentation in the seed coat; as hard or soft, based on the resistance of the seed to crushing; and as winter or spring types, depending on their suitability to temperature and environmental conditions (Uthayakumaran and Wrigley, 2010).

The main applications for wheat exploit the reserves of starch and protein in the endosperm (Wan et al., 2008). One of the major features of wheat is its unique ability to form gluten, which makes it well suited for flour used in bread production. However, while bread-making flour requires hard wheats with high protein (gluten) levels, these are very different from those that are suitable for Scotch whisky production, for which soft wheats with low levels of protein (and gluten) and high levels of starch are essential requirements (Brosnan et al., 1999). In general, distilling wheat varieties are closer in quality to feed and biscuit-making types.

Winter wheat is the highest yielding cereal in Britain and comprises the largest acreage sown. In Scotland, winter wheat is sown in late September to November and is harvested the following September. Nitrogen applications are carefully controlled during crop growth, as nitrogen applied to crops nearer harvest will accumulate in the grain and will give an increased final nitrogen content (Taylor and Roscrow, 1990), which is not favourable in wheat for distilling, but which is desirable in bread wheat (David Cranstoun, SAC, personal communication). The harvest weather in Scotland tends to produce wheat that is inherently low in nitrogen (Brown, 1990) and hence is generally better suited for whisky production. In addition to being important for the production of potable spirits such as grain whisky, vodka, and gin, wheat is also an important feedstock for the current generation of bioethanol plants.

Distillers have been using white soft winter wheat (*Triticum aestivum*) as the major starch source for whisky production since 1984 (Brown, 1990). Originally, the predominance of wheat was mainly due to economic factors (Palmer, 1986), but more recently the difficulty in sourcing maize from reliable, certified non-GM sources has also been a factor. However, the use of wheat is now firmly established by the industry, with five out of seven Scotch whisky grain distilleries currently using wheat as their principal cereal raw material (Brosnan et al., 2010). The industry's reliance on locally produced feedstocks is an important underpinning feature of the UK cereals supply chain, as it provides a significant market for Scottish winter wheat, accounting for at least somewhere in the region of 50% of the crop. This equates to approximately 5% of the total UK wheat production, which was around 16.2 million tonnes in 2019 (DEFRA, 2019).

Many aspects of the biochemistry and physiology of wheat have for many years been reasonably well understood and are well reported in the literature. The biggest change in recent times has been the significant increase in the numbers of new distilling wheat varieties that have been developed by plant breeders for different markets and which are now progressing successfully through official UK trials in response to improved industry understanding of their properties and suitability for their processes. This has allowed the distilling industry to input directly into the UK selection systems for modern distilling wheat varieties (Bringhurst et al., 2008).

More recently, there have been major advances in wheat science with the unravelling of the genetic factors controlling the characteristics of wheat and which highlight the suitability of certain types of wheat for different end-user applications (Sylvester-Bradley et al., 2010; Wan et al., 2008). This knowledge has been observed feeding its way into the development of new varieties for both grain distilling and for bioethanol production (Bringhurst et al., 2012a; Weightman et al., 2010).

Biochemistry and physiology of wheat

The wheat grain is the seed or fruit of the wheat plant. This contains about 85% carbohydrate (Uthayakumaran and Wrigley, 2010), of which starch is the main reserve carbohydrate. The amount of starch in wheat is generally inversely related to the nitrogen content (Taylor and Roscrow, 1990). Structurally, the kernel can be divided into three distinct parts: the endosperm, which comprises around 83% of the kernel; the bran, which accounts for 14%; and the germ or embryo, which makes up about 3% of the kernel (Bushuk, 1986). The endosperm is the most important part in terms of alcohol production and consists of more than 80% carbohydrate (mostly starch), approximately 12% protein, and 2% fat, with 1% minerals and other constituents (Bushuk, 1986). Wheat also contains smaller amounts of nonstarch carbohydrate comprising mono-, di-, and oligosaccharides and fructans (about 7%), as well as cell wall polysaccharides, including arabinoxylans, β-glucans, celluloses, and glucomannans (around 12%) (Uthayakumaran and Wrigley, 2010). The cell wall polysaccharides have important implications for distillery processing, and high levels of these can result in excessive viscosity, which can potentially give serious problems, particularly with the recovery of co-products.

The hardness of wheat is an important factor in determining the suitability of wheat for different end-user applications. Hardness is often measured by a standard milling test, such as the Perten Single-Kernel Characterisation System (SKCS) (Pearson et al., 2007), in which samples that are more resistant to milling are classified as the hardest and those that are easier to mill are classified as soft. The distinction between hard and soft wheat is well resolved at the genetic level (Pearson et al., 2007). The hardness or softness of grain texture is linked to the way in which the starch is bound up in the protein matrix and is thus also related to the nitrogen content. Differences in hardness appear to relate to adhesion between starch granules and storage proteins (Wrigley and Bietz, 1988).

The inverse relationship between grain nitrogen levels and alcohol yield is well known and is related to the relatively low starch content of higher nitrogen wheats (Taylor and Roscrow, 1990). This has been confirmed in distilling trials, where it has been shown that there is a direct link between low nitrogen content and high spirit yield (Brosnan et al., 1999). This relationship has been found to be stronger in some harvest years more than in others (Scotch Whisky Research Institute unpublished data), but the nitrogen content is considered a reliable index of distilling quality regarding its influence on the expected yield of alcohol.

Studies by Brosnan et al. (1999) suggest that other factors that are relatively independent of nitrogen content, such as the proportion of A and B starch granules, and the way they can be densely packed together may also have an important role to play in determining the maximum alcohol yield of wheat. This emphasises that simply determining the amount of starch in the grain, on its own, may not be a good indicator of alcohol yield. This is primarily because of the limitations of the starch assays that are used in vivo (Kindred et al., 2008), and it is considered that the nitrogen (protein) in the grain gives the best indication of its potential for alcohol production. This is because this parameter is related to the accessibility of the starch within the grain. As a result, the best predictive measures of alcohol yield are those that take the protein content into account (Agu et al., 2008b; Kindred et al., 2008; Swanston et al., 2007). Swanston et al. (2007) indicated that wheat alcohol yield is determined by a mixture of genetic and environmental factors, with the effect of variety having a major impact on the amount of alcohol produced and the level of nitrogen assimilated by the grain. The importance of both variety and environment is emphasised by the annual assessment of new wheat varieties carried out by the Scotch Whisky Research Institute (Bringhurst et al., 2008), which look closely at the performance of individual varieties over a range of sites at different locations in the UK, including Northern and Central Scotland, the Scottish Borders, Northern England (Yorkshire), and Eastern England (Lincolnshire, East Anglia) in which trial varieties are grown at a range of different nitrogen levels.

The types of wheat best suited for alcohol production are soft winter wheat varieties. Durum and hard red spring wheats are generally not suitable for alcohol production, due to the lower starch content and resultant low yield of alcohol (Stark et al., 1943). Distillers prefer to use soft wheat varieties because they tend to give fewer processing problems in distilleries (Brosnan et al., 1999; Brown, 1990; Riffkin et al., 1990; Taylor and Roscrow, 1990). Hard wheats are generally associated with higher-viscosity worts than soft wheats (irrespective of protein content) (Brown, 1990) and tend to increase problems in important rate-determining areas of the process, such as the transfer of worts, and in co-products streams (centrifugation and/ or evaporation of spent wash). Because of the high degree of production integration and co-product recovery streams, significant spent wash viscosity can effectively stop the process, resulting in serious and expensive downtime.

Wheat and maize have different processing characteristics, and these have important implications for the handling of distillery co-products, such as spent grains and spent wash. Some grain distillers dry their spent grains and spent wash, then combine them to produce animal feed. Co-products deriving from maize dry more easily and efficiently than wheat, as maize spent wash is less viscous than that from wheat. As a result, wheat has been reported to cause more processing problems because spent grains and spent wash deriving from wheat can cause a residue build-up on heat exchangers and evaporators, which reduces efficiency and may cause processing down time (Newton et al., 1995). Wheat is also associated with process viscosity problems, which are believed to be due to a number of different factors, including the gluten content and pentosan polymers such as arabinoxylans. Wheat endosperm cell walls contain around 75% arabinoxylans, while maize contains about 25% (Newton et al., 1995). In contrast to maize, certain wheat types and cultivars (varieties) are associated with high viscosity, particularly hard wheat varieties, and even individual soft wheat varieties have different viscosity characteristics that can affect their suitability for processing in a grain distillery.

Some of the wheat properties are influenced primarily by genetic factors and are inherent in the varieties of wheat produced. Others are governed by environmental factors, such as soil fertility, rainfall, and temperature, during both the growing season and at harvest (Orth and Shellenberger, 1988).

Although the change from maize to wheat has been generally considered not to have affected the quality of the final spirit, some processing adjustments have been necessary (Nicol, 1990), and some distillers believe that wheat spirit is lighter bodied than that distilled from maize.

AHDB maintains a Recommended List for winter wheat and other cereals (AHDB, 2020) that is designed to encourage farmers to grow a specific range of wheat varieties, including those that are suitable for distilling. Scotland's Rural College (SRUC) provides a similar annual list of winter wheat varieties and other cereals (SRUC, 2020) that are of economic and agronomic importance in Scotland. Together, these provide guidance in terms of both agronomic performance and end-user quality and indicate the suitability of particular varieties for individual end-user markets, such as for distilling. As a result, distillers can be confident that wheat of suitable quality will continue to be available from the market. In the UK, winter wheat varieties are officially classified into four groups by the National Association of British and Irish Millers (nabim) (Table 10.3). Group 1 and 2 hard wheats are targeted at milling and baking end users and are considered unsuitable for distilling. Groups 3 and 4 contain the varieties that are more suited for distilling; however, only certain varieties within these groups will give acceptable distilling performance. Group 3, which is soft wheat, is mainly targeted at the biscuit and cake-making market, although several important distilling varieties are included in this category. This is important because the usefulness of a particular variety in a number of markets will contribute to the overall attractiveness of this variety to growers, as well as end users. Group 4 is split between hard and soft varieties, and distillers will only show interest in soft varieties in this group, which is mainly comprised of feed varieties but also includes some of the more high-performance soft grain distilling varieties, such as Viscount and more recently LG-Skyscraper.

TABLE 10.3   National association of British and Irish millers classification for UK winter wheat.

NABIM group	Type	Principal applications	Notes
1	Hard	Milling and baking	Premium quality, consistent milling, and baking performance. Meets specific quality requirements for specific weight (76 kg/hL), protein (13%), and HFN (~250). Varieties not interchangeable; depend on end-user requirements
2	Hard	Milling and baking	Variable premium breadmaking potential. Some consistent but not as good as Group 1. Others perform inconsistently or are suited to specialist flours. Lower protein than Group 1
3	Soft	Biscuit, cake, distilling	Soft varieties for biscuit, cake, and other flours; will also include some distilling varieties. Quality requirements include soft milling characteristics, low protein, good extraction rates, and an extensible but not elastic gluten
4s/4h	Soft/ hard	Distilling (*soft only*), general purpose, feed	Nonpremium varieties grown mainly as feed varieties but may be used by millers for some general-purpose grists. Distillers will use certain *soft* varieties (will not usually use *hard* varieties)

Adapted from Bringhurst, T.A., Agu, R.C., Brosnan, J.M., 2012a. Creating better cereal varieties for the sustainability of the distilling industry. In: Walker, G.M., Goodall, I., Fotheringham, R., Murray, D. (Eds.), Distilled Spirits: Science and Sustainability. Nottingham University Press, Nottingham, UK, pp. 77–86.; Bringhurst, T.A., Brosnan, J.M., Thomas, W.T.B., 2012b. New approach to barley breeding. AGOUEB—the association genetics of UK elite barley project. Brewer Dist. Int. 8 (1), 25–28.

Wheat specifications

In general, distillers require inexpensive soft winter wheat, which is easy to process and which will give an acceptable alcohol yield. In broad terms, the requirement for distilling wheat would be similar to a specification for feed wheat. This is in contrast to that for bread production, where millers require high-quality, hard wheat with high nitrogen and gluten content and relatively low levels of starch, and are prepared to pay a premium for this. Wheat supplied at a bread-making specification would normally be unsuitable for grain distillers, as this could cause stickiness problems in distilleries, thus reducing process efficiency and potentially giving a low alcohol yield. This means that distillers' requirements for wheat are not in competition with those of bread millers and distillers can target the 'best of the rest' (Brown, 1990). This allows Scotch whisky distillers to buy suitable wheat at more economically favourable commodity prices.

The general requirements of wheat for Scotch whisky grain distilling are well established (Agu et al., 2006, 2008b; Bringhurst et al., 2003; Brosnan et al., 1999), and these are primarily soft endosperm texture, high levels of starch, and low total nitrogen (protein) and high alcohol yield. Several other parameters have been identified as being important in defining distilling quality, including moisture, high specific weight, corn size, hardness, and screenings. The most important defining factor is the strong inverse relationship between alcohol yield and protein content. In most cases, variety is not specified, although some distillers might have a preferred list of varieties. However, the agronomic system has developed in such a way that the grower's market in Scotland strongly reflects the end-user requirements of distillers.

Soft low-nitrogen wheat for grain distilling is normally traded as a commodity; for this reason, quality specifications are kept to a minimum. Supplies for distilleries are normally obtained from quality-assured sources, such as those covered by a traceable on-farm assurance passport, such as from Scottish Quality Crops (SQC), either from seed merchants or in some cases directly from local farms. Scotch whisky distilleries largely obtain their supplies from within Scotland, and can use more than 70% of the Scottish soft winter wheat crop (Brosnan et al., 2010). Because grain distilleries are often located in the central and southern areas of Scotland, significant amounts of wheat will also be sourced from the north of England. The relative proportions of these can change from year to year, depending on availability and cost of suitable supplies.

Individual distillery companies may differ slightly with regard to their specific delivery specifications but, in general, the main quality parameters that are assessed are moisture and total nitrogen, both of which are well suited to rapid near-infrared (NIR) analysis at intake. In some cases, NIR may also be used to screen wheat supplies for predicted alcohol yield (PAY). Distillers will also analyse other parameters such as specific weight, hardness, and screenings that will all impact alcohol yield performance and process efficiency. They will also assess for soundness (e.g. sprouting), other contamination, and unwholesome off-notes and will also carry out visual checks for moulds (e.g. fusarium red, ergot black). Distillers will also arrange (or request) routine due diligence analysis to confirm the absence of mycotoxins and other undesirable trace elements.

Specific weight is the weight of a known volume of grain (expressed in kg/hL). This was originally referred to as the bushel weight and is broadly similar in concept to the thousand corn weight used for barley. The higher the specific weight, the more starch (and protein) it contains. Grain distilleries usually set specifications at no lower than 72 kg/hL (Brown, 1990), but this may vary depending on availability and ambient harvest conditions. The size of kernels or screenings can also be specified, as this may be important in some distillery processes. If the grains are too small, they can pass through mills and enter the process whole, which could mean that not all of the available starch will be extracted, possibly leading to decreased alcohol yields. There also appears to be a relationship between corn size and alcohol yield (Agu et al., 2008b), which favours wheat supplies with larger kernel sizes.

Advances in modern (non-GM) genetics have identified some of the genes associated with certain phenotype traits that are important to end users, and now this information is used actively by plant breeders to accelerate this process using classical non-GM breeding techniques (Sylvester-Bradley et al., 2010). One of the aims of modern research is to develop new varieties that will require less nitrogen and agronomic treatments, which will reduce the economic and agronomic costs of growing and supporting wheat varieties that are better suited for the UK market and provide support for the long-term sustainability of the cereals supply chain (Sylvester-Bradley et al., 2010). These initiatives are now resulting in the emergence of more new high-quality wheat varieties with strong potential for use by grain distillers (Bringhurst et al., 2012a) and are attractive to growers as a result of better agronomic performance and improved disease resistance. These developments make a strong positive contribution to sustainability, both for the distilling industry and within the cereal supply chain, and will help to stimulate plant breeders to provide continuing improvement of modern wheat varieties.

Selection of new distilling wheat varieties

For many years, the wheat cultivar Riband was the distilling wheat market leader (Lea, 2001) but was succeeded by agronomically improved varieties with acceptable distilling properties. The variety Viscount proved very popular as a distilling variety for its consistently high alcohol yield over many seasons but has now been overtaken by, at the time of writing, newer varieties such as LG-Skyscraper. This process is designed to assess and identify new wheat varieties that are better suited to the UK market in terms of agronomic performance and end-user quality.

Each year, the AHDB, in conjunction with the British Society of Plant Breeders, the National Institute for Agricultural Botany (NIAB), and Scotland's Rural College, conducts a series of trials to evaluate new wheat varieties, not just for distillers but also for other end users, including flour millers, bread makers, and biscuit manufacturers, as well as for use as animal feed. Plant breeders initially submit new wheat varieties for the UK market into National Level (NL) trials over 2 years (NL1 and NL2). If these trials are successful, then named, new varieties will move on to full Recommended List (RL) trials, normally 1 year later. Wheat trials are carried out in various sites every year, in many different locations throughout the UK. This allows new wheat varieties to be tested in a range of environments, which adds to the robustness of the trials process and also helps identify any particular regional or market suitability. In the final stages of the recommendation process, certain varieties will have gained provisional or full approval, and other varieties will be rejected. New varieties are not approved until they have completed at least 3 years of trials.

The AHDB Wheat Committee uses the outcome of the RL trials to select the most promising new wheat varieties for various end-user groups and place them on the annual AHDB and SRUC Recommended Lists. The most important parameters influencing inclusion on the Recommended Lists are related to agronomic factors such as agronomic yield, tendency to sprouting damage (i.e. Hagberg falling number, or HFN), disease resistance, and straw strength (lodging). However, end-user quality data and suitability for particular applications (milling, biscuit/cake making, distilling) have an important role to play in the recommendation process. Every year, distillation performance data (based on alcohol yield and viscosity) are supplied by the Scotch Whisky Research Institute to identify those new varieties that are suitable for distilling.

These trials have shown, on the basis of their performance in distilling quality tests, that certain wheat varieties have a consistent tendency to give higher potential alcohol yield results than other varieties. Under the current quality ratings for distilling wheat (Bringhurst et al., 2012a), established soft wheat varieties consistently providing good distilling performance are rated as 'good', those with acceptable but average laboratory results (or distillery performance) would be rated as 'medium', and those that are significantly below average performance or show potential viscosity issues are rated as 'poor'. The 'poor' category includes all hard wheat and any soft wheat varieties carrying the 1B1R rye gene, which are both associated with high viscosity and poor processing performance and which can give low alcohol yields.

The key attribute of a 'good' soft wheat variety for distilling is its consistency in producing low-nitrogen grain, which provides a high alcohol yield relative to other varieties coupled with good agronomic performance (Brosnan et al., 1999). Ideally, distillers prefer to target varieties with a 'good' distilling rating, such as Viscount, which normally gives very high alcohol yields and good processing characteristics. However, they can be willing to accept 'medium' varieties at lower alcohol yield if the 'good' ones are in short supply and the price is right, provided distillery processing performance is not seriously compromised.

Consultations between the distilling industry and plant breeders to find new varieties for distilling are continuing, and research to evaluate the alcohol yield potential of new varieties of wheat as they process through the AHDB recommendation system enables Scotch whisky distillers to inform plant breeders and farmers about which new varieties are suitable for distilling in addition to, crucially, identifying those that are undesirable for distillers. The winter wheat variety Warrior is a good example of a Recommended List variety that is unsuitable for distilling. Although this variety has agronomic advantages, it is consistently associated with both poor alcohol yield and high viscosity, which are strong negative factors in distillery processing. Warrior was particularly poor but other soft wheat varieties show consistently depressed alcohol yield and would be best avoided by distillers seeking the maximum efficiency.

One of the major benefits of the RL trials system is that in recent years there has been a much wider choice of varieties that are better suited to distillers' (and other end users') requirements, and the resulting increased diversity can help to address the supply chain sustainability agenda by spreading the risks for farmers and growers, as well as supporting end-user requirements. Recent collaborative research between scientists at Rothamsted Research, the Scotch Whisky Research Institute, and the plant breeder Limagrain have investigated ways to improve distilling wheat quality in the future by lowering viscosity to enable easier processing (BBSRC, 2018).

Cereals processing

Before looking at some of these processes in more detail, it is important to be aware of some of the basic biochemistry of cereals processing and how this is applied to the processing of raw materials in the distillery. The main areas of interest are the structure and composition of starch, starch gelatinisation and retrogradation, starch-degrading enzymes, and their action. All of these have a direct impact on the way that cereals are treated during processing and the subsequent conversion of starch into fermentable sugars.

Starch structure

The efficient conversion of starch into ethanol is the major determinant of distillery efficiency, as the primary objective of grain distilling is to produce as much alcohol as possible from the raw materials. The cost of the raw materials is a major component of the overall costs of running a distillery (Nicol, 1990), and maximising spirit yield is of fundamental importance to distillers. In order to achieve the maximum potential of the raw materials, distillers must have a good understanding of the structure and properties of cereal starch and the implications these have for processing this material into a fermentable substrate that can be converted into alcohol.

The general principles governing the structure and functions of starch have been well understood for many years, and in the present work, there is no intention to provide a comprehensive review of what is a very broad subject area, as this has been well documented and reviewed in many fundamental sources, such as Whistler et al. (1984), Pomeranz (1988), and MacGregor and Bhatty (1993). Kelsall and Piggott (2009) have also provided a concise, recent summary of the aspects that are relevant to distilleries. The purpose of this section is to provide a very brief overview of aspects of starch that are particularly relevant in defining processing characteristics of cereals used in the production of Scotch whisky. In the last decade, there have been some advances, particularly in our understanding of the fine structure of starch and in the detailed taxonomy of the groups of enzymes that degrade starch into fermentable sugars, but, in general, the principles that were outlined remain valid. Pérez and Bertoft (2010) have provided a fairly recent comprehensive review of starch structure and its components, and Bathgate and Bringhurst (2011) have presented an overview of the aspects that are relevant to the production of Scotch whisky.

It has already been discussed that the major sources of starch used for making Scotch grain whisky are wheat and maize (corn). The composition and structure of starch can vary for different cereals, which has important implications for the ways that these cereals are processed in the distillery, both in terms of conversion to fermentable sugars and maximising the efficiency of cereal throughput.

After cellulose, starch is the most abundant form of carbohydrate produced in plants. Starch is a condensation polymer of glucose found in all the major organs of plants and is the primary form of storage carbohydrates, providing a reserve food supply for periods of dormancy (Swinkels, 1985). In cereals such as wheat, barley, and maize, reserve starch is mainly stored in the starchy endosperm, where it is embedded in a protein matrix. Starch is laid down as granules (size range, 2–200 μm), which accumulate in organelles called *amyloplasts* in the endosperm cells. Wheat starch is comprised mainly of two major populations of starch granules: large lenticular A-type granules (20–35 μm) and small spherical B-type granules (2–10 μm). Normally, the number of small granules is very much greater than the large granules. However, the large granules, although they only account for about 12%–13% of the total number of granules, contain more than 90% of the total starch (Bathgate and Palmer, 1973; Shannon and Garwood, 1984). In contrast to wheat, maize starch granules are irregular and polyhedral in shape and are generally smaller, averaging up to 15 μm in diameter (Lynn et al., 1997). A range of starch granule sizes of 3–26 μm in maize has been reported in the literature (Swinkels, 1985). However, maize starch granules have a single size distribution (Cochrane, 2000), rather than the bimodal distribution associated with wheat.

The size and configuration of the starch granules have an important influence on parameters, such as the gelatinisation temperature, which determines the temperature required to efficiently process the cereal. The starch in small B-type granules is more tightly bound, has a higher gelatinisation temperature, and thus requires more vigorous conditions to be fully extracted compared to the larger A-type granules, where the starch is more accessible. The small granules tend to remain ungelatinised at normal mashing temperatures (Bathgate et al., 1973).

Although the small granules contain only a relatively small amount of starch, these are still significant in terms of overall yield, and it is essential that these are utilised efficiently to obtain acceptable alcohol yields. The relatively small and crystalline granules in maize require higher temperatures to gelatinise and release the starch from maize.

Starch granules also contain noncarbohydrate components such as proteins and lipids (Cochrane, 2000), which have potential implications for cereals processing. It has been suggested that the presence of lipids can reduce the susceptibility of the starch to amylolytic breakdown (Palmer, 1989). The presence of lipids is also an important factor influencing the propensity for starch to retrograde after cooking (Swinkels, 1985).

Starch itself is composed of two major fractions. Amylose is a linear molecule comprising long chains of α-(1,4)-linked glucose units and generally accounts for about 15%–37% of the total starch. Amylopectin, which has a highly branched structure, is comprised of a large number of relatively short α-(1,4)-linked chains linked to α-(1,6)-linked branches and makes up the bulk of the starch. Amylose is largely considered to have a regular left-handed helical structure, whereas amylopectin takes a largely crystalline form (French, 1984; Lineback and Rasper, 1988). The properties of starch are influenced by the relative amounts of both amylose and amylopectin (Fredriksson et al., 1998).

Amylose in wheat has a molecular weight of 105–106 Da with a chain length of about 2000 glucose units (Barnes, 1989). Amylose contains a relatively small number of α-(1,6) branches (about two to four chains per 1000 glucose units; two to eight branch points per molecule) (Hoover, 1995) but shows behaviour that is characteristic of a linear polymer (Lineback and Rasper, 1988).

On the other hand, amylopectin has one of the highest molecular weights associated with naturally occurring polymers (up to about 108 Da), and this is of the order of 1000 times that of amylose (Barnes, 1989). Amylopectin has a highly branched structure made up of a large number of relatively short glucose chains (10–60 units), with an average length of about 20–25 units (Cochrane, 2000), but overall the total chain length can be as high as 2×10^7 glucose units. The branch points account for about 5% of the total glucose units that are present

(Swinkels, 1985), accounting for about one branch point for every 20–25 residues (Hoover, 1995). Amylopectin is considered to be a major factor in determining the physical and chemical properties of starch (Tester, 1997).

The relative amounts of amylose and amylopectin are relatively constant for starches from a single source. Most starches contain 20%–30% amylose and 70%–80% amylopectin (Jane et al., 1999). However, different sources of starch have varying amylose-to-amylopectin ratios, with cereals such as wheat, maize, and sorghum having a significantly higher ratio of amylose (about 28%) compared with starches deriving from tubers and roots (potato, tapioca, and arrowroot), which contain about 20% amylose. Some starches, such as waxy maize, contain little or no amylose. More specialised cereals such as amylomaize can contain as much as 80% amylose (Swinkels, 1985).

The relative amounts of amylose and amylopectin and the distribution of A- and B-type granules have a strong influence on the physical and chemical properties of starch and are the important factors affecting the processing characteristics of cereals, such as gelatinisation temperature, viscosity, and the tendency for retrogradation or recrystallisation to occur as a result of high levels of amylose. These can have a serious effect on processing efficiency, as well as on alcohol yield.

The composition of the starch has an important influence on its breakdown by starch-degrading enzymes, such as α- and β-amylase, which by themselves are unable to degrade α-(1,6) glycoside links, and the breakdown of amylopectin by these enzymes results in the presence of α- and β-limit dextrins, as well as fermentable sugars such as maltose and maltotriose. The α-(1,6) links in α- and β-limit dextrins are degraded by limit dextrinase.

Gelatinisation

After deposition in starch granules, the starch is partly crystalline and is largely insoluble in water. In order to utilise the starch, it is necessary to disrupt the granular structure so it can absorb water (Evers and Stevens, 1985). Gelatinisation has been defined by Zobel (1984) as the process of swelling and hydration of starch granules so the starch can be solubilised. Normally, this is achieved by slurrying the starch in water and heating until the starch begins to melt. Ultimately, this results in a suspension containing amylose and amylopectin fragments, which are then amenable to the action of amylolytic (starch-degrading) enzymes, which can convert the solubilised starch into fermentable sugars (Palmer, 1986).

Gelatinisation takes place in several stages. Initially, when dry starch granules are exposed to excess water at low temperatures (0–40°C), they undergo limited reversible swelling (amorphous phase). As more heat is applied the crystalline starch begins to lose its integrity (melting phase) and, after combining with the noncrystalline fraction, undergoes irreversible swelling and hydration (French, 1984). This is associated with a substantial increase in viscosity, which has been attributed to the leaching of amylose from the granules. The swelling is accompanied by the disruption of the molecular order inside the starch granule, which is manifested as a loss of birefringence (Atwell et al., 1988; MacGregor and Fincher, 1993). This is a measure of the degree of order, or crystallinity, of starch granules when viewed under polarised light (Cochrane, 2000). The degree of swelling has been largely attributed to the effect of amylopectin (Fredriksson et al., 1998).

Gelatinisation of starch granules results in the dissociation and uncoiling of the helical regions of amylose and break-up of the crystalline structure of amylopectin, allowing the hydration and swelling of liberated amylopectin side chains (French, 1984). This causes the starch granule to swell, first allowing linear amylose to diffuse out of the granule and then ultimately resulting in a complete disruption of the granule structure.

There appears to be a negative correlation between the proportion of amylose and the onset of gelatinisation (Fredriksson et al., 1998); thus, as the relative amount of amylose increases, the temperature at the onset of gelatinisation decreases. Fig. 10.2 gives a visual impression of the physical changes that take place as starch granules begin to gelatinise; moving from left to right, as the temperature rises, the starch granules become less crystalline as they begin to swell and become disrupted as they undergo gelatinisation.

In the past, the standard method for measuring the gelatinisation temperature of cereals has been differential thermal analysis (Zobel, 1984) or differential scanning calorimetry (Atwell et al., 1988). However, the progress of gelatinisation has also been studied using various other techniques, such as the Brabender Micro Visco-Amylo-Graph (Zobel, 1984), which measures changes in viscosity (pasting) as a cereal slurry is subjected to a programmed temperature cycle. Pasting has been defined as the granular swelling and exudation of molecular components following gelatinisation (Atwell et al., 1988). A more modern variant of the Micro Visco-Amylo-Graph® is the Rapid Visco™ Analyser (RVA), which has been developed by Calibre Control, Inc. Fig. 10.3 shows an idealised RVA amylogram, or pasting curve, which is designed to illustrate the main stages in the gelatinisation (pasting) process.

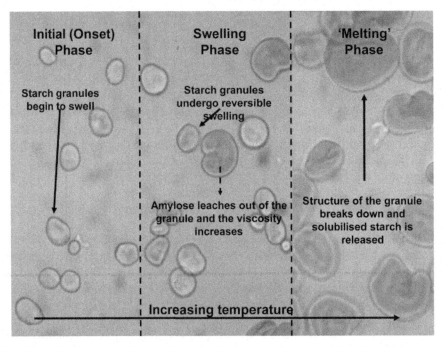

FIG. 10.2 Composite photograph showing the phase changes that occur in cereal starch as they undergo gelatinisation. As the temperature rises, they begin to swell and become less crystalline as they become disrupted. Prepared from a set of photographs supplied by M. P. Cochrane and used with permission.

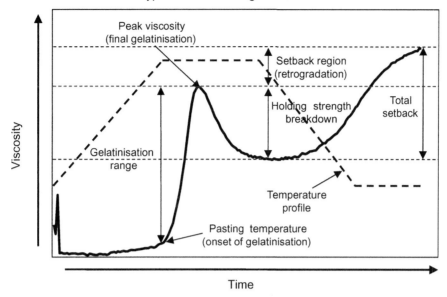

FIG. 10.3 A schematic representation of the main features of a pasting curve produced by a Rapid Visco Analyser. Figure contains some terminology from Dengate (1984).

The increase in viscosity observed as the temperature is increased to about 95–100°C shows that starch granules do not gelatinise at the same rate but do so over a large temperature range. This is dependent on the degree of crystallinity of areas within each granule, which results in considerable variation between different granules. In addition, small granules appear to gelatinise at a higher temperatures and over a wider range (MacGregor and Fincher, 1993). Jane et al. (1999) suggested that amylopectin, with its longer branch chain lengths, gives an increase in gelatinisation temperature. As the temperature is maintained, the viscosity begins to fall as the gelatinised starch is solubilised. When the temperature is reduced, the viscosity begins to rise to a much higher level than the previous peak as the gelatinised starch begins to agglomerate and recrystallise to form a resistant gel. This phenomenon is known as *setback* or *retrogradation*.

Gelatinisation temperatures for cereals are highly dependent on the method that is used to measure them (MacGregor and Fincher, 1993); however, the gelatinisation temperatures of wheat and maize are considerably different, with that for maize (70–80°C) being higher than that for wheat (52–54°C) (Palmer, 1989). This has important implications for the production of Scotch grain whisky as high-temperature conditions are required to process maize efficiently (Bathgate, 1989).

Starch hydrolysis

Because starch in unmalted cereals is generally not accessible, it is necessary to convert it into a useable form by heating or cooking it to gelatinise the starch. It is also possible to utilise starch effectively at lower temperatures by applying mechanical energy, such as by finely grinding or hammer-milling the cereal. In both cases, this facilitates access to the starch by malt enzymes contained within the mash.

In a malt distillery, the enzymes in malted barley, primarily α- and β-amylases, are able to convert starch and other oligosaccharides to fermentable substrates that can be converted by yeast to alcohol during fermentation (Robyt, 1984). Grain distillers use a small proportion of malted barley containing a high level of these starch-degrading enzymes, collectively known as dextrinising power (DP) (a mixture of α- and β-amylase), and diastatic units (DU) (α-amylase), converting unmalted cereal starch into fermentable sugars, including glucose, fructose, sucrose, maltose, and maltotriose, as well as a mixture of linear and branched dextrins and limit dextrins that are largely nonfermentable.

The α- and β-amylase degrade the α-(1,4) links in the α-glycoside chains of starch. The first of these, α-amylase, is the most important of the starch-degrading enzymes and is largely responsible for the degradation of starch into lower molecular weight dextrins and sugars (Muller, 1991). The major form of this enzyme is heat stable up to 70°C, although its activity declines at temperatures above 67°C (Briggs et al., 1981).

The second enzyme, β-amylase, is essential in hydrolysing unfermentable dextrins and oligosaccharides into fermentable sugars, primarily maltose and, to a lesser extent, maltotriose. The β-amylase is less heat stable than the α-amylase and is denatured at normal mashing temperatures; its activity completely disappears after 40–60 min at 65°C (Briggs et al., 1981). A temperature of 67°C will leave substantial levels of dextrins instead of efficiently converting them to maltose (Palmer, 1989). These facts underline the importance of careful temperature control during the conversion stage of both grain and malt distillery production.

The α-amylase is an endoenzyme that rapidly degrades the α-(1,4) bonds of starch molecules at random within the chains, producing a large number of progressively smaller oligosaccharides and dextrins. The α-amylase enzyme can attack ungelatinised starch granules, but will only do so very slowly. The linear smaller oligosaccharides and dextrins are in turn degraded into maltose by β-amylase, which is an exoenzyme that degrades starch residues by the stepwise release of maltose units from the nonreducing ends of the chains.

Fig. 10.4 illustrates the action of α- and β-amylase on starch. Neither of these enzymes can degrade the large number of α-(1,6) branch points associated with amylopectin, and significant amounts of branched residues or limit dextrins remain in the mash (MacGregor et al., 1999). These limit dextrins are degraded by a third enzyme, limit dextrinase, which specifically attacks the α-(1,6) branch points of amylopectin and branched oligosaccharides to produce smaller more linear components that can be further degraded by α- and β-amylase. Although it is considered that limit dextrinase will only degrade amylopectin itself very slowly and to a limited extent (Stenholm, 1997), the enzyme acts rapidly on limit dextrins produced by the action of α-amylase (MacGregor et al., 1999). More recent work (Walker et al., 2001) suggests that the bulk of the limit dextrinase passes through the mashing stage in an inactive bound form and is released later in the process during the early stages of fermentation, when the active enzyme is released and is available to hydrolyse branched dextrins (Bringhurst et al., 2001; Bryce et al., 2004).

The three major enzymes—α-amylase, β-amylase, and limit dextrinase—work together at mashing temperatures (62–65°C) to progressively degrade starch, first into large oligosaccharides that are in turn degraded to a mixture of fermentable sugars, primarily maltose and maltotriose, and (mainly) branched dextrins. High levels of unconverted dextrins in the wort

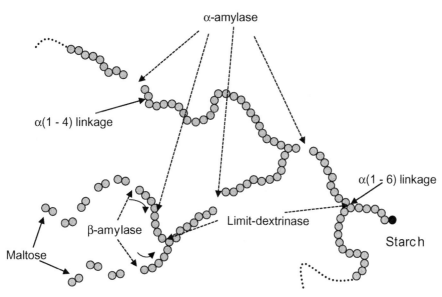

FIG. 10.4 Degradation of starch by α- and β-amylase and limit dextrinase.

normally indicate problems with enzyme hydrolysis. A fourth enzyme, α-glucosidase, is generally associated with the mobilisation of starch during germination (Sun and Henson, 1992), but may also have a minor role in mashing (Agu and Palmer, 1997). This enzyme can release glucose units from a variety of other α-glucosides and small dextrins (Fincher and Stone, 1993); however, the overall contribution of this to the degradation of starch during mashing/ conversion has not yet been fully established (MacGregor, 1991).

In contrast to brewing, the wort is not boiled and the starch-degrading enzymes are able to survive into fermentation, where they have an important role in the further hydrolysis of dextrins during fermentation, thus maximising the amount of fermentable substrate that is potentially available for conversion into alcohol (Bringhurst et al., 2001; Bryce et al., 2004).

Proteolysis

Another major function of conversion is that of proteolysis, by which proteins are broken down into amino acids and other low-molecular-weight nitrogen-bearing protein fragments (Boivin and Martel, 1991). These supply a source of essential nutrients that can be readily utilised by yeast, allowing it to grow and operate efficiently during fermentation.

Proteolytic enzymes, particularly in the area of endopeptidases, are probably the least well-understood part of the mashing (conversion) process (Bamforth and Quain, 1989). However, it is generally accepted that the formation of soluble nitrogenous materials in wort is primarily a result of the action of heat-stable endoproteinases working in conjunction with heat-stable carboxypeptidases (Boivin and Martel, 1991; Briggs et al., 1981), which are able to tolerate the temperatures associated with mashing and conversion (up to 65°C) (Bamforth and Quain, 1989).

The proteolytic enzymes include a mixture of proteinases that are endoenzymes and peptidases that are exoenzymes. Endoproteinases (endopeptidases) break the internal peptide links of polypeptides (proteins) at random, to produce smaller molecules, which are in turn broken down by carboxypeptidases, which are exoenzymes that remove amino acids stepwise from the carboxyl end of the chains (Bamforth and Quain, 1989). However, many proteolytic enzymes, such as aminopeptidases, as well as some endoproteinases, are heat labile above 55°C and are inactivated by the conditions encountered during mashing (Briggs et al., 1981; Jones and Marinac, 2002) and would not contribute significantly to the overall soluble nitrogen content of the wort. Hence, it is generally considered that the bulk of the free amino nitrogen (FAN) in grain distilling wort originates from malt proteolysis, because under normal cooking and conversion conditions the proteolytic enzymes are inactivated and will not operate on the proteins in the unmalted cereal.

Malt distillery processing

The malt distillery

The malt distillery is the fundamental production unit for Scotch malt whisky. Malt distilleries can vary considerably in both size and technology, although all of them share the main basic features. Bathgate (2003) highlighted the traditional nature of the process and emphasised that the basic technology underlying the malt distillery process has changed little since the nineteenth century. Of course, there have been many refinements and improvements to both the process and equipment since then, which have resulted in better process control and higher efficiency. Fig. 10.5 shows the general layout of a typical Scotch malt whisky distillery highlighting the main features of the process. The main inputs to the process are water,

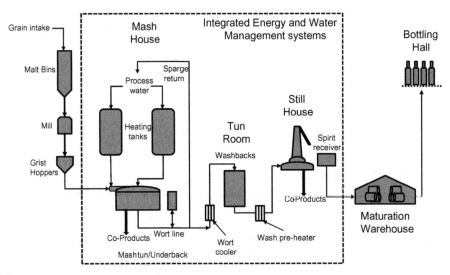

FIG. 10.5   General layout of a malt distillery.

energy, and barley malt. In modern practice, energy and water management systems are to some extent integrated within the distillery process. These may range from very small and basic heat and water recovery systems in older and smaller distilleries to fully integrated energy and water management systems in larger and more modern plants.

The main parts of the malt distillery process are the *grain intake*, where the grain is delivered, cleaned, and distributed to the malt bins in readiness for the production process; *milling*, where the malted barley is crushed to give a suitable grist for the mashing process; and *mashing*, where the grist is mixed with hot water and extracted and filtered to give a liquor (wort) that is cooled, pitched, with yeast, and added to the fermenters (washbacks). After fermentation, the wash is distilled into low wines and then into spirit, as described elsewhere in this publication. After distillation, the new-make spirit is matured in oak casks in a maturation warehouse for at least 3 years, more typically 8–12 years, but sometimes much longer (>20 years) for a premium malt whisky. After maturation, the finished whisky is shipped to a bottling hall, where the strength is reduced to bottling strength (normally 40% or 43% abv) and is then bottled as a single-malt whisky, married with other malt whiskies as a blended malt, or blended with grain whisky and other malt whiskies as a blended whisky prior to being placed on the domestic or export market.

Although the processes in each malt distillery are very similar, there is a very wide range of malt distilleries. Some are very small (500 LPA), and others are very large (>20 million LPA) (Gray, 2019). Some distilleries have very traditional features, such as floor maltings and traditional mash tuns, while others have more modern state-of-the-art equipment, such as full lauter tuns and highly integrated energy transfer/recovery systems. Some distilleries are old and traditional, retaining many of the original craft features, which are interesting from a historical perspective. These contribute to the iconic imagery traditionally associated with Scotch whisky, even though the production levels in such a unit may be small. Some more traditional distilleries still use a simple boiler house producing steam to provide energy and power to heat and power the still house.

Modern energy recovery and management systems that are fully integrated with the distillery process allow the distillery to operate more efficiently by recovering and moving heat (energy) from one part of the process to another, in addition to reducing water usage and limiting discharge into the environment. Examples of energy recovery systems are high-temperature heat pumps in the still house. Distillation is an energy-intensive process, with 90% of the thermal energy in a malt distillery being used in the still house (Ludford-Brooks and Davies, 2015). Thermal vapour recompression and mechanical vapour recompression are the primary methods currently available that can be used to aid in the recovery of energy from hot distillate in the condenser.

Processing of malted barley

Grain intake

Most malted barley is supplied to malt distilleries by commercial malting companies (sales maltsters) and is shipped by the lorry load (~20–30 tonnes per load) to the distillery, where it is weighed, visually checked, and tested for quality using a range of simple and rapid tests before discharging and conveying, either by conveyer belt or grain elevator

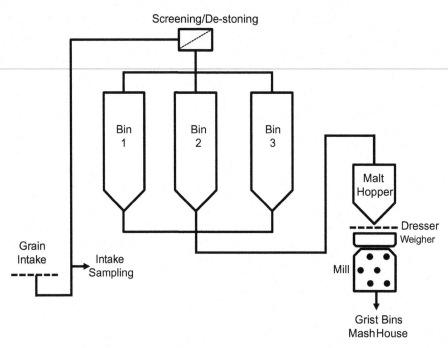

FIG. 10.6   Malt distillery grain intake.

through the malt intake system (Fig. 10.6). Before entering the malt bins, the grain is passed through a screening and sieving system to remove dust, small corns, and other fine debris. An important part of this system is a vibrating de-stoner, which separates stones and other heavy debris from the grains. Metallic debris is normally removed by passing the grain stream through a strong magnetic field. If such debris is not removed, there is a high probability of damage to the plant and a serious risk of fire or explosion as the materials pass through the process.

When barley malt is delivered, each load is sampled, either manually using sampling probes or mechanically using a grab or automatic trickle sampler. Multiple representative samples are subjected to a series of quality checks. Before accepting delivery, each batch is analysed using simple, rapid tests to ensure that it is wholesome, of acceptable delivery quality, and free from taints. The most important intake tests include moisture and nitrogen (protein) content, which are often measured using near-infrared (NIR) spectrophotometric analysis, and thousand corn weight and screenings (2.8-, 2.5-, 2.2-mm sieves). These tests are used to accept or reject the delivery.

If acceptable, the load is discharged into the grain intake system and distributed to the next available malt bin. Unless the distiller wants to segregate an individual variety, quality grade, or peating level, depending on the available storage capacity, loads may simply be added to the existing content of the malt bin. Depending on availability, quality, and price, distillers may receive supplies of a single variety or a mixture of varieties and malt batches blended by the maltster to meet a particular quality specification.

At intake, additional material samples are taken and sent to the laboratory for further analysis to ensure that the material reflects the required quality specification. Normally, samples are retained for a specified period at the distillery for future reference in case of a quality dispute. If there is a significant deviation from the agreed specification, the distiller may seek compensation.

Malt processing

Cereal processing in a malt distillery can be divided into three main stages: milling, mashing/conversion, and wort separation. In milling, the barley malt is crushed into a grist in which the composition is suitable for processing in the distillery process. Various distilleries and mash tuns require different grist compositions. During mashing, the grist is mixed with hot mashing liquor (sparge), which is usually recovered from the previous mash and transferred to a mash tun (traditional, semilauter, or lauter tun) where the starch is gelatinised and converted to fermentable sugars by starch-degrading enzymes present in the malt at the standard mashing temperature (63.5–64°C). The final stage is wort separation, where the sweet wort is filtered from the grains in the mash tun and then pumped to the fermenter. The remaining grains in the mash tun are then extracted with second and third waters at higher temperatures—typically 70–75°C (second water) and 80–90°C (third water and/or fourth water)—to solubilise the remaining starch and sugars. The second worts are normally used to fill the washback, while the third water is returned as sparge to the heating tank for the next mash. At the end of the mashing cycle, the spent grains are discharged. These are used for animal feed, and the mash tun is generally cleaned with water prior to the next mash. In modern practice, in some cases, the spent grains may be further processed to provide renewable fuels—for example by using an anaerobic digestion system. In addition, research continues to look for alternative uses that can add value to this co-product (Harrison and Brosnan, 2017).

Because the mashing and fermentation stages are not carried out under sterile conditions, effective cleaning is a critical part of the distillery process, as it prevents the build-up of microbial infections that can result in a significant loss of potential alcohol yield. Periodically, usually once or twice a week, the mash tun and ancillary pipework is thoroughly cleaned with a caustic solution or a similar cleaning agent and the mash tun plates are lifted for cleaning. A more modern approach is to use self-contained, automatic cleaning in place (CIP) systems to regularly and thoroughly clean the vessels and pipework. These normally include effective spraying systems, which will give more thorough coverage for surfaces that otherwise are normally difficult to access. These may include an underplate spray system for cleaning the lower surfaces of the mash tun plates (see Chapter 22 for more on sanitation).

Principles of milling

Milling is a critical part of the malt distillery process, and the composition of the grist plays an important role in the efficiency of wort separation during mashing. Although no unbroken grains should survive milling, it is essential to minimise damage to the husk fraction, as this provides the main filter medium in the mash tun. An intact husk will help maximise the extraction efficiency by reducing the wort run-off time. The composition of the coarse and fine fractions is important in determining both the extraction efficiency and wort separation, as high levels of flour will result in a potential loss of extract by allowing material to escape through the bottom of the mash tun, and excessive levels of fines can tend to block the mash tun plates and reduce the separation from the spent grains. Hence, it is essential to determine the optimum or best practical grist composition for the particular mashing system used in each distillery. Briggs et al. (2004) emphasise that optimal milling is not easy to achieve and that it is sometimes necessary to adjust the mill settings when changing between different types or batches of malt.

It is important to routinely check the grist composition if there are significant changes in the supply of malt to the distillery or when changing between batches of barley malt. This may require some adjustment to the mill settings, particularly if using barley from a number of different suppliers, which may vary in their degree of modification. Undermodified malt requires fine grinding compared with well-modified malts, which will readily shatter into smaller fragments. Seasonal and environmental variations, as well as the use and availability of different varieties, are also important factors as these can affect the malting characteristics of the barley and the resulting modification of the malt.

Grist composition is largely monitored manually using a grist box or some other sieving system. Multiple samples of grist are normally obtained from the bottom of the mill, along the length of the rollers, while the mill is running. These are usually analysed individually by shaking in a grist separation box (1.98-mm and 0.212-mm mesh) (Stevenson Reeves, Ltd., Edinburgh, Scotland) (or similar set of sieves), and the weight of each of the sieve fractions gives the relative proportions of husk, middles, and flour fractions. Dolan (2003) suggested an optimum grist for a well-modified distilling malt of 20% husk, 70% grits (middles), and 10% flour for a traditional mash tun, while that for a semilauter or lauter tun would be somewhat finer. However, mash tuns have different characteristics, and the actual grist specification will depend very much on the scope of the equipment being used. Excessive levels of fine flour may give problems with wort separation.

Some authorities cite a wider range of different sieve seizes than the simple grist separator that has traditionally been used by distillers. Briggs et al. (2004) highlighted the EBC/Pfungstadt sieving system that is typically used in brewing and which gives a much more precise range of mesh sizes associated with conventional and lauter tuns. Table 10.4 gives a comparison of the differences and shows the typically finer grind used for a lauter tun. Comparative data for a (brewing) mash filter show a much finer grist. The arrangement of the sieves (Table 10.4) gives a more precise indication of the composition of the grist from the mill. Although there is some overlap between these values and the traditional distiller's grist separation box, which separates only the coarse, middle, and flour fractions, it is difficult to correlate them precisely. Roughly, the coarse/husk fraction would be similar to that held on sieve number 1, while the middles would be the combined total of sieves 2, 3, 4, and 5, and the flour would approximate the material passing the smallest sieve (5).

However, there is no specific correspondence between the two systems, and it is important for the distiller to apply one or the other consistently to understand the necessary performance limits of the individual distillery and the distillers' own experience in matching the grist composition to the operational requirements of a particular mash tun. Doing so will allow the distiller to respond to changes in the malt supply in an effective way, ensuring that the distillery mash tun is operating efficiently and consistently within the confines of the available plant.

TABLE 10.4   Grist composition associated with mash tuns and a conventional mash filter.

Sieve number	Mesh (mm)	Fraction	Mash tun	Lauter tun	Mash filter
1	1.27	Husk	27	18–25	8–12
2	0.26	Coarse grits	9	8–<10	3–6
3	0.15	Fine grits I	24	35	15–25
4	0.07	Fine grits II	18	21	35–45
5	0.04	Flour	14	7	8–11
Tray	–	Fine flour	8	11–<15	12–18

Adapted from Briggs, D.E., Hough, J.S., Stevens, R., and Young, T.W. (Eds), 1981. Malting and brewing science. Vol. 1. Malt and Sweet Wort, second ed. Chapman & Hall, London; Briggs, D.E., Boulton, C.A., Brookes, P.A., Stevens, R., 2004. Brewing Science and Practice. Woodhead Publishing, Cambridge, UK.

The milling process

Prior to entering the distillery process, barley malt is transferred from the malt bins into a smaller malt hopper, from which it is dispatched to the mill after passing through a further de-stoner and screens to remove any remaining stones, debris, culms, and dust. This is performed via a weighing trap, which discharges repeatedly by tipping a fixed weight of malt (typically around 25–50 kg) into the mill feed. Normally, the distiller will set a fixed number of 'tips' (or 'coups') to give the required weight of grist for the mash. Typically, this would be around 10 tonnes per mash, but this can vary considerably depending on the size and capacity of the equipment in an individual distillery. Typical throughput through a four-roller mill would be around 2–6 kg/h/mm roll length (Briggs et al., 2004).

In a roller mill, cereal grains are compressed as they pass between sets of rollers. The use of a roller mill provides a relatively gentle separation of the grist and leaves the husk fraction relatively undamaged, which renders them particularly suitable for use in processes requiring the separation of wort in a traditional or lauter tun, as the husk is able to act as a filter bed during mashing (Kelsall and Lyons, 1999).

Milling equipment can vary for different distilleries, but variations on the four-row type shown in Fig. 10.7 are in common use. Some distilleries can use six-row mills, which are more versatile and can deliver a finer grist and are more suited to larger distilleries (and breweries). Dolan (2003) suggests that the latter are not essential for handling modern well-modified malt.

In the four-row mill, the malt enters the mill via a feed roll, which directs the flow so that individual grains are presented 'end on' through a pair of adjustable spring-loaded rollers (normally fluted), generally with a diameter of around 250 mm and a maximum working length of about 1500 mm. The main powered rollers normally operate between 250 and 500 rpm. These are designed to crush the grain along its length without seriously damaging the husk fraction. The space between the first set of rollers is set to give the appropriate initial grind (1.3–1.9 mm) to crack the malt kernels; the rollers should be checked regularly for wear and to ensure that they are parallel. Sometimes both rollers are driven, but in some cases one may be a follower driven by friction between the grain and the powered roller. The rollers may operate at different speeds in order to provide a shearing force, to give more efficient grinding of the grain.

4-Roll Mill

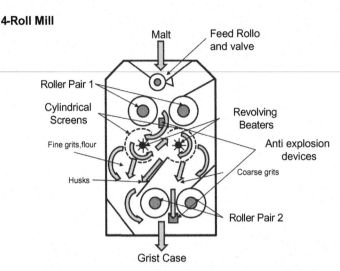

FIG. 10.7   Four-row mill. Adapted from Briggs, D.E., Boulton, C.A., Brookes, P.A., Stevens, R., 2004. Brewing Science and Practice. Woodhead Publishing, Cambridge, UK and Kelsall, D.R., Lyons, T.P., 2003. Grain dry milling and cooking procedures: extracting sugars in preparation for fermentation. In: Jacques, K.A., Lyons, T.P., Kelsall, D.R. (Eds.), The Alcohol Textbook, fourth ed. Nottingham University Press, Nottingham, UK, p. 11.

From the first pair of rollers, the cracked grist is partly separated into grist and husk fractions before passing through a second pair of rollers, which are set more closely together (0.3–1.0 mm). In some designs, the cracked grist is passed through a set of beaters, which direct the grist fractions through screens that separate the flour and fine grits from the coarse grits, which are then directed through the second pair of rollers for further crushing. The husk fractions fall more or less intact through the beaters, bypassing the second set of rollers and hence minimising any further damage before they join the rest of the grist in the grist case.

The screens are arranged so that the proportion of coarse and fine grits passing through them to the second set of rollers can be varied to fine-tune the final grist composition. In some mills, oscillating flatbed screens are used to provide similar results (Briggs et al., 2004). The mill also contains anti-explosion devices, which are shelves or dams set below the rollers designed to smother any sparks and minimise the risk of explosion from the dust generated as the grain is passing through the mill. As it is milled, the grist is collected in the grist case, from which it is transferred to the mashing machine and eventually into the mash tun.

Conditioning

In some cases, conditioning may be used to soften and reduce the brittleness of the husk material by dampening it for a short time (30 s to 1 min) with a small amount of low-pressure steam just prior to reaching the mill. This carefully controlled treatment can improve the survival of the husk fraction without affecting the milling characteristics of the starchy endosperm. It can also give other benefits, such as improving wort run-off times and improving the performance characteristics of a lauter tun by allowing it to be loaded more deeply, as it results in reduced bed density and increased porosity (Briggs et al., 2004). It has also been suggested that conditioned malt will give improved performance with better extract yield and attenuation and faster saccharification (Briggs et al., 2004). The use of conditioned malt requires the mill to be adjusted to give a closer roller gap to achieve the best extract. Conditioning will also increase the volume of the spent grains.

Hammer milling

Impact milling, using equipment such as a hammer mill, is rarely used by (conventional) malt distillers, who generally use relatively coarse grists. However, these are used more extensively in grain distilleries, where it is essential to have a relatively fine grind to handle unmalted cereals. One application would be in a malt distillery system where a mash filter is used. This system requires a very fine grist to operate efficiently. Currently, these are very rarely used in the production of Scotch malt whisky, but at the time of writing there are at least two distilleries operating a mash filter.

Mashing

Mashing is the stage of the malt distillery process where the malt grist is mixed with liquor, hydrated, and infused at a specific temperature (typically 63.5–64°C), so that the starch is gelatinised and malt enzymes are able to efficiently hydrolyse it into fermentable sugars. These are solubilised and extracted into the wort, which is separated from the grains, cooled, and transferred to the fermenter after pitching with yeast. After the first water has been drained off, the mash is extracted with further waters at a range of higher temperatures (typically 70–90°C).

Types of mash tun

Mashing is a batch process carried out using one of three main types of mash tun. The traditional mash tun, which is still sometimes found in older distilleries, is a cylindrical vessel often made from cast iron or steel (Bathgate, 2003) (Fig. 10.8). This relatively crude system, which originally would not be insulated or have a lid, featured large rotating paddles to mix the mash and a plough to remove the spent grains (draff) after mashing. In some very old traditional mash tuns, there was no plough, and the draff would have to be discharged by hand using a shovel. The mash tun would have a false bottom, or deck, with movable slotted gun-metal or steel plates or wedge wire supported above the base to support the mash bed. Over the years, the original open mash tuns were covered with a copper or steel lid and at least partially insulated to retain more heat. Originally, a traditional tun would not have any sparging system (Dolan, 2003), but eventually rotating sparge arms with relatively crude sprinkling systems were installed so the mash could be sparged with hot water.

A traditional tun would use an infusion mashing system, where the mash would stand for a fixed time period (normally up to 1 h) before draining to dryness. The grains would then be extracted with second water at a higher temperature (typically 70–75°C) and again with third water at >80°C and drained to dryness. Some traditional mash tuns used a fourth water at about 90°C (or above). In some cases, the third and fourth waters would be combined to give a composite third water at a relatively high temperature (>85°C).

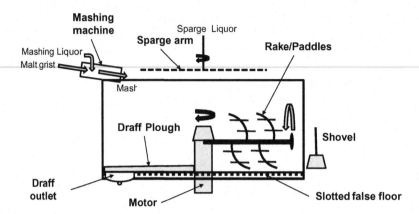

Traditional Mash Tun

FIG. 10.8   Traditional mash tun.

In general terms, the amount of each water can vary considerably, depending on the amount of barley malt used and the target original gravity, but the approximate ratio of the first, second, third, and fourth waters has been considered to be 4:2:2:4 m^3 per tonne of malt mashed (Bathgate, 1989; Dolan, 2003). The main drawback with a traditional mash tun is that the mashing process is relatively slow and inefficient, because the wort run-off is interrupted to allow for the addition of each water and subsequent mixing to remake the mash bed. The traditional mash tun was originally designed to work at relatively low original gravities, ranging from 1040 to 1050° IOB (10–13° Plato) (Dolan, 1976). In modern use, higher wort original gravities are more typical, and modern lauter and semilauter tuns are capable of working routinely at much higher gravities. An original wort gravity of up to 1080° (20° Plato) is achievable using the latest lauter technology. Higher gravities still could be achieved; however, downstream factors such as yeast temperature tolerance, and whether fermentation cooling is possible, become limiting factors. Though now uncommon, traditional mash tuns are still important in certain distilleries, as they are often considered to provide a unique character and contribute to the traditional image of Scotch malt whisky production.

Modern lauter and semilauter systems are much more efficient and flexible. These attributes are particularly important for expanding production as they allow distillers some flexibility to increase production levels without adding additional distillery capacity. The malt distillers' concept of a full lauter system is different from that of brewers, for which it generally refers to a system where the mash is mixed and infused in a separate vessel and then transferred to a lauter tun to separate the wort (O'Rourke, 2003). In a malt distillery, a full lauter tun is a high-performance mash tun in which the technology in the vessel is flexible enough to efficiently provide a very high extract and good run-off characteristics.

The most important type of mash tun currently used by malt distillers is probably the semilauter tun (Fig. 10.9), which Dolan (2003) defined as 'using a lauter tun with a traditional

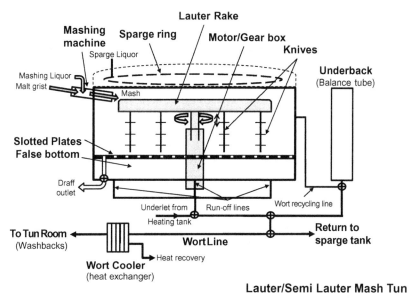

FIG. 10.9 Schematic of a lauter/semilauter mash tun.

infusion mash protocol'. However, in reality, the range of semi-lauter tuns is much broader than this and effectively covers everything between the traditional tun and the full lauter system. The main difference between a lauter (or semilauter) tun and a traditional tun is that the lauter system is designed to deliver a filtered wort faster and more efficiently than a traditional tun. The bed depth in a lauter tun is normally shallower, in the region of 0.5–1.0 m, compared with a traditional tun, where the bed depth is much higher (1.0–1.5 m). Instead of the traditional rakes, the lauter tun is fitted with a set of knives with angled vanes (shark fins) set at right angles to each blade; they can exert additional, gentle pressure on the mash bed, which helps to extract more of the liquid wort from the mash bed. The rake can be raised and lowered to maintain the rate of run-off. More advanced versions may have several forward and reverse gears to increase the pressure or if necessary help lift the bed if there are severe drainage problems. Where the traditional mash tun has slotted gun-metal or steel plates or wedge wire, more modern semi-lauter (and lauter) tuns use stainless-steel plates with lauter slots. In some cases, these are fixed in place, requiring some form of underplate cleaning to maintain them free from contamination.

Sparging is carried out through a series of high-efficiency sparge nozzles set within the sparge ring. These are designed to deliver a fine spray of very hot water that will help to elute soluble material, such as fermentable sugars, starch, oligosaccharides, and dextrins, efficiently from the mash.

Extraction efficiency can be similar on all mashing systems—between 0.5% and 1.0% residual extract in the spent grains is typical (D. Murray, 2020, personal communication, 12/08). However, because the liquid removal is much quicker using modern semilauter and lauter tun designs, the mash is cooled more rapidly. There is, therefore, more opportunity for the preservation of enzymes for further conversion of larger extracted saccharides to fermentable sugars during fermentation.

Mashing with a semilauter tun

Malt distilleries generally have two main heating tanks, one of which will contain clean water (i.e. a water tank) and the other containing sparge (i.e. a sparge tank), which is the weak (third) worts from the previous mash. Some distilleries keep each tank separate, while others interchange them. It is essential that these are cleaned regularly to avoid a build-up of bacterial contamination, which can have negative effects on production efficiency.

The weak worts are normally returned to the sparge tank at high temperature, directly from the wort line from the mash tun/underback. The sparge is normally held at a relatively high temperature (>70°C) to maintain it free from infection and prevent deterioration of the liquor. Because the sparge is recycled back to the tank at a temperature close to the final mash temperature, which may be around 90°C, it must be cooled before it is sent back to the mash tun. This can be accomplished by adding cold water to the heating tank or, more efficiently, by passing the sparge through a heat exchanger to reduce the temperature as it is returned to the sparge tank. This allows better control of the temperature, minimises the amount of additional water added to the tank, and allows the distillery to operate at a higher gravity. The heat removed from the sparge can then be used elsewhere in the process. In order to save energy, some distilleries may operate at lower (final) sparge temperatures, which means that there is less requirement for cooling before starting the next mash. The temperature in the sparge tank is normally monitored. This can be used to calculate the temperature of the mash when the malt is added, so the correct temperature is achieved in the mash as it enters the mash tun.

Underletting

The first stage in mashing is known as underletting and consists of adding hot mashing liquor to the mash tun to cover the surface of the plates. This has the effect of preheating the mash tun plates and helps minimise any cooling of the mash as it comes into contact with the plates. The presence of a thin layer of liquor also helps to prevent fine material from passing through the mash floor during mashing-in. The volume of liquor that is added as underlet will vary with the design and size of each individual mash tun.

Mashing-in

Many malt distilleries still use a variation on the traditional Steel's mashing machine to mix the malt grist with the mashing liquor (Fig. 10.10). Ultimately, more modern variants such as the vortex mixer (Fig. 10.11), which has no moving parts, will find increasing use in malt distilleries as well as in breweries.

In the steel's mashing machine, the grist is added from the grist case through a slide valve and mixed with hot liquor entering through the hot liquor inlet. When the grist strikes the hot liquor, it is thoroughly mixed by passing along a mixing screw, with a series of beaters, which drives the mash slurry along to the spout at the end of the mashing machine. The striking temperature and the final temperature of the slurry are critical in ensuring that mashing is carried out correctly. In a vortex mixer (Fig. 10.11), the same basic principle is used, and the grist is mixed with the mash liquor that is injected to provide a vortex, which effectively mixes the mash as it passes through the vessel.

Steel's Mashing Machine

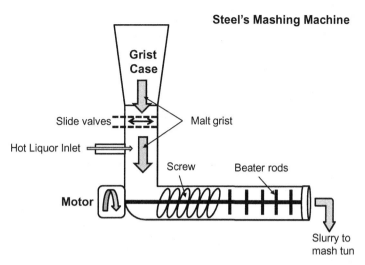

FIG. 10.10   Steel's mashing machine. Adapted from Briggs, D.E., Boulton, C.A., Brookes, P.A., Stevens, R., 2004. Brewing Science and Practice. Woodhead Publishing, Cambridge, UK

Vortex Mixer

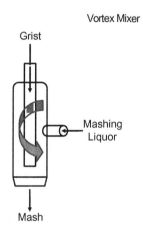

FIG. 10.11   Vortex mixer.

Originally, the temperature of the hot liquor was controlled manually by adding cold water to adjust the striking temperature (normally 68–70°C) to achieve the correct mash-in spout temperature (normally 63.5–64°C), which would be displayed from a sensor set at the end of the mashing machine. In more modern distilleries, this is more likely to be controlled automatically by using a feedback loop. Some distilleries may aim to minimise the input of cold water by adjusting the sparge liquor directly to the correct striking temperature using a heat exchanger. It is critical that the mash temperature remains constant at 63.5–64°C throughout mashing-in. Temperatures above 65°C will result in the rapid deactivation of important malt enzymes and will give suboptimum hydrolysis of starch and dextrins, resulting in incomplete fermentations. When the mashing-in is complete, the operator normally records the final mash temperature and the level in the mash tun. Normally, the sparge tank will be more or less empty when the mashing-in is complete and will then be filled up with fresh water and heated to the third water temperature.

Mashing-in takes between 20 and 30 min, depending on the amount of grist added and the size of the mash tun. Initially, the mash is not raked until the end of the mashing-in process, as raking too early will result in fine particles being forced through the mash tun slots, which may result in problems with subsequent wort run-off.

When the mashing-in is compete, the rakes are rotated quickly to make up the bed, which is then allowed to settle for about 10 min before carefully opening the valve to the underback and slacking to the underback (or hydrostatic balance tube), which will then fill by gravity until the liquor level balances that in the mash tun. The underback maintains the hydrostatic balance of the mash, preventing it from being drawn onto the mash tun plates as the wort is drained. The wort level in the under-back is normally balanced to reflect the level in the mash tun and any differential is a good indicator of any problems with wort separation.

Because a much finer grist can be used in a lauter tun, fine particles can escape though the lauter slots, and some distillers prefer to remove them by carefully recycling a small portion of the wort from the underback to the mash tun (vorlaufing). When this is complete, run-off commences and the wort is gently pumped though the wort cooler to cool it to the correct pitching temperature (normally around 16–19°C), prior to yeast inoculation, and then to the washback in the tun room, where it is fermented. The yeast may be injected into the wort line before it enters the washback, which reflects modern practice with creamed and/or propagated yeast, or it may be added to the fermentation vessel as bagged (pressed) or dried yeast.

When run-off is commenced, the mash is raked slowly. Initially, the mash floats around gently with the rakes but gradually slows down as the floating bed comes into contact with the plates. The knives slowly and gently cut through the bed, gradually squeezing out the liquor from the mash.

When the surface of the mash bed begins to show, the operator will normally begin to add the second water through the sparge nozzles. This is relatively time-consuming compared with simply adding it through the mashing spout but offers the advantage of evenly eluting the wort through the mash bed while gradually raising the mash temperature. Different distilleries may vary how they add the second water but it is important that the temperature of the mash is not increased too quickly before there is sufficient wort in the fermenter (washback) to supply enough enzymes to fully convert the dextrins in the wort into fermentable sugars and ensure that fermentation is complete. During mashing, it is essential to maintain a careful balance between the run-off rate, the level of wort in the washback, and the time available to achieve the required production levels. In practice, distillers will use their experience to develop a system that is suited to the constraints of the distillery and gives a consistent and reproducible procedure. This may simply be based on a fixed run-off time and/or flow rate. In more modern plants, the flow from the mash tun will be monitored and the second water will be added at a set point, once a specific volume of wort has been collected. Precision regarding the point of addition of the second water is essential to achieve consistent original gravity from mash to mash. Also, the timing and rate of addition of the second water are critical to allowing a gradual rise in the mash temperature as the hot liquor elutes through the mash.

If a proper lauter system is being used, draining will normally proceed without interruption while the second water is added and will then continue until the washback is full. The second water is normally clean hot water (85–95°C) from the heating tank, but, depending on the relative size of the vessels, it may contain sparge left over from the mashing-in stage. The third water is normally clean water added at a high temperature (85–95°C) from the hot water tank; it is often added to the mash as soon as the washback has been filled to the required level, when the flow of wort is simply diverted back to the hot liquor tank. The constant sparging with hot water retains the heat in the bed and allows the mash temperature to rise steadily to its maximum, which may be as high as 95°C, although some distilleries may choose to run at a lower temperature (e.g. 85°C) to save energy. In the later stages, the speed of the rakes will be increased. Together with the increase in mash temperature, this has the effect of efficiently eluting the extract from the mash and also helps to increase the run-off rate. Each mash tun will have its own protocol that has been adapted to the capacity and production requirements of the individual distillery.

In some semilauter systems, as with the traditional tun, draining is stopped while the second water is added and commences again when the mash bed is remade. In this case, the second water is added at a lower temperature (normally around 70–75°C), and draining recommences until the washback is filled. The third water (and sometimes fourth water) is added after stopping run-off when the wash back is full. Because sparging can be relatively slow, to save some time some distilleries may add the final water either partially or wholly through the mashing spout, but this generally requires the mash bed to be remade. The third water run-off is directed to the (sparge) heating tank after any temperature adjustment; the heating tank is then filled to the required level for the next mash.

When the mash is complete, draining is stopped and the spent grains (draff) are discharged from the mash tun, either by using a retractable plough or by adjusting the raking knives to direct them through the draff outlets. The spent grains are generally sold to local farms as animal feed, sent off to a central co-products processing plant to be incorporated into dark grains, or added to anaerobic digestion feed stock or some other waste recovery treatment.

A key consideration during mashing is to balance the liquor used during the mash to avoid fluctuations in wort concentration and to ensure that there are no shortages of liquor or that there is no excess that needs to be discarded.

Mash filter

Mash filtration is currently used routinely in brewing, but has not found general favour in malt distilleries, although two Scotch whisky distilleries are now known to be operating a mash filter. Bathgate (2003) mentioned that this technology has been used in the past but had shown no major advantages over conventional mashing systems using traditional and semilauter tuns. However, the technology continues to develop, so it would be expected that it could deliver further improvements in performance.

A mash filter is used together with a separate mashing/conversion vessel in which the mash is mixed and held at normal mashing temperature (64–65°C) to allow the malt enzymes to convert the starch into fermentable sugars before being transferred to the filter for the wort separation stage. Unlike a mash tun, the husk fraction is not necessary to provide a filter bed, and the use of modern high-pressure mash filters allows the use of finer hammer-milled flour. This gives the mash filter additional flexibility to handle barley malt that would otherwise be

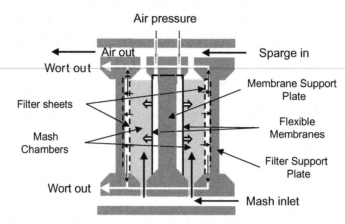

FIG. 10.12   Diagram showing the main features of a typical mash filtration module.

difficult to process. In addition, because the filter layers are normally very narrow (4–6 cm), the filtration rate is faster, and wort collection times can be shorter than those for a mash tun. However, for a mash filter to work efficiently, it must be completely filled and can only handle a fixed volume of mash, unless cumbersome blanking plates are fitted.

Individual mash filters vary, but generally they are composed of two main unit types (membrane supports and filter plates) contained within a chamber module (Fig. 10.12). Central to this is the membrane frame, which is a thin, grooved plate supporting two elastic diaphragms, usually of polypropylene or rubber, on either side of the plate which can be inflated or deflated by controlling the air pressure. The membrane frames are sandwiched between two plates supporting the filter sheets, with the space between them providing a narrow chamber within which the mash is retained. The membranes can be inflated to apply additional pressure to the mash to drive the wort through the filter sheets. The wort is filtered through fine-pore polypropylene filter sheets, which retain the grains as the wort is collected. A large number of these individual modules (typically 10–60, depending on the mash capacity) are compressed together to make up the full mash filtration unit.

O'Rourke (2003) and Briggs et al. (2004) described the operation of a typical mash filter in brewing. The overall principle is the same as that for Scotch whisky production. The filter is preheated and the mash is filled from the bottom to reduce mash aeration, allowing the mash chambers to fill evenly and venting the air. Mash delivery continues until all the chambers are filled, when the vents are closed and the wort outlets opened. Air pressure applied to the space between the diaphragm/membrane and its support allows the diaphragm to expand and squeeze the wort through the filter sheets, and the liquid is collected at the wort outlets at the top and bottom of the filter. The wort streams can then be combined, cooled, and pumped to the fermenters in the normal way. As filtration proceeds, the pressure to the membranes is gradually released, and the mash is sparged by pumping second water from the top of the unit at a higher temperature (usually 75–78°C) so that it is flushed through the grains, displacing the remaining worts, and collected as weak worts after the filter. Once sparging is complete, the mash is further compressed to collect the final portion of the wort. Normally, the strong (first) worts go to the fermentation process, while the weak (second) worts are returned to a separate tank or the mashing vessel for the next mash. A typical 60-plate wort filter (used in brewing) can accommodate about 10.5 tonnes of barley malt (Briggs et al., 2004). The first worts are highly concentrated, with a specific gravity of about 1085–1090° (21–22° Plato).

When mashing is complete, the filter is opened and the grains are discharged to the normal co-product stream, usually via a screw conveyor. The filtration process is very efficient and will result in much drier spent grains that will have less loading on effluent handling systems. The cycle time is usually around 2h (Briggs et al., 2004). As the filter sheets become dirty, resistance to the flow of wort increases and the filters need to be cleaned regularly by rinsing with clean water, periodically using caustic cleaning agents and possibly hypochlorite (Briggs et al., 2004). If the grains are sticky, the residues must be physically removed from the filter sheets (O'Rourke, 2003).

Two advantages of a mash filter are that it provides additional flexibility to handle malt that would otherwise be difficult to process and its ability to handle worts deriving from finely milled flours enables the process to be adapted for a wider variety of materials, such as hull-less barley malt. The mash filter can give highly efficient extract recovery—about 99.5% of the laboratory extract (Briggs et al., 2004)—together with a faster turnaround. In addition, mash filters offer more scope for automation and can deliver reduced water usage with less effluent production.

On the other hand, the system can be more expensive to operate compared to a lauter tun (Briggs et al., 2004), and there is a requirement for separate mashing vessels and additional plant to mix and transfer the mash to the filter. The relatively small capacity of some mash filters, compared with a large mash tun, may also require more than one cycle to fill a washback. In general, although they have fewer moving parts, mash filters require regular maintenance, cleaning, and replacement of filter sheets and membranes. Sometimes the sequence of production runs needs to be interrupted for cleaning (Briggs et al., 2004).

Although the spent grains are much drier than conventional draff, which has some advantages, the co-products may be more difficult to handle. In addition, because of the way in which the air is expelled during the process, the worts have less aeration and a lower fatty acid content compared with conventional worts (Briggs et al., 2004). This may have an impact on fermentation efficiency and affect the quality of the final spirit but can be overcome in brewing, as the wort en route to the fermentation is usually oxygenated.

The technology that has been discussed so far shows the main components of the processes that are used in nearly all of the current malt distilleries. Much of this process remains traditional, but lessons have been learned from technology companies with regard to modern ways of integrating the process with water and energy recovery systems and co-product utilisation, particularly with modern expansions and new-build distilleries. This is particularly apparent when considering recent developments in the design and operation of grain distilleries.

Grain distillery processing and principles

Developments in the basic process and technology of grain spirit production have occurred over the years in response to requirements for improved efficiency. Changes include raw materials and environmental protection legislation as well as modern initiatives to achieve sustainability targets such as reduced carbon footprints and water and energy use, together with lower environmental emissions. One approach to reducing energy requirements has been the use of low-temperature processing. Closely coupled to this is increased integration of energy recovery systems and co-product streams within the production process. Other major changes include increases in wort original gravity, sometimes to >1080° (20° Plato) to reduce water usage and to increase throughput, as well as a move from pressed to cream yeast. The latter has resulted in increased capital spending, but has resulted in improvements in yeast management and better control of yeast pitching rates.

Grain distilling, the processing of unmalted cereals, has two main objectives: (1) to release the starch from the grain and (2) to convert this starch into fermentable sugars. Generally, the first of these aims is achieved by cooking the cereals at high temperatures or pressures. This gelatinises the starch so it can be released and solubilised. Enzymes from high-enzyme malted barley can then convert the starch to fermentable sugars, which are in turn fermented to alcohol by yeast. In the low-temperature process, the starch is not gelatinised in the same way; instead, the granules are damaged by fine hammer milling to make them more accessible to α-amylase, which can degrade native starch and reduce the gelatinisation temperature (Lynn et al., 1997, 1999).

The flow chart in Fig. 10.13 illustrates the main features of what might be regarded as a typical grain distillery process. In general, the main features of the process apply to both high-temperature and low-temperature processes. Fig. 10.13 represents a generalised overview of cereal processing in grain distilleries as used in the Scotch whisky industry; in practice, there is a wide range of process options for each part of the process. Table 10.5

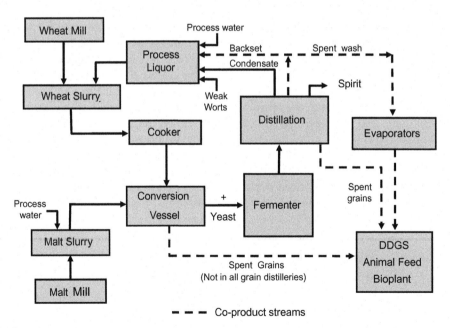

FIG. 10.13   Chematic flow chart of a typical grain distillery.

TABLE 10.5   Processing options for Scotch whisky grain distilleries.

Process steps	Range of options
Process	Batch, semicontinuous
Cereal	Wheat, maize
Milling	Hammer mill (coarse/fine), disintegrator, roller mill (coarse cracking)
Process liquor	Water, backset (stillage), condensate, weak worts
Cooking	High temperature/pressure, high temperature atmospheric, low temperature
Conversion	Dry (white) high diastatic power (DP) malt, green malt (commercially produced)
Wort separation	Generally none, coarse wort screens
Original gravity	1060°–1085° (15–21° Plato)

summarises some of the options that are currently being used in Scotch whisky grain distilleries and illustrates the wide range of processes that are employed in Scotch grain whisky production.

Each of these options has its own advantages and disadvantages, and the system in each distillery is designed to fit with individual company aims and objectives. These have mainly evolved as a result of their own experience with plant design and technology in the context of their production requirements. In some cases, these may be modelled on features from other grain spirit production processes that have been developed internationally.

Most grain distilleries continue to operate batch cooking processes, mainly using relatively high cooking temperatures (95–145°C). However, in recent times, with the expansion in whisky production, there has been a tendency to move towards larger batches or semicontinuous operation. More recent new-build grain distilleries have been designed to operate at much lower processing temperatures, but some features of the traditional high-temperature process have been retained, at least in part, to maintain the overall energy balance of the distillery and to allow heat recovery and transfer systems to operate efficiently.

Wheat continues to be the main cereal used by Scotch whisky grain distillers, although maize (corn) still has a major role to play in at least one distillery. Generally, most grain distillers will hammer mill their cereals, particularly when using a low-temperature cooking process. Technically, there is no wort separation stage, and nearly all grain distilleries now operate by processing unfiltered worts. In certain cases, the wort may be passed through a rough screening system to remove any coarse debris, which could have downstream effects later in the process.

One fundamental and increasingly important change is the emergence of highly integrated water and energy management systems to reduce overall water and energy usage. These are managed by reusing heat-exchanged process water in various parts of the process and recycling certain liquid co-product streams as process liquor (for example backset). Another aspect of this is that in order to reduce water usage and increase cereal throughput, some grain distilleries are gradually moving to a higher gravity operation, with original gravities rising from around 1060° (16° Plato) to as high as 1075–1085° (19–21° Plato) in some cases.

Although there have been many changes in the detail of the processes used by grain distillers, in general the principles behind grain distillery processing outlined in the first edition (Bringhurst et al., 2003) are still relevant.

The cooking process

The cooking process has been defined by Kelsall and Lyons (1999) as a process that begins with mixing the grain with liquor and ends with delivery of the mash to the fermenter. In practice, this is made up of four components: milling, cooking, blowdown, and conversion.

Milling

When cereals are processed in a grain distillery, the first stage after intake, before cooking, is normally to mill the grains, although on occasion whole grains may be processed. The choice of milling has been largely due to the balance between its cost and the energy saved through reduced cooking times (Piggott and Conner, 1995), and today hammer milling of cereals is the norm rather than the exception. Kelsall and Piggott (2009) reviewed some of the milling options used in grain distilleries, and in general, the principles have not changed.

The main purpose of milling is to break the structure of cereal grains in order to facilitate water penetration of the cereal endosperm during subsequent cooking (Kelsall and Lyons, 1999). Fine milling also has the effect of mechanically damaging starch granules, which promotes the absorption of water (Evers and Stevens, 1985), facilitates the mechanical release of starch from the protein matrix of the grain, and reduces the gelatinisation temperature (Lynn et al., 1997). Milling also helps to break down gums (such as arabinoxylans and β-glucans) and other cell wall materials and promotes the solubilisation of proteins later in the process.

In the Scotch whisky industry, as already described, two main forms of mills are used: roller mills and hammer mills. Pin mills (disintegrators) have also been used. In some instances, wet milling can be employed but is mainly used for processing green malt as described below. Roller mills are usually used in the production of malt whisky, but can be also used in the context of grain distilling; they are particularly suited to the grinding of small grain cereals such as barley malt and wheat. In a roller mill, cereal grains are compressed as they pass between sets of rollers, usually three sets of two (for a six-roll mill).

Hammer mills are normally used in grain distilleries, as these can break the grains into a homogeneous flour, which can be handled relatively easily. Hammer milling also enables grain distillers to use short-term cooking and mashing processes and is particularly suited to continuous processes (Wilkin, 1983).

In a hammer mill, whole cereal grains (maize, wheat) are fed into a grinding chamber and crushed to a uniform flour by a number of rotating hammers. Control of the grist size is achieved by using a fixed-size retention screen (typically 0.5–2.5 cm), which retains larger particles until they are broken into a uniform size (Kelsall and Lyons, 1999). When using a hammer mill, it is important not to grind the grain too finely. This can result in *balling*, which allows small amounts of unprocessed starch to pass through the process. In addition, grinding too finely will have an adverse effect on the solids content of the postdistillation stillage (spent wash) and puts an extra load on evaporators, giving rise to potential downstream processing problems.

The fineness of the grind also has an impact on the yield of alcohol from the process. If the grind is too coarse, there is greater potential for the starch to be incompletely gelatinised during processing, and an increase in coarseness of about 0.2 cm has been reported to result in a reduction in spirit yield of about 7.5% (Kelsall and Lyons, 1999).

Grain distillers generally prefer to mill cereals prior to processing, and hammer milling is now normal practice. Although the use of whole grains in the past was effective (Bathgate, 1989), there was a cost penalty for extended cooking times. Using whole grains, however, has the advantage of potentially reducing the degree of browning reactions, which can result in improved alcohol yield (Bathgate, 1989).

Modern low-temperature processes require the cereals to be relatively finely milled, as this damages the starch granules sufficiently to allow the starch to gelatinise at the lower temperature (Lynn et al., 1997). In addition, the use of a small proportion of premalt and backset may be used to facilitate enzyme hydrolysis and ensure that there is enough free α-amino nitrogen (FAN) present to support fermentation.

The principles of cooking

The main function of the cooking process is to break the hydrogen bonds linking starch molecules and to separate the starch from the protein matrix, thus breaking its granular structure and converting it into a colloidal suspension (Kelsall and Lyons, 1999). In cereals such as maize, the gelatinisation temperature is substantially higher than the temperatures at which the enzymes involved in the conversion of starch to fermentable sugars are able to function (62–67°C) (Palmer, 1986). Thus, before the starch can be utilised, the cereal generally has to be cooked (Wilkin, 1989). The degree of cooking is very much dependent on the cereal used. This is generally determined by the gelatinisation temperature. Maize, which has a substantially higher gelatinisation temperature than wheat, must be cooked under more rigorous conditions (Bathgate, 1989). True solubilisation of starch molecules occurs when a starch paste is cooked at 100–160°C (Swinkels, 1985).

Although maize is still occasionally used in some distilleries, since 1984 the major cereal used for the production of Scotch grain whisky has been wheat (Brown, 1990). In theory, wheat would appear to require little or no additional cooking, but in practice the experience of distillers has indicated that the cooking of wheat gives improved access to the starch and more complete disintegration of the grain (Bathgate, 1989). In addition, because the economics of the process could conceivably, in the future, promote a general return to the large-scale processing of maize, many distillers have maintained the capacity to process maize as well as wheat and have thus, at least in part, retained their traditional cooking processes rather than making substantial changes. By not fully implementing cold cooking processes as suggested by Wilkin (1989) and Newton et al. (1995), these distilleries can respond to changes in the raw materials market. In addition, the modern design of grain distilleries integrates heat recovery and effluent reduction systems, which are intrinsically linked to the technology designed for high-temperature cooking of cereals.

One of the problems with wheat has been that it contains substantial amounts of cell wall materials such as arabinoxylans, which can present serious problems with a distiller's throughput, during both processing and recovery of spent grains/wash after distillation. The practical opinion of distillers is that this problem can be alleviated by cooking. However, the emergence of modern co-product handling systems, such as anaerobic digestion (AD) and other biogas generation systems that do not require conventional evaporation systems, may also provide a partial solution to these problems.

Although unmalted barley has, on occasion, been used in grain distilleries because it can be less expensive than other cereals, it causes very severe problems with viscosity due to the presence of high levels of β-glucans during cooking and in the recovery of co-products (Bathgate, 1989; Brown, 1990). For this reason, unmalted barley has not been used in recent times.

Cooking in the grain distillery

The technology used for cooking cereals in a grain distillery is continuously evolving, and some distillers have shown a strong interest in adapting some of these methods when upgrading or modernising their processes. Hence, there is now a relatively wide spectrum of cooking options for the production of Scotch whisky. These range from batch to continuous processes, from pressure cooking to atmospheric and low-temperature processing, from hammer-milled cereals to unmilled grain, and from wheat to maize. Each of these has its own individual features and its own particular advantages and disadvantages in relation to the available technology. Publications describing cooking for grain whisky production include Pyke (1965), which still provides the definitive account of the batch pressure cooking of maize. Rankin (1977) described the evolution of grain distillery processing from more traditional processes to more modern ones, similar to those in use today. Wilkin (1983) provided an account of both batch and continuous processes and described the use of both wheat and maize. Collicut (2009) primarily focused on North American grain whiskies, but also reviewed the raw materials and technologies used globally in a wide range of distilleries, including Scotch grain whisky. Kelsall and Piggott (2009) have also described a range of cooking processes, some of which are similar to those currently used in the production of Scotch whisky.

In recent years, there has been increased interest in utilising low-temperature cooking systems and at least one modern grain distillery has now adopted this technology. Cereal flour (or unmilled grains) is generally mixed with process liquor in a slurry tank. The concentration of slurry can vary depending on the process used and the target original gravity, but it typically contains about 2.5 litres of liquor per kilogram of cereals (Piggott and Conner, 1995; Wilkin, 1989). The process liquor is usually water but can also include recycled stillage (backset), weak worts, or sparge recovered from mash filtration or separation systems. Backset is a recycled portion of the stillage from the distillation when most of the solid matter has been removed, either by centrifugation or by screening (Travis, 1998), and is used in certain cases as a supplement to the process liquor. Although backset can be quite acidic, when it is used properly it is considered to provide important benefits, particularly during fermentation (Travis, 1998). Backset can provide nutrients that are essential for yeast growth, but too much can result in an oversupply of certain minerals and ions (such as sodium and lactate), which can suppress fermentation (Kelsall and Lyons, 1999). Backset is now considered an essential requirement for the successful operation of low-temperature cooking processes and is an important aspect of the integration of processing, water management, co-products, and energy recovery systems.

Normally, the contents of the slurry tank are mechanically mixed thoroughly to avoid balling of the grist, which can result in lost extract and the presence of unconverted starch, which in addition to resulting in lost alcohol yield can also cause problems later in the process. The initial slurry can be carried out at ambient temperature but is often at about 40°C (or higher), using waste heat from the process. The higher temperature helps to hydrate and condition the grist as well as reduce the energy input required for cooking. In some processes, particularly in continuous processes, a small amount of barley malt is added as a premalt (Collicut, 2009). The purpose of this is for enzymes in the malt (amylases, proteinases, and β-glucanases) to partially hydrolyse starch, protein, and gums such as β-glucans in order to reduce viscosity and facilitate pumping of the slurry through the process.

The slurry is then passed forward into the cooker, which is generally a cylindrical pressure vessel, normally fitted with stirring equipment (Pyke, 1965). Mixing during cooking is essential to avoid sticking and subsequent burning (caramelisation). Steam is then injected into the cooker to heat the slurry to the required temperature to gelatinise, liquefy, and release the starch until the cereal is cooked. The cooker is usually programmed to operate over a fixed cycle, which has been optimised for a particular process and cereal, and cooking temperatures and times can vary for different distilleries. In practice, the cooking temperature is programmed to ramp up to the maximum (usually 130–145°C) and is maintained there for only a relatively short time. Longer times at high temperature and pressure are necessary for whole grains. Some distilleries may operate several cookers in parallel to maintain sufficient, semicontinuous production capacity to support distillation in continuously operating Coffey or patent stills.

Normal batch cooking is energy-intensive and requires relatively long cooking times, which can result in excessive browning reactions and can give reduced yield if the process is not controlled correctly. However, with batch cooking the wort is more likely to be sterile. The batch process is also more adaptable for use with a wide range of cereals, including wheat and maize.

Overall, the cooking process represents a delicate balance between gelatinisation, the release of starch, and its thermal degradation into undesirable products. If the temperature is too low, some starch granules will remain intact and the starch will not be fully gelatinised, resulting in lost alcohol yield. On the other hand, if the temperature is too high or the starch is cooked for too long a period, browning, or Maillard, reactions will take place between carbohydrates and protein-derived amino acids and peptides. This results in a loss of alcohol yield.

Table 10.6 gives an indication of the range of cooking temperatures operated at different distilleries.

Low-temperature continuous processing

In the 1980s, continuous cooking was envisaged to have enormous potential in the production of Scotch grain whisky (Wilkin, 1989). However, for various reasons, such as process delays, energy efficiency concerns, and changes in the whisky market, continuous processes

TABLE 10.6 Cooking conditions at Scotch grain whisky distilleries.

Cooking process	Maximum temperature
Low temperature	68–72°C
Atmospheric	95–105°C
Pressure cooking (batch)	125–150°C

have not generally found favour in the Scotch whisky industry, although there has been an increased interest recently. With current production schedules, several modern processes could be regarded as being semicontinuous, as they are constrained only by the capacity of the fermentation process (Palmer, 2012). Because fermentation is still a batch process, continuous cooking is particularly vulnerable to interruptions in the process resulting from downstream process problems. Low-temperature processing has several advantages, as the system is less energy-intensive and reduces the extraction of viscous cell wall materials. One of the main benefits of low-temperature processing is that the system has the potential to give significant increases in alcohol yield (Agu et al., 2008a) because the substrate is less damaged by the process conditions. However, modern low-temperature processes require careful integration with energy recovery and co-product streams to ensure that they are operating at maximum efficiency. The main drawback with low-temperature cooking is that the wort may not be completely sterile, so the system has to be carefully managed to avoid the build-up of microbial infection, which can cause problems during the fermentation stage.

At present, only one grain distillery is using a true low-temperature (68–72°C) cooking process, but it is still reliant on a significant proportion of high-temperature (>140°C)/pressure production to support overall production levels. Maintaining the high-temperature process at a certain level is considered to be a critical component of the energy balance of the distillery, and it provides essential back-up to ensure production efficiency is maintained.

The adoption of a low-temperature cooking/mashing process requires some important changes to the traditional grain distillery process, particularly with the introduction of backset, which is now regarded as an essential component of this type of process. Without backset, it would be difficult to get consistently good alcohol yields from the low-temperature process. This is primarily because when backset is returned to the process, it can provide additional levels of soluble nitrogen and FAN to support fermentation, which can be slower than normal. Additionally, adding a proportion of high-temperature cooked mash to each fermenter also provides some benefits, as the high-temperature process also contributes to the soluble nitrogen in the wort.

In modern low-temperature continuous cooking, wheat is hammer-milled and the grist is screened through a 2.5-mm screen to give a flour that can be easily hydrated when mixed with water. The wheat flour and a small proportion of malt (premalt) are slurried with process liquor at 45–55°C and heated to about 68–72°C on the way to the mashing/conversion vessel, where it is mixed with malt slurry at the normal mashing temperature (63–64°C). The parallel use of multiple cooking and conversion vessels allows a continuous production stream to be maintained. After mashing/conversion, the in-grains wort is cooled and added to the fermenter.

After distillation, the spent wash is centrifuged to remove the remaining solids and the clear centrate is returned to the process as backset. The backset is closely monitored to ensure that the pH is suitable (approximately pH 4) and added to the cereal slurry prior to the next mash. The backset usually comprises around 40–50% of the low-temperature cooking process liquor.

The key aspect of the combined low- and high-temperature processes is the way in which the heat recovery systems are integrated within the process and optimised to support the energy balance of the distillery to reduce external inputs and to minimise the carbon footprint.

Maillard reactions

Maillard or browning reactions are a highly complex series of reactions that take place between sugars and amino acids or proteins most prominently under high-temperature conditions, producing a range of products. These reactions have been extensively reviewed in the literature (Briggs et al., 2004; Hough et al., 1982; O'Brien et al., 1998), and there is no attempt to cover this topic in any detail in this chapter; however, Fig. 10.14 gives an idea of the complexity of the processes involved.

Mlotkiewicz (1998) described the main stages in Maillard reactions. Initially, reducing sugars combine with amino acids to form products (Amidori or Heynes rearrangement products) that are in turn degraded through a complex series of reactions into a large number of flavour intermediates and other flavour active compounds. One of the key routes is Strecker degradation, where amino acids react with dicarbonyl compounds called *reductones* to form products that are also ultimately converted into brown-pigmented polymeric materials. The final products of Maillard reactions are a mixture of low-molecular-weight colour compounds containing two to four linked rings and melanoidins, which have much higher molecular weights. Additionally, a large number of other flavour and aroma active products, such as furans, pyrroles, and cyclic sulphur compounds, are also produced. The development of Maillard products is enhanced by increasing temperature, heating time, and pH (particularly above pH 7).

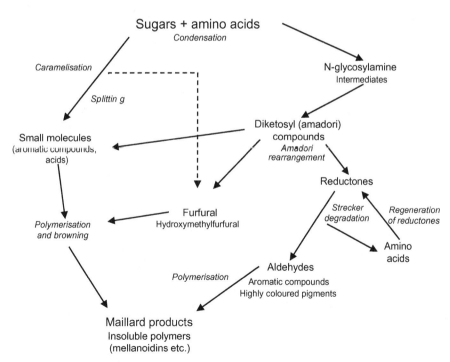

FIG. 10.14 Summary of the processes involved in Maillard (browning) reactions. Adapted from Adrian, J., Legrand, G., Frangne, R., 1988. Dictionary of Food and Nutrition. Ellis Horwood, Chichester, pp. 120–121.

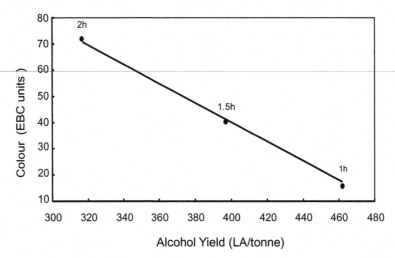

FIG. 10.15 Laboratory data showing the inverse relationship between wort colour and alcohol yield for extended times for wheat (1, 1.5, and 2h at 145°C) (Scotch Whisky Research Institute data).

The occurrence of Maillard reactions during cooking is important because they result in the uptake and degradation of both fermentable carbohydrate and amino acids and could have a significant impact on spirit yield (Fig. 10.15). They also have implications for downstream processing and the potential to cause problems with co-products.

The optimum cooking time and temperature depend very much on the cereal used and the process employed. Maize generally requires higher temperatures and/or longer cooking times than wheat. The degree of milling is also important. Finely milled cereals require shorter cooking times than unmilled cereals, which can require as long as 2h (Brown, 1990).

The pressure cooking of unmilled cereals can be both efficient and cost-effective, provided it is economical to use longer cooking times (Bathgate, 1989). In addition, when the pressure is released on blowdown, the grain endosperm is thoroughly disintegrated as it is passed forward to the conversion stage. It has been suggested that the whole-grain processing stage is less prone to problems with overcooking or browning. Huskless cereals such as wheat and maize are more suitable for whole-grain cooking, because they disintegrate more thoroughly after cooking (Walker, 1986).

Blowdown and retrogradation

When the cooked cereal slurry is discharged from a pressure cooker, generally into a flash cooling vessel or expansion tank, there is a rapid decrease in pressure. This is known as blowdown, which has the effect of mechanically releasing any remaining tightly bound starch from the grain matrix (popcorn effect). When cooked whole grains are discharged, the pipework associated in the transfer from the cooker is instrumental in ensuring the disintegration of the grains (Walker, 1986).

Blowdown is a critical part of the cooking process and is normally associated with rapid, but carefully controlled, cooling. Poor temperature control will cause serious problems with retrogradation (setback), which will result in the cooked slurry forming a gel that is resistant to enzymic breakdown (MacGregor and Fincher, 1993) and will lead to subsequent processing problems related to viscosity (Swinkels, 1985), filterability, and alcohol yield (Jameson et al., 2001).

Retrogradation has been defined as a change from a dispersed amorphous state to an insoluble crystalline condition (Swinkels, 1985), which occurs when heated gelatinised starch begins to reassociate upon cooling (Atwell et al., 1988), resulting in gel formation and precipitation. The process of retrogradation is very complex (Swinkels, 1985) but is considered to be predominantly influenced by the relative amount of amylose present in the starch (Sasaki et al., 2000). Other important factors involved in retrogradation are the cooking conditions, starch concentration, cooling procedure, and pH. Both wheat and maize starch, each containing 26%–28% amylose, are prone to retrogradation. Waxy maize, which is effectively all amylopectin, is much less likely to retrograde.

The length of the amylose chain is considered to have a major influence on the processes that take place during retrogradation. This can affect parameters such as the propensity to form precipitates or gels and the gel strength. Longer-chain lengths (> 1100 units) show a stronger trend towards gel formation, due to the alignment and crosslinking of adjacent chains to form ordered structures (Hoover, 1995).

Normally, retrogradation occurs as a result of hydrogen bonding between chains of adjacent amylose molecules, which become bound together irreversibly to form aggregates. This crosslinking probably involves a number of regions within a single amylose chain leading to the formation of a macromolecular network (Hoover, 1995). The aggregated material entraps liquid within a network of partially associated starch molecules and leads to the formation of a gel (Swinkels, 1985). The rate of retrogradation is highest between pH 5 and pH 7. Retrogradation does not occur above pH 10 and only proceeds slowly below pH 2. Retrograded amylose is not readily degraded by α-amylase (Miles et al., 1985a). Thus, when starch has retrograded, it is extremely difficult to solubilise again. Retrogradation of amylose is considered to be irreversible, even at high temperatures (greater than 100°C) (Miles et al., 1985b). Fig. 10.16 illustrates the main features of retrogradation.

Amylopectin is much less susceptible to retrogradation than amylose, and the presence of this polymer has been regarded as a moderating influence on this process (Swinkels, 1985). However, under extreme conditions such as high starch concentration and very low temperatures, amylopectin can also undergo retrogradation, but this can be, to some extent, reversible (Miles et al., 1985b). Amylopectin from maize has a higher propensity for retrogradation than either barley or wheat and retrogrades more quickly. This has been attributed to the higher degree of crystallinity of maize amylopectin and a greater proportion of chains of DP 15–20. Wheat amylopectin is less susceptible to retrogradation than that from barley (Hoover, 1995).

Normal retrogradation occurs when starch is cooled, but it can also take place at high temperatures (75–95°C) when starch solutions are stored. This takes the form of a precipitate of regularly sized particles. This phenomenon appears to be related to the presence of lipid or fatty acid material forming complexes with amylose. These complexes are not formed above 95°C (Swinkels, 1985). Jameson et al. (2001) observed that extended holding of cooked maize or wheat in a grain distillery at temperatures up to 100°C prior to mashing can result in the formation of resistant starch, which can lead to subsequent problems in the distilling process.

During blowdown, the temperature is dropped very quickly to about 70°C, which is very close to the normal striking temperature of malt. This allows the malt to be added in such a

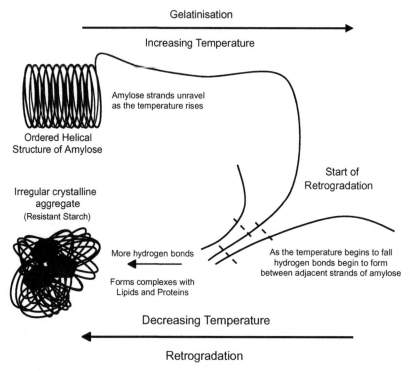

FIG. 10.16   Retrogradation of starch.

way that the malt enzymes begin to rapidly hydrolyse the cooked starch before retrogradation takes place. This represents a very delicate balance, and the process requires careful control. If the malt is added too late or the striking temperature is too low, the cooked starch will begin to retrograde and become inaccessible. If the malt is added too early, the striking temperature will be too high and the malt enzymes will be damaged, also resulting in lost yield.

Problems with cooking

There are several main sources of problems during the cooking process. Undercooking occurs when the cooking time is too short or there is insufficient heat to gelatinise the starch properly. This gives significant losses in alcohol yield because the ungelatinised starch is not readily accessible to enzyme hydrolysis. This can be a particular problem with continuous processing, but can also be related to variations in cereal quality, such as wheat nitrogen content and milling performance. When the slurry is overcooked, excessive browning (Maillard) reactions occur, which remove both fermentable substrate and proteinaceous material from the wort, resulting in losses of alcohol yield. Overcooking can also cause physical problems with sticking and burning as the cooked slurry caramelises. Higher levels of cell wall material and gums such as β-glucans and arabinoxylans associated with wheat can also cause problems with viscosity during cooking, but these tend to have more serious downstream effects, particularly on the recovery of co-products after distillation. Retrogradation can be a problem if the cooking and blowdown process is not controlled carefully. When the starch retrogrades, it becomes insoluble and inaccessible to enzyme hydrolysis and will not be available to produce fermentable sugars, resulting in significant losses in alcohol yield.

Conversion

Principles of conversion

When the cooker is discharged, the cooked slurry is then added to a malt slurry, which has been held at a particular temperature (often about 40°C but sometimes higher and closer to mashing temperatures). In some cases, the cooked cereal is discharged directly into a vessel containing the malt (drop tank) after cooling to a suitable striking temperature; in other cases, the malt slurry is added to the cooked cereal either directly into the process stream (continuous process) or to the conversion vessel. The actual procedure used depends on the individual process. Each of these has advantages and disadvantages. The choice is largely dictated by the technology that is in place and the economics of the individual process.

The main function of conversion is the breakdown of gelatinised starch into fermentable sugars and small dextrins (starch hydrolysis). It is also essential to degrade proteins into amino acids and low-molecular-weight, nitrogen-rich fragments, which are necessary to provide nutrients for the yeast during fermentation. Because no additives are permitted in the production of Scotch grain whisky, both of these processes must be accomplished entirely by the endogenous enzymes from the malt. This is in contrast to the production of neutral spirit, where the use of added commercially produced food-grade enzymes is permitted.

The malt used for conversion is normally commercially produced high-enzyme malt that has been germinated and carefully dried (kilned) to produce high levels of α- and β-amylase (DU/DP). Some green (unkilned) malt is still used, but its use in most grain distilleries has been superseded by kilned malt. Green malt is normally less expensive to produce than kilned malt and can potentially be used at a lower dosage (malt inclusion rate); however, it has a shorter shelf life and higher transportation costs than dried malt (Walker, 1986).

Prior to mashing, the malt is milled, usually using a hammer mill, into a fine flour, which is then slurried with process liquor, usually water. This slurry is conditioned for a short period (20–30 min) at a suitable temperature (around 40°C) prior to mixing with the cooked cereal slurry. It is important that the residence time in the slurry tank be kept to a minimum, to avoid the build-up of microbial infection levels in the malt slurry as that could affect both the efficiency of the mash and fermentation performance. If used, green malt is normally wet-milled into a slurry before being mixed with the cooked cereal. Typically, the amount of malted barley used in a Scotch whisky grain distillery (malt inclusion rate) is around 10% of the overall grain bill (on a dry weight basis), although this can vary for different distilleries and can be lower than 10% or as high as 15% in some cases.

It is essential to add the malt as soon as possible after the cooker is discharged, so enzymes (mainly α-amylase) from the malt can begin to degrade the solubilised starch, reducing the viscosity, before it begins to retrograde. As described above, extended storage of cooked starch will result in retrogradation and the formation of resistant starch, which will not be efficiently degraded by the enzymes.

In many cases, the malt slurry and cooked cereal slurry are mixed together in a mash tun, or some other conversion vessel, and held at mashing temperatures (62–65°C) for up to 30 min to convert the starch into fermentable sugars, mainly maltose. In a continuous-tube converter, the residence time can be considerably shorter due to the increased heat transfer and the more intimate contact between the malt enzymes and the cereal slurry. Robson (2001) described a continuous conversion vessel in a grain distillery, where on average the residence time is about 20 min. The conversion time should be sufficient for the complete conversion of starch to fermentable sugars.

The main problem associated with conversion is the incomplete conversion of starch into fermentable sugars, which is manifested by high levels of oligosaccharides and dextrins in the wort. This can be a result of poor temperature control during conversion, insufficient conversion time, or the presence of insufficient enzymes from the malt because the malt inclusion rate was too low or the malt was of poor quality. In addition, the pH levels could be too low for optimum enzyme performance (for example as result of using too high a proportion of backset). During conversion, mash pH levels are usually around 5.0–5.5, which facilitate the activity of α- and β-amylase. One of the features that distinguishes grain distilling from brewing is that the wort is not boiled prior to fermentation. This allows the enzymes to maintain their activity during fermentation. During the early stages of fermentation, the pH falls to 4.2–4.4, which will result in the release and action of the active form of the debranching enzyme limit dextrinase (Bryce et al., 2004; Walker et al., 2001). The malt enzymes are still active and will continue to slowly degrade dextrins during fermentation (Bringhurst et al., 2001), but they will be unable to degrade any substantial quantities of remaining nondegraded starch if the conversion process has not been carried out effectively.

Wort separation

Traditionally, in grain distilleries, the worts were separated in a mash tun after conversion and sparged several times with liquor at increasing temperatures to provide a clear, filtered wort for fermentation, much in the same way as malt distilleries (Rankin, 1977). However, this process is complex and time-consuming in relation to the production demands of a modern grain distillery, and wort separation has largely been eliminated in modern grain distilleries. After conversion with malt, the whole mash is normally transferred directly to the fermenter after cooling and pitching with yeast. This allows for the use of higher original gravities (up to 1085°, or 21° Plato) than would otherwise be the case.

The use of unfiltered worts, however, can cause problems with fouling of stills during distillation and may have serious downstream effects, particularly on the efficient operation of the evaporators used to collect co-products. Problems with the evaporators can lead to substantial process delays, which have serious effects on the efficiency of the whole grain distillery process. These problems can, to some extent, be alleviated by removing solid matter from the spent wash after distillation, such as by centrifugation.

In some grain distilleries, as a result of processing constraints, it has been considered necessary to have some sort of separation process before pumping the wort to the fermenter. The main advantage of using a wort separation process is that a largely liquid wort is pumped to the fermentation vessel, which will cause fewer problems during subsequent distillation and processing of co-products. However, the filtration process itself can lead to delays in the production cycle, which can have serious implications for plant efficiency (Jameson et al., 2001). There is also a greater probability of losing residual starch in the spent grains, which could lead to significant losses in the alcohol yield. Additionally, because only about 10%–15% of the total protein is solubilised during mashing (Boivin and Martel, 1991), the bulk of the remaining unhydrolysed protein remains with the spent grains. This means that there may be a potential shortage of free amino nitrogen to sustain an adequate level of yeast fermentation.

Several different means of separation have been employed in Scotch whisky grain distilleries. Originally, worts were filtered using a lauter type of filtration system (Pyke, 1965) but these have long been superseded by other filtration systems that are more adapted to more or less continuous operation. In such systems, the mash is filtered through a series of filters or sieves (hydrosieves) (Robson, 2002), and the grains are sparged to remove soluble material. First and second worts are pumped to the fermenters, and the recovered sparge (weak worts) is recycled to the process. Largely these operate in a similar way to the traditional malt distillery wort separation process, although the equipment used is completely different. The requirement for wort separation has largely been determined by the efficiency and economics of the particular process and by individual companies' experience with the technology. In general, grain distillery processes are much more complex than those in the malt distillery, and there are greater opportunities for technical innovation in terms of improved process efficiency and better management of energy and water resources, which are also important features of distillers' sustainability and environmental objectives.

Water as a raw material

Water is one of the key raw materials and is covered separately in Chapter 11.

Future trends

Malt distillery processes tend to be more traditional and less adaptable than grain distilleries, and the main areas of future innovation are likely to be behind the scenes. Improved process control will provide more efficient, integrated energy utilisation and heat recovery systems, as well as reduced water usage and environmental emissions in line with the industry climate change and sustainability objectives. Successful collaborative projects have brought together groups of stakeholders, including plant breeders, agronomists, academics, maltsters, brewers, and distillers to look closely at potential ways of improving distillery performance, while maintaining their commitments to sustainability.

One area that is looking very promising is the development of modern, high-yielding barley varieties that are better suited for malt distillery processing. This is primarily due to the development and improvement of the genetic tools that are used by plant breeders to identify promising new materials and which will help to accelerate the classical (non-GM) breeding approach to developing new barley varieties. We already had a good genetic marker for glycosidic nitrile (EPH) (Swanston et al., 1999), but the use of modern approaches has led to improvements in reliability and throughput (JHI, 2018). Markers for other economically important agronomic and end-user traits are now in development and will help to meet the future requirements of distillers, maltsters, and brewers (Bringhurst et al., 2012b). A similar approach has also been used to improve our understanding of wheat genetics that can be applied to both Scotch grain whisky production and bioethanol production (BBSRC, 2018; Bringhurst et al., 2012b; Sylvester-Bradley et al., 2010).

From a technical perspective, grain distillery processes are rather more adaptable to changes in technology and lend themselves more easily to technical innovation than malt distilleries, within the constraints of the legal definition of Scotch whisky. This degree of flexibility allows the use of a wider range of potential raw materials, which are essentially driven by market forces. In addition, the economics of grain distilling can support the diversification from grain whisky to other neutral spirits for use as vodka or gin, in order to supply as wide a range of products as possible. Essentially, the technology for producing these products is identical to that for grain whisky, with the possible exception of the distillation stage, but production is not constrained by the limitations imposed by the legal definition of Scotch whisky. In addition, for neutral spirits, it is also possible to use commercial starch-degrading enzymes, as an alternative to malt, which is relatively expensive, for products deriving from neutral spirits such as vodka and gin. However, in distilleries producing diverse streams of products, there should be a strict separation of the Scotch whisky production stream from that for other spirits to avoid contamination of the Scotch whisky process, thus maintaining the integrity of Scotch grain whisky.

So, what are the likely developments for the future production of Scotch grain whisky? At present, the technology that is used is largely determined by the scope of existing processes and the economics of how these can be developed in terms of the relative capital cost of modifying the existing process and that of adopting new technology, in response to market pressures defining the products in demand as well as the raw materials which are available.

The main raw material of choice for grain whisky production is, and is likely to continue to be, UK soft winter wheat., However, future local uncertainty on climate change impacts with more variable harvest conditions coupled with potential fluctuation in global wheat markets could affect the supply of wheat used by Scotch whisky producers. This is one reason why grain distilleries, where possible, may wish to retain the capability to process maize, as well as potentially other cereals.

As process technology continues to develop, a much wider range of process options is now available to Scotch whisky producers, both in the context of the traditional batch and continuous cooking processes. For example, the introduction of highly efficient cooking systems such as steam jet cookers, which can operate at higher temperatures (160°C) and can also be easily integrated with modern energy and water recovery systems, which can be adapted to both batch and continuous cooking processes (Collicut, 2009).

Because of the relatively low gelatinisation temperature of wheat in relation to maize (Bathgate, 1989), it is possible to process wheat at lower temperatures than maize, and it has been shown that this has the potential to provide higher alcohol yields and reduced spent wash viscosity (Agu et al., 2008a). There is now a wider interest from Scotch grain whisky producers in perfecting and applying this technology more widely. A modern low-temperature process can use a combination of fine milling and temperatures in the range of 70–85°C, to allow sufficient gelatinisation of the wheat starch to provide high alcohol yield and production throughput.

Some no- or cold cooking processes were reviewed by Wilkin (1989). These were used with varying degrees of success as long ago as 1980–84. In these processes, cereal and tuber

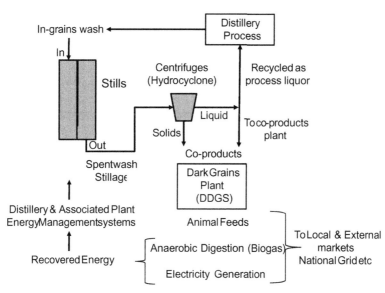

FIG. 10.17   Integrated co-product handling systems in a typical grain distillery.

starches were used to produce substantial alcohol yields at temperatures as low as 55°C. However, the raw materials were required to be milled ultrafinely, which used significant amounts of the energy saved by reducing the cooking temperature. Additional problems were encountered with the incomplete release and saccharification of the starch granules, which could only be resolved by resorting to the addition of exogenous enzymes such as proteases and hemicellulases and, as such, could not be used for the manufacture of Scotch whisky as it is currently defined.

An obvious direction for Scotch grain whisky would be to modify the process in order to make it truly continuous (Rankin, 1990). Whitby (1995) predicted that in the longer-term distilleries would eventually become fully integrated with the maximum production of alcohol by the more efficient utilisation of raw materials and minimum use of energy, and it is clear that the current trend is for individual distilling companies to move steadily in this direction.

At this stage, it is difficult to say whether the Scotch whisky industry is prepared to move fully in the direction of whole crop utilisation, as described by Petersen and Munck (1993) and Bekers et al. (1997), where the distillery production process is a part of a closed biotechnology system or biorefinery (Audsley and Sells, 1997; Lyons, 1999), allowing the complete utilisation of raw materials to produce a wide range of marketable co-products, as well as ethanol (whisky). However, ultimately the future direction of distilling will probably be as processors of grain rather than exclusively as alcohol producers (Lyons, 1999). Some of these aims were primarily aspirational at the time, but they are now central to current initiatives for addressing the long-term sustainability objectives of the industry.

The adoption of partially integrated co-product handling systems, similar to that illustrated in Fig. 10.17, shows how far the Scotch whisky industry has moved in this direction. It is now an important objective for the industry to refine how the overall distillery process is organised to provide an efficient and sustainable utilisation of both the raw materials and co-products in conjunction with modern energy recovery systems to allow them to achieve their environmental and sustainability objectives and reducing the carbon footprint, by adopting renewable energy sources and recycling water. Applications that are now in operation include the use of biomass to generate energy, improvements in water management, and the use of aerobic and anaerobic digestion and membrane biofiltration, as well as improved heat recovery systems (Jappy, 2010).

Potential future technological developments for the production of Scotch grain whisky are largely already in place in other sectors and are currently in use for the production of a wide range of other cereal, fermentation, and distillation products for other industries ranging from brewing to pharmaceuticals to fuel alcohol. However, a pipeline of new barley and wheat varieties with good distillery processing and robust agronomic characteristics is a key priority.

The adoption of new technologies for cereal processing into Scotch grain whisky is still dependent on market pressures and other future economic factors facing the Scotch whisky industry. The challenge facing the Scotch whisky industry will continue to be in its ability to tap into new technology while maintaining its market position, but without losing the distinctive quality of its products.

References

Agu, R.C., Palmer, G.H., 1997. a-Glucosidase activity in sorghum and barley. J. Inst. Brew. 103, 25–29.

Agu, R.C., Bringhurst, T.A., Brosnan, J.M., 2006. Production of grain whisky and ethanol from wheat, maize and other cereals. J. Inst. Brew 112 (4), 314–323.

Agu, R.C., Bringhurst, T.A., Brosnan, J.M., Jack, F.R., 2008a. Effect of process conditions on alcohol yield of wheat, maize and other cereals. J. Inst. Brew 114 (1), 39–44.

Agu, R.C., Swanston, J.S., Bringhurst, T.A., Brosnan, J.M., Jack, F.R., Smith, P.L., 2008b. The influence of nitrogen content and corn size on the quality of distilling wheat cultivars. In: Bryce, J.H., Piggott, J.R., Stewart, G.G. (Eds.), Distilled Spirits: Production. Technology and Innovation. Nottingham University Press, Nottingham, UK, pp. 67–74.

Agu, R.C., Bringhurst, T.A., Brosnan, J.M., 2012. Effect of batch-to-batch variation on the quality of laboratory and commercially malted Oxbridge barley. J. Inst. Brew. 118 (1), 49–56.

AHDB, 2020. Recommended Lists for cereals and oilseeds (RL). https://ahdb.org.uk/rl.

Atwell, W.A., Hood, L.F., Lineback, D.R., Varriano Marston, E., Zobel, H.F., 1988. The terminology and methodology associated with basic starch phenomena. Cereal Foods World 33, 306–311.

Audsley, E., Sells, J., 1997. Determining the profitability of a whole crop biorefinery. In: Campbell, G.M., Webb, C., McKee, S.L. (Eds.), Cereals: Novel Uses and Processes. Plenum Press, New York, pp. 191–203.

Badr, A., Muller, K., Schafer-Pregl, R., El Rabey, H., Effgen, S., Ibrahim, H.H., 2000. On the origin and domestication history of barley (Hordeum vulgare). Mol. Biol. Evol. 17 (4), 449–510.

Bamforth, C.W., Quain, D.E., 1989. Enzymes in brewing and distilling. In: Palmer, G.H. (Ed.), Cereal Science and Technology. Aberdeen University Press, Aberdeen, Scotland, pp. 326–366.

Barnes, P.J., 1989. Wheat in milling and baking. In: Palmer, G.H. (Ed.), Cereal Science and Technology. Aberdeen University Press, Aberdeen, pp. 367–412.

Bathgate, G.N., 1989. Cereals in Scotch whisky production. In: Palmer, G.H. (Ed.), Cereal Science and Technology. Aberdeen University Press, Aberdeen, Scotland, pp. 243–278.

Bathgate, G.N., 1998. The recipe for Scotch whisky. In: Beach, R.H.B. (Ed.), Brewing Room Book 1998-2000. Pauls Malt, Ltd, Bury St. Edmunds UK, pp. 81–84.

Bathgate, G.N., 2003. History of the development of whisky distillation. In: Russell, I. (Ed.), Whisky: Technology. Production and Marketing. Academic Press, London, pp. 2–24.

Bathgate, G.N., Bringhurst, T.A., 2011. Letter to the editor: update on knowledge regarding starch structure and degradation by malt enzymes (DP/DU and Limit dextrinase). J. Inst. Brew. 117 (1), 33–38.

Bathgate, G.N., Cook, R., 1989. Malting of barley for Scotch whisky production. In: Piggott, J.R., Sharp, R., Duncan, R.E.B. (Eds.), The Science and Technology of Whiskies. Long- man Scientific & Technical, Harlow, UK, pp. 19–63.

Bathgate, G.N., Palmer, G.H., 1973. The in vivo and in vitro degradation of barley and malt starch granules. J. Inst. Brew. 79, 402–406.

Bathgate, G.N., Clapperton, J.F., Palmer, G.H., 1973. The significance of small starch granules. European Brewery Convention Proceedings 14, 183–196.

BBSRC, 2018. New wheat variety to boost UK whisky production. https://bbsrc.ukri.org/research/impact/new-wheat-variety-boost-uk-whisky-production/.

Bekers, M., Laukevics, J., Vedernikovs, N., Ruklisha, M., Savenkova, L., 1997. A closed biotechnological system for the manufacture of non-food products from cereals. In: Campbell, G.M., Webb, C., McKee, S.L. (Eds.), Cereals: Novel Uses and Processes. Plenum Press, New York, pp. 169–176.

Boivin, P., Martel, C., 1991. Proteolysis during malting. Ferment 4, 182–186.

Briggs, D.E., Hough, J.S., Stevens, R., Young, T.W. (Eds.), 1981. Malting and Brewing Science. Vol. 1. Malt and Sweet Wort, second ed. Chapman & Hall, London.

Briggs, D.E., Boulton, C.A., Brookes, P.A., Stevens, R., 2004. Brewing Science and Practice. Woodhead Publishing, Cambridge, UK.

Bringhurst, T.A., Broadhead, A.L., Brosnan, J.M., Pearson, S.Y., Walker, J.W., 2001. The identification and behaviour of branched dextrins in the production of Scotch whisky. J. Inst. Brew. 107, 137–149.

Bringhurst, T.A., Brosnan, J.M., Fotheringham, A.L., 2003. Grain whisky: Raw materials and processing. In: Russell, I. (Ed.), Whisky: Technology. Production and Marketing. Academic Press, London, pp. 77–115.

Bringhurst, T.A., Agu, R.C., Brosnan, J.M., Fotheringham, A.L., 2008. Wheat for Scotch whisky production: Broadening the horizon. In: Bryce, J.H., Piggott, J.R., Stewart, G.G. (Eds.), Distilled Spirits: Production. Technology and Innovation. Nottingham University Press, Nottingham, UK, pp. 51–66.

Bringhurst, T.A., Agu, R.C., Brosnan, J.M., 2012a. Creating better cereal varieties for the sustainability of the distilling industry. In: Walker, G.M., Goodall, I., Fotheringham, R., Murray, D. (Eds.), Distilled Spirits: Science and Sustainability. Nottingham University Press, Nottingham, UK, pp. 77–86.

Bringhurst, T.A., Brosnan, J.M., Thomas, W.T.B., 2012b. New approach to barley breeding. AGOUEB—the association genetics of UK elite barley project. Brewer Dist. Int. 8 (1), 25–28.

Brosnan, J.M., Makari, S., Paterson, L., Cochrane, M.P., 1999. What makes a good grain distillery wheat? In: Campbell, I. (Ed.), Proceedings of the Fifth Aviemore Conference on Malting, Brewing and Distilling. Institute of Brewing, London, pp. 225–228.

Brosnan, J.M., Bringhurst, T.A., Agu, R.C., 2010. Growing sustainability in the wheat supply chain. In: Walker, G.M., Hughes, P.S. (Eds.), Distilled Spirits: New Horizons: Energy, Environment and Enlightenment. Nottingham University Press, Nottingham, UK, pp. 27–32.

Brown, J.H., 1990. Assessment of wheat for grain distilling. In: Campbell, I. (Ed.), Proceedings of the Third Aviemore Conference on Malting, Brewing and Distilling. Institute of Brewing, London, pp. 34–47.

Bryce, J.H., Stewart, G.G. (Eds.), 2004. Distilled Spirits: Tradition and Innovation. Nottingham University Press, Nottingham, UK.

Bryce, J.H., McCafferty, C.A., Cooper, C.S., Brosnan, J.M., 2004. Optimising the fermentability of wort in a distillery: the role of limit dextrinase. In: Bryce, J.H., Stewart, G.G. (Eds.), Distilled Spirits: Tradition and Innovation. Nottingham university press, Nottingham, UK, pp. 69–78.

Bryce, J.H., Piggott, J.R., Stewart, G.G. (Eds.), 2008. Distilled Spirits: Production, Technology and Innovation. Nottingham University Press, Nottingham, UK.

Bryce, J.H., Goodfellow, V., Agu, R.C., Brosnan, J.M., Bringhurst, T.A., Jack, F.R., 2010. Effect of different steeping conditions on endosperm modification and quality of distilling malt. J. Inst. Brew. 116 (2), 125–133.

Buglass, A.J., 2011. Handbook of Alcoholic Beverages. John Wiley & Sons, Chichester.

Bushuk, W., 1986. Wheat: chemistry. Cereal Foods World 31 (3), 218–226.

Cochrane, M.P., 2000. Seed carbohydrates. In: Black, M., Bewley, J.D. (Eds.), Seed Technology and Its Biological Basis. Sheffield Academic Press, Sheffield, UK, pp. 85–120.

Collicut, H., 2009. Whisky: Grain mashing and fermentation. In: Ingledew, W.M., Kelsall, D.R., Austin, G.D., Kluhspies, C. (Eds.), The Alcohol Textbook, fifth ed. Nottingham University Press, Nottingham, UK, pp. 413–430.

Cook, R., 1990. The formation of ethyl carbamate in Scotch whisky. In: Campbell, I. (Ed.), In: Proceedings of the Third Aviemore Conference on Malting Brewing and Distilling. Institute of Brewing, London, pp. 237–243.

DEFRA, 2019. Structure of the agricultural industry in England and the UK at June (https://www.gov.uk/government/statistical-data-sets/structure-of-the-agricultural-industry-in-england-and-the-uk-at-june).

Dengate, H.N., 1984. Swelling, pasting, and gelling of wheat starch. In: Pomeranz, Y. (Ed.), Advances in Cereal Science and Technology. vol. VI. American Association of Cereal Chemists, St. Paul, MN, pp. 49–82.

Dolan, T.C.S., 1976. Some aspects of the impact of brewing science on Scotch malt whisky production. J. Inst. Brew. 82, 84–86.

Dolan, T.C.S., 2003. Malt whiskies: Raw materials and processing. In: Russell, I. (Ed.), Whisky: Technology. Production and Marketing. Academic Press, London, pp. 27–74.

European Brewery Convention, 2010. Analytica-EBC. Fachverlag Hans Carl, Nürnberg, Germany. European Parliament and Council of the European Union. (2008). Regulation (EC) No. 110/2008 of the European Parliament and of the Council of 15 January 2008 on the definition, description, presentation, labelling and the protection of geographical indications of spirit drinks and repealing Council Regulation (EEC) No. 1576/89. Off. J. Eur. Union February 13, 16–54.

Evers, A.D., Stevens, D.J., 1985. Starch damage. In: Pomeranz, Y. (Ed.), Advances in Cereal Science and Technology. vol. VII. American Association of Cereal Chemists, St. Paul, MN, pp. 321–349.

FAOSTAT. (2012). Food and Agriculture Organization of the United Nations (FAOSTAT) data, www.fao.org.

Fincher, G.B., Stone, B.A., 1993. Physiology and biochemistry of germination in barley. In: MacGregor, A.W., Bhatty, R.S. (Eds.), Barley: Chemistry and Technology. American Association of Cereal Chemists, St. Paul, MN, pp. 247–295.

Fredriksson, H., Silverio, J., Andersson, R., Eliason, A.-C., Åman, P., 1998. The influence of amylose and amylopectin characteristics on gelatinisation and retrogradation properties of different starches. Carbohydr. Polym. 35, 119–134.

French, D., 1984. Organisation of starch granules. In: Whistler, R.L., BeMiller, J.N., Paschall, E.F. (Eds.), Starch: Chemistry and Technology, second ed. Academic Press, London, pp. 183–247.

Goodall, I., Fotheringham, R., Murray, D., Speers, R.A., Walker, G.M. (Eds.), 2015. Distilled Spirits: Future Challenges, New Solutions. Context Publishers, Nottingham, UK.

Gray, A.S., 2019. The Scotch Whisky Industry Review, forty-second ed. Pagoda Scotland, Aberdeen.

Harrison, B.M., 2012. Impact of peat source on the flavour of Scotch malt whisky. In: Walker, G.M., Goodall, I., Fotheringham, R., Murray, D. (Eds.), Distilled Spirits: Science and Sustainability. Nottingham University Press, Nottingham, UK, pp. 19–26.

Harrison, B.M., Brosnan, J.M., 2017. Technical opportunities and challenges in distilled spirit production. In: Walker, G.M., Abbas, C., Ingledew, W.M., Pilgrim, C. (Eds.), The Alcohol Textbook (sixth edition). Lallemand Biofuels and Distilled Spirits, Duluth, USA, pp. 573–582.

Harrison, B.M., Ellis, J., Broadhurst, D., Reid, K.J.G., Goodacre, R., Priest, F.G., 2006. Differentiation of peats used in the preparation of malt for Scotch whisky production using Fou rier transform infrared spectroscopy. J. Inst. Brew. 112 (4), 333–339.

Hesketh-Laird, J., Hutcheon, G., Rae, D., 2012. Sustainable Scotch: Will science deliver the industry's sustainability strategy or will the strategy deliver the science. In: Walker, G.M., Goodall, I., Fotheringham, R., Murray, D. (Eds.), Distilled Spirits: Science and Sustainability. Nottingham University Press, Nottingham, UK, pp. 323–329.

Hoad, S.P., Russell, J., Looseley, M., Bayer, M., Bull, H., Ramsay, L., Thomas, W., Booth, A., Morris, J., Hedley, P., Hess, L., Brosnan, J., 2017. A Genome Wide Analysis of Key Genes Controlling Diastatic Power Activity in UK Barley(DPGENES). (PR583 ed.). Agriculture and Horticulture Development Board. https://pure.sruc.ac.uk/en/publications/a-genome-wide-analysis-of-key-genes-controlling-diastatic-power-a.

Hoover, R., 1995. Starch retrogradation. Food Reviews International 11, 331–346.

Hough, J.S., Briggs, D.E., Stevens, R., Young, T.W., 1982. Malting and brewing science. In: Hopped Wort and Beer, second ed. vol. 2. Chapman & Hall, London.

Hume, J.R., Moss, M.S., 2000. The Making of Scotch Whisky, second ed. Canongate Books, Edinburgh.

Ingledew, W.M., Kelsall, D.R., Austin, G.D., Kluhspies, C. (Eds.), 2009. The Alcohol Text-Book, fifth ed. Nottingham University Press, Nottingham, UK.

IOB, 1997. IOB Methods of Analysis. Analytical, Methods, vol. 1 Institute of Brewing, London.

Jack, F., Dabrowska, D., Davies, S., Garden, M., Maskell, D., Murray, D. (Eds.), 2018. Distilled Spirits: Local Roots, Global Reach. Context Publishers, Nottingham, UK.

Jacques, K.A., Lyons, T.P., Kelsall, D.R. (Eds.), 2003. The Alcohol Textbook, fourth ed. Notting-ham University Press, Nottingham, UK.

Jameson, R.P.M., Palmer, G.H., Spouge, J., Bryce, J.H., 2001. Resistant starch formation in the distillery. J. Inst. Brew. 107, 3–10.

Jane, J., Chen, Y.Y., McPherson, A.E., Wong, K.S., Radosavljievic, M., Kasemsuwan, T., 1999. Effects of amylopectin branch chain length and amylose content on the gelatinization and pasting properties of starch. Cereal Chem. 76, 629–637.

Jappy, M., 2010. New build distilleries—Challenges of sustainability. In: Walker, G.M., Hughes, P.S. (Eds.), Distilled Spirits. Vol. 3. New Horizons: Energy, Environment and Enlightenment. Nottingham University Press, Nottingham, UK, pp. 105–110.

JHI, 2018. New genetic marker to identify potential EPH non-producing barley varieties (https://www.huttonltd.com/news/new-genetic-marker-identify-potential-eph-non-producing-barley-varieties).

Jones, B.L., Marinac, L., 2002. The effect of mashing on malt endoproteolytic activities. J. Agric. Food Chem. 50, 858–864.

Kelsall, D.R., Lyons, T.P., 1999. Grain dry milling and cooking for alcohol production: designing for 23 percent alcohol production. In: Jacques, K., Lyons, T.P., Kelsall, D.R. (Eds.), The Alcohol Textbook, third ed. Nottingham University Press, Nottingham, UK, pp. 7–24.

Kelsall, D.R., Piggott, R., 2009. Grain dry milling and cooking for alcohol production: Designing for the options in dry milling. In: Ingledew, W.M., Kelsall, D.R., Austin, G.D., Kluhspies, C. (Eds.), The Alcohol Textbook, fifth ed. Nottingham University Press, Nottingham, UK, pp. 161–175.

Kindred, D.R., Verhoeven, T.M.O., Weightman, R.M., Swanston, J.S., Agu, R.C., Brosnan, J.M., Sylvester-Bradley, R., 2008. Effects of variety and fertiliser nitrogen on alcohol yield, grain yield, starch and protein content, and protein composition of winter wheat. J. Cereal Sci. 48, 46–57.

Lea, A., 2001. There's something about Riband. In: Arable Farming, p. 29. May, 26.

Lineback, D.R., Rasper, V.F., 1988. Wheat carbohydrates. In: Pomeranz, Y. (Ed.), Wheat Chemistry and Technology, third ed. vol. 1. American Association of Cereal Chemists, St. Paul, MN, pp. 277–372.

Ludford-Brooks, J., Davies, S., 2015. Get your energy back. Brew. Dist. Int. (April), 36–40.

Lynn, A., Prentice, R.D.M., Cochrane, M.P., Cooper, A.M., Dale, F., Duffus, C.M., Ellis, R.P., Morrison, I.M., Paterson, L., Swanston, J.S., Tiller, S.A., 1997. Cereal starches: Properties in relation to industrial use. In: Campbell, G.M., Webb, C., McKee, S.L. (Eds.), Cereals: Novel Uses and Processes. Plenum Press, New York, pp. 69–79.

Lynn, A., Smith, R., Cochrane, M.P., Sinclair, K., 1999. Starch granule enzymic digestion. In: Campbell, I. (Ed.), Proceedings of the Fifth Aviemore Conference on Malting. Brewing and Distilling. Institute of Brewing, London, pp. 46–54.

Lyons, T.P., 1999. Thinking outside the box: Ethanol production in the next millennium. In: Jacques, K., Lyons, T.P., Kelsall, D.R. (Eds.), The Alcohol Textbook, third ed. Nottingham University Press, Nottingham, UK, pp. 1–6.

MacGregor, A.W., 1991. Starch degrading enzymes. Ferment 4, 178–182.

MacGregor, A.W., Bhatty, R.S. (Eds.), 1993. Barley: Chemistry and Technology. American Association of Cereal Chemists, St. Paul, MN.

MacGregor, A.W., Fincher, G.B., 1993. Carbohydrates of the barley grain. In: MacGregor, A.W., Bhatty, R.S. (Eds.), Barley: Chemistry and Technology. American Association of Cereal Chemists, St. Paul, MN, pp. 73–130.

MacGregor, A.W., Bazin, S.L., Macri, L.J., Babb, J.C., 1999. Modelling the contribution of alpha amylase, beta amylase and limit dextrinase to starch degradation during mashing. J. Cereal Sci. 29, 161–169.

MAGB, 2014. The Assured UK Malt Technical Standard, Version 3.5. Maltsters Association of Great Britain, Newark. www.ukmalt.com.

MAGB, 2020. The MBC Approved Malting Barley List for 2021 Harvest. Maltsters Association of Great Britain, Newark. www.ukmalt.com.

Miles, M.J., Morris, V.J., Ring, S.G., 1985a. Gelation of amylose. Carbohydr. Res. 135, 257–269.

Miles, M.J., Morris, V.J., Orford, P.J., Ring, S.G., 1985b. The roles of amylose and amylopectin in the gelation and retrogradation of starch. Carbohydr. Res. 135, 271–281.

Mlotkiewicz, J.A., 1998. The role of the Maillard reaction in the food industry. In: O'Brien, J., Nursten, H.E., Crabbe, M.J.C., Ames, J.M. (Eds.), The Maillard Reaction in Foods and Medicine. Royal Society of Chemistry, Cambridge, UK, pp. 19–27.

Muller, R., 1991. The effects of mashing temperature and mash thickness on wort carbohydrate composition. J. Inst. Brew. 97, 85–92.

Newton, J., Stark, J.R., Riffkin, H.L., 1995. The use of maize and wheat in the production of alcohol. In: Campbell, I., Priest, F.G. (Eds.), Proceedings of the Fourth Aviemore Conference on Malting, Brewing and Distilling. Institute of Brewing, London, pp. 189–192.

Nicol, D.A., 1990. Developments in distillery practices in response to external pressures. In: Campbell, I. (Ed.), Proceedings of the Third Aviemore Conference on Malting, Brewing and Distilling. Institute of Brewing, London, pp. 117–137.

O'Brien, J., Nursten, H.E., Crabbe, M.J.C., Ames, J.M. (Eds.), 1998. The Maillard Reaction in Foods and Medicine. Royal Society of Chemistry, Cambridge, UK.

O'Rourke, T., 2003. Mash separation systems. The Brewer Int. 3 (2), 57–59.

Orth, R.A., Shellenberger, J.A., 1988. Origin, production and utilization of wheat. In: Pomeranz, Y. (Ed.), Wheat: Chemistry and Technology, third ed. vol. 1. American Association of Cereal Chemists, St. Paul, MN, pp. 1–14.

Palmer, G.H., 1986. Adjuncts in brewing and distilling. In: Campbell, I., Priest, F.G. (Eds.), Proceedings of the Second Aviemore Conference on Malting. Brewing and Distilling. Institute of Brewing, London, pp. 24–45.

Palmer, G.H., 1989. Cereals in malting and brewing. In: Palmer, G.H. (Ed.), Cereal Science and Technology. Aberdeen University Press, Aberdeen, Scotland, pp. 61–242.

Palmer, I., 2012. Building a brand new grain whisky distillery. Brew. Dist. Int. 8 (12), 18–23.

Pearson, T., Wilson, J., Gwirtz, J., Maghirang, E., Dowell, F., McCluskey, P., Bean, S., 2007. Relationship between single wheat kernel particle-size distribution and Perten SKCS 4100 hardness index. Cereal Chem. 84 (6), 567–575.

Pérez, S., Bertoft, E., 2010. The molecular structures of starch components and their contribution to the architecture of starch granules: a comprehensive review. Starch/Stärke 62, 389–420.

Petersen, P.B., Munck, L., 1993. Whole crop utilization of barley, including potential new uses. In: MacGregor, A.W., Bhatty, R.S. (Eds.), Barley: Chemistry and Technology. American Association of Cereal Chemists, St. Paul, MN, pp. 437–474.

Piggott, J.R., Conner, J.M., 1995. Whiskies. In: Lea, A.G.H., Piggott, J.R. (Eds.), Fermented Beverage Production. Chapman & Hall, Glasgow, pp. 247–274.

Piggott, J.R., Sharp, R., Duncan, R.E.B. (Eds.), 1989. The Science and Technology of Whiskies. Longman Scientific & Technical, Harlow.

Pomeranz, Y., 1988. Chemical composition of cereal structures. In: Pomeranz, Y. (Ed.), Wheat Chemistry and Technology, third ed. vol. 1. American Association of Cereal Chemists, St. Paul, MN, pp. 97–158.

Pyke, M., 1965. The manufacture of grain whisky. J. Inst. Brew. 71, 209–218.

Rae, W.D., 2008. Sustainability in the cereals supply chain. In: Bryce, J.H., Piggott, J.R., Stewart, G.G. (Eds.), Distilled Spirits: Production. Technology and Innovation. Nottingham University Press, Nottingham, UK, pp. 2–6.

Looseley, M.E., Ramsay, L., Bull, H., Swanston, J.S., Shaw, P.D., Macaulay, M., Booth, A., Russell, J.R., Waugh, R., Thomas, W.T.B., 2020. Association mapping of malting quality traits in UK spring and winter barley cultivar collections. Theor. Appl. Genet. 133, 2567–2582.

Rankin, W.D., 1977. New production methods in distilling. The Brewer 63, 90–95.

Rankin, W.D., 1990. Future trends in plant for maltings breweries and distilleries. In: Campbell, I. (Ed.), Proceedings of the Third Aviemore Conference on Malting. Brewing and Distilling. Institute of Brewing, London, pp. 138–148.

Riffkin, H.L., Bringhurst, T.A., McDonald, A.M.L., Hands, E., 1990. Quality requirements of wheat for distilling. Asp. Appl. Biol. 25, 29–40.

Robson, F., 2001. Cameronbridge: a distilling giant. The Brew. Int. 1 (4), 16–19.

Robson, F., 2002. Roof with a view. The Brew. Int. 2 (1), 42–47.

Robyt, J.F., 1984. Enzymes in the hydrolysis and synthesis of starch. In: Whistler, R.L., Be-Miller, J.N., Paschall, E.F. (Eds.), Starch: Chemistry and Technology, second ed. Academic Press, London, pp. 87–123.

Sasaki, T., Yasui, T., Matsuki, J., 2000. Effect of amylose content on gelatinization and pasting properties of starches from waxy and non-waxy wheat and their seeds. Cereal Chem. 77, 58–63.

Shannon, J.C., Garwood, D.L., 1984. Genetics and physiology of starch development. In: Whistler, R.L., BeMiller, J.N., Paschall, E.F. (Eds.), Starch: Chemistry and Technology, second ed. Academic Press, London, pp. 25–86.

SRUC, 2020. Scottish Recommended Lists for Cereals 2020/21. https://www.sruc.ac.uk/cerealslist.

Stark, W.H., Kolachov, P., Willkie, H.F., 1943. Wheat as a raw material for alcohol production. Ind. Eng. Chem. 35 (2), 133–137.

Statutory Instrument 2009, 2009. Scotch Whisky Regulations 2009 (citation 2009, No. 2890). HMSO, London.

Stenholm, K., 1997. Malt Limit Dextrinase and its Importance in Brewing. VTT Publications, Espoo, Finland.

Sun, Z., Henson, C.A., 1992. Extraction of α-glucosidase from germinated barley kernels. J. Inst. Brew. 98, 289–292.

Swanston, J.S., Thomas, W.T.B., Powell, W., Young, G.R., Lawrence, P.E., Ramsay, L., Waugh, R., 1999. Using molecular markers to determine barleys most suitable for malt whisky distilling. Mol. Breed. 5, 103–109.

Swanston, J.S., Smith, P.L., Gillespie, T.L., Brosnan, J.M., Bringhurst, T.A., Agu, R.C., 2007. Associations between grain characteristics and alcohol yield among soft wheat varieties. J. Sci. Food Agric. 87, 676–683.

Swinkels, J.J.M., 1985. Sources of starch its chemistry and physics. In: Van Beynum, G.M.A., Roels, J.A. (Eds.), Starch Conversion Technology. Marcel Dekker, New York, pp. 15–46.

Sylvester-Bradley, R., Kindred, D., Weightman, R., Thomas, W.T.B., Swanston, J.S., et al., 2010. Genetic Reduction of Energy Use and Emissions of Nitrogen through Cereal Production: GREEN Grain, HGCA Project Report No. 468. Home Grown Cereals Authority, Agriculture and Horticulture Development Board, Kenilworth, UK.

Taylor, B.R., Roscrow, J.C., 1990. Factors affecting the quality of wheat grain for distilling in northern Scotland. Asp. Appl. Biol. 25, 183–191.

Tester, R.F., 1997. Properties of damaged starch granules: composition and swelling properties of maize, rice, pea and potato fractions in water at various temperatures. Food Hydrocoll. 1, 293–301.

Travis, G.L., 1998. American bourbon whiskey production. Ferment 11, 341–343.

Uthayakumaran, S., Wrigley, C.W., 2010. Wheat: characteristics and quality requirements. In: Wrigley, C.W., Batey, I.L. (Eds.), Cereal Grains, Assessing and Managing Quality. Woodhead Publishing, Cambridge, UK, pp. 59–111.

Walker, E.W., 1986. Grain spirit—Which cereal? In: Campbell, I., Priest, F.G. (Eds.), Proceedings of the Second Aviemore Conference on Malting, Brewing and Distilling. Institute of Brewing, London, pp. 375–380.

Walker, G.M., Hughes, P.S. (Eds.), 2010. Distilled Spirits. Vol. 3. New Horizons: Energy, Environment and Enlightenment. Nottingham University Press, Nottingham, UK.

Walker, J.W., Bringhurst, T.A., Broadhead, A.L., Brosnan, J.M., Pearson, S.Y., 2001. The survival of limit dextrinase during fermentation in the production of Scotch whisky. J. Inst. Brew. 107 (2), 99–106.

Walker, G.M., Goodall, I., Fotheringham, R., Murray, D. (Eds.), 2012. Distilled Spirits: Science and Sustainability. Nottingham University Press, Nottingham, UK.

Walker, G.M., Abbas, C., Ingledew, W.M., Pilgrim, C. (Eds.), 2017. The Alcohol Textbook, sixth ed. Nottingham University Press, Nottingham, UK.

Wan, Y., Poole, R.L., Huttly, A.K., Toscano-Underwood, C., Feeney, K., et al., 2008. Transcriptome analysis of grain development in hexaploid wheat. BMC Genomics 9, 121.

Weightman, R., Sylvester Bradley, R., Kindred, D., Brosnan, J., 2010. Growing Wheat for Alcohol/Bioethanol Production, HGCA Information Sheet 11. Home Grown Cereals Authority, Agriculture and Horticulture Development Board, Kenilworth, UK.

Whistler, R.L., BeMiller, J.N., Paschall, E.F., 1984. Starch: Chemistry and Technology, second ed. Academic Press, London.

Whitby, B.R., 1995. Designing a distillery for the future. In: Campbell, I., Priest, F.G. (Eds.), Proceedings of the Fourth Aviemore Conference on Malting, Brewing and Distilling. Institute of Brewing, London, pp. 100–111.

Wilkin, G.D., 1983. Raw materials: milling, mashing and extract recovery. In: Priest, F.G., Campbell, I. (Eds.), Current Developments in Malting Brewing and Distilling. Institute of Brewing, London, pp. 35–44.

Wilkin, G.D., 1989. Milling cooking and mashing. In: Piggott, J.R., Sharp, R., Duncan, R.E.B. (Eds.), The Science and Technology of Whiskies. Longman Scientific & Technical, Har-low, pp. 64–88.

Wrigley, C.W., Bietz, J.A., 1988. Proteins and amino acids. In: Pomeranz, Y. (Ed.), Wheat Chemistry and Technology, third ed. vol. 1. American Association of Cereal Chemists, St. Paul, MN, p. 201.

Zobel, H.F., 1984. Gelatinization of starch and mechanical properties of starch pastes. In: Whistler, R.L., BeMiller, J.N., Paschall, E.F. (Eds.), Starch: Chemistry and Technology, second ed. Academic Press, Orlando, FL, pp. 285–309.

Further reading

Adrian, J., Legrand, G., Frangne, R., 1988. Dictionary of Food and Nutrition. Ellis Horwood, Chichester, pp. 120–121.

SWA, 2012. Scotch Whisky Industry Environmental Strategy Report 2012. Scotch Whisky Association, London. www.scotch-whisky.org.uk.

11

Water: An essential raw material for whisk(e)y production

Tom A. Bringhurst, Barry M. Harrison, and James Brosnan

The Scotch Whisky Research Institute, Edinburgh, United Kingdom

Process water

A plentiful, wholesome supply of water is an essential requirement for any distillery process (Dolan, 2003), and the distribution of water through the various parts of the process is a major aspect of distillery production, which involves repeated stages of adding and removing water. Water is added during malting and mashing and removed during distillation and coproduct treatment. It is added again following maturation as reducing water, bottling, and ultimately by consumers, who will eventually return it to the environment! Dolan (2003) has made the important point that it is relatively cheap to add (cold) water to the process but expensive to remove it.

Water is necessary at various stages in the production process in both distilling and brewing (Palmer and Kaminski, 2013). The water source and quality will often depend on the process requirements. Often, a distillery will require access to more than a single water source. Broadly, water usage in a distillery falls into three main categories: (1) water used during the process itself; (2) cooling water, which is used in fermenters and in condensers during distillation; and (3) ancillary and service supplies, which are used to supply the boiler house and heat recovery systems, as well as fulfil other site requirements. In practice, the boundaries between these categories are not fixed, and there may be considerable overlap between them, depending on availability and the individual requirements of the distillery. In Scotch whisky production, process water is covered by the current Private or Public Water Supplies (Scotland) Regulations depending on the source. The exception to this is water used for one or a combination of the mashing process or for washing plant, and for no other purpose. This exception applies where the quality of the water has no influence, either directly or indirectly, on the health of any person consuming the final product. Any further addition of water to the spirit, as reducing water, must be of potable quality according to the aforementioned regulations.

The main sources of distillery water supplies are boreholes, surface or springwater, and public town water. The composition of surface waters such as streams, rivers, lochs, and canals may be subject to organic influences, such as peat and vegetation, whilst groundwater from springs or boreholes is more affected by the underlying geology. Town (mains) water is generally more consistent but can be more expensive (Wilson et al., 2010). Distillers often have exclusive use of their own private water supplies, which gives them more control over the supply and helps to minimise any problems with external contamination, although, of course, that cannot always be guaranteed. In some cases, the distillery water supply may be some distance from the distillery and may have to be piped to the distillery (e.g. from a dam). In some cases, particularly in isolated areas, drought conditions in the summer can result in water shortages that can limit production. This is one reason why some distilleries traditionally have a silent season during the summer months. As a result of climate change, such drought conditions are likely to occur more frequently in the future. Increasing water efficiency, by investing in monitoring and auditing measures, for example, has therefore become an important focus for distillers (SWA, 2020).

Normally, malt distilleries in Scotland use soft water, the lower pH of which is favourable to the yeast. In other locations, depending on local availability and geology, hard water is used. For example, American distillers of bourbon and some Canadian distillers extol the virtues of their hard limestone water and its positive effect on the production of certain congeners. In order to overcome the alkalinity of limestone water, a sour mash process is often employed and 'thin stillage' can be added to the mash to make it more acidic. Palmer and Kaminski (2013) have produced a very comprehensive and easy to understand guidebook for treating water for beer production, with excellent chapters ranging from understanding the chemistry of water to wastewater treatment, topics that are also applicable to a wide range of spirit production.

As far as possible, water supplies should be reliable and available at all times. Dolan (2003) summarised the main quality characteristics that are required for the water supply to a Scotch malt whisky distillery. Generally, the water supply should be clean and wholesome, with a clear appearance that is free from taints. The exception to these criteria would be certain island or peaty areas where the water can have distinctive colouring and flavour characteristics at certain times. Process water does not strictly need to be of potable quality, provided it is free from contaminants and otherwise meets the mineral and salt requirements for mashing. Distillery mashing waters from different sources can vary greatly in their organic and mineral content (Wilson et al., 2010).

Although traditionally Scotch whisky production does not use water treatment prior to its use in the process, this is not the case in all countries. The purpose of water treatment is to remove unwanted components from the water prior to use and to add desirable components missing from the water (Eumann and Schildbach, 2012). Treatment procedures will not be discussed in detail but can be summarised as follows:

- Activated carbon filtration is commonly used to purify ground and surface water containing moderate concentrations of dissolved organic matter. Activated carbon improves water taste, colour, and odours. It also filters residual suspended solids and chlorine. In addition, carbon filtration removes organic pollutants such as pesticides, phenols, and chlorinated compounds, including trihalomethanes (THMs). Excessively turbid water requires clarification prior to carbon treatment. Activated carbon is very effective in removing excessive chlorine residues from municipal water supplies. Periodic steam or hot water treatment is essential to sterilise the carbon. This treatment is usually preceded by a water backwash.

- Microbiological control of water has advanced during the past few decades. The identification methods and the major contaminating microorganisms that can cause problems are discussed in Chapter 13. The following techniques are used to eliminate contaminating microorganisms used in distilling:
  - Ultraviolet (UV) light
  - Heat
  - Chemicals
  - Sterile filtration
- Water clarification is necessary in some instances prior to carbon filtration. The following techniques have been employed:
  - Sand filtration
  - Gravity settling
  - Coagulants
  - Flocculating polymers
  - Oxidation plus filtration
- Contemporary methods used for removal of ions and organic molecules include the following:
  - Ion exchange resins (IER)
  - Reverse osmosis (RO)
  - Nanofiltration (NF) membranes
  - Crossflow filtration

Process water is often considered to make a major contribution to spirit quality and, traditionally, malt distillers would not generally use any water treatments as doing so might affect the flavour characteristics of the wort and the spirit. However, if there are problems with the available water supplies, the process water might be sterilised by filtration or, for example, by ultraviolet light to remove any microorganisms that could potentially have a negative effect on production, either through infection, which could impact mashing and fermentation performance, or through spoilage by introducing unpleasant or undesirable flavour notes.

In addition to impacting flavour, the salts present in process water can affect the process pH, which can influence the efficiency of the malt enzymes and provide essential trace elements for the yeast necessary to maintain good fermentation performance. The summary by Dolan (2003) of the impact of the most important components commonly found in water sources is a useful guide to selecting sources of process water. Calcium, magnesium, and zinc are all essential ions for yeast. Calcium is also an important cofactor for the activity of starch-degrading enzymes (α-amylase). Excess levels of sulphates can reduce the mash pH, which can also have a negative impact on mashing and subsequent fermentation performance. High levels of carbonates can increase pH and will increase the risk of scale formation on heating surfaces. The presence of significant levels of nitrates and/or nitrites indicates surface water or sewage contamination, which should be avoided. A typical analysis of soft water used in a Scotch whisky distillery is shown in Table 11.1.

TABLE 11.1 Typical analysis of soft water used in a Scotch whisky malt distillery.

Component	Concentration (mg/L)
Dissolved solids	46
Calcium	5
Magnesium	6
Bicarbonate	5
Sulphates	14
Nitrate	–
Chloride	27

Service water used for steam generation and other ancillary purposes may require more extensive treatment to avoid problems with the boiler and ensure that heating surfaces and heat exchangers and associated pipework are kept free from deposits and scales. Cleaning water from a normal standard water supply is used for general hygiene and general cleaning purposes.

Some distilleries require the use of cooling towers when there is insufficient cold water available to efficiently recover the heat from the process or to comply with local environmental discharge consent conditions. Compliance with such conditions may be more challenging where climate change causes ambient temperatures to increase. Cooling towers must be regularly monitored and carefully maintained to avoid the presence of *Legionella*, a bacterium that can result in significant pathogenic effects in humans. The system must be maintained in a clean condition and any bacterial infection controlled by regular decontamination and sterilisation.

Any water used in the plant must be discharged according to the current environmental regulations and may be passed through a wastewater treatment system to ensure that it meets satisfactory consent conditions before it is returned to the environment. Dolan (2003) outlined general good practice in managing water supplies to avoid waste and reduce overall water usage and highlighted the importance of using automatic cleaning in place (CIP) to reduce discharges of wastewater and prevent the dumping of cleaning agents and sterilants into the environment. Cooling water should be recovered and reused in the most energy-efficient manner. The mashing process should be carefully controlled to avoid excessive drainings from the mash tun and reduce the amount of flushing necessary to accomplish product transfer into the fermenter. The distillery plant should be well maintained and supervised to avoid losses of steam condensate and other materials from cooling systems. Any leaks in the system (condensers, heat exchangers, valves, pipework) should be diligently addressed.

When considering waste streams from a distillery, in addition to wastewater, air emissions from alcohol and volatile organic compounds (VOCs) are also an environmental concern in the spirit industry but are not covered in this book. The reader is referred to local regulations for their particular country; these currently vary greatly from very little regulation to very strict guidelines. One example of regulations for air emissions for distilled spirit production and storage in warehouses can be found on the US Environment Protection Agency website (http://www.epa.gov/ttnchie1/ap42/ch09/final/c9s12-3.pdf).

Water treatment

The purpose of water treatment is to remove unwanted components from the water prior to its use in mashing and subsequent processes and, in some circumstances where legally allowed, to add desirable components missing from the water. For wastewater disposal, distilling is characterised by the fact that the raw materials employed, such as malt and unmalted cereal, are organic and biodegradable. Consequently, process discharges are predominantly biodegradable. Removal of copper from wastewater is a unique issue and will be discussed separately in this chapter. An excellent resource book for managing water treatment is *The NALCO Water Handbook* (2009). At over 1200 pages, it is a very comprehensive guide on the use and conditioning of water and wastewater in an industrial facility.

Due to regulations and the local requirements, distilleries increasingly have their own wastewater treatment plants or have an agreement to pay for wastewater treatment performed by the municipal treatment plant. Two main treatment systems exist for distillery wastewater: anaerobic and aerobic. During anaerobic treatment, organic matter is converted to methane and carbon dioxide; however, during aerobic treatment, organic matter is converted to carbon dioxide and biomass.

Anaerobic treatment of wastewater means that the degradation process takes place in the absence of oxygen. Microorganisms degrade organic material such as starch, lipids, and proteins into smaller molecules such as fatty acids, amino acids, and sugars and subsequently into hydrogen, carbon dioxide, and methane. Different microorganisms take care of these processes. The result is that the organic material is converted predominantly into biogas, and only a limited amount of biomass is produced. A major advantage of anaerobic systems is that biogas can be used as a fuel, thereby saving the purchase of oil, coal, or natural gas (Kormelinck, 2008). In turn, as a renewable energy source, the production of biogas lowers the carbon footprint of the process by reducing reliance on the delivery of fossil fuels.

There are several types of aerobic digestion. The most well-known and widely used system employs activated sludge. Other types are the oxidation bed, rotating disc reactor, and fluidised bed reactor (Kormelinck, 2008). The activated sludge system consists of a pretreatment step to remove large particulate substances by screening, an aeration basin, a sedimentation basin, and, finally, sludge handling facilities. The biomass or sludge absorbs the organic material, increases the amount of cell material, and, as a consequence, produces more sludge. A large part of the dissolved organic material, between 40% and 80%, will be converted into solid material, which has to be disposed of usually into landfill. The rest of the BOD provides the energy for cell maintenance and is released into the atmosphere as carbon dioxide. Table 11.2 provides a comparison of the anaerobic and aerobic processes.

Whether an aerobic or an anaerobic wastewater treatment system is employed depends mainly on the demands of the local authorities, the cost of onsite treatment, and the costs for treatment by the local authorities. Local circumstances that might lead to a combined anaerobic-aerobic wastewater treatment plant include the following:

- Limited building area is available.
- Distillery is located in a residential area and odour treatment is required.

TABLE 11.2 Comparison of the anaerobic and aerobic processes.

Aspect	Anaerobic process	Aerobic process
BOD removal	70% to 85%	95% to 99%
Space requirements	Small area, 20% of aerobic process requirement	Large surface area
Investment	Low	High
Power consumption	Low	High
Sludge produced	20% of aerobic	High
Biogas	Yes	None
Running costs	Low	High
Bioreaction process time	Low, approximately 2h	24h
Robustness	Poor	Good

- The first part of this combined process is anaerobic and produces methane. This is followed by aerobic treatment producing an effluent that can be discharged into the public sewer system.
- Since this treatment system requires less oxygen, it results in lower energy costs. In addition to the need for a relatively small building area and near odourless treatment, the entire process is also relatively rapid.

Treatment of copper in the effluent

Copper pot stills are used in all-malt whisky distilleries, primarily because of their efficient heat-transferring properties. Copper is also present in the continuous (Coffey) stills. In addition, the copper plays a very important role in spirit quality. When the spirit vapour comes in contact with a copper still, it reduces the less desirable sulphur compounds in the vapour such as dimethyl trisulphide (DMTS). Copper picked up at this stage of the distilling process is in a soluble form. Although some can be found in the final malt whisky (385 to 480ng/L), most of it will find its way into the discharge effluent. Typically, copper levels of 10mg/L are found at this stage in the process.

Distilleries can discharge their effluent into town sewage as well as the freshwater and marine environment. For Scotch whisky distillers, environmental quality standards (EQS) for copper are based on the latest scientific understanding of the UK Technical Advisory Group (UKTAG) for the Water Framework Directive (WFD). Benefitting from the science, current EQS better reflect the bioavailability of copper to aquatic wildlife, thus placing a greater emphasis on developing a more ecologically relevant approach to assessing water quality (SEPA, 2020). The EQS are calculated differently for fresh and marine environments given the differences, such as salinity and tidal currents in marine environments. One factor contributing to copper bioavailability in both environments is the level of organic material, measured as dissolved organic carbon (DOC). Higher levels of DOC are associated with a lowering of the bioavailability of copper as the copper binds to the organic material.

Though the changes to the EQS have tended to make it easier to meet copper discharge limits, there is still an interest in technologies available to extract copper from effluent. For this purpose, much work has been performed looking at a variety of techniques based on precipitation, adsorption, ion exchange, or filtration principles (Walker, 2012; Jack et al., 2014). Currently, there is an interest in methods that can be applied that allow the copper to be recovered and recycled as a valuable product. In the past, electrolysis has been used in conjunction with ion exchange to recover copper, but more recently the process of bioremediation has attracted attention as a green and cost-effective approach to produce valuable copper nanoparticles (Kimber et al., 2020).

Water conservation and the future

Water conservation is now an important factor in the design and operation of any distillery, from the perspective of increased efficiency as well as in terms of long-term sustainability. This requires careful consideration of water usage to minimise waste and reduce environmental impacts. Systems that integrate water usage throughout the distillery, where water from one part of the process can be used effectively elsewhere in the distillery, are now important features in the design of modern and new-build distilleries. The capability to meter water use and a clear definition of the different categories of water, such as process water or cooling water, are important to accurately calculate water efficiency. Older plants may be retrofitted to meet new targets for reduced water usage and improved energy efficiency. Careful treatment of the water at all stages of the process is key to a successful distillery today.

References

Dolan, T.C.S., 2003. Malt whiskies: raw materials and processing. In: Russell, I. (Ed.), Whisky: Technology, Production and Marketing. Academic Press, London, pp. 27–74.

Eumann, M., Schildbach, S., 2012. 125th Anniversary review: water sources and treatment in brewing. J. Inst. Brew. 118, 12–21.

Jack, F., Bostock, J., Tito, D., Harrison, B., Brosnan, J., 2014. Electrocoagulation for the removal of copper from distillery waste streams. J. Inst. Brew. 120, 60–64.

Kimber, R.L., Parmeggiani, F., Joshi, N., Rakowski, A.M., Haigh, S.J., Turner, N.J., Lloyd, J.R., 2020. Synthesis of copper catalysts for click chemistry from distillery wastewater using magnetically recoverable bionanoparticles. Green Chem. 21, 4020–4024.

Kormelinck, V.G., 2008. Wastewater treatment and energy recovery go hand in hand in the distilling industry. In: Bryce, J.H., Piggott, J.R., Stewart, G.G. (Eds.), Distilled Spirits: Production, Technology and Innovation. Nottingham University Press, Nottingham, UK, pp. 257–264.

Nalco Chemical Company, 2009. The Nalco Water Handbook, third ed. McGraw-Hill, New York.

Palmer, J., Kaminski, C., 2013. Water: A Comprehensive Guide for Brewers. Brewers Publications, Boulder, CO.

SEPA, 2020. Supporting Guidance (WAT-SG-53): Environmental Quality Standards and Standards for Discharges to Surface Waters. https://www.sepa.org.uk/media/152957/wat-sg-53-environmental-quality-standards-for-discharges-to-surface-waters.pdf.

SWA, 2020. Environmental Strategy Report 2020. https://www.scotch-whisky.org.uk/media/1728/environmental-strategy-report-2020.pdf.

Walker, S., 2012. Cost effective solution for copper removal in distillery effluents. In: Walker, G.M., Goodall, I., Fotheringham, R., Murray, D. (Eds.), Distilled Spirits: Science and Sustainability. Nottingham University Press, Nottingham, UK, pp. 289–294.

Wilson, C.A., Jack, F.R., Priest, F.G., 2010. The role of water composition on malt spirit quality. In: Walker, G.M., Hughes, P.S. (Eds.), Distilled Spirits. Vol. 3. New Horizons: Energy, Environment and Enlightenment. Nottingham University Press, Nottingham, UK, pp. 267–275.

Distilling yeast and fermentation

Inge Russell[a] and Graham G. Stewart[a,b]

[a]International Centre for Brewing and Distilling (ICBD), Heriot-Watt University, Edinburgh, United Kingdom [b]GGStewart Associates, Cardiff, Wales, United Kingdom

Introduction

The selection of a suitable yeast culture is very important during fermentation in any distillery. Traditionally, little thought had been devoted to yeast selection, and a locally sourced spent brewing yeast culture was employed for whisk(e)y fermentations. This yeast culture was traditionally available from breweries located in the vicinity of the distillery. It was inexpensive and convenient to obtain and use. Currently, more specialised distilling yeasts that possess an improved tolerance to ethanol and that can employ a wider range of fermentable substrates have become more popular as replacements for spent brewing yeast cultures. These are sometimes (not often) blended with a spent brewing yeast culture.

Pure culture distilling yeast strains are now commercially available that possess an additional genetic spectrum. This allows many such strains to ferment larger sugar molecules such as maltotetraose (G4) and smaller dextrin materials (G5 and larger), and small peptides (Lekkas et al., 2009). They usually have an ability to better withstand various fermentation stresses (Cheung et al., 2012; Pauley and Maskell, 2017; Gibson et al., 2017)—details later.

Important yeast parameters for the distiller include tolerance to alcohol, osmotic pressure effects, temperature, and reduced pH. Sugar uptake and flocculation properties and culture storage, culture viability, and culture vitality characteristics are key (Stewart, 2020). In addition, the fermentation velocity of the strain, its fermentation lag period when initially pitched into wort, and appropriate congener (metabolite) formation for particular distilled products are all relevant attributes.

Unlike brewers, Scotch and other whisky producers do not recycle their yeast—details later. Also, the regulations (see Chapter 10) do not permit the addition of nutrients such as yeast foods or enzymes to the mash or to the all-malt wort fermentation, making the selection of the correct yeast strain for the metabolism of the appropriate substrate(s) even more important. However, there are different regulations for neutral spirit production, where these additions are allowed (Noguchi et al., 2008). A final ethanol yield of over 90% theoretical conversion efficiency is the goal (Chapter 10).

Sourcing the yeast

The primary yeast culture can be propagated in a distillery from its own starter culture or supplied by a commercial yeast manufacturer in a number of forms such as dry yeast, wet cake yeast, or as a stabilised liquid yeast culture (Walker and Stewart, 2016). These yeast cultures can have the following composition: dried (~95% solids content), compressed (~35% solids content), or creamed (~18% solids content), and typically the yeast has been propagated aerobically. A secondary yeast (traditionally spent ale brewer's yeast) can be used for additional flavour development and, in some cases, this addition also results in a higher final ethanol yield.

There are a number of commercial suppliers of high-quality distilling yeast cultures. The use of spent brewer's yeast has been phased out due to issues of supply, varying quality, and problems with maintaining a consistent product flavour profile. Consistency of flavour associated with a particular product is an important factor as the flavour of the spirit can be affected by which yeast culture is selected and by the wort composition and fermentation conditions employed. This latter yeast property is not fully appreciated (Wanikawa, 2020). For many years, it was considered that the only function of a distilling yeast culture was ethanol production. However, this is not the case!

Distillers have standard microbiological quality criteria for the yeast they purchase. The requirements for the culture include the following: high viability (>95%) and low levels of contaminating bacteria and wild yeasts (usually $<1 \times 10^4$ total per gram of dried yeast). Acetic acid bacteria should total $<1 \times 10^3$ per gram of dried yeast, with no pathogenic microorganisms present—details in Chapter 13.

A rapid start to wort fermentation is key, particularly in terms of avoiding bacterial contamination. Some yeast strains include a killer factor in their genetic make-up, which can help protect the fermentation from contamination by some wild yeast strains (Chapter 13). However, yeasts containing the killer factor are encountered more commonly in the wine industry (Niël van Wyk et al., 2020).

Yeast nomenclature

The nomenclature of yeast used for traditional whisky fermentations is a confusing subject and not uniformly or extensively considered in the literature. Baker's yeast is traditionally a strain of the species *Saccharomyces cerevisiae* (Pyke, 1958) and most whisky strains are *S. cerevisiae*, although other whisky yeast species are sometimes used in conjunction with this particular yeast species (Suomalainen and Lehtonen, 1979).

Lager yeast cultures from breweries [correctly designated *S. pastorianus* rather than its original (traditional) name of *S. carlsbergensis* (Hittinger, 2013), a name some brewers still prefer to use for lager yeast strains)] will struggle to grow at temperatures above 34°C

(Stewart and Russell, 2009). Since lager yeasts can tolerate the lower growth temperatures, it makes them ideal for use at the fermentation temperatures used for lager beer production, which is always conducted at lower temperatures (8–18°C) than that employed for ale fermentations (20–26°C) and typically for whisky fermentations (28–34°C).

S. pastorianus is a hybrid of the ale yeast *S. cerevisiae* and a closely related yeast species *S. eubayanus*. Sequencing studies with *S. bayanus*, a wine yeast found in wine fermentation environments, has shown that it is a complex hybrid of *S. eubayanus, S. uvarum,* and *S. cerevisiae* (Hittinger, 2013).

For industrial processes, baker's yeast is usually a strain of *S. cerevisiae*, lager yeast is *S. pastorianus*, ale yeasts include *S. cerevisiae* and rarely some *S. bayanus* strains, rum is fermented primarily with *S. cerevisiae* and the fission yeast species *Schizosaccharomyces pombe* (also with various wild yeasts) (also in Chapter 25). The wine industry mostly uses *S. cerevisiae* and/or *S. bayanus*, together with various wild yeasts from a plethora of other genera such as *Kloeckera, Saccharomycodes, Schizosaccharomyces, Hansenula, Candida, Pichia,* and *Torulopsis,* which are found naturally on grapes. Various natural species of the *Saccharomyces* genus have been recognised, including *S. cerevisiae, S. paradoxus, S. mikitae, S. kudriavzevii, S. aboricola, S. bayanus,* and *S. cariocanus* (Naumov et al., 2013; Stewart, 2017; Niël van Wyk et al., 2020). Multiple combinations of these *Saccharomyces* species have led to the industrial hybrids illustrated in Fig. 12.1. Various species can be further classified into different strains, and a considerable number of different strains of *S. cerevisiae* have been recognised. Hybridisation is common between the domesticated yeasts used for alcohol production. Unravelling the complex history of these strains continues to evolve with the use of advanced molecular techniques now available to this area of yeast taxonomy (Walther et al., 2014; Stewart, 2017; Libkind et al., 2020).

The yeast *S. cerevisiae* is an intensively researched organism because it has been used as a model species by scientists to explore eukaryotic systems [eukaryotic cells include mammalian, fungal, and plant cells, while prokaryotic cells encompass bacteria and cyanobacteria (blue-green algae)] (Pedersen, 2019). In 1996, *S. cerevisiae* (haploid strain S288c) was the first eukaryotic genome to be completely sequenced (Goffeau et al., 1996), and over 6000 protein-coding genes have now been identified in the *Saccharomyces* genus. There are a number of databases that contain a wealth of information regarding the genes and proteins in this yeast species. The Saccharomyces Genome Database (SGD) is located at http://www.yeastgenome.org/, and this site not only provides information on the yeast but has provided research and analytical

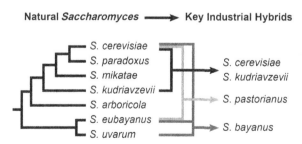

FIG. 12.1   Relationships of *Saccharomyces* strains. Adapted from Hittinger, C.T., 2013. *Saccharomyces* diversity and evolution: a budding model genus. Trends Genet. 29(5), 309–317.

tools that allow researchers to explore the data in detail. There is now even an iPhone App that has been introduced to assist researchers in conveniently navigating the vast database of *Saccharomyces* in order to monitor the research on the genes of interest (Wong et al., 2013).

Understanding the structure of the yeast cell

Yeast, including the genus *Saccharomyces*, is a unicellular fungus. The *Saccharomyces* yeast cell varies in size, but typically, it is roughly 5–10μm in diameter (1μm=10^{-4}cm). Fig. 12.2 is a diagrammatic representation of a yeast cell illustrating this eukaryotic cell's main features.

Individual yeast cells are invisible to the naked eye and need to be viewed under a light microscope. With unstained cells, it is difficult to see any intracellular details other than the larger vacuoles situated in the cell's cytoplasm (Fig. 12.3).

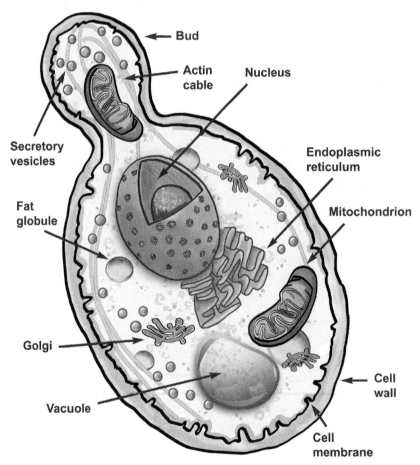

FIG. 12.2 Main features of a typical *Saccharomyces* yeast cell. The organelles inside the cell are not drawn to scale but are enlarged for illustration purposes.

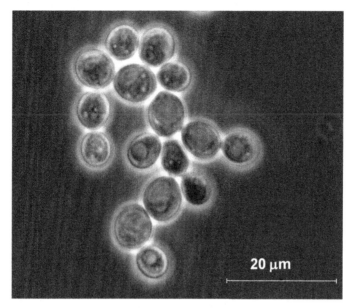

FIG. 12.3   Phase-contrast microscopic image of distiller's yeast cells.

The cell wall

The yeast cell wall is a rigid structure that provides the cell with physical protection, particularly against the stresses of the 'outside world', in addition to other roles such as osmotic homeostasis, cell shape maintenance, and a scaffold function for proteins. Glucan, mannoprotein (previously called mannan), and chitin make up over 90% of the cell wall (Fig. 12.4). The cell wall constitutes 10%–25% of the cell's dry weight, with the mannoprotein fraction making up 30%–50% of the wall, the β-glucan 35%–55%, and the chitin 1.5%–6%. The cell wall thickness varies depending on the environment, from ~100 to 200nm. When the cell reaches the stationary growth phase, it thickens.

The plasma membrane

Although the wall gives the yeast cell its overall shape, it is the plasma membrane that is the key to the cell's survival. The plasma membrane is a physical and chemical barrier between the outside world and the inside of the cell. It is a lipid bilayer, only 7nm thick, with proteins (some are glycoproteins) inserted or traversing this layer, all with various functions (Fig. 12.4). The plasma membrane provides selective permeability and controls what enters and leaves the cell, such as carbohydrates (sugars), nitrogen (amino acids and peptides), ions, and metabolites. The cell membrane allows gases and water to pass in and out of the cell while also controlling the passage of many other chemicals, particularly ethanol, carbon dioxide, glycerol, esters, carbonyls, and organosulphur compounds.

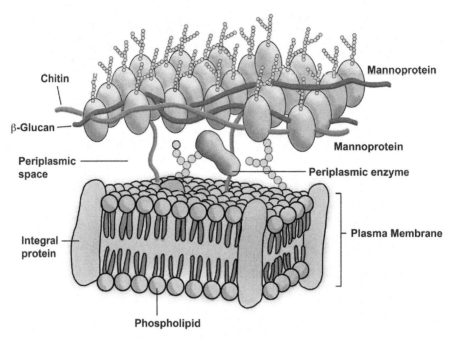

Chitin

β-Glucan

Periplasmic
space

Integral
protein

Mannoprotein

Mannoprotein

Periplasmic enzyme

Plasma Membrane

Phospholipid

FIG. 12.4   Architecture of the yeast cell wall and the plasma membrane.

The periplasmic space

The periplasmic space is the area between the cell wall and the plasma membrane. It is here that secreted proteins (mannoproteins), which are unable to permeate the cell wall, are located. These proteins fulfil essential functions such as enzymes for hydrolysing some sugars, for example, sucrose, and melibiose. These sugars cannot cross the plasma membrane into the cell. The enzyme invertase, located in the periplasmic space, converts sucrose into glucose and fructose, and these two sugars (monosaccharides) can enter the cell through the plasma membrane. In a similar manner, in lager yeast, the enzyme melibiase in the periplasmic space hydrolyses melibiose into glucose and galactose (monosaccharides) for subsequent uptake (Stewart and Russell, 2009).

The cell nucleus

The cell's nucleus is the structure that contains most (over 98%) of the cell's deoxyribonucleic acid (DNA). For the genus *Saccharomyces*, the cell's nuclear DNA is arranged into 16 chromosomes, which contain over 6000 genes. These encode most of the proteins (enzymatic and structural) synthesised in the cell. The double membrane of the nucleus is perforated at intervals with pores, and the yeast cell nucleus remains intact throughout the cell cycle. At the opposite poles are the spindle pole bodies (SPBs), which are essential for cell division.

The mitochondria

The mitochondria contain the enzymes for the reactions required during aerobic respiration. Under aerobic conditions (in the presence of oxygen), yeast mitochondria are involved in ATP synthesis coupled to oxidative phosphorylation. Mitochondria are easily recognisable in an electron micrograph of an aerobically grown yeast cell as spherical or rod-shaped (sausage) structures, surrounded by a double membrane (Fig. 12.5). They contain cristae, formed by the folding of the inner membrane. Most of the enzymes of the

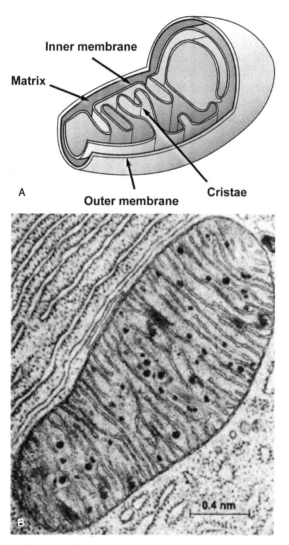

FIG. 12.5 (A) Diagram of the internal structure of a yeast mitochondrion (B) Electron micrograph of one mitochondrion.

tricarboxylic acid (TCA)/Krebs cycle are present in the matrix of the mitochondria. Enzymes involved in electron transport and oxidative phosphorylation are associated with the inner membrane cristae. Cells grown anaerobically in the absence of lipids have very simple incomplete mitochondria consisting of an outer double membrane but lack cristae. Mitochondria possess their own DNA (mtDNA), which is less than 2% of the cell's total DNA (Stewart, 2014).

The cell cytoplasm and its constituents

The yeast cell's cytoplasm is where many enzymes reside and is where reactions involved in respiration occur. The cytoplasm also contains a system of membranes known as the endoplasmic reticulum (ER), and some of these membranes are associated with ribosomes (mRNA translation). The endoplasmic reticulum is the site of protein biosynthesis and the modification of proteins that are to be exported into the cytoplasm. From the endoplasmic reticulum, proteins are directed *to* the Golgi apparatus by vesicles, and proteins are delivered *from* the Golgi to various destinations within the cell or to the cell's exterior via different secretory vesicles.

The largest organelle inside the yeast cell is the vacuole (Fig. 12.3). The vacuole is a dynamic structure, and it can rapidly modify its morphology depending on the environmental conditions. Vacuoles are primary storage compartments for basic amino acids, polyphosphates, certain metal ions (Ca^{++}, Zn^{++}, Mg^{++}, and Mn^{++}), and specific enzymes, including many proteinases. They are the key yeast organelles that are involved in the intracellular trafficking of proteins and osmoregulation. The vacuoles, which can become quite large, can be seen in the yeast's cytoplasm under a light microscope (Stewart, 2017) (Fig. 12.6).

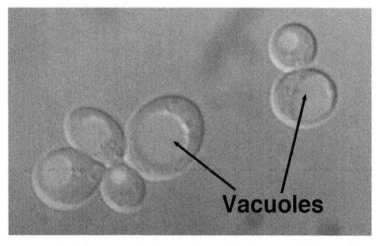

FIG. 12.6   Yeast cell vacuole stained with methylene blue (wet mount under a light microscope).

The vacuole space is encircled by the vacuolar membrane, which is called the tonoplast. The membrane contains specific membrane-bound proteins and several permeases. The tonoplast plays a vital role in metabolic processes associated with the vacuole.

Multiplication of yeast—Bud scars and birth scars

A distinguishing feature of a growing population of *Saccharomyces* cells is the presence of budding yeast cells. The daughter cell begins as a small bud, which increases in size throughout the cell cycle until it is almost a similar size to the mother cell, at which stage the bud and the mother cell separate. The cell separation site is marked on the mother cell by a bud scar (a chitin-containing ring) and on the daughter cell by a birth scar (Fig. 12.7).

These scars can be stained with a fluorescent stain such as calcofluor and then easily visualised and counted using a fluorescent microscope. No two buds arise at the same site on the yeast cell wall, and each time a bud is produced, a new bud scar forms on the cell wall of the mother cell. Counting the number of bud scars can be used to determine the age of a cell (i.e. the number of times less one that it has budded) (Mortimer and Johnston, 1959).

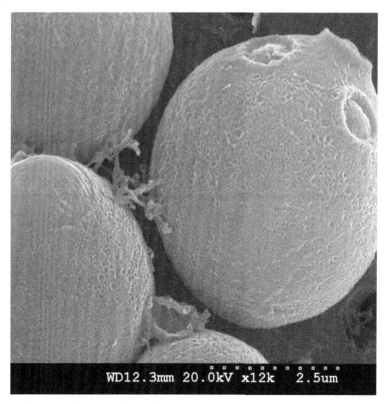

FIG. 12.7 Electron micrograph of a yeast cell with bud scars. Courtesy of A. Speers—Dalhousie University, Halifax, NS, Canada.

Biochemistry of the yeast cell and nutrient uptake from wort

When the yeast is first pitched into wort, it is introduced into a very complex environment consisting of simple sugars, dextrins, amino acids, peptides, proteins, vitamins, minerals, and many other constituents (Stewart and Russell, 2009). Without the correct nutritional components being present in the wort, at a suitable temperature and pH, the yeast cell will struggle to grow and survive!

Carbohydrate metabolic pathways

Saccharomyces cells use wort carbohydrates as their primary source of energy, in particular glucose (not exclusively), maltose, and maltotriose. The glycolysis pathway (anaerobic) converts glucose into pyruvate, producing energy in the form of ATP, which is then coupled to the formation of metabolic intermediates and reducing power in the form of NADH for the cell's various biosynthetic pathways. Yeast is the only microorganism with the ability to efficiently alternate between the cell's aerobic (respiration) pathway and anaerobic (fermentation) pathway. The two pathways are illustrated in Fig. 12.8.

Once the glucose is inside the cell, it is converted via what is called the *glycolytic pathway* or the *Embden-Meyerhof-Parnas (EMP) pathway* into pyruvate. The *EMP pathway (glycolysis)* takes place in the cytoplasm and does not require oxygen. It generates NADH and precursors such as pyruvate; however, the final step, the *tricarboxylic acid (TCA) cycle* (*citric acid cycle* or *Krebs cycle*—named after the primary discoverer of the pathway—Hans Krebs) and *oxidative phosphorylation*, is an **aerobic** pathway. This step yields biomass, CO_2,

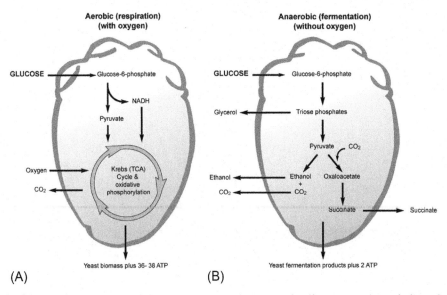

FIG. 12.8 *Saccharomyces* yeast (A) aerobic (respiration) and (B) anaerobic (fermentation) metabolic pathways.

and ATP, as shown in Fig. 12.8A. The **anaerobic** pathway (fermentation) for pyruvate production is illustrated in Fig. 12.8B.

The overall equation for the fermentation of glucose to ethanol is given below.

$$C_6H_{12}O_6 + 2H_2PO_4^- + 2ADP = 2CH_3 - CH_2 - OH + 2CO_2 + 2ATP$$

Fig. 12.9 illustrates the detailed steps in the yeast's metabolic pathway from the entry of glucose into the cell to pyruvate production and then the last steps to ethanol production (Pronk et al., 1996).

An alternate pathway to the one described earlier, which the yeast cell also uses, is called the *pentose phosphate cycle*. This cycle occurs in the cytosol and provides the cell with pentose (5-carbon) sugars and cytosolic NADPH, which are needed for biosynthetic reactions to produce primary nucleic acids and amino acids.

What happens inside of the yeast cell?

With so many metabolic pathways and so many different names for the same pathway, the various metabolic cycles of yeast can become very confusing! During aerobic conditions (in the presence of oxygen), there are a series of reactions in the *TCA cycle* (*citric acid cycle* or

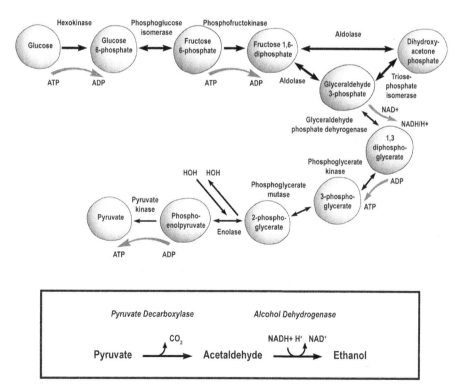

FIG. 12.9   Yeast pathway and production of pyruvate.

the *Krebs cycle*) and in the electron transfer chain (also called the *respiratory cycle* or *oxidative phosphorylation*) (Fig. 12.10).

The *TCA cycle* is a series of chemical reactions used by all aerobic organisms to generate energy and compounds for biosynthesis. This cycle feeds into the *respiratory cycle* (*electron transport chain*), which is also located in the mitochondria. The *respiratory cycle* produces a large amount of ATP and consumes oxygen. If oxygen is not present, the *respiratory cycle* cannot function, which then shuts down the *TCA cycle* (for this reason, the *TCA cycle* is called an aerobic pathway for energy production).

The *TCA cycle* begins with pyruvate (Pronk et al., 1996), which is the end product of glycolysis, and it is the first step in all types of cell respiration (and biosynthetic reactions). The *TCA cycle* takes place in the matrix of the mitochondria and uses the reactions of NAD^+ and FAD, which carry high-energy electrons to the *electron transport systems* in the inner mitochondrial membrane. The reactions in this organelle are responsible for producing most of the ATP by

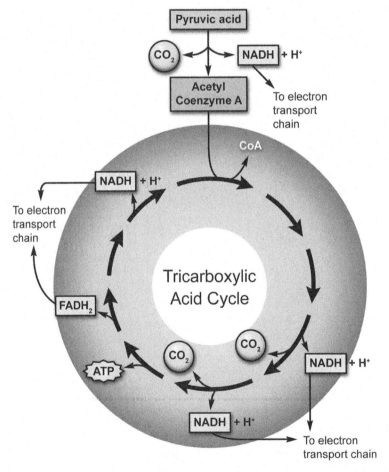

FIG. 12.10 Simplified diagram of the tricarboxylic acid (TCA) cycle (citric acid cycle/Krebs cycle) illustrating the pathway to produce energy for the cell under aerobic conditions in the mitochondria.

chemiosmosis (the movement of ions across a selectively permeable membrane, down their electrochemical gradient). This pumping establishes a proton gradient, and the energy of the protons generates ATP, using ADP and phosphate ions as the starting points.

As a result of the *TCA cycle* reactions, the *electron transport system* and *chemiosmosis*, 36 molecules of ATP are produced from each glucose molecule metabolised during *cellular respiration*. In addition, two extra ATP molecules are produced for every molecule of glucose metabolised through *glycolysis* (*EMP pathway* Fig. 12.8B). Consequently, in total, 38 molecules of ATP are produced.

Yeast and nitrogen metabolism

Yeast cells, in general, have relatively simple nutritional needs. However, they do require a source of nitrogen to grow, and they are very selective as to how they take up specific nitrogen compounds (Lekkas et al., 2007; Hill and Stewart, 2019). They can also utilize a variety of nitrogen compounds, including small peptides, ammonia, urea, and a number of amino acids (Fig. 12.11). Small peptides can be taken up by the yeast cell, and once inside the cell,

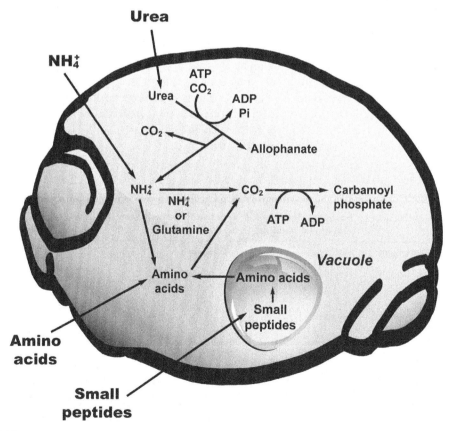

FIG. 12.11   Nitrogen uptake into the yeast cell.

they can be cleaved into amino acids. *S. cerevisiae* can store amino acids in vacuoles until they are needed by the cell. Zhang et al. (2018) extensively reviewed nitrogen regulation in *Saccharomyces* in terms of sensing, transportation, and catabolism.

During whisky fermentations, similar to brewing wort fermentations, the main sources of nitrogen are the amino acids, small peptides, and ammonium ions produced during the hydrolysis of the proteins released from the barley, during the malting and mashing processes (Lekkas et al., 2014). Yeast needs this nitrogen as a nutrient to be assimilated into proteins, RNA, DNA, and other crucial cellular nitrogen-containing compounds. *Saccharomyces* yeast can synthesise all of the building blocks it needs for nitrogenous macromolecules from these nitrogen sources. Amino acids are not usually limiting in an all-malt wort but can become limiting in a grain wort, produced mainly from either corn (maize) or wheat.

When good nitrogen sources, such as glutamic and aspartic acid, are present in the medium (wort), the transcription of some genes involved in the uptake and utilisation of the poorer nitrogen sources, such as ammonia and tryptophan, is repressed (this is called *nitrogen catabolite repression*). Three permeases are responsible for the import of ammonium ions into the yeast cell, and these permeases are subject to repression by other nitrogen sources in the medium. This limits the yeast's ability to use poor nitrogen sources in the presence of a preferred nitrogen source.

The amino acids taken up by the yeast cell are used as the building blocks of proteins, and a set of amino acid permeases enable their transport and regulation. Yeast has the capacity to synthesise virtually all of its amino acids from simple carbon skeletons plus a nitrogen source. For the sulphur-containing amino acids (cysteine, methionine, and homocysteine), the yeast can assimilate the sulphur from sulphate present in the wort.

The yeast will take up amino acids present in the wort in a preferred order. This order has been modified with only minor variations since first being described by Jones and Pierce (1964) (Table 12.1) and has been slightly revised by Lekkas et al. (2007). For example, methionine uptake has been moved from group B to group A.

TABLE 12.1   Order of wort amino acid uptake by an ale brewing yeast.

Group A Absorbed rapidly early in the fermentation	Group B Absorbed slowly at the start of fermentation	Group C Absorbed slowly later in the fermentation	Group D Little or no absorption during fermentation
Aspartic acid/asparagine	Histidine	Alanine	Proline
Glutamic acid/glutamine	Tryptophan	Glycine	
Lysine			
Arginine, serine, threonine	Tyrosine		
Methionine	Valine		
Isoleucine Leucine	Phenylalanine	(Ammonia)	

Source: Jones, M., Pierce, J., 1964. Absorption of amino acids from wort by yeasts. J. Inst. Brew. 70, 307–315., revised by Lekkas, C., Stewart, G.G., Hill, A.E., Taidi, B., Hodgson, J., 2007. Elucidation of the role of nitrogenous wort components in yeast fermentation. J. Inst. Brew. 113, 183–191.

Yeast has nitrogen sensing and signalling mechanisms that involve two different sensors, a cell surface amino acid sensor (related to amino acid permeases) and cytoplasmic nitrogen sensors (Lekkas et al., 2007). In order for the nitrogen sources to be taken up by the yeast, they must be able to traverse the plasma membrane, which is mostly impermeable to these compounds. Dedicated transporter proteins embedded in the plasma membrane assist with this transport, bringing amino acid residues and small peptides into the cell coupled with a hydrogen ion that is later expelled by the cell. These proteins (permeases) are highly regulated, and there is much about the operation of the nitrogen pathway(s) that is not entirely understood.

Distiller's yeast and oxygen

It has been discussed earlier in this chapter that distilling yeast strains can employ both respiration (aerobic) and fermentation (anaerobic) metabolic pathways. Yeast has a strong tendency to use the fermentation pathway, and it will only use the respiration pathway when the concentration of fermentable sugars is low and oxygen is present (Crabtree, 1928). Thus, growing yeast for its biomass formation requires low sugar feeding with a considerable amount of oxygen present (Pasteur and Faulkner, 1879).

When high concentrations of fermentable sugar are present (higher than ~1%), even when oxygen is present, the yeast will prefer to operate in the fermentation pathway mode. It is speculated that this could ensure that the cell does not become overloaded with the high amounts of ATP produced during respiration.

A distilling or brewing yeast cannot multiply indefinitely under anaerobic conditions. After two or three generations, the yeast requires a nutritional supplement if there is no oxygen present for the yeast to be able to maintain cell membrane health. Sterols are essential lipid components of yeast cell membranes (Quain et al., 1981). With every cell division, membrane sterols are diluted between mother and daughter cells. When the membrane lacks sufficient sterols, oxygen or lipids must be added to the wort. The addition of one of these allows the cell to synthesize membrane sterols and then continue to form new cells. When the membrane is no longer in good condition, the weakened cell becomes more susceptible to stress. 'Stuck' (incomplete) wort fermentations and the production of unwanted off-flavours can result (Stewart, 2020).

Yeast cells can take up sterols from the wort and synthesise ergosterol in the endoplasmic reticulum and store it in lipid droplets. It can then be transported to the plasma membrane as needed. The direct addition of unsaturated fatty acids and sterols to a distiller's wort fermentation during whisky production is against some countries' regulations (such as the Scotch Whisky Regulations) (Gray, 2013). However, wort aeration (oxygenation of wort) is legal in all countries. The pitching yeast for a typical whisky fermentation is commonly grown aerobically on an appropriate substrate (usually molasses). This yields a healthy yeast to carry out the subsequent wort fermentation without difficulty. This yeast is pitched at a much higher rate than for a beer fermentation, and it is not reused (recycled) in contrast to the situation in brewing fermentations.

The commercial cultivation of distiller's yeast

There are numerous manufacturers of distilling yeasts, and a large number of specialised strains are now available, depending on the particular fermentation and organoleptic profile desired in the fermented wort prior to distillation. The stock culture is propagated by the yeast manufacturer, through a succession of fermentation vessels, with slow increases in the size of the vessels employed during scale-up, with the goal being rapid growth and healthy cells (Nielsen, 2010; Cheung et al., 2012).

The culture medium (often molasses based) is supplemented with ammonium salts and other necessary nutrients. The yeast is grown with vigorous aeration (oxygenation) and careful temperature control to obtain a product that exhibits maximum cell viability and vitality. Sugar is fed into the fermentation at a level of approximately 0.5% (w/v) to maintain the yeast in a respiration mode (rather than allowing it to switch to a fermentation mode) as the goal is to accumulate yeast biomass not to produce ethanol (Stewart, 2017)!

The yeast can then be harvested by a number of methods such as rotary vacuum filtration (compressed yeast), collected and sold as a cream yeast (for easy delivery by tanker trucks and for automated pitching), centrifuged yeast, or yeast dried under a partial vacuum. By using an inert gas packaging method (e.g. nitrogen), the shelf life of the product can be enhanced.

Manufacturers of yeast for distilleries aim for minimal alcohol production during the yeast production process. Sugar is carefully fed, and higher oxygen levels are needed to produce yeast that can survive storage well in whatever form of storage is selected before use in a distilling fermentation. Table 12.2 illustrates the difference in yield between a yeast scaled up in a pure respiration mode versus a propagation in a wort with an oxygen limitation.

Depending on the yeast packaging/storage method used, a liquid reactivation step is usually employed before the yeast is pitched, especially if it has been dried prior to shipment. Each yeast manufacturer carefully details the propagation method for the specific strain being employed to ensure that rapid wort fermentation will commence during pitching. This reactivation process is called 'bubbing' or 'livening' by distillers, and it helps the yeast culture adapt to a wort environment (Stewart, 2010).

If a brewing yeast culture is used as a secondary yeast for fermentation (unusual currently), extra care must be taken in terms of yeast shipment from the brewery to the distillery and the yeasts' consequent storage in order to maintain cell viability and to avoid bacterial contamination. Yeast washing, using the method of cold phosphoric acid at pH2.2 as described by

TABLE 12.2   Comparison of oxygen consumption on two propagation modes.

	Yield factor	Oxygen consumption
Pure respiration propagation	0.54g yeast dry solids[a] per g carbohydrate	0.74g oxygen per g yeast dry solids
Propagation in a 12° Plato wort	0.10g yeast dry solids per g carbohydrate	0.12g oxygen per g yeast dry solids[b]

[a] Assumption that 1g dry matter yeast per litre is equivalent to approximately 40 million cells.
[b] Equivalent to 10–20mg oxygen per g dry yeast per hour (depending on temperatures in the range of 12–20°C).
Adapted from Nielsen, O., 2005. Control of the yeast propagation process 'glycolysis' how to optimise oxygen supply and minimize stress. Tech. Q. Master Brew. Assoc. Am. 42(2), 101–106; Nielsen, O., 2010. Status of the yeast propagation process and some aspects of propagation for re-fermentation. Cerevisia 35, 71–74.

Simpson and Hammond (1989), can be employed to lower the bacterial load. However, acid washing will not remove any wild yeast cells that may be present in the culture, but it will remove bacteria (Cunningham and Stewart, 1998).

How much yeast is required to pitch a typical distilling fermentation?

The pitching rate for a distilling fermentation is the weight of pressed yeast expressed as a percentage of the weight of malt and other cereals employed. For malt distillery fermentations, a rate of 1.8% (w/w) is typical, whereas, for grain distilleries, 1.0% (w/w) is typical. It will very much depend on the particular yeast used and the wort gravity employed. For a malt distillery fermentation, a minimum of ~18kg **dry weight** of a distilling yeast per tonne of malt (or 5g of **pressed weight** of yeast per litre of wort) is typical. The objective is a small amount of yeast growth, a rapid fermentation, and maximum theoretical alcohol production. When brewing yeast is also used as half of the inoculum, the pitching rate would be increased to ~22kg **dry weight**/tonne. The goal is to achieve a pitching rate of around $3-4\times10^7$ cells/mL. Unlike in a brewing fermentation, the distilling yeast will only divide a few times once pitched into wort and the yeast concentration at the end of fermentation will be around 2×10^8 cells/mL.

When a brewer's yeast is used as a secondary pitching yeast, it has been credited with assisting in increasing the wort pH and, consequently, assisting in the maintenance of enzyme activity. Pyruvic acid, excreted into the wort by the aerobically grown yeast, is taken up by the anaerobically grown brewer's yeast, thereby raising the wort pH. This allows the amylase enzymes to hydrolyse the starch into fermentable sugars and conduct their activity for a longer period before a low wort pH inactivates them. However, as already discussed, the use of spent brewer's yeast in Scotch whisky distilleries is currently uncommon. With the demise of local breweries (particularly ale breweries) in Scotland, this situation will continue.

Fermentation congeners

Fermentation congeners are the chemical compounds produced during fermentation and maturation. Congeners include higher alcohols, esters, acids, aldehydes, ketones, and sulphur compounds. The role of yeast and its fermentation products in the final aroma of an alcoholic beverage is well known and undisputed (Suomalainen, 1971). Quantities of the congeners vary and depend on the exact distilled product. They form the largest and possibly most important group of fermentation aroma compounds and are key in contributing to the products' organoleptic characteristics.

The higher alcohols and fusel alcohols are largely produced by yeast during fermentation by a series of biochemical reactions with amino acids and carbohydrates. Abundant yeast growth results in more higher alcohols being produced (Quain and Duffield, 1985). These can be produced via catabolic routes (e.g. isobutanol from valine) or via anabolic routes (e.g. n-propanol). The organoleptic compound most often identified in whisky is isoamyl alcohol, followed by isobutanol, amyl alcohol, and propanol (Suomalainen, 1971). Isoamyl alcohol is the primary fusel alcohol in whisky, and it can comprise 40–70% of the total fusel alcohol concentration. It has a pleasant fruit odour at low levels, but the aroma is unpleasant at high levels.

The esters can impart a flowery and fruit-like aroma to spirits, and over 90 distinct fermentation esters have been identified. Ethyl acetate is the principal whisky ester in terms of the quantity present. Other esters with lower aroma thresholds also significantly contribute to whisky flavour. Ethyl acetate and many of the long-chain esters are formed by yeast, principally by enzymic reactions during fermentation. Chemical reactions to form esters between ethanol and the corresponding acids can be of importance during ageing. Fig. 12.12 illustrates the formation of esters from various routes (Suomalainen, 1971; Suomalainen and Lehtonen, 1979). Esters are the largest group of aroma compounds present in whisky (Stewart, 2008), and the major components are formed by the ethyl esters of fatty acids. The number of esters is increased even further by the esterification of the same acids with fusel alcohols (Suomalainen, 1971). In addition to ethyl acetate (fruity/solvent aroma), some of the other whisky esters, to name just a few, include the following: ethyl caproate (apple/aniseed), ethyl caprylate (apple-like aroma), ethyl caprate (sweet, waxy, and fruity), ethyl laurate (apple, sweetish, and fruity), isoamyl acetate (banana/apple), and β-phenylethyl acetate (roses/honey).

The acids produced during fermentation can impact the overall aroma profile of the distillate. This includes acetic, propionic, isobutyric, butyric, and isovaleric acids. The fatty acids include caprylic, caproic, capric, and lauric acids. Of the acids present, isovaleric acid, with its strong, pungent cheesy smell, contributes the most to the odour profile of the spirit.

The sulphur compounds in a distillery fermentation can be formed as by-products from the biosynthesis of sulphur-containing amino acids and from the reduction of sulphate salts in the wort during fermentation (Stewart and Ryder, 2019). Copper in the still offers remedial effects in terms of neutralising the flavour of some of the sulphur compounds and meaty notes due to the formation of insoluble copper salts (e.g. copper sulphide) (Harrison et al., 2011) (see Chapters 11 and 14 for additional details).

Sulphur compounds have very low odour thresholds and are of concern since they can have large negative effects on the final aroma of the spirit. Much of the hydrogen sulphide (rotten egg aroma) and the sulphur dioxide (burnt match aroma) formed during fermentation is purged by the evolution of carbon dioxide during the fermentation. The remaining

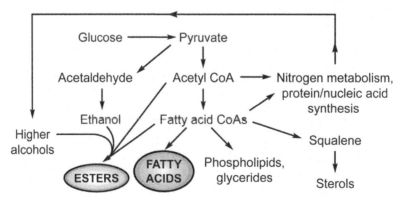

FIG. 12.12   Formation of esters and fatty acids as by-products of yeast growth.

hydrogen sulphide in the fermentation can form ethanethiol during distillation and can be oxidised to the very unpleasant garlic-like aroma of diethyl disulphide. Some of the sulphur compounds, such as diethyl disulphide, will be reduced over the years of maturation in oak barrels (casks), while other sulphur compounds remain, or new ones are formed during maturation (Stewart and Ryder, 2019).

The carbonyl compounds formed during fermentation are the side products of amino acid synthesis. In whisky, they include acetaldehyde, propionaldehyde, furfural, 2,3 pentanedione, isobutyraldehyde, n-butyraldehyde, isovaleraldehyde, n-valeraldehyde, and diacetyl. Although the carbonyl compounds constitute only a small portion of the aroma components, they contribute to the odour intensity with a sharp and poignant aroma (Salo et al., 1972).

Diacetyl is a carbonyl compound (butter, butterscotch, stale milk, and movie popcorn aroma/flavour) and a by-product of yeast nitrogen metabolism, as well as a product of bacterial metabolism. A small amount of diacetyl in a distilled product can be very pleasant, whereas a large amount is usually regarded as a serious flavour defect (Krogerus and Gibson, 2013). The precursors of diacetyl can be rapidly converted to diacetyl during the early stages of distillation when heat is applied. This compound has a low aroma threshold and a similar volatility to ethanol. Consequently, once present, it is not a compound that can be easily (or totally) removed. A diacetyl rest period at the end of fermentation (as performed by many brewers) can be used to allow the yeast to remove excess diacetyl from the fermented wort. There is a general rule in the industry that says, 'distilling a bad wash will never make a good whisky'. It is always essential to start with a good wash!

Phenol compounds occur in minor amounts in whisky, and the majority originate from the raw materials rather than being fermentation by-products.

Although all of the aroma compounds already discussed can be formed during fermentation, during the final distillation stage, the composition of the aroma compounds present will vary depending on the distillation techniques employed and how the product is subsequently matured (Suomalainen and Lehtonen, 1979). It is also important to remember that the presence of bacteria during the main fermentation and the rapid growth of lactic acid bacteria at the end of fermentation will affect the flavour (and pH) of the final spirit.

Wort fermentation

There are many similarities between the fermentation of a brewer's wort and a distiller's wort, but there are also significant differences. Again, it must be emphasised that distiller's wort is not boiled for sterility before fermentation, as is the case with brewer's wort (Noguchi et al., 2008). Therefore, the amylase enzymes will not have been inactivated in a distiller's wort. These enzymes will continue to hydrolyse the larger carbohydrate molecules in the wort during the fermentation into smaller fermentable units that the yeast can metabolise (Bathgate and Bringhurst, 2011). As a result, a beer fermentation will contain residual unfermentable dextrins, which give beer mouthfeel, calories, and body. In contrast, a distillery wort will ferment to a much lower final gravity and yield a higher alcohol concentration.

The fermentation vessels (washbacks) used vary greatly in size, overall geometry, and construction. Malt distillery fermenters usually have a much smaller capacity than typical brewery fermenters (up to 30,000L), and traditional wooden fermentation vessels are not uncommon. Grain distillery fermenters typically have a much larger capacity (250,000L or larger) with temperature control, in-place cleaning, stirring capability, and are usually constructed of stainless steel.

The concentration of a typical Scotch whisky malt wort is in the gravity range of 1.060–1.070 (15–17°Plato), yielding about ~7.5% (w/v) alcohol, with the grain wort being higher, at approx. 1.080 (20°Plato). Higher gravity wort fermentations are becoming more common with the goal of reaching increased alcohol levels, as there is a greater understanding of what is needed to conduct high-gravity fermentations in terms of yeast viability and the handling of the mash (Stewart, 2010, 2020). High-gravity worts positively affect overall plant capacity; however, the resulting higher alcohol concentrations that occur in the final wort can contribute to distillation problems (details in Chapter 14). Final gravities in a distillery fermentation fall below 1.000 and are referred to as 'degrees under'. Consequently, 2.5 'degrees under' (or 0.975 on the hydrometer scale) is a typical reading at the end of a fermentation.

Fermentation generates significant amounts of heat. Cooling during a beer fermentation is carefully monitored to maintain yeast viability and vitality, to ensure that the yeast remains healthy for reuse and to maintain the beer flavour character. Distillery fermentations usually place less emphasis on the control of the temperature compared to brewery fermentations. The wort is pitched (inoculated) at an ambient temperature (~20°C), and cooling is accomplished through the use of an external cooler to maintain the fermentation temperature at a level where the yeast can still multiply and ferment. The temperature will increase to 28–32°C, depending on the local ambient temperature. The ideal fermentation temperature is a compromise. Too high a temperature will negatively affect yeast viability. There are some commercial yeasts with temperature tolerances up to 34°C and higher, which can be employed. Above 38°C, most potable distilling yeasts struggle to reproduce or ferment. The fermentation progresses rapidly during the first 30h, and the specific gravity decreases to about 1.000 and the starting wort pH of 5.2–5.3 decreases to about pH4.2. It can rise later during fermentation as a result of bacterial metabolism.

Because the wort has not been boiled, some contamination and bacterial growth can be expected (Chapter 13). However, the wort mashing temperature of 63–65°C will reduce the bacterial load considerably. Contamination by bacteria late in the fermentation can contribute positively to the spectrum of flavour congeners (see Chapter 13). However, a heavy bacterial infection early in the fermentation is undesirable, primarily due to resultant loss in spirit yield if the sugars are utilised by the bacteria for growth rather than by the yeast for alcohol production.

The wort carbohydrate composition continues to change during the fermentation of a distiller's wort. Fig. 12.13 illustrates the progress of a typical distillery fermentation with active yeast growth occurring during the first 24h of fermentation (log phase).

The temperature increases during the initial stages of fermentation, and sugars continue to be taken up by the yeast until all of the fermentable sugars have been utilised. Nitrogen is utilised along with the sugars, but only during the active phase of yeast growth. After the active growth phase, nitrogen is no longer taken up, and the temperature does not increase, even though alcohol continues to be produced. The activity of the amylolytic enzymes (α and β-amylase, limit dextrinase—all obtained from the malting barley) continues, and starch is converted to fermentable sugars. This hydrolysis continues throughout the fermentation (enzyme activity is not inactivated as the wort was not boiled). A typical malt distillery

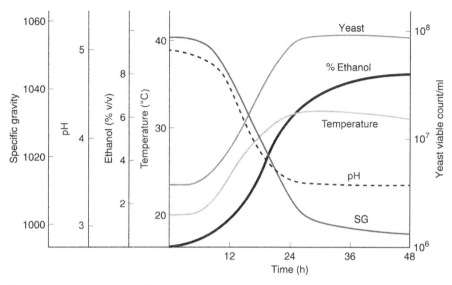

FIG. 12.13   Progress of a typical distillery fermentation.

fermentation can be completed in 48h. It can be longer if bacterial fermentation is a part of the process. However, to accommodate the necessary distillery schedule, grain distillers tend to operate all week and thus favour more uniform fermentation schedules. The enzyme activity during a distilling fermentation (called secondary conversion) can account for up to 15% of the spirit yield. Amino acids are not limiting in all-malt worts, but they can become limiting in a typical grain wort.

Sugar uptake from a distiller's wort

The yeast takes up the fermentable sugars from the wort fermentation cycle in order to produce ethanol, CO_2, glycerol, and a plethora of flavour congeners. The excretion of these flavour metabolites into the wash (fermented wort) gives the spirit much (not all) of its distinct flavour. The primary fermentable sugars in both brewer's and distiller's wort are glucose, maltose, and maltotriose (and sometimes minor amounts of fructose and sucrose). In addition, the wort contains dextrins [maltotetraose (G4) and larger dextrins]. With a wort produced for beer production, the sugars are taken up by the brewer's yeast in a distinct order: glucose, fructose (when present), maltose, and maltotriose (Stewart, 2017). The dextrins are **not** taken up by a typical brewer's yeast. The uptake of wort sugars from a distiller's wort for a whisky fermentation is more variable and complex (Fig. 12.14). As already discussed, distiller's wort is unboiled (unlike brewer's wort, which is boiled). Therefore, it contains active enzymes, particularly amylases, proteinases, and contaminating microorganisms, which come from the raw materials (barley malt and unmalted cereals) and distillery equipment—mash mixers, fermenters, etc. Consequently, during fermentation, fermentable sugars are continuously being produced from the dextrins and starch, which are taken up by the yeast

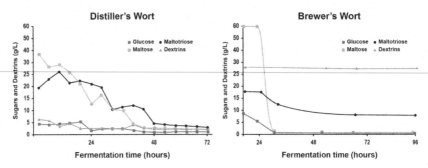

FIG. 12.14 Carbohydrate utilisation during a typical whisky all-malt fermentation and a typical 12°Plato beer fermentation.

and to a lesser extent by the contaminating bacteria such as *Lactobacillus* (and other bacterial species) (Chapter 13). This means that the sugar uptake patterns during distilling fermentations are more variable, complex, and dynamic than in brewer's wort fermentations. The sugar uptake profiles in distilling fermentations are more poorly characterised than brewing fermentations, and this is an area that requires considerable further attention.

The approximate sugar composition of a typical malt distilling wort at yeast pitching is sucrose 2%, fructose 1%, glucose 10%, maltose 50%, maltotriose 15%, maltotetraose 10%, and dextrins 10% (Bathgate, 2016). The dextrin levels in wort continue to decrease during fermentation because of continuing enzymatic activity in order to break them into smaller molecules. The sugars present are not utilised simultaneously. Glucose is immediately available to the yeast. It is taken up first by the yeast as all of the required transport systems for its uptake are present and available. Maltose and maltotriose are present in the largest amounts, but they are not initially utilised by the yeast. Their transport systems require induction, and the enzymes are only synthesised in the presence of the particular sugar and not until the glucose concentration has significantly decreased. During fermentation, smaller fermentable sugars are continuously being produced from the larger dextrins due to the amylase activity. In addition, glucose is produced from the maltose and maltotriose that is already present in the wort as one of the fermentable sugars produced during mashing. As a result, sugar uptake patterns in distilling fermentations are very different from those observed in brewing wort fermentations, where the sequence of sugar uptake progresses in an orderly and consistent manner.

Summary

It has already been discussed that distilling yeasts have many characteristics similar to brewing yeast strains. However, there are also a number of differences between these two types of yeast. The most important difference has a direct bearing on the way that distilling cultures are managed. Distilling cultures are not recycled, and they are only used once. In contrast, brewing yeast cultures are recycled through a number of separate fermentation cycles. In addition, as discussed previously, a distiller's yeast culture is usually propagated by an independent yeast supplier, whereas a brewer's yeast culture is typically cultured on site in the brewery or supplied from another production plant within the brewery.

The objectives of wort fermentation by distiller's yeast are to consistently metabolise wort constituents (primary, but not only, sugars and nitrogenous materials) into alcohol, glycerol, and numerous other fermentation products to produce a fermented wort prior to its distillation and to yield an immature whisky, which ultimately as a result of appropriate maturation in oak casks, will possess exceptional flavour, quality, and drinkability.

References

Bathgate, G.N., 2016. A review of malting and malt processing for whisky distillation. J. Inst. Brew. 122, 197–211.

Bathgate, G.N., Bringhurst, T.A., 2011. Letter to the editor: update on knowledge regarding starch structure and degradation by malt enzymes (DP/DU and limit dextrinase). J. Inst. Brew. 117, 33–38.

Cheung, A.W.Y., Brosnan, J.M., Phister, T., Smart, K.A., 2012. Impact of dried, creamed and cake supply formats on the genetic variation and ethanol tolerance of three *Saccharomyces cerevisiae* distilling strains. J. Inst. Brew. 118, 152–162.

Crabtree, H.G., 1928. The carbohydrate metabolism of certain pathological overgrowths. Biochem. J. 22, 1289–1298.

Cunningham, S., Stewart, G.G., 1998. High gravity brewing and acid washing in brewer's yeast. J. Am. Soc. Brew. Chem. 56, 12–18.

Gibson, B., Geertman, J., Hittinger, C.T., Krogerus, K., Libkind, D., Louis, E.J., Magalhães, F., Sampaio, J.P., 2017. New yeasts—new brews: modern approaches to brewing yeast design and development. FEMS Yeast Res. 17 (4). https://doi.org/10.1093/femsyr/fox038.

Goffeau, A., Barrell, B.G., Bussey, H., Davis, R.W., Dujon, B., Feldmann, H., Galibert, F., Hoheisel, J.D., Jacq, C., Johnston, M., Louis, E.J., Mewes, H.W., Murakami, Y., Philippsen, P., Tettelin, H., Oliver, S.G., 1996. Life with 6000 genes. Science 274, 563–567.

Gray, A.S., 2013. The Scotch Whisky Review, thirty-sixth ed. Sutherlands, Edinburgh, Scotland.

Harrison, B., Fagnen, O., Jack, F., Brosnan, J., 2011. The impact of copper in different parts of malt whisky pot stills on new make spirit composition and aroma. J. Inst. Brew. 117, 106–112.

Hill, A.E., Stewart, G.G., 2019. Free amino nitrogen in brewing. Fermentation 5 (1), 22. https://doi.org/10.3390/fermentation5010022.

Hittinger, C.T., 2013. *Saccharomyces* diversity and evolution: a budding model genus. Trends Genet. 29 (5), 309–317.

Jones, M., Pierce, J., 1964. Absorption of amino acids from wort by yeasts. J. Inst. Brew. 70, 307–315.

Krogerus, K., Gibson, B.R., 2013. 125th anniversary review: diacetyl and its control during brewery fermentation. J. Inst. Brew. 119, 86–97.

Lekkas, C., Stewart, G.G., Hill, A.E., Taidi, B., Hodgson, J., 2007. Elucidation of the role of nitrogenous wort components in yeast fermentation. J. Inst. Brew. 113, 183–191.

Lekkas, C., Hill, A.E., Taidi, B., Hodgson, J., Stewart, G.G., 2009. The role of small wort peptides in brewing fermentations. J. Inst. Brew. 115, 134–139.

Lekkas, C., Hill, A.E., Stewart, G.G., 2014. Extraction of FAN from malting barley during malting and mashing. J. Am. Soc. Brew. Chem. 72, 6–11.

Libkind, D., Peris, D., Cubillos, F.A., Steenwyk, J.L., Opulente, D.A., Langdon, Q.K., Rokas, A., Hittinger, C.T., 2020. Into the wild: new yeast genomes from natural environments and new tools for their analysis. FEMS Yeast Res. 20. https://doi.org/10.1093/femsyr/foaa008.

Mortimer, R.K., Johnston, J.R., 1959. Life span of individual yeast cells. Nature 183 (4677), 1751–1752.

Naumov, G.I., Lee, C.-F., Naumova, E.S., 2013. Molecular genetic diversity of the *Saccharomyces* yeasts in Taiwan: *Saccharomyces arboricola*, *Saccharomyces cerevisiae* and *Saccharomyces kudriavzevii*. Antonie Van Leeuwenhoek 103, 217–228.

Niël van Wyk, A.R.C., von Wallbrunn, C., Swiegers, J.H., Pretorius, I.S., 2020. Biotechnology of wine yeasts. In: Reference Module in Life Sciences., https://doi.org/10.1016/B978-0-12-819990-9.00007-X.

Nielsen, O., 2010. Status of the yeast propagation process and some aspects of propagation for re-fermentation. Cerevisia 35, 71–74.

Noguchi, Y., Urasaki, K., Yomo, H., Yonezawa, T., 2008. Effect of new-make spirit character due to the performance of brewer's yeast. In: Bryce, J.H., Piggott, J.R., Stewart, G.G. (Eds.), Distilled Spirits—Production, Technology and Innovation. Nottingham University Press, Nottingham, UK, pp. 109–122.

Pasteur, L., Faulkner, F., 1879. The physiological theory of fermentation I. On the relations existing between oxygen and yeast. In: Studies on Fermentation. Kessinger Legacy Reprints. Macmillan & Co, London, pp. 235–336.

Pauley, M., Maskell, D., 2017. Mini-review: the role of *Saccharomyces cerevisiae* in the production of gin and vodka. Beverages 3 (1), 13. https://doi.org/10.3390/beverages3010013.

Pedersen, T., 2019. Prokaryotic vs eukaryotic cells: what's the difference? In: Live Science. July 11, 2019 https://www.livescience.com/65922-prokaryotic-vs-eukaryotic-cells.html.

Pronk, J.T., Yde Steensma, H., van Dijken, J.P., 1996. Pyruvate metabolism in *Saccharomyces cerevisiae*. Yeast 12 (16), 1607–1633.

Pyke, M., 1958. The technology of yeast. In: Cook, A.H. (Ed.), The Chemistry and Biology of Yeasts. Academic Press, New York, pp. 535–586.

Quain, D.E., Duffield, M.L., 1985. A metabolic function for higher alcohol production by yeast. In: Proc. 20th Congress of the Eur. Brew. Conv. Congr. IRL Press, Oxford, pp. 307–314.

Quain, D.E., Thurston, P.A., Tubb, R.A., 1981. The structural and storage carbohydrates of *Saccharomyces cerevisiae*: changes during fermentation of wort and a role for glycogen catabolism in lipid biosynthesis. J. Inst. Brew. 87, 108–111.

Salo, P., Nykänen, L., Suomalainen, H., 1972. Odour thresholds and relative intensities of volatile aroma components in an artificial beverage imitating whisky. J. Food Sci. 37 (3), 394–398.

Simpson, W.J., Hammond, J.R.M., 1989. The response of brewing yeast to acid washing. J. Inst. Brew. 96, 347–354.

Stewart, G.G., 2008. Esters – The most important group of flavour-active compounds in alcoholic beverages. In: Bryce, J.H., Piggott, R.J., Stewart, G.G. (Eds.), Distillers Spirits, Production, Technology and Innovation. Nottingham University Press, Nottingham, UK, pp. 243–250.

Stewart, G.G., 2010. High-gravity brewing and distilling - past experiences and future prospects. J. Am. Soc. Brew. Chem. 68, 1–9.

Stewart, G.G., 2014. Yeast mitochondria – their influence on brewer's yeast fermentation and medicinal research. Tech. Q. Master Brew. Assoc. Am. 51, 3–11.

Stewart, G.G., 2017. Brewing and Distilling Yeasts. Springer International Publishing, Cham, Switzerland.

Stewart, G.G., 2020. Stresses imposed on yeast during brewing fermentations and their effect on cellular activity. Tech. Q. Master Brew. Assoc. Am. 57, 1–8.

Stewart, G.G., Russell, I., 2009. The IBD Blue Book on Yeast, second ed. Series III: Brewer's Yeast, The Institute of Brewing and Distilling, London.

Stewart, G.G., Ryder, D.S., 2019. Sulphur metabolism during brewing. Tech. Q. Master Brew. Assoc. Am. 56, 38–46.

Suomalainen, H., 1971. Yeast and its effect on the flavour of alcoholic beverages. J. Inst. Brew. 77, 164–177.

Suomalainen, H., Lehtonen, M., 1979. The production of aroma compounds by yeast. J. Inst. Brew. 85, 149–156.

Walker, G.M., Stewart, G.G., 2016. *Saccharomyces cerevisiae* in the production of fermented beverages. Beverages 2 (4), 30. https://doi.org/10.3390/beverages2040030.

Walther, A., Hesselbart, A., Wendland, J., 2014. Genome sequence of *Saccharomyces carlsbergensis*, the world's first pure culture lager yeast. Genes, Genomes, Genetics 4 (5), 783–793.

Wanikawa, A., 2020. Flavors in malt whisky: a review. J. Am. Soc. Brew. Chem. 78 (4), 260–278.

Wong, E.D., Karra, K., Hitz, B.C., Eurie, L., Hong, E.L., Cherry, J.M., 2013. The yeast genome app: the Saccharomyces genome database at your fingertips. Database (Oxford). https://doi.org/10.1093/database/bat004.

Zhang, W., Du, G., Zhou, J., Chen, J., 2018. Review—regulation of sensing, transportation, and catabolism of nitrogen sources in *Saccharomyces cerevisiae*. Microbiol. Mol. Biol. Rev. 82. https://doi.org/10.1128/MMBR.00040-17, e00040-17.

Further reading

Nielsen, O., 2005. Control of the yeast propagation process 'glycolysis' how to optimise oxygen supply and minimize stress. Tech. Q. Master Brew. Assoc. Am. 42 (2), 101–106.

Contamination: Bacteria and wild yeasts in whisky fermentation

Nicholas R. Wilson

Whyte & Mackay, Invergordon Grain Distillery, Invergordon, United Kingdom

During whisky production, the yeast *Saccharomyces cerevisiae* converts fermentable carbohydrates into ethanol, carbon dioxide, and other metabolites, many of which contribute to whisky quality. Distillery wort is not boiled; malt enzymes are preserved, and a variety of contaminating microbes are allowed to persist. This chapter deals with the diversity of these microbes and their impact on distillery performance and spirit character.

Sources and types of contamination

The lack of a boiling stage in distillery fermentations makes microbial contamination inevitable (Table 13.1). Contaminating microbes are introduced through raw materials: grain, water, and yeast, and are allowed to proliferate depending on length of fermentation, washback material (wood or stainless steel), and the efficacy of distillery cleaning procedures. The grain supply harbours thermotolerant microbes, such as lactic acid bacteria (LAB), wild yeasts, and acetic acid bacteria. These microorganisms are able to survive mashing temperatures and inhabit the resulting wort. Porous wooden washbacks provide shelter for any microbes present in previous fermentations, allowing for contamination in subsequent fermentations. These vessels are virtually impossible to sterilise (Dolan, 1976), but are cleaned using steam and hot water. Stainless steel washbacks can be more thoroughly cleaned using high-temperature caustic solution (2% sodium hydroxide). Contaminated process water used for cleaning and the yeast supply itself may contain very low levels of wild yeasts and LAB.

The osmotic stress imposed upon contaminating microbes by wort has some antimicrobial activity. However, it is not until fermentation is underway that the hostility of the resulting environment begins to compromise all but the most tolerant microbes. Acetic acid bacteria, such as *Acetobacter* spp. and aerobic yeasts, which are part of the malt microflora, are relatively unscathed by the mashing stage but are unable to survive during increasing anaerobic conditions. Enterobacteria, which may be introduced via contaminated process water, cannot tolerate the

TABLE 13.1 Potential contaminants of distillery fermentations.

Contaminant	Source	Location	Issues
Lactic acid bacteria	Malt, grain dust, yeast supply	Throughout, but particularly in late fermentation	Yield reduction, acid production, off-flavours
Acetic acid bacteria	Plant material, process water	Wort, initial stages of fermentation, yeast supply	Acidic off-flavours
Enteric bacteria	Plant material, process water	Wort, initial stages of fermentation, yeast supply	Sulphide and diacetyl production, off-flavours
Wild yeasts	Malt, grain dust, yeast supply	Fermentative yeast occur throughout, aerobic yeast only in initial stages	Yield reduction, higher alcohol and diacetyl production, off-flavours

fall in pH that occurs due to acid production in distillery fermentations. The brewery practice of repitching yeast between fermentations is not observed in the distilling industry and, as such contamination with *Obesumbacterium* spp., common enteric contaminants of brewery fermentations (Priest and Barker, 2010), is not generally regarded as a problem. Furthermore, the acidic pH of distillery fermentations prevents the proliferation of such organisms.

Modern yeast supply is typically of good hygienic quality; low numbers of contaminating microbes are of little significance once the dominant yeast population has established itself during fermentation. However, secondary pitching of brewing yeast can provide another source of contamination. The use of ale yeast can introduce *Zymomonas* bacteria, which are responsible for rotten egg and fruity off-odours in beer (van Vuuren and Priest, 2003), and would have a negative impact on produced spirit. *Pediococcus* spp. may be introduced if secondary brewing yeasts are used and are known to confer a so-called Sarcina sickness to beer through the production of acids and diacetyl, and produce an extracellular slime responsible for 'ropiness' in infected fermentations (van Vuuren and Priest, 2003). Although both *Zymomonas* and *Pediococcus* have the potential to be serious contaminants in distillery fermentations where secondary yeasts are used, the presence of alcohol, the lack of available nutrients, and the low pH of such fermentations are often sufficient to restrict the effects of infection.

While contamination arrives at distilleries through the raw materials, production and processing, idiosyncrasies can result in differing procedures for managing contamination depending on distillery type. The use of 100% malt, no sterilising step, and wooden washbacks in many malt distilleries can make them more susceptible to contamination than larger, industrial-scale grain distilleries, in which pressure-cooking wheat or maize provides a sterilising step and contributes as much as 90% of the fermentable substrate. The decreased malt content and the use of stainless steel washbacks reduce the number of potential causes of infection. However, the large-scale industrial nature of such operations offers alternative routes of contamination. Issues relating to cleaning-in-place (CIP) systems, recycling of process water, blockages in heat exchangers and piping, and larger, more numerous washbacks can all allow for contamination to proliferate in a grain distillery.

Pitched *S. cerevisiae* rapidly establishes itself as the dominating microbe during fermentation by altering the environment for its exclusive benefit; accumulating ethanol and acidic pH combined with a lack of oxygen and a rapidly depleting nutrient supply excludes all but the most adaptable contaminating microbes, primarily represented by LAB, which are part of the malt microflora and are tolerant to a greater or lesser extent to the conditions discussed earlier (van Beek and Priest, 2000). The range of LAB strains and species present in a distillery is typically stable but is subject to fluctuations in malt supply and distillery hygiene practices (Simpson et al., 2001).

Lactic acid bacteria (LAB)

The LAB are classified according to the products of their carbohydrate metabolism. Obligate homofermenting LAB, including *Lactococcus*, *Pediococcus*, *Streptococcus*, and some *Lactobacillus* species, use glycolysis to produce lactic acid as the sole end product (Fig. 13.1). Obligate heterofermenting LAB, such as *Leuconostoc* and some *Lactobacillus* species, metabolise carbohydrates using the 6-phosphogluconate/phosphoketolase (6-PG/PK) pathway, producing lactic acid, carbon dioxide, and either ethanol or acetic acid depending on wort oxygen content (Fig. 13.2). Facultative heterofermenters, comprised primarily of various *Lactobacillus* species, are able to utilise both glycolysis and the 6-PG/PK pathway depending on environmental conditions (Axelsson, 2004). As LAB are in competition with yeast for carbohydrates during whisky fermentations, the concentration and to a lesser extent the type of LAB present during fermentation are important.

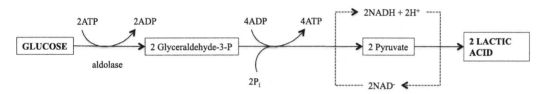

FIG. 13.1   Homolactic fermentation.

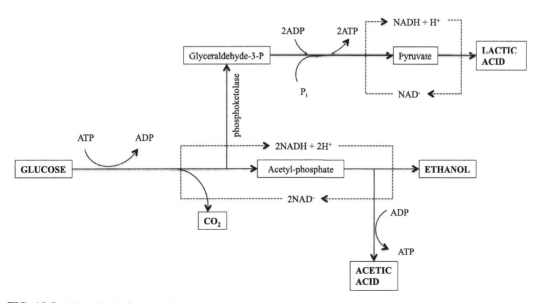

FIG. 13.2   Heterolactic fermentation.

The LAB population increases in concentration, but declines in diversity, over the course of fermentation and can be split into three stages (Fig. 13.3) (van Beek and Priest, 2003). During the initial stage of fermentation (0 to 30–40h), bacterial growth is inhibited by rapid yeast growth and ethanol accumulation (Thomas et al., 2001; van Beek and Priest, 2002). During this stage, the bacterial flora is primarily comprised of heterofermentative LAB, such as *Leuconostoc*, *Saccharococcus*, and *Streptococcus*, as well as *Lactobacillus brevis* and *Lactobacillus fermentum* (van Beek and Priest, 2002) and *Bacillus coagulans*, which has been observed in the mash (Cachat, 2005). *Leuconostoc* spp. are less tolerant of high levels of ethanol and consequently do not persist much beyond the initial stages of fermentation. During the second phase of fermentation (30–40 to 70h), the yeast population enters a stationary phase and begins to decline, at which point ethanol production is between 80 and 90% complete. As the yeast population dies, lactobacilli such as *L. fermentum*, *L. paracasei*, and *L. ferintoshensis* proliferate, resulting in lactic and acetic acid accumulation, hastening yeast decline, and ultimately allowing for the dominance of lactobacilli (van Beek and Priest, 2002). During the final stage (70h onward), bacterial populations, comprised primarily of homofermenters such as *L. acidophilus*, *L. delbrueckii*, and *L. paracasei*, peak and then decline. Due to the lack of fermentable sugars at this stage, these bacteria are believed to survive on nutrients liberated from dying and autolysing yeast cells and residual oligosaccharides. Such diversity is less apparent in grain whisky fermentations, with *L. plantarum* and *L. fermentum* dominating (Cheung et al., 2015), likely due to the prevalence of stainless steel washbacks and the ease with which these vessels can be cleaned.

The presence of LAB in whisky fermentations can be regarded as both detrimental and beneficial, depending on the degree of contamination and how that contamination is managed.

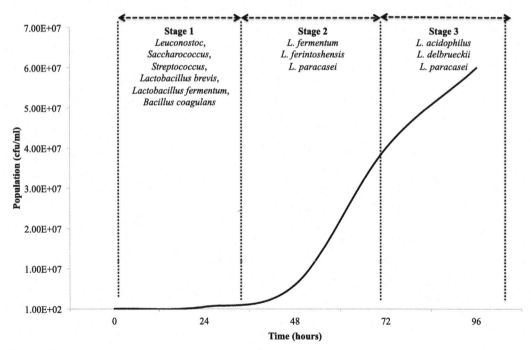

FIG. 13.3   Evolution and growth of the LAB population throughout fermentation.

In Scotch whisky distilleries, the LAB population is passively managed through fermentation time and cleaning procedures, while distilleries in the USA and elsewhere may encourage LAB growth through the use of 'backset', in which a volume of spent mash is inoculated into a fresh fermentation (Buglass, 2011). High initial levels of LAB ($> 10^6$ cells/ml) are known to reduce ethanol yield, resulting in a loss of revenue (Makanjoula et al., 1992) and is the primary contamination concern in grain distilleries that produce highly rectified spirit. Moreover, the presence of a large bacterial population at the start of a fermentation can result in the production of negative flavour characteristics (van Beek and Priest, 2002).

The primary method by which LAB inhibit ethanol production by yeast is through the production of lactic acid and, to a lesser extent, acetic acid. Lactic acid production diverts fermentable substrate away from alcohol production, as each sugar molecule used by LAB results in the loss of two ethanol molecules. The secondary method of inhibition is the accumulation of lactic acid and the effects of low pH (< 3.8) on yeast physiology and metabolism, which is further compounded if acetic acid is present (Narendranath et al., 2001). Fatty acid production by LAB during distillery fermentations may have an inhibitory effect on yeast metabolism, particularly during the latter stages of fermentation (Lowe and Arendt, 2004).

A low initial LAB population (10^3–10^5 cells/ml) will be kept in check by extensive yeast growth and ethanol accumulation (Thomas et al., 2001) until the yeast population begins to decline and ethanol production ceases. At this point, the LAB population may be allowed to bloom (Fig. 13.4). This secondary fermentation does not interfere with ethanol production and allows for the accumulation of desirable flavour and aroma compounds (van Beek and Priest, 2002)

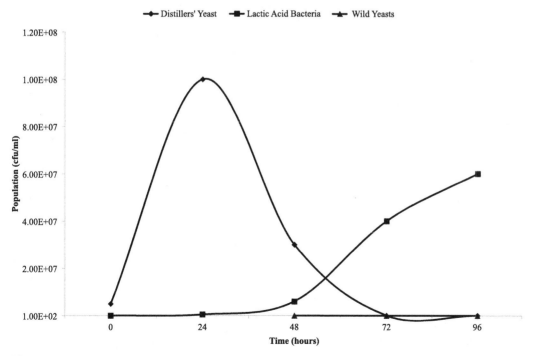

FIG. 13.4   Microbial profile of a typical distillery fermentation.

derived from the products of homolactic and heterolactic metabolism, as well as other metabolic pathways involved in the decarboxylation of substituted cinnamic acids (van Beek and Priest, 2000), the hydroxylation of fatty acids (Wilson, 2008), and the production of succinic acid (Dudley and Steele, 2005; Reid et al., 2020).

Wild yeasts

Although LAB are very much dominating contaminant microbes during whisky fermentations, a background population of wild yeasts has been observed (Neri, 2006). The dominance of a distilling strain of *S. cerevisiae* keeps the wild yeast population at a low level (10^1–10^4 cells/ml), which does not appear to evolve during the course of fermentation (Neri, 2006).

The contaminating wild yeasts include various nondistilling strains of *S. cerevisiae*, *Pichia membranifaciens*, *Candida* spp., *Issatchenkia orientalis*, and *Torulaspora delbrueckii*. Wild strains of *S. cerevisiae* have been observed as brewery contaminants, with no perceived adverse effects on fermentation for limited populations. However, their biochemical similarity to distilling yeast means that competition for sugars and the production of potential off-flavours cannot be discounted (Campbell, 2003). Variants of *S. cerevisiae* observed during the later stages of fermentation may represent mutations in the distilling strain (e.g. respiratory deficient yeast) as a result of the hostile conditions that occur during prolonged fermentations (approximately 50 h). Significant *P. membranifaciens* contamination can lead to serious problems due to its aerobic metabolism in high ethanol concentrations (Campbell and Msongo, 1991). Such yeasts are common brewery contaminants (Campbell, 2003) and may be introduced to distillery fermentations through the use of secondary yeasts. *I. orientalis*, which is an aerobic yeast known to produce off-flavours and to form a pellicle (scum-like layer) on the surface of contaminated beer fermentations, is a potential contaminant in distilleries using secondary yeasts (Campbell, 2003).

As with wild strains of *S. cerevisiae*, *T. delbrueckii*, a fermentative yeast, competes with distilling yeast for nutrients due to their biochemical similarity. It is a common contaminant in brewery fermentations, but few deleterious effects have been observed as a result of its presence. In high enough numbers, however, it is likely to contribute to off-flavour development (Campbell, 2003). It has been suggested that this background population may contribute to positive flavour profiles associated with sweet and creamy notes if present during the late lactic fermentation through the detoxification of hydroxylated fatty acids via conversion to γ-lactones (Neri, 2006; Wanikawa et al., 2000; Wilson, 2008).

Effects on spirit composition and quality

The primary congeners in new-make spirit, other than ethanol, are higher alcohols, esters, aldehydes, ketones, organic acids, carbonyls, phenols, and sulphur compounds (Palmer, 1997), as well as fatty acids and lactones (Wanikawa et al., 2000; Wanikawa, 2020). It has been extensively documented that high levels of LAB contamination can be detrimental to the perceived quality of the resulting spirit through a reduction in ethanol yield and the presence of off-notes (Makanjoula et al., 1992) such as through acrolein production, which can lead to a rancid, bleach-like aroma, and poses a potential health risk due to its toxicity. The primary concern for grain distilleries is ethanol yield; thus, loss of ethanol and revenue is the main impact that LAB have on grain spirit quality, as fermentations are finished once ethanol accumulation has concluded, and the products of excessive bacterial metabolism are generally removed during the rectification stage of distillation. While reduced yield due to excessive LAB contamination is also a concern for malt distilleries, many distilleries also rely on the resident LAB population to impact spirit quality and character.

In fermentations that are allowed to continue beyond the conclusion of ethanol production (> 50h), LAB can contribute to positive flavour notes during the subsequent late lactic fermentation, if followed by batch rather than continuous column distillation. This stage is characterised by falling pH due to lactic and acetic acid accumulation, increasing concentration of ethyl lactate and ethyl acetate in the later stages of fermentation and during distillation, contributing to desirable fruity notes in produced spirit (Campbell, 2003; van Beek and Priest, 2002). The amounts and relative proportions of these esters are dependent on the type of carbohydrate metabolism exhibited by the dominating LAB population, with elevated concentrations of ethyl lactate in fermented wash reflecting its concentration in spirit (Wilson, 2008). The use of backset or the addition of specific LAB species to wort to promote bacterial growth either during or before yeast fermentation produces a fruitier spirit (Takatani and Ikemoto, 2002), with noted increases in ethyl lactate, ethyl acetate, and phenylethyl acetate (Reid et al., 2020). However, wort acidified in this manner is less fermentable, has compromised ethanol yield, and has reduced higher alcohol content (Reid et al., 2020), which may in turn reduce the perception of green/grassy aromas. Furthermore, hydration/hydroxylation of unsaturated fatty acids by LAB during lactic fermentation provides increased quantities of γ-lactone precursors, imparting sweet and creamy aromas to the resulting spirit due to increased concentrations of γ-decalactone and γ-dodecalactone (Wilson, 2008). Unsaturated fatty acids implicated in the perception of nutty character (Webb, 2015) may in turn be removed by LAB activity, reducing the perception of nutty aroma.

The impact of wild yeast on spirit character is less well documented, possibly due to their innocuous presence in whisky fermentations (Fig. 13.4). However, the development of green/grassy aromas may be influenced by the presence of wild yeasts, such as *T. delbrueckii*, which produce higher alcohols and aldehydes (Wanikawa et al., 2002; Wanikawa, 2020). Wild yeasts may also impact the development of sweet and buttery aromas through significantly elevated production of the vicinal diketones diacetyl and pentanedione, contributing a butterscotch-like aroma to new-make spirit (Wilson, 2008). The use of different yeast strains for whisky production is not generally restricted by law, as such there exists a potential for modification of spirit character through the use of yeasts, either as the primary fermenting yeasts or as the secondary yeasts, which produce higher levels of organic acids, vicinal diketones, or higher alcohols/aldehydes. The growing popularity of organic acid-producing yeasts in craft brewing (Reid et al., 2020) may be followed by their use in craft distilling to produce spirits of fruitier and sweeter character.

While the presence of LAB (*L. brevis*, *L. paracasei*, and *L. plantarum*) and wild yeast (*T. delbrueckii*) populations in whisky fermentations may separately impact spirit quality, either positively or negatively depending on how the contamination is managed, their combined presence during the later stages of fermentation can confer positive effects on spirit quality greater than each population individually. Such mixed fermentations show increased concentrations of ethyl lactate and γ-dodecalactone (Wilson, 2008) and are likely to contribute to the perception of fruity, sweet, and buttery aromas.

References

Axelsson, L., 2004. Lactic acid bacteria: classification and physiology. In: Salminen, S., von Wright, A., Ouwehand, A. (Eds.), Lactic Acid Bacteria: Microbiological and Functional Aspects. Third and Expanded Edition Marcel Dekker, New York, pp. 1–66.

Buglass, A.J., 2011. Whiskeys. In: Handbook of Alcoholic Beverages: Technical, Analytical and Nutritional Aspects. John Wiley & Sons, Ltd, Chichester UK, pp. 515–534.

Cachat, E., 2005. Diversity of Lactic Acid Bacteria in the Malt Whisky Distillery and Their Impact on the Process. PhD thesis, Heriot-Watt University, Edinburgh.

Campbell, I., 2003. Wild yeasts in brewing and distilling. In: Priest, F.G., Campbell, I. (Eds.), Brewing Microbiology, third ed. Kluwer Academic/Plenum Publishers, New York, pp. 247–264.

Campbell, I., Msongo, H.S., 1991. Growth of aerobic wild yeast. J. Inst. Brew. 97, 279–282.

Cheung, A., Walker, J., Lawrence, S., Phister, T., Brosnan, J.M., Smart, K., 2015. Identification of lactic acid bacteria from grain whisky fermentation. In: Goodall, I., Fotheringham, R., Murray, D., Speers, R.A., Walker, G.M. (Eds.), Worldwide Distilled Spirits Conference—Future Challenges. New Solutions. Context Products Ltd, Packington UK, pp. 91–95.

Dolan, T.C.S., 1976. Some aspects of the impact of brewing science on Scotch malt whisky production. J. Inst. Brew. 82, 171–181.

Dudley, E.G., Steele, J.L., 2005. Succinate production and citrate catabolism by Cheddar cheese nonstarter lactobacilli. J. Appl. Microbiol. 98, 14–23.

Lowe, D.P., Arendt, E.K., 2004. The use and effects of lactic acid bacteria in malting and brewing with their relationships to antifungal activity, mycotoxins and gushing: a review. J. Inst. Brew. 110, 163–180.

Makanjoula, D.B., Tymon, A., Springham, D.G., 1992. Some effects of lactic acid bacteria on laboratory-scale yeast fermentations. Enzym. Microb. Technol. 14, 350–357.

Narendranath, N.V., Thomas, K.C., Ingledew, W.M., 2001. Acetic acid and lactic acid inhibition of growth of Saccharomyces cerevisiae by different mechanisms. J. Am. Soc. Brew. Chem. 59, 187–194.

Neri, L., 2006. The Involvement of Wild Yeast in Malt Whisky Fermentations. PhD thesis, Heriot-Watt University, Edinburgh.

Palmer, G.H., 1997. Scientific review of Scotch malt whisky. Ferment 10, 367–379.

Priest, F.G., Barker, M., 2010. Gram-negative bacteria associated with brewery yeasts: reclassification of Obesumbacterium proteus biogroup 2 as Shimwellia pseudoproteus gen. nov., sp. nov., and transfer of Escherichia blattae to Shimwellia blattae comb. nov. Int. J. Syst. Evol. Microbiol. 60, 845–849.

Reid, S.J., Speers, R.A., Willoughby, N., Lumsden, W.B., Maskell, D.L., 2020. Pre-fermentation of malt whisky wort using Lactobacillus plantarum and its influence on new-make spirit character. Food Chem. 320, 126605.

Simpson, K.L., Pettersson, B., Priest, F.G., 2001. Characterization of lactobacilli from Scotch malt whisky distilleries and description of Lactobacillus ferintoshensis sp. nov., a new species isolated from malt whisky fermentations. Microbiology 147, 1007–1016.

Takatani, T., Ikemoto, H., 2002. Contribution of bacterial microflora in malt whisky quality. In: Bryce, J.H., Stewart, G.G. (Eds.), Distilled Spirits—Tradition and Innovation. Nottingham University Press, Nottingham, pp. 197–207.

Thomas, K.C., Hynes, S.H., Ingledew, W.M., 2001. Effect of lactobacilli on yeast growth, viability and batch and semi-continuous alcoholic fermentation of corn mash. J. Appl. Microbiol. 90, 819–828.

van Beek, S., Priest, F.G., 2000. Decarboxylation of substituted cinnamic acids by lactic acid bacteria isolated during malt whisky fermentation. Appl. Environ. Microbiol. 66, 5322–5328.

van Beek, S., Priest, F.G., 2002. Evolution of the lactic acid bacterial community during malt whisky fermentation: a polyphasic study. Appl. Environ. Microbiol. 68, 297–305.

van Beek, S., Priest, F.G., 2003. Bacterial diversity in Scotch whisky fermentations as revealed by denaturing gradient gel electrophoresis. J. Am. Soc. Brew. Chem. 61, 10–14.

van Vuuren, H.J.J., Priest, F.G., 2003. Gram-negative brewery bacteria. In: Priest, F.G., Campbell, I. (Eds.), Brewing Microbiology, third ed. Kluwer Academic/Plenum Publishers, New York, pp. 219–245.

Wanikawa, A., 2020. Flavours in malt whisky: a review. J. Am. Soc. Brew. Chem. 78, 1–19.

Wanikawa, A., Hosoi, K., Takise, I., Kato, T., 2000. Detection of γ-lactones in malt whisky. J. Inst. Brew. 106, 39–43.

Wanikawa, A., Hosoi, K., Kato, T., Nakagawa, K.-I., 2002. Identification of green note com- pounds in malt whisky during multidimensional gas chromatography. Flavour Fragance J. 17, 207–211.

Webb, D., 2015. The impact of unsaturated lipids on nutty character. In: Goodall, I., Fotheringham, R., Murray, D., Speers, R.A., Walker, G.M. (Eds.), Worldwide Distilled Spirits Conference—Future Challenges. New Solutions. Context Products Ltd., Packington UK, pp. 239–242.

Wilson, N.R., 2008. The Effect of Lactic Acid Bacteria on Congener Composition and Sensory Characteristics of Scotch Malt Whisky. PhD thesis, Heriot-Watt University, Edinburgh.

14

Batch distillation

Denis A. Nicol

Allied Domecq (Retired), Dumbarton, United Kingdom

Introduction

Scotch malt whisky production is derived by the double or triple distillation of a fermented mash derived from pure malt, water, and yeast as the only permissible raw materials defined by law. This chapter reviews the history of stills, differences in still design, the construction of stills and their ancillary equipment, the operation of wash and spirit stills, product quality and efficiency, potential problems to be avoided during distillation, and the role of copper in the construction of pot stills and quality of the final product.

History

Distillation was carried out from earliest times using pot stills, which were initially made of ceramic or glass and eventually of copper. These early stills were directly heated by open fires in a furnace or hearth. Heat-sensitive materials to be distilled could be heated by means of a water- or sand-filled bath, called a *bain marie*, invented by a first-century alchemist known as Mary the Jewess. Distillates were originally air-cooled condensers, which had tapering lye pipes delivering the product to glass or clay vessels. Worm tubs and eventually the shell-and-tube condenser superseded the primitive condensers. Alcohol was not recovered by distillation in any quantity until the 12th century, when stills of a crude design, caulked with clay and straw, were improved upon by using a close-fitting pot and head and lye pipe to improve the recovery of alcohol from inferior, unpalatable beers and wines.

The production of alcohol was the preserve of monks in monasteries, within whose hallowed cloisters alchemy was practised in the vain search for the philosopher's stone, which was believed to be key to the transmutation of base metals into gold. An 'elixir of life' was also sought but to no avail; instead, alcohol filled this niche, apparently being prescribed for all manner of ills.

The Reformation of the Church saw the dissolution of the monasteries in England and Scotland. The knowledge accumulated by the monks was dispersed throughout the land, being acquired by individuals seeking to learn a trade or profession: brewers, distillers, alchemists, apothecaries, or barber–surgeons. The monks were instrumental in establishing the medical sciences and the early brewing and distilling industries. Thus, whisky (*uisge beatha*, or 'water of life') was first noted by Henry II in 1170, when he and his army invaded Ireland and witnessed the natives of that land engaged in making such a drink from a mash of cereals called *usquebaugh*. The first written account of the making of whisky dates back to the Scottish Exchequer Rolls of 1494, which refer to *aqua vitae*, the tantalising and unattainable elixir of life.

Distillery design

All malt whiskies, derived from low-nitrogen barleys, are produced in copper pot stills (Nicol, 1989) using a time-honoured traditional design (Fig. 14.1). The following principal elements comprise a simple pot still, whether for distilling wash or low wines and feints:

- *Heating source*—Direct fire (coal or liquid petroleum gas; see Fig. 14.2) or indirect fire (steam coils, kettles, pans, or external heat exchanger, where the steam is raised by an oil- or a gas-fired boiler; see Figs. 14.3–14.5)
- *Pot*, which contains charge to be distilled
- *Shoulder*
- *Swan neck*
- *Head*
- *Lyne arm, lye pipe,* or *vapour pipe*
- *Worm tub* or *condenser*
- *Tail pipe*
- *Spirit safe*

Fig. 14.6 illustrates the overall layout of a typical malt distillery.

Heating source

Several fuels are available to heat the stills (Watson, 1989). With direct-fired stills, the pot must be designed to withstand the rigours of direct firing, and the copper crown and flue plates must be made of a sufficient gauge (16-mm) copper to withstand the intense local heating. Where copper is not expected to be exposed to intense heat, the gauge can be reduced (10 mm). The base of the direct-fired still is convex, resembling an inverted saucer, the rim facilitating discharge. The hearth or furnace upon which the still rests is of brick or steel construction, lined with suitable firebricks to protect the supporting structure from the heat.

Whether gas or coal fired, the exhaust gases must be ducted to a flue stack or chimney made of brick or steel. With more than one still, the flue gas can be led into a manifold, the flue gas being individually controlled by dampers. Where coal is used as the heat source, each hearth is equipped with a chain grate stoker with automatic solid fuel feed and ash removal.

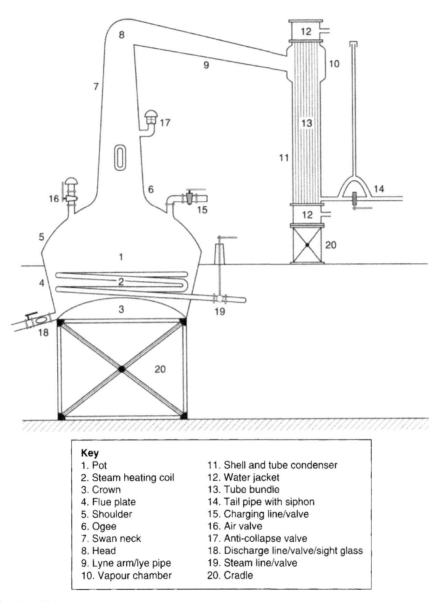

FIG. 14.1 Pot still design.

Key

1. Pot	11. Shell and tube condenser
2. Steam heating coil	12. Water jacket
3. Crown	13. Tube bundle
4. Flue plate	14. Tail pipe with siphon
5. Shoulder	15. Charging line/valve
6. Ogee	16. Air valve
7. Swan neck	17. Anti-collapse valve
8. Head	18. Discharge line/valve/sight glass
9. Lyne arm/lye pipe	19. Steam line/valve
10. Vapour chamber	20. Cradle

A damper is fitted to the flue to control the heat input. With gas firing, the burner can be modulated by controlling the gas flow. In Scotland today, coal is no longer, or rarely, used as a primary heat source for distilling.

For indirect heating by coil, pan, or kettle, steam is supplied by an oil- or gas-fired boiler and is transferred by a steam manifold from the central boiler at the crown pressure.

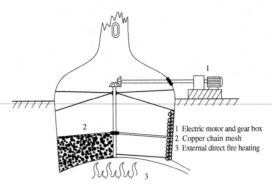

FIG. 14.2 Direct-fired still with rummager.

1 Electric motor and gear box
2 Copper chain mesh
3 External direct fire heating

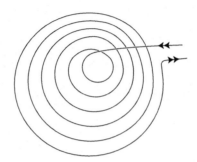

FIG. 14.3 Plan view of steam coil.

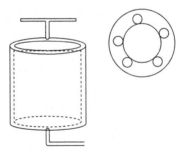

FIG. 14.4 Steam pan.

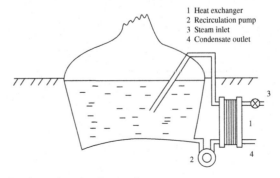

1 Heat exchanger
2 Recirculation pump
3 Steam inlet
4 Condensate outlet

FIG. 14.5 External steam heating using plate heat exchanger.

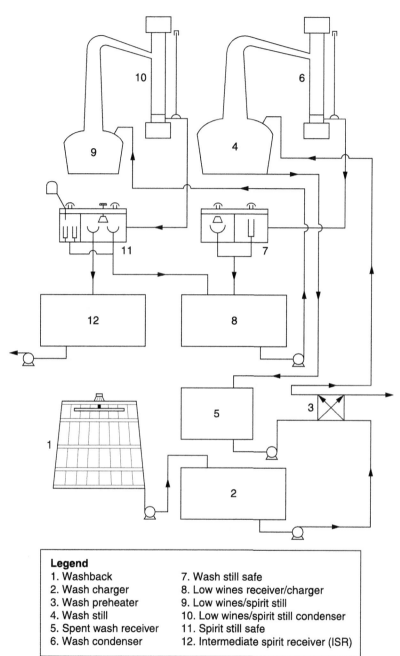

Legend

1. Washback
2. Wash charger
3. Wash preheater
4. Wash still
5. Spent wash receiver
6. Wash condenser
7. Wash still safe
8. Low wines receiver/charger
9. Low wines/spirit still
10. Low wines/spirit still condenser
11. Spirit still safe
12. Intermediate spirit receiver (ISR)

FIG. 14.6   Typical malt distillery layout.

The pressure is reduced to the desired operating pressure for the individual still heating elements. The heating elements must be designed to be totally immersed in the relevant charge to be distilled, at the beginning and end of the distillation cycle. A low wines and feints still can be fitted with an extra coil to be used to distil the middle cut gently, thus ensuring good reflux. Steam traps are strategically positioned to remove steam condensate that might otherwise waterlog the steam lines. Condensate returns, after heating, are fed back to the boiler feedwater tank via a condensate manifold as an energy-saving step. The steam demand is calculated to provide sufficient heat to bring the still to a boil and into the safe within an hour or less. The calculation should include the maximum demand for steam when heating hot liquor tanks and several stills simultaneously. The boiler output is then designed based on this demand and sized accordingly, building in extra capacity.

Direct firing of wash stills requires a rummager (a flail resembling chain mail) made of copper or brass, suspended from a rotating geared shaft, which as it rotates scours the flue plate and base of the wash still. This reduces the amount of charring, thus maintaining heat transfer.

To conserve energy, the external flue plates of an indirectly fired still can be insulated to prevent radiant heat loss through the sides of the pot. The shoulder and all the other parts of the still should not be insulated as this would adversely impact on the reflux, which is essential for spirit character. Wash preheating, when used, can considerably reduce total distillation times, thus saving energy. A treatise on energy-saving techniques can be found in a government report titled *Drying, Evaporation and Distillation* (Energy Technology Support Unit, 1985).

Condensers can now be run hot (> 85°C), but this can only be accomplished by using a subcooler prior to distillates entering the spirit safe to protect the delicate safe instruments. The hot water thus recovered can be incorporated in the first water for mashing purposes.

Pot

The pot (Fig. 14.7) can assume many shapes (e.g. onion, plain, straight, and ball), provided that sufficient volume and surface area are maintained to ensure that the heating elements remain totally immersed at the end of distillation (Whitby, 1992). Such a problem does not exist with direct-fired stills. The pot is equipped with close-locking air, charging, discharging, and safety valves. If manually operated, a special interlocking valve key that is shared among the air, discharge, and charging valves is used. It can only be used in sequence, ensuring that the air valve followed by the discharge valve can be opened to prevent the collapse or accidental discharge of the still. The reverse is applicable on charging, with sequential closing of the discharge valve followed by opening of the charging valve and finally closure of the air valve on the completion of charging and application of heat. Such requirements are redundant with today's use of automatic remote controls through programmable logic controllers (PLCs) with sequencing operations. It is now possible to operate distillation plants with a minimum of personnel. The pot connects to the swan neck via the ogee. Access to the pot is provided by a lockable brass man-door.

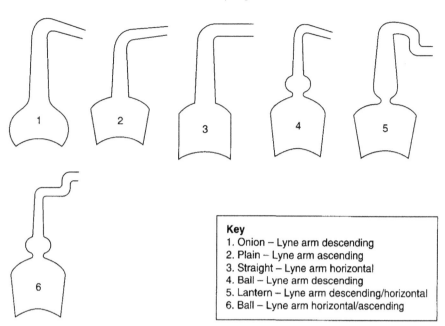

FIG. 14.7   Shapes of stills.

Swan neck

The swan neck has the greatest influence on the final character of new-make spirit (Plain British Spirit, or PBS). A good example of the importance of the swan neck is Hiram Walker's Lomond still, developed by Alastair Cunningham, which incorporated sieve plates in the neck of the Inverleven malt still in Dumbarton, Scotland. This type of still is still in use in certain distilling plants in the industry today. It enables 'tunes' to be played with the distillates, providing distillates of different congener ratios. The neck of the still can vary from short to long. It can be tapered, straight sided, or severely swept in to the head. At the base of the neck, it can assume the shape of a lantern glass, be ball shaped, or just be directly connected to the pot, thus resembling an onion. The neck is provided with two oppositely placed sight glasses, so when the still comes in, any foaming can be seen, especially in a wash still, thus demanding a reduction in heat. A light can be attached to the rear sight glass to illuminate the still internally. A cold finger can be installed at the top of the neck. Cold water can be used to help prevent overfoaming of the still into the safe (i.e., foul distillation). A vacuum relief valve or seal pot (Fig. 14.8) is fitted well above the boiling and foaming line in a wash still to prevent seizure of the valve by dextrins and solids.

Head

The head is a curved extension of the neck, connecting to the lyne arm or lye pipe. The head can be fitted with a thermometer to indicate the imminent arrival of the hot distilling vapours. The length or height of the head will dictate the degree of reflux within the still.

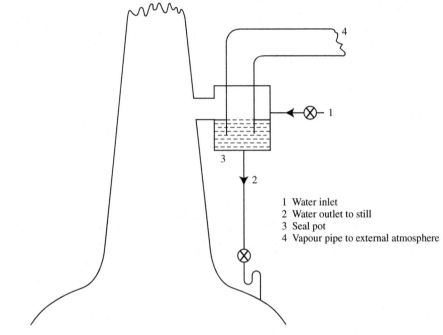

FIG. 14.8   Anti-collapse seal pot.

Lyne arm or lye pipe

The lyne arm, lye pipe, or vapour pipe is of cylindrical construction and connects the head to the worm tub and shell-and-tube condenser. The attitude of the lyne arm has an important bearing on the spirit character. It can be designed to be horizontal, ascending, or descending to the condenser or worm tub. The angle of ascent or descent is computed to be shallow. Such permutations of the lyne arm will affect the organoleptic nature of the new spirit. The lyne arm can be interrupted by a purifier (Fig. 14.9), a device fitted with baffles and cooled by an external water jacket or internal coil. Its use is to encourage heavy oils (higher fatty acid esters, C15+) to return to the body of the still during distillation. The purifier returns the heavy oils to the still via a U-bend.

Worm tub or condenser

The worm tub, of ancient origin and design, is a large coopered, water-filled wooden vessel containing the worm (a long coiled, tapering copper tube), which is an extension of the lyne arm, beginning with a diameter equivalent to that of the lyne arm and reducing to about 76mm, before being led to the spirit safe. Cold water is fed to the base of the tub and exits through an overflow pipe at the top. At certain times of the year, when water is in short supply, the exiting heated cooling water can be chilled by means of a cooling tower or a seawater-cooled plate heat exchanger at coastal distilleries, before returning to the cooling system. The successor to the worm tub is the shell-and-tube condenser or even a plate heat

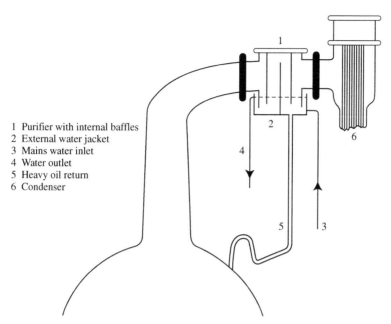

1 Purifier with internal baffles
2 External water jacket
3 Mains water inlet
4 Water outlet
5 Heavy oil return
6 Condenser

FIG. 14.9   Purifier on low wines and spirits still.

exchanger with copper plates to condense the vapours. If a condenser is used in a system that recovers sensible heat from the outflowing water at an elevated temperature (>80°C), a subcooler postcondenser will be required to chill the spirit to less than 20°C to protect and enable spirit-safe instruments to record reliable readings. The cooling technique can impinge on spirit quality, as worms can produce a product imbued with the aroma of sulphur compounds; this will be discussed further later.

Spirit safe

The spirit safe, which is under lock and key to prevent illicit sampling (first introduced circa 1823 at the Port Ellen distillery on Islay), is used to monitor the cut point, strength, and temperature of the outflowing distillates prior to delivery to the relevant receivers. Traditionally, by means of spirit hydrometers, the strength of low wines from a wash still or foreshots, middle cuts, and feints from a spirit still can be ascertained. The wash still distillation is monitored using a hydrometer, calibrated at 20°C in the full range of 0% to 75% alcohol by volume (abv), or with a narrow-range hydrometer reading 0% to 10% abv, to determine the completion of wash distillation.

The spirit still is controlled by two hydrometers, the tailpipe outflow being directed to one or the other of two collecting bowls, one for foreshots and feints and the other for the potable spirit. The bowl receiving the spirit may on occasions have a muslin cloth placed across it to act as a filter for verdigris particles emanating from the condenser. Such a piece of muslin is referred to historically as a *Hippocratum*. By means of a swivelling spout, the distillate flow can be directed to the foreshots and feints receiver or to the intermediate spirit receiver, as indicated by the hydrometer readings.

A small reservoir for cold water is provided for carrying out the demisting test. This test is used to distinguish between foreshots and true potable spirit. By mixing foreshots with water to a strength of 46% abv, a time is reached when the normally milky/turbid mixture becomes clear at this strength and pure spirit is flowing. It is this test that determines the primary cut point, while the second cut point is chosen according to the desired bouquet and strength of the final collected spirit, usually 68% to 70% abv. The demisting test hydrometer jar is equipped with a three-way sampling valve. Distillate overflows from the hydrometer jar, draining via a small bore pipe to the low wines and feints receiver. In some instances, the demisting test has been abandoned, and the collection of new-make spirit commences immediately when the distillate flow enters the safe. This is driven by the desire to capture all of the flavour-enhancing congeners.

Construction of stills and ancillary equipment

Any reputable coppersmith or distillery engineer should be capable of manufacturing the necessary equipment, from stills to receivers. As previously mentioned, stills are totally constructed from copper, including the condensers and worms. The outer shell of a shell-and-tube condenser can be fashioned from stainless steel and the tubes from copper. Man-doors and valves can be made from brass or stainless steel. The pipework, 50–76 mm in diameter, can be made from suitable grade stainless steel, with flanges sealed using food-grade, alcohol-resistant gaskets. The safe, as already described, is made of brass with plate-glass windows and a lockable lid for security. The safe is fitted with air vents and capped with mushroom domes, and the vents can be extended to the external atmosphere through flame arresters.

Vessels for collecting charges—wash charger, low wines, foreshots and feints receivers, and intermediate spirit receivers—are now usually made of stainless steel. Coopered oak wood, epoxy-lined COR-TEN steel, and glass-lined vessels have also been used for collecting distillates. For excise purposes, vessels are gauged and fitted with dipsticks and striking pads for measuring wet dip. Volumes are calculated using the relevant gauging tables.

Valves are constructed of stainless steel and may be of ball, gate, butterfly, or diaphragm design. When attached to gauged vessels, the valves must be lockable. Vessels containing spirit are fitted with agitators or rousing devices (compressed air or mechanical screw agitation) to achieve good mixing prior to taking account. Venting is also necessary in spirit, receiving, and charging vessels, with the vents leading to the external atmosphere via flame arresters.

All electrical equipment must conform to current flameproofing practice under Health and Safety Regulations. Pipe runs should be constructed to avoid dead legs and should be angled slightly towards the receiving vessels to provide for complete drainage.

Wash still operation

A number of texts provide details regarding wash and spirit still distillation (see, for example, Lyons, 2003; Lyons and Rose, 1977; Nicol, 1989, 1997; Piggott and Connor, 1995;

Whitby, 1992). Fermented wash, with an original gravity (OG) of 1050–1060° (12.5–15° Plato), is pumped or gravity fed for wash preheating to the wash charger, preheating being one of the energy-saving techniques. The charge volume (two-thirds of the wash still capacity, usually lipping the base of the man-door) is heated by either direct or indirect firing. Preheated wash, heat-exchanged with hot discharging pot ale within a few degrees of boiling, is gently brought in. Although the initial heat can be applied vigorously as soon as the flap on the spout in the safe indicates the arrival of expanding air and the condensed distillate will not be long in arriving, the heat should be reduced to prevent the still contents from boiling over, resulting in a foul distillation. As previously mentioned, this should be avoided at all costs.

Before charging the still, the discharge valve is checked to see that it is closed, while the charging and air valves are open. The anticollapse valve should be checked to see that it is moving freely and cleaned if necessary. When the still has been charged with the required volume, the air and charging valves are closed. The man-door, if open, should also be closed, as the still contents expand with the heat and with the evolution of the dissolved carbon dioxide. With the advent of the programmable logic controllers, the sequential opening and closing of valves manually has given way to automation. With manual systems, an interlocking valve key arrangement is used to prevent the accidental opening of valves out of sequence.

The sight glass on the side of the still indicates the degree of frothing that can occur during the initial distillation stages, depending on the age of the fermented wash. Using this indicator, the amount of heat applied to the still can be controlled to prevent a foul distillation. When the frothing subsides, the heat can be increased to allow a steady and uniform flow of low wines to be collected.

The progress of the distillation is followed by hydrometry in the safe until the hydrometer reading indicates about 1% abv, when the distillation can be deemed complete. The end point of distillation at 1% abv ensures that time and fuel are not wasted in recovering a small amount of very weak spirit. A distillation cycle can last from 5 to 8 h, in parallel with the mashing and fermentation cycle.

When the distillation is complete and the low wines have been collected, the air valve is opened to equilibrate the internal still pressure with the external atmospheric pressure. Failure to carry out this vital but very necessary procedure may result in a collapsing still, should the anticollapse valve fail to open.

A record is kept of the original dip in the low wines receiver and of the final dip after distillation. The temperature and low wines strength, corrected to 20°C in the hydrometer jar, are recorded every 15 min over the period of the distillation cycle. Increasing distillation time is indicative of charring of wash on the internal still heating surfaces. This charring can be minimised by ensuring that excessive heat is not applied to the cold wash, thus maintaining a minimal temperature differential. Wash preheating reduces this charring effect, which is equivalent to a cold protein break. Deteriorating heat transfer indicates the need for caustic cleaning. Caustic soda (1%–2% w/v) is boiled up within the still to strip deposits from the heating surfaces. Should the charring be intractable and fail to respond to the caustic solution, it may be necessary to manually scrub the heating surfaces or to use an alternative sequestered caustic cleaning agent.

The manufacturer's instructions should be followed and all safety precautions observed. On no account should pearl or powdered caustic soda be added directly to hot water, as this will result in an exothermic blow back, with the resultant solution boiling violently.

Under prevailing health and safety regulations, work permits should be issued for both caustic cleaning and entry into a confined space. Adequate protective clothing, including chemical-resistant gauntlets and goggles, must be provided, and the alkaline washing residues must be neutralised with acidic effluent prior to downstream processing.

Cleaning-in-place (CIP) systems are now available and are preferred to manual cleaning, as they greatly reduce the risk factors involved in handling aggressive and dangerous chemicals; another benefit is the ability to recycle the cleaning agent. Rinsing out with water is necessary to prevent the wash from being contaminated with cleaning agent residues.

The wash still is emptied via the discharge valve, and the discharging pot ale is heat-exchanged with the incoming wash. Hot pot ale or hot preheated wash can be held prior to charging the empty still.

In some distilleries, the condenser water is regulated to outflow at around 80°C. This hot condenser water can be pumped through a mechanical vapour compressor or a steam ejector, with the flash steam produced being used to drive the still. Such techniques require the use of a spirit subcooler to ensure that the low wines are collected at a temperature no higher than 20°C.

Wash that is not fully fermented is at risk of causing foul distillations. This exacerbates any potential ethyl carbamate problem among other flavour problems; hence, distillation of short fermentations should be avoided. The volume of low wines collected is approximately one-third of the original wash charge volume.

Heat recovery in wash stills using thermal vapour recompression (TVR)

As a means of saving energy, thermal vapour recompression (TVR) has been introduced to achieve significant energy savings in wash still distillation arrangements.

Some of the heat from the vapour stream in a wash still condenser is reused to heat the pot still. The ethanol vapour on the shell side is used to evaporate the condensate flowing through the tube side under vacuum.

The condensate is pumped around the condenser, entering a flash vessel, which is operated under vacuum. The liquid from the bottom of the flash vessel is recirculated to the condenser, and the level in this vessel is maintained automatically by condensate top up. The vapour from the separator vessel is drawn under vacuum into a thermocompressor, which is mixed with high-pressure live steam. The outlet stream from the thermocompressor feeds directly into an external plate heat exchanger, which is used to heat the wash for the pot still distillation. The content of the still (i.e., the wash) is recirculated through the heat exchanger via pump. Energy savings of 25%–30% can be achieved using this type of TVR arrangement (Fig. 14.10).

Spirit still operation

Due to the increased risk of alcohol loss, low wines and feints are not normally preheated, although the discharge of spent lees may be treated similarly to that of pot ale as a source for preheating. As in a wash still, the spirit is charged to a level not exceeding two-thirds of

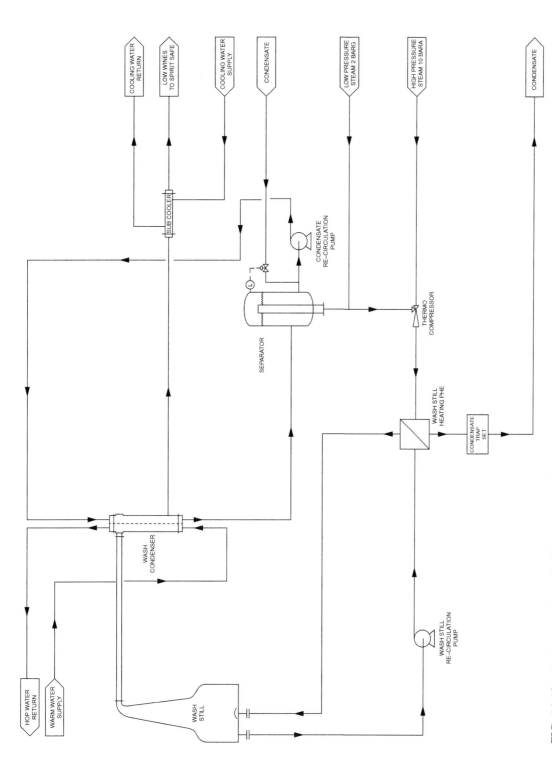

FIG. 14.10 Heat recovery in wash stills using thermal vapour recompression.

the working capacity of the spirit still. The precautions for charging are the same as in wash still distillation. The charge ingredients, a mixture of foreshots, low wines, and feints, are of greater excisable value and therefore require assiduous handling. Any loss of the charge will weigh heavily against the distiller. The low wines and feints receiver is dipped before and after distillation. The spirit receiver into which the new potable spirit will flow is also dipped, both at commencement of the new spirit run and at its completion, when as feints the flow is directed towards the low wines and feints receiver. A spirit distillation is divided into three fractions:

- Foreshots
- Middle cut
- Feints

The foreshots are the first runnings of the spirit distillation. In most cases, they are not deemed worthy of collection as potable spirit, as they contain highly volatile and aromatic compounds such as ethyl acetate. The time on foreshots is usually about 15–30 min, when the incoming strength of the distillate (~85% abv) drops to 75% abv.

Normally, a demisting test is carried out that involves mixing foreshots with water in a hydrometer jar in the safe and reducing the strength of the foreshots to 45.7% abv (old Sykes proof 80°). Initially, the mixture is turbid, with a milky appearance not unlike the reaction between anis and water. This turbidity is caused by displacement of the water-insoluble, long-chain fatty acids and esters (C14 and above) that have remained attached as a film to the inner surfaces of the still and in the residual subpool at the bottom of the spirit still condenser from the previous distillation. Being soluble in the high-strength incoming foreshots, they are flushed into the hydrometer jar. When the mixture of foreshots and water clears at the stated strength, the spirit is deemed potable. The flow of foreshots is redirected from the low wines and feints receiver to the spirit receiver by means of the swivelling spout and is collected as new spirit.

Some blenders and distillers have abandoned the time-honoured demisting test, preferring to collect the foreshots as new spirit after a timed run, with no resort to the demisting test, regardless of the potability of the spirit. Such final distillates are high in fatty acid ester concentration, making future chill proofing of mature whisky more difficult. Regardless of the way in which the spirit is deemed potable, the collection of new spirit lasts for about 2½ to 3h, during which time the strength drops from 72% to 60% abv, depending on the chosen final cut point.

The amount of heat applied to the still (as foreshots distil) and during the spirit distillation affects spirit quality. Too harsh an application of heat will result in a fiery spirit that has not benefited from a gentle natural reflux on the sides of the swan neck. To avoid adverse flavour notes, both foreshots and middle cut collections should be subjected to the delicate action of heat. On the other hand, feints can be treated like a wash distillation, following the initial collapse of the froth. The feints can be driven hard, reaching a distillation end point of 1% abv, and the resulting residue (spent lees) can be discharged while observing the safety procedures adopted for discharging the wash still. Chemical cleaning of the heating surfaces of a spirit still is rarely necessary to avoid disrupting the internal patina, the disruption of which is implicated in flavour reactions in the still.

Sulphur compounds present in the distillate vapour (as with wash stills) are highly volatile, and these odorous substances take their toll on the copper, forming sulphides. The carbon dioxide in the wash encourages the formation of copper carbonate, which manifests itself as verdigris. As mentioned earlier, these solids are the origin of the use of the muslin gauze filter, or Hippocratum, placed over the collecting bowl in the safe. The attack by carbon dioxide, sulphur, and solids (in wash still) also thins the copper, so eventually areas subject to this attack (above the boiling line, the shoulder, the swan neck, the lyne arm, condenser tubes, and start of worm) erode, requiring patching or replacement. A still affected by erosion emulates the breathing of a dog, with the shoulders rising and falling in a rhythmic pattern called *panting*. Such a condition renders the still more vulnerable to collapse, and the offending pot should be replaced.

Like a wash distillation, a spirit distillation should last from 5 to 8h, paralleling the wash distillation time. Ethyl carbamate precursors, being more soluble in aqueous solution, exit via the spent lees (Riffkin et al., 1989).

The alcoholic strength of the charge of combined foreshots, feints, and low wines should not exceed 30% abv; strengths in excess of this lead to blank runs when the demisting test fails to indicate potable spirit. In such circumstances, the demisting test protects the previously collected spirit from an influx of nonpotable spirit, which, with its high concentrations of higher fatty acid esters and long-chain saturated carboxylic acids, would impart a 'feinty' note to the spirit. The demisting test should always be available, even if foreshots are collected on a time basis.

Low wines and feints receivers and chargers act as separating vessels. The last runnings of a spirit distillation contain heavy oils and esters that are not readily soluble in water. Such oils have an affinity for alcohol, especially at high strength. At a strength of less than 30% abv, these compounds undergo a phase separation, such that the esters float on top of the aqueous layer while a small concentration is dissolved in the low-alcohol aqueous layer. If the concentration of the lower alcoholic strength aqueous layer is allowed to exceed 30% abv, these floating surface oils will migrate into this layer, completely dissolving. This effect eventually impacts not only the demisting test but also the entire spirit distillation—potable spirit cannot be collected, as the low wines and feints charge contains a disproportionate concentration of heavy oils, making it impossible to have a turbidity-free demisting test result.

With low wines and feints charges at less than 30% abv, it is still possible to suffer from distillation problems. The presentation of the floating surface layer of heavy oils or higher fatty acid esters as a charge to the still (by emptying the contents of the charger into the still) will result in an episode when the collection of potable spirit (as determined by the demisting test) is unachievable. The entire spirit distillation system will have been contaminated by these esters, and it can take several distillations before satisfactory spirit is again obtained.

To avoid such scenarios, when the low wines and feints appear to be approaching higher strengths (or have even reached this situation), the charge can be diluted with water, aiming for a combined strength of less than 30% ABV and thus stimulating hydroseparation. The surface phase in the low wines and feints charger must not be allowed to enter the still on charging. Adherence to these principles will ensure a consistent product, with regard to both nose and analysis. The low wines and feints components will reach a steady concentration state and maintain equilibrium during subsequent distillations.

Product quality

A competent distiller ensures that the distillery staff is fully aware of the parameters that must be controlled to provide a high-quality spirit. To ensure consistency, the plant and equipment must be designed in a balanced manner and techniques, borne of tradition, strictly observed along with modern improvements allied to that tradition.

First, the wash (whether traditionally derived from a wort's recipe producing an OG of 1050° or 1060° using high-gravity brewing techniques) should be fully attenuated, fermenting for at least 48 h. It has been shown that short fermentations of less than 40 h adversely affect the congener spectrum, producing an inferior spirit. Prolonged fermentations, exceeding 48 h in duration, undergo malolactic fermentation, the products of which, when distilled, produce a superior, more mellow spirit. Even at 48 h, such a secondary fermentation is unlikely to have occurred, as it relies on the autolysis of yeast cells with the spilling of the cell contents to provide nutrients for the lactic acid bacteria.

Fermentations should not be less than 2 days long to avoid an excessively gassy wash. A gassy lively wash is difficult to distil and has much frothing, leading to the risk of foul distillation and producing a final spirit with an unacceptable ethyl carbamate concentration. The wash (still) charge should ideally not exceed two-thirds the working capacity of the still, thus reducing the risk of foul distillation and subsequent poor spirit quality.

Without wash preheating, the wash still contents should be heated gently to minimise charring on the heating surfaces, which are more susceptible to burn-on of proteins and dextrins in the early stages of distillation if too much heat is applied (i.e., too great a temperature differential between the still contents and the heating elements, coils, pans, or kettles). With wash preheating, the temperature difference is much reduced and charring is less prevalent.

An adequate supply of cooling water must be provided to prevent hot, uncondensed vapours from reaching the spirit safe, causing irreparable damage to safe instruments, jars, bowls, and the final distillate quality. Spirit-filled thermometers have now replaced all mercury-filled thermometers for obvious health and safety reasons.

The bouquet of the final spirit is influenced initially by the raw materials—particularly the variety of barley malted and the pitching yeast. Water also influences spirit character. Process parameters, including mashing temperatures, washback setting temperatures, length of fermentation, and variable new spirit cut points, will impact the flavour characteristics of the spirit. Such production variations can be ameliorated by collecting several days' distillations in one large vat. Variability in individual production quality will then be eliminated by averaging congener concentrations.

With peated malts, the spirit produced reflects a marked concentration of steam volatile phenols, whose arrival in the final spirit tends to concentrate towards the end of the middle cut collection as the alcohol-to-water ratio changes with decreasing spirit strength, favouring phenol entrainment. To enhance the phenol concentration in the new spirit, the strength of the second cut point in the middle cut can be reduced, but not at the expense of producing a feinty spirit. A cut point of not less than 60% abv would be acceptable.

The rate of distillation is critical. Too rapid a distillation will result in an unpalatable spirit, fiery in aroma and taste and lacking a refined congener balance. Foreshots and the middle cut should be carefully and gently distilled to ensure adequate reflux, with foreshots completely purging the oily residues of the previous distillation. As mentioned earlier, it is these residues that demand the demisting test. Slow spirit distillation ensures the production of a clean balanced spirit devoid of aroma and flavour blemishes.

It must be emphasised that an adequate supply of cold water to condensers or worm tubs should be maintained. Inadequate cooling (> 20°C) will lead to spirit endowed with an aroma reflecting higher concentrations of compounds associated with the feints. This is also true of forced or too rapid distillation. Warm weather, with resultant warmer cooling condenser water, demands that the distillation rate be reduced to allow the spirit to be collected at the desired temperature (20°C). Prolonged distillation times will have an adverse effect on production schedules (e.g., mashing and fermentation). Distilling spirit, during times of high ambient temperatures, will increase losses due to evaporation. Thus, the practice of malt distilling requires low ambient air and water temperatures, as experienced during late autumn, winter, and early spring. Distillation in past centuries was, like curling, essentially a winter sport, a time during which the floor malting of barley was easily controlled, providing fully modified malt unaffected by high summer temperatures. Summer was the 'silent season', when distillery staff were engaged in the maintenance of plant and buildings, harvesting barley, and bringing the peats home.

Chemical contaminants that can bedevil the malt distiller are nitrosamines, ethyl carbamate, methanol, pesticide residues, haloforms, polycyclic aromatic hydrocarbons, and pesticide and herbicide residues, all of which are monitored in compliance with government regulations and procedures. Some of these contaminants are derived from the raw materials and others from processing malt or during distillation. Genetically modified cereals and yeast come under scrutiny, as the definition of Scotch whisky demands the use of only pure water, yeast, and cereals derived from natural sources.

Copper has already been mentioned as a silent contributor to spirit quality as it removes highly volatile sulphur compounds. It is also implicated in the formation of esters. Copper catalyses the formation of ethyl carbamate from the cyanogenic glycosides derived from the original barley. Stainless steel is not recommended for the construction of distillation equipment to avoid compromising quality, but it can be used for ancillary pipework and vessels. The original gravity (OG) of the wash impacts on spirit quality, and it has been determined that OG values in the range of 1045–1050° (11.3–12.5° Plato) encourage the formation of esters, thus imparting a fruity, sweet aroma to the finished product. The role and impact of copper on the quality of potable spirit are discussed further in Chapter 11.

Efficiency and production yield

Following the design and construction of a well-balanced distillery, where milling, mashing, fermentation, and distillation are in harmony, it is fairly easy to establish in-tandem programmes of mashing. This is achieved by ensuring that the time cycles for mashing, fermentation, and distillation are in step, with the week being divided into set periods reflecting the mashing cycle. Consequently, if mashing takes 6 h to complete (with washbacks individually filled within this time period), the maximum distillation time from charging to discharging should not exceed 6 h. This enables four mashes per day to be performed. The distillery may be fully automated, removing the human element and its uncertainty.

One tonne of malted barley, fully modified and efficiently mashed, should ensure complete extraction of available fermentable sugars, resulting in an overall distillery yield approaching 425 L of pure alcohol. Without complete extraction, it will be impossible to achieve the potential spirit yield determined by laboratory analysis. Mashing efficiency is vital in achieving the maximum possible spirit yield.

In the distillery, it is essential that the integrity of the pipework, vessels, and stills is maintained without leaks. Spirit can be lost through insidious, invisible vapour leaks that are not easily detected. The worm or tube bundle in a shell-and-tube condenser is under constant attack by sulphur compounds and carbonic acid emanating from the vapour phase, which eventually corrodes the copper. Condenser or worm leaks are noticeable by the entry of cooling water into the product side, reducing the strength of the distillate (as detected by the safe hydrometers) and most definitely producing a water flow into the safe when the still is off. Such a scenario requires that the distillation be stopped, with the offending tubes spilled or worms patched under strict safety conditions, before continuing further distillation. In the case of shell-and-tube condensers, several tubes may be affected, and the condenser should be pressure tested for further possible tube weaknesses. In the event of multiple tube failures, the condenser will require retubing.

Thinning copper on the shoulder of the pot, due to erosion at the boiling surface, swan neck, or lyne arm, can lead to pinhole leaks as the copper becomes spongy. Such leaks are rectified by soldering or the temporary use of molecular metal. Soldering requires the use of a blow torch, and flammable vapours must be purged from the system by blanking off receivers to prevent explosion and fire. Such repairs require the complete cessation of distilling operations. Leaks are not acceptable and must be dealt with as soon as is practically possible.

Other losses occur via the pot ale or spent wash and spent lees, when the distillation end point is not accurately observed. Again, these end points would indicate that energy is being wasted if the distillation is continued beyond 1% abv, as indicated by the safe hydrometers. Premature ceasing of distillation will result in significant detectable ethanol being present in the pot ale or spent lees. Permissible spirit losses are as follows:

• Pot ale	< 0.03% abv
• Spent lees	< 0.03% abv
• Condensate	< 0.0001% abv
• Condenser water	< 0.0001% abv

Where pot ale is evaporated for syrup, residual ethanol interferes with the evaporation efficiency.

The distillery yield is calculated from the weekly production figures. It takes account of the weight of malted barley used and the amount of spirit remaining in the low wines and feints receiver, the intermediate spirit receiver (ISR), and the final spirit receiver/warehouse (W/H) vat. Depending feints, carried forward from the previous week, are deducted from the total sum of spirit produced, expressed in litres of absolute alcohol (LAA).

Calculation of distillery yield

(a) Depending feints	= 15,500 LAA (carried forward from previous week)
(b) Spirit produced in ISR	= 7200 LAA
(c) Spirit receiver W/H vat	= 30,300 LAA
(d) Feints remaining	= 13,500 LAA
(e) Spirit produced	= (b) + (c) + (d)—(a)
	= 7200 + 30,300 + 13,500–15,500
	= 35,500 LAA
(f) Tonnes of malt mashed	= 85.54 t
(g) Distillery yield	= Spirit produced (LAA) ÷ malt used (t)
	= 35,500 LAA ÷ 85.54 t
	= 415 LAA per tonne

Her Majesty's Revenue and Customs (HMRC) is able to calculate the projected amount of spirit produced from a given amount of malt through the attenuation charge.

Calculation of attenuation charge and percentage over attenuation

As an example, the attenuation across several fermentations comprising a week's production is 57° gravity from an OG of 1055° (13.75° Plato) and a final gravity (FG) of 998° or −02 (02 under). For 10 fermentations with a total volume of 488,500 L of wash and an average attenuation of 57°, the attenuation charge is calculated as follows:

$$\text{Attentuation charge} = \frac{\text{Litres of wash} \times \text{Average attenuation}}{8 \times 100}$$
$$= \frac{485,000 \times 57}{800}$$
$$= 34,556 \text{ LAA}$$

HMRC calculates the percentage over-attenuation from the following formula:

$$\left(\frac{\text{Spirit produced}(\text{LAA}) \times 100}{\text{Attenuation charge}} - 100\right) \times 100$$
$$= \left(\frac{35,500\,\text{LAA} \times 100}{34,556\,\text{LAA}} - 100\right) \times 100 = 2.73\%$$

Hence, an acceptable percentage over-attenuation has been achieved at 2.73%. A value of 3.0% is deemed acceptable. Any departure from this figure by more than 1% to 2% either way demands an investigation.

Under-declaring the OG of worts collected will inflate the figure, while over-declaring will reduce the figure. Recourse to laboratory analysis for OG determinations is necessary. The law demands that at least six declared washbacks per week should be analysed to find the true declarations, with the distiller adjusting the saccharometer readings to allow for the amount of work conducted relating to the temperature and gravity of the wash at the time of declaration.

Triple distillation

Within the Scotch malt whisky industry, there are at least two distilleries that practise triple distillation. This technique ensures a lighter final spirit at higher natural strength than double-distilled whiskies and is primarily carried out in lowland distilleries. It is similar to the distilling practice in Ireland. In principle, there is a wash still from which two fractions are derived—strong low wines and weak low wines—and are separately collected. A second still, the low wines still, is charged with the weak low wines. From this low wines still, two fractions are similarly collected: strong feints and weak feints (tails). The strong feints are presented to the third still, the spirit still, and the weak feints are redistilled in the low wines still.

The distillates from the spirit still are divided into three collected fractions: the foreshots or heads, the new spirit, and the tails (which, with the heads, are collected and returned for redistillation in the spirit still). This recycling of the various fractions derived from the low wines and spirit stills impacts on the final bouquet and strength of the new spirit. This is collected at a strength in excess of that of normal double-distilled products, which are usually in the region of 68 to 72% abv. The triple-distilled product can approach a strength of 90% ABV. The Irish distillers boast very large pot stills in comparison with the double-distilling techniques of their Scottish counterparts.

Dealing with distillation problems

As with the all manufacturing processes, problems can occur that impact the quality of the finished product if not rapidly addressed. Such problems can occur during the mashing, fermentation, and distillation stages. Declaration problems related to over- or under-attenuation attract the unwelcome attention of Revenue and this is best avoided!

Atypical over-attenuation percentages can be attributed to false declarations when determining the OGs in a set washback. As previously discussed, this can be overcome by recourse to laboratory determinations of OG in at least six declared washbacks, calculating the allowance for loss in gravity due to fermentation and temperature. The average calculated OG and the difference between what was observed at the time of declaration and the true OG from the laboratory checks are the figures that must be added to the tun room declaration. Saccharometers and thermometers must be regularly checked for accuracy against standard solutions and thermometers. The addition for wort and temperature is then not suspect guesswork but a true reflection of the genuine OG.

All worts and final wash should be attempered to 20°C prior to taking readings. Product losses, with low percentages of over-attenuation accompanied by low yields when compared with the potential spirit yields obtained by analysis, are invariably caused by one or a combination of the following problems:

- Poor mash tun extraction
- Infection (Geddes, 1985)
- High wash fermentation temperatures
- Physical losses (worts, wash, low wines, feints, spirit)
- Human error

To satisfy Revenue, any loss should be accountable to avoid the possibility of excise penalties, as long as the loss is accidental.

To improve mash tun extraction efficiencies, it is necessary to examine sparge-to-grist ratios as well as mashing temperatures, especially the first water. Overloading a mash tun with goods is definitely counterproductive, exacerbating fermentable sugar losses in the draff.

Bacterial infection, which competes with yeast for fermentable sugars, is alleviated by paying close attention to cleaning regimes, concentrating on mashing, fermentation equipment, and pipe work; any dead legs should be eliminated (see Chapter 22).

Attention should be paid to setting proper fermentation temperatures depending on prevailing atmospheric conditions. Fermentations that exceed 33°C contribute to evaporation losses and may also contribute to the growth of strains of *Lactobacillus*, which will be responsible for off-notes and the production of acrolein during distillation. Physical losses of process materials can be accidental due to plant or human failure, demanding investigation to satisfy Revenue.

The nature of the wash, with dissolved carbon dioxide, can lead to foul distillations when the still boils over into the safe, damaging the safe instrumentation and glasswork. Low wines and feints, thus contaminated with wash, can be a source of increased ethyl carbamate formation. Overzealous application of heat, an overloaded wash still, young lively wash, or blocked condenser tubes can also lead to this problem. Sensory or organoleptic problems, presenting themselves as feinty spirit, are caused by the overrun of the middle cut. Safe spirit hydrometers and high condensate temperatures should be suspected and the necessary checks made. It is possible to be plagued with blank spirit runs when no potable spirit is collected, producing adverse effects on fuel usage. This is caused by weak low wines and feints charges or by an increasing charge strength ($>30\%$ abv), when higher concentrations of fusel oils dissolve in the alcoholic aqueous layer. The problem can be overcome by reducing the strength of the charge ($<30\%$ abv) by adding water.

Discharging the total content of the low wines and feints charger into a spirit still is a recipe for blank runs, as the surface layer of fusel oils contaminates the still, making it impossible to have a successful demisting test when applied to the foreshots. Increasing time lengths to clear foreshots can indicate an increasing concentration of fusel oils in the foreshots. A balanced distillation regime can help to alleviate these problems.

The collection of weak spirit, resulting in a congener imbalance, can be attributable to poor reflux, poor cooling, or a leaking condenser. To overcome this effect, the distillation rate can be lowered to improve reflux and cooling. Leaking condenser tubes manifest themselves by presenting a water flow to the safe when the distillation is complete. The condenser tubes should be checked by pressure testing with water. Any offending tubes should be temporarily blanked off by spiling, and a constant check should be maintained for future potential tube failure. One leaking tube indicates that neighbouring tubes should be treated with suspicion.

Wash stills are prone to fouling of direct or indirectly fired heating surfaces due to the nature of the wash with its solids and unfermentable sugars. Fouling impedes heat transfer, resulting in increasing distillation times and energy wastage. A 1 to 2% caustic soda solution treatment of the still or heating elements should be sufficient to solve this problem.

High ambient water and air temperatures, especially in summer, combine to elevate the temperature of distillates entering the safe. This demands that the distillation should be slowed so that the low wines, feints, and spirit are collected at a slower rate and at temperatures as close to 20 °C as possible. Thus, evaporation losses and detrimental organoleptic effects are reduced but unfortunately at a slower production rate. However, some distillation losses are difficult to detect. An almost inaudible high-pitched hissing noise emanating from a steam coil indicates a steam leak, which will not only dilute the charge but also result in the charge entering the condensate system, with resultant losses and condensate contamination. Most indirect heating material is constructed from stainless steel and less so from copper. Nevertheless, regardless of the material of construction, the integrity of all flanges should be examined, and, in the case of copper, the existence of any cracks or pinholes should be determined.

Pinhole vapour leaks occurring above the charge line in the still can be readily detected when external verdigris blue–green staining occurs on any of the surfaces of the still components, from neck to head and lyne arm. If not dealt with quickly, these insidious and unsightly leaks, especially if occurring at inaccessible places, can rapidly grow. Passing valves, air and anticollapse valves, and flanges on pipe runs can all contribute to product loss. Management must therefore maintain a high level of inspection for physical losses due to an ageing plant, especially copper, fundamental to the quality of the final product. When the body of a copper still begins to thin above the charge line, like an old dog it begins to pant, visibly heaving up and down. Such a still is reaching the end of its useful days and will require either patching or, in extreme cases, replacement.

The role of copper in the quality of new whisky

Copper has been mined as an ore since the Bronze Age (~3500 BCE). As a metal, it has many uses relating to its properties. It is soft and easy to work in its annealed state. Its hardness and tensile strength can be doubled by working it cold through hammering, beating, rolling, or drawing. For distillery purposes, its properties related to its malleability, thermal conductivity, and resistance to corrosion make it an ideal metal for the manufacture of a distillation apparatus.

It was not until the introduction of stainless steels that a more subtle property of copper manifested itself. It was known that a copper distillation apparatus had to be replaced periodically, as it was subject to wear. Certain distillers believed that to prolong the life of the distillation equipment stainless steel would make a suitable substitute. Plans were made to replace the copper condenser tubes and even the stills with longer lasting stainless steel, as American distillers had done in their distilleries.

It was not until the new spirit took on a sulphury odour that the impact of stainless steel on the bouquet of new spirit was realised. New spirit derived from worm tubs reflected a similar note but of less intensity. It was therefore back to the drawing board! To focus on the area most likely to have an impact on this aroma, a pair of Quickfit glass stills was assembled under laboratory conditions. In the first glass still, copper turnings were placed immediately above the condenser, in the splash head, and also in the connection between the glass still and the condenser. The other glass still was assembled without copper (unpublished results, 1968). Low wines were obtained from a distillery that boasted stainless steel condensers in the wash stills. Using the glass laboratory apparatus, the low wines were distilled through both small stills, one with and the other without copper turnings.

The products from each still were examined organoleptically. The low wines distilled over the copper produced a noticeably clean spirit compared with the still without copper. Further trials were carried out to find out which part of the apparatus containing copper had the most impact on the bouquet. The conclusion was that hot spirit vapours reacted readily with copper in the tube bundle at the top of the condenser and to a lesser extent on the copper surface in the lyne arm. Where the reaction was greatest at the top of the tube bundle, it was associated with the erosion of copper, requiring eventual retubing of the condenser with new copper. However, little had been known about the substances contained within the low wines vapour, except that such aromas were derived from sulphur compounds associated with foul smells. Stills with all-copper condensers, including the tube bundles, produced an aroma with a lesser contribution from sulphur, and stills with worm tubs possessed a hint of a sulphury aroma. It was not until recently that the impact of the offending sulphur compounds was identified through work carried out by a team at the Scotch Whisky Research Institute (Harrison et al., 2011).

The primary offending compound, dimethyltrisulphide (DMTS), was identified as having the most meaty and sulphury aroma. It was also concluded that sulphury or meaty aromas were established in the wash still condensation phase on the copper condenser and in the low wines or spirit still on the inner copper surface of the body of the still, reflecting the results derived using the Quickfit apparatus with and without copper turnings. This is a scientifically interesting area that demands further investigation with regard to the mechanisms involved.

One of the recently constructed distilleries has incorporated two condensers for each wash still, one of copper and the other of stainless steel. They were being shared by means of a bifurcate valve arrangement in the lyne arm, so that the still can produce either a sulphury or sulphur-free spirit. The organoleptic contribution of highly volatile sulphur compounds to new spirit decreases upon maturation, unless it has been obtained from casks treated with burning sulphur sticks prior to purchase.

Miniature pot stills

Over the past few years, there has been an explosion of single miniature pot stills, primarily engaged in the production of gin. These stills come in many sizes compatible with the size of premises they are housed in. Under the latest HMRC laws relating to distillation in the UK, a compounder or rectifier's licence is required. The raw material used, rectified spirit (97.1% alcohol), is already duty paid.

To manufacture whisky in the UK requires a distiller's licence, which at one time had cost only £10.75. HMRC does not permit whisky production in stills of less than 18 hectolitres in size (40 gal) capacity. This size stems from the time when illicit distilling was carried out in small stills. The larger size of stills required bigger premises, thus discouraging this nefarious secret practice. Today, these miniature stills can be made of stainless steel or of the more traditional copper. For the purpose of this chapter, it is not necessary to go into great detail as all of the required information on where to purchase such stills, modes of operation, and recipes for gin production can be found on the Internet. As far as the whisky production is concerned, there has been the introduction of automation, removing the need for manual control of certain aspects of the distillation process.

The future

Environmental, energy, fiscal, and health and safety pressures continue to be experienced by distillers due to greater public awareness of diminishing wildlife species and oil and gas supplies, profligate governmental schemes and banking mismanagement, and ever-present health and safety issues that demand greater monetary contributions through taxation. With the introduction of computerised controls in almost every distillery department—from malt intake to milling, mashing, fermentation, distillation, and cask filling—the industry has reduced its production labour force as a cost-cutting exercise; one person can now operate a distillery, but law requires that person to have a working companion. This trend towards automation is set to continue. Raw materials (barley, malt, yeast, and water) will continue to come under scrutiny during the search for ways to improve distillery yields so whisky, the elixir of life, will continue to dominate the spirits market. Other countries are attempting to reproduce the success of Scotch whisky by producing products similar to Scotch. By retaining age-old but modernised techniques and a standard recipe, the continued success of whisky is assured.

References

Energy Technology Support Unit, 1985. Whisky distilling. In: Drying, Evaporation, Distillation: The Potential for Improving Energy Efficiency in Twelve Industrial Sectors. Vector Publishing, Farnborough, U.K, pp. 34–41.

Geddes, P.J., 1985. Bacteriology in the whisky industry. J. Inst. Brew. 91, 56–57.

Harrison, B., Fagnen, O., Jack, F., Brosnan, J., 2011. The impact of copper in different parts of malt whisky pot stills on new make spirit composition and aroma. J. Inst. Brew. 117 (1), 106–112.

Lyons, T.P., 2003. The production of Scotch and Irish whiskies: their history and evolution. In: Jacques, K.A., Lyons, T.P., Kelsall, D.R. (Eds.), The Alcohol Textbook, fourth ed. Nottingham University Press, Nottingham, U.K, pp. 193–222.

Lyons, T.P., Rose, A.H., 1977. Whisky. In: Rose, A.H. (Ed.), Economic Microbiology. vol. 1. Alcoholic Beverages Academic Press, New York, pp. 635–692.

Nicol, D.A., 1989. Batch distillation. In: Piggott, J.R., Sharp, R., Duncan, R.E.B. (Eds.), The Science and Technology of Whiskies. Longman Scientific & Technical, Harlow, U.K, pp. 118–147.

Nicol, D.A., 1997. Distilling, past, present and future. Ferment 10 (6), 382–391.

Piggott, J.R., Connor, J.M., 1995. Whiskies. In: Lea, A.G.H., Piggott, J.R. (Eds.), Fermented Beverage Production. Blackie Academic & Professional, London, pp. 247–274.

Riffkin, H.L., Wilson, R., Bringhurst, T.A., 1989. Ethyl carbamate formation in the production of pot still whisky. J. Inst. Brew. 95, 115–119.

Watson, J.G., 1989. Energy management. In: Piggott, J.R., Sharp, R., Duncan, R.E.B. (Eds.), The Science and Technology of Whiskies. Longman Scientific & Technical, Harlow, pp. 327–359.

Whitby, B.R., 1992. Traditional distillation in the whisky industry. Ferment 5 (4), 261–267.

Further reading

Smith, G.D., 1997. A–Z of Whisky. Neil Wilson Publishing, Glasgow.

CHAPTER

15

Grain whisky distillation

Douglas Murray

Institute of Brewing & Distilling, London, United Kingdom

Introduction

Throughout the history of Scotch whisky production, cereals such as wheat, oats, and barley and various distillation methods have been used to make raw spirit. By the early 19th century, the two main Scotch whisky types, grain and malt, had become separated in terms of both production technology and geographical location. In the central belt of Scotland, the increasing population and demand from other countries required large volumes of relatively inexpensive distilled spirit, which led to a rise in the manufacture of grain whisky. The manufacture of this grain whisky evolved into a more efficient distillation method with regard to labour, energy cost, and throughput. This evolution from simple distillation culminated in the design of continuous stills by leading distillers of the day such as Robert Stein in 1827 and Aeneas Coffey, who patented his design in 1830. The introduction of these continuous stills led to a lightly flavoured whisky that contrasted with the stronger flavoured whiskies made by batch pot still distillation. Over time, this type of whisky was recognised as a valuable component in the emerging blended Scotch whisky market. In 2013, there were seven major grain distilleries in operation in Scotland (Fig. 15.1) providing the range of flavours required by the Scotch whisky industry, and the product was used mainly for blending with malt spirit. There are also several small distilleries that make a range of spirits including some grain spirit. The theory and basic technology are no different to what is contained in the batch and continuous distillation chapters of this book. For example, Diageo's Teaninich distillery can process cereal through its use of a mash filter.

Feedstock for distillation

The Scotch whisky regulations are exactly the same for grain whisky production as for malt whisky production. In contrast to malt whisky, the term 'grain' implies that any cereal can be

---

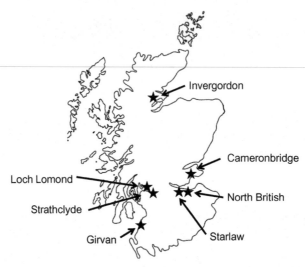

FIG. 15.1 Geographical location of grain whisky distilleries. Courtesy of Diageo PLC, London.

used, providing the starch content of the cereal can be converted into sugar using the endogenous enzymes contained in the cereal or in the malted barley. It should be remembered that the regulations on Scotch whisky production require the whole grain to be processed. This prevents the use of cereals where the seeds have been processed or polished. Historically, as mentioned above, many different cereals were used, but in more modern times the main cereals are wheat and maize (corn), with the occasional use of raw barley. It is normal practice to use only one cereal type at a time due to the different processing requirements of each type. The key factors in selecting the cereal type are availability, price, starch content (related to alcohol production), energy requirement prior to distillation, and ease of processing through the plant. In addition, each cereal imparts a slightly different chemical composition to the fermented wash, which can provide a flavour nuance for each distillery's product, an attribute that is much desired by the blenders.

Factors affecting energy use are the ease of milling, or indeed the need to mill, the volume of water required, and the gelatinisation temperature, which, when high, requires greater energy input to release the starch into solution. This has led to grain distilleries using different cereal preparation methods ranging from pressure cooking to lower temperature hydration/gelatinisation mashing. The ease of processing relates to the cost of the equipment required, such as pressure vessels, increased pump sizes due to viscosity, wear and tear on the pipes, foam control during fermentation, and a build-up of cereals at the top of the fermenter into a thick crust. For example, maize contains oil, which is a natural antifoam, while the high glucan and husk levels in barley can cause significant fouling and crust formation, thus affecting process efficiency.

Table 15.1 shows a comparison of percentage starch, gelatinisation temperature, and ethanol yield for cereals of the quality used by the whisky industry. Traditionally, wort was prepared in a manner similar to that for malt whisky, in a mash tun with separation of the solid material from the liquid occurring through a false bottom. This practice has been superseded in most, but not all, grain distilleries by the use of a mash mixer, in which all of the grain residue is transferred to the fermenter, known as an 'all grains in' process. In this pro-

TABLE 15.1 Comparison of starch content, potential yield, and gelatinisation temperature for potential raw materials.

Cereal	Typical starch range (%)	Typical ethanol yield (as is) (LA/t)	Gelatinisation temperature range (°C)
Maize (corn)	70–74	400	70–80
Rice	74–80	430	70–80
Wheat	60–65	390	52–55
Barley	65–70	350	60–62
Sorghum	62–68	388	70–80

Source: Data courtesy of Diageo PLC, London, and Palmer, G.H. (Ed.), 1989. Cereal Science and Technology. Aberdeen University Press, Aberdeen, Scotland, p. 154, Table 24.

cess, the non-fermentable portion of the cereal is not removed prior to distillation. The distillation equipment used requires modification to allow solids to pass through the still without significant fouling. The ability for the distiller to change the feed material and to process the slightly differing levels of solids from each cereal type is one of the main advantages of continuous distillation. The higher level of rectification, providing it is less than 94.8% (the legal upper limit for Scotch and Irish whiskies), and different still designs add flexibility. These factors have a double benefit and allow control of consistency and being able to change the final composition of the spirit and, consequently, quality if required.

To facilitate continuous distillation, a wash charger vessel is usually placed between the fermenter and the still and is used to store fermented wash (the liquid produced at the end of fermentation) to provide a constant feed to the still. This provides the link between the batch fermentation process and a continuous still operation. The concentration of ethanol in this vessel depends on the fermentation conditions and feed material but is typically in the range of 8%–12% (v/v). Higher levels of ethanol can be generated using the next generation of yeasts, but these may not have the typical congener profile associated with (Scotch) grain whisky.

Theory of continuous distillation

The fermented wash produced is a mixture of ethanol, water, desirable flavour components (congeners), undesirable flavour compounds, and solid material (mainly unfermented cereal and yeast). The act of distillation carries out the basic separation of ethanol from the water and solid material, thus increasing the concentration of ethanol. At the same time, congeners need to be recovered and undesirable flavour compounds removed. All of these processes are carried out by the still to ensure a consistent flavour profile and to maintain quality.

Ethanol has a boiling point of 78°C, and water has a boiling point of 100°C. A simple mixture of the two will boil at a temperature between these two values. However, as ethanol is more volatile, the vapour produced will contain ethanol at a higher concentration than in the boiling liquid. This relationship is shown in Table 15.2 and in graphical form in Fig. 15.2.

The bubble point line defines the point where evaporation takes place. Under this line, the ethanol and water exist only in the liquid state. Above the dew point line, the mixture exists only as a vapour. Between these lines, liquid and vapour coexist, and it is in this region that the science of distillation operates. As shown in Fig. 15.2, a mixture of 35% ethanol (X) is being heated. Between points X and Z, say point Y, the mixture is entirely liquid. As the temperature is increased above point Z, vapour is being formed. At point K, both vapour and liquid exist (i.e., it is a two-phase equilibrium mixture). Drawing a line parallel to the x-axis gives the relative composition where it intersects the dew and bubble point lines. Point L is the composition of the vapour and point M the composition of the liquid. The amount of each is defined by the relative lengths of the lines from points M, K and points K, L. It can be seen that the composition of the vapour is increased in the relatively lower boiling point ethanol. In continuous distillation, the constant feed allows a steady state to be established (Campbell, 2003).

The point at which the two curves meet is at a temperature of 78.15°C and a concentration of 97.2% v/v ethanol and represents the azeotropic point. At this point, further boiling (distillation) does not change the concentration of ethanol. Continuous distillation uses this principle. A simple distillation column and its components are represented in Fig. 15.3.

In simple terms, liquid introduced into the still boils and the enriched vapour passes up to the column through the plate (see later section on plates for more detail). The liquid is depleted of ethanol and passes down to the plate below. If the liquid enters the column near

TABLE 15.2 Vapour–liquid equilibrium data for ethanol and water mixture at a constant pressure of 1 bar (101.3 kPa or 101.3 kN/m^2).

Temperature (°C)	Mole fraction of ethanol in		% w/w		% v/v	
	Liquid	Vapour	Liquid	Vapour	Liquid	Vapour
78.2	0.8943	0.8943	95.6	95.6	97.2	97.2
78.4	0.7472	0.7815	88.4	89.3	92.1	92.8
78.7	0.6783	0.7385	84.2	87.8	88.9	91.7
79.3	0.5732	0.6841	77.4	84.7	83.3	89.2
80.7	0.3965	0.6122	62.7	80.1	70.3	85.6
82.3	0.2608	0.5580	47.4	76.3	55.2	82.4
84.1	0.1661	0.5089	33.7	72.6	40.5	79.2
86.7	0.0966	0.4375	21.5	66.5	27.5	73.8
89.0	0.0721	0.3891	16.6	61.9	20.4	69.6
95.5	0.0190	0.1700	4.7	34.4	5.9	41.2
100	0.0000	0.0000	0.0	0.0	0.0	0.0

Source: As calculated from Seader, J.D., Kurtyka, Z.M., 1984. Distillation. In: Perry, R.H., Green, D.W. (Eds.), Perry's Chemical Engineer's Handbook, 6th ed. McGraw-Hill, New York, pp. 13–21.

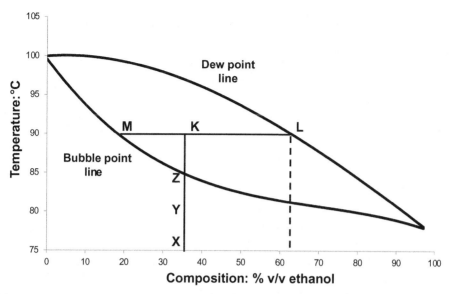

FIG. 15.2   Vapour–liquid equilibrium data for ethanol and water mixture at a constant pressure of 1 bar (101.3 kPa or 101.3 kN/m²) as calculated from Seader and Kurtyka (1984).

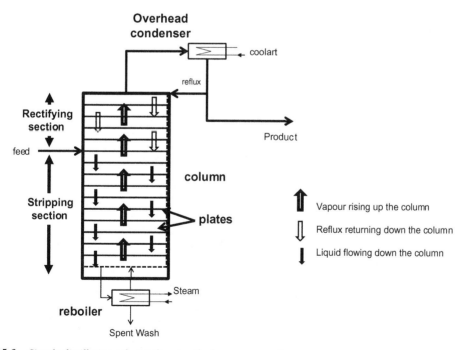

FIG. 15.3   Simple distillation column showing the key components.

its boiling point, then all of the heat being supplied by the steam will be transferred to the vapour. As the liquid passes downwards from plate to plate, it is boiled on each plate by contact with the ascending vapour. The vapour produced on each plate going down the column becomes progressively reduced in ethanol concentration. In a similar manner, vapour rising through the column condenses on the next plate above and the enriched liquid, which has a lower boiling point, will be reboiled. The liquid on the plate is then allowed to flow downwards. This downward flow of liquid is called *reflux* and is critical to the distillation process. This results in the boiling point of the liquid increasing as you go down the column and the ethanol concentration increasing as you go up the column. This is shown in Fig. 15.4. Below the feed entry point is an area called the *stripping* section, and the area above is termed the *rectification* section.

The design of the column requires that the steam and wash input matches the spirit and spent wash output. If the volume of ethanol being taken off is reduced, then the ethanol balance is disrupted, and as the ethanol in the feed is fixed, the extra ethanol will leave the still in the spent wash and an ethanol loss will result.

When the industrial application is considered, the design is not identical to that of the theoretical design. The number of actual plates required is greater than the number predicted by the theoretical model, and this makes the column very tall. To accommodate the required number of plates, while keeping building capital costs low, the still is normally constructed as two individual columns making up the rectification and stripping sections. The overhead condenser cannot condense all of the vapour, as some has to be vented for quality reasons. In addition, the feed is not a simple two-component (ethanol and water)

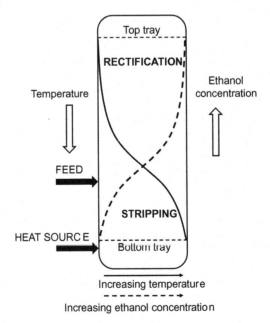

FIG. 15.4   Temperature and ethanol concentration profile in a single column still.

mixture but rather contains hundreds of other compounds. The build-up of highly vol-atile components at the very top of the column results in an unacceptable level of these components (e.g. methanol) being present in the reflux; therefore, the spirit is drawn off several plates below the top, rather than as proportion of the condenser exit line. The actual strength of the spirit collected is between 93.5% and 94.5% to ensure the correct level of desirable flavour compounds. Finally, care has to be taken when using a reboiler, as the steam on the inside of the tubes may cause solid material to become burnt onto the outsides of the reboiler tubes. In most cases, direct injection of steam is used to prevent this from happening and also to prevent the production of unpleasant flavour compounds from the burning effect.

The vapour–liquid equilibrium of ethanol and water is usually expressed differently than what was shown in Fig. 15.2; this is shown in Fig. 15.5 using the term *percentage molar*. This method uses the fortunate property that the molar latent heats of most compounds in whisky distillation have similar values. The latent heat of evaporation of ethanol is 854 kJ/kg, and multiplying by its molecular weight (46) gives a molar latent heat figure of 39,290 kJ/kg mol (854 × 46). Water has a molar latent heat of 40,525 kJ/kg mol (2250 × 18). This is only 3% different than ethanol, and an average figure of 40,000 kJ/kg mol is used. If molar mass units are used, then a mass balance is the same as a heat balance, and the heat

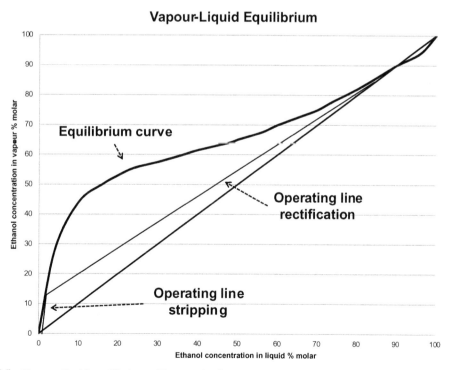

FIG. 15.5 Vapour–liquid equilibrium with operating line.

quantities do not have to be considered. Molar units and concentration are therefore used for distilling calculations.

In Fig. 15.5, a line is drawn from the x-axis from the point representing the ethanol in the spent wash (a very low value) up to the point where the wash ethanol vapour line meets the equilibrium curve. A second line is drawn from this point to the point on the x,y line, where the ethanol concentration in the liquid is the desired spirit strength. These are the two operating lines. The mathematics behind the derivation of the operating line is known as the McCabe–Thiele method (McCabe and Thiele, 1925). For those who wish to learn more about this aspect, further information can be found in *Perry's Chemical Engineer's Handbook* (Perry and Green, 1984). The operating line represents the relationship between the liquid composition on a plate and the vapour passing through the liquid from the plate below. This is made up of these two straight lines. The first section is the stripping line, and the other is the rectification line. In practice, the operating line is derived from the required reflux ratio, which is typically 4:1. This can vary depending on the amount of energy the distiller is prepared to inject. In practice, the reflux ratio is a compromise between energy use and capital cost of the additional plates needed in the still. By drawing horizontal lines from the x,y line to the equilibrium line, the number of steps or theoretical plates can be determined (Fig. 15.6). In Fig. 15.6A, at the top of the rectification line these theoretical plates get closer and closer on the graph as they reach the azeotrope. In Fig. 15.6B, the stripping zone has been expanded to show the large number of theoretical plates in this area.

Continuous still design and operation

Within the Scotch whisky industry, each grain distiller employs a unique distillation regime to give the desired spirit quality. In terms of still design, these can be categorised into three groups:

- Traditional or Coffey stills
- Atmospheric column stills
- Vacuum column stills

All three types have the following features in common:

- Method of preheating the feed prior to entry into the still
- Initial stripping column (wash, beer, or analyser column)
- Rectification column
- Reflux condensing system
- Fusel oil recovery system
- Energy input system
- Stripped feed removal system

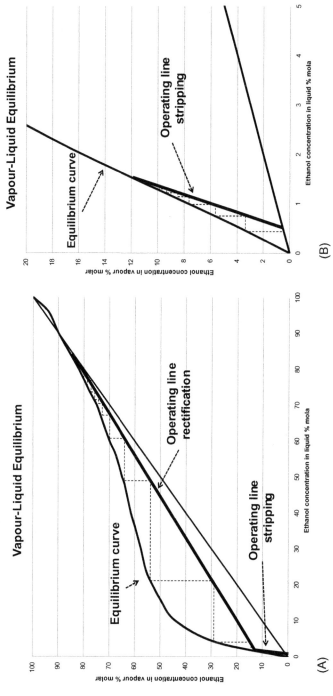

FIG. 15.6 (A) Vapour–liquid equilibrium with theoretical plates, and (B) vapour–liquid equilibrium with stripping operating line area magnified. Source: Diageo PLC, London.

Distillation design

Traditional or Coffey still (also referred to as patent still)

Fig. 15.7 shows the schematic of a traditional Coffey still. In the Coffey still, the feed stream is heated by passing a coil containing the feed through the space above the plates in the rectifier. The energy in the vapour is transferred to the feed, and, at the same time, the vapour condenses to form a reflux. The preheated feed is then introduced high up the wash or analyser column and allowed to flow down the still. The liquid flows across the plates by providing downpipes, or downcomers, at alternative ends of the plate. Steam is introduced at the base, usually through a thermocompressor, or *lurgi*, which removes heat, in the form of flash steam, from the spent wash at the base of the still to reduce energy costs. The steam boils the liquid and fractionation begins. Vapour from the top of the analyser is directed through vapour pipes into the base of the rectifier and continues to rise up the column, where it meets the relatively cool wash coil and begins to condense. The spirit condensed on the spirit plate is drawn off as the product stream. Fusel oil is removed towards the base of the rectifier. Vapour that condenses above the spirit plate runs back down to the spirit plate and is removed with the main product stream. Incondensable gases such as carbon dioxide are vented from the top of the still through the *ether pipe*. The liquid at the base of the rectifier is collected as hot feints and pumped back to the top of the analyser to ensure that the reflux is kept balanced. The stripped feed (spent wash) collects at the base of the analyser and is removed. This material is

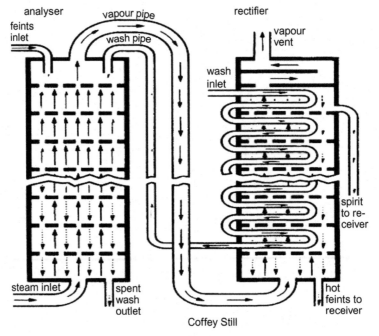

FIG. 15.7   Schematic of Coffey still. Courtesy of Diageo PLC, London.

disposed off by further processing it into animal feed (main route), using it as a carbon source for methane generation in a bioreactor, or through other approved routes, such as long sea outfall (least preferred route).

Atmospheric column still

Fig. 15.8 is a simple double distillation column still, in which the feed is preheated by passing through the heads condenser on the wash column. The preheating system usually also incorporates a degassing device to remove the dissolved carbon dioxide contained in the feed. For simplicity, the number of plates shown is much less than the typical 25 to 50 in the wash and 40 to 80 in the rectification columns.

The still pictured has an analyser and rectifier column, like the Coffey still, but it does not have the coil to preheat the feed. In this design, the steam can be injected into the base of both stills. More energy-efficient versions utilise a reboiler, as shown, at the base of the column. The bottom product is pumped through a heat exchanger using live steam, thereby utilising the energy in the bottom product and reducing energy costs. This process can be fouled with the solid material in the wash column, and care is required in still operation. Vapour at the top of the stills is condensed using an external condenser. A proportion of this liquid is returned as reflux to the top of the column, while the rest is transferred to become the feed for the second column.

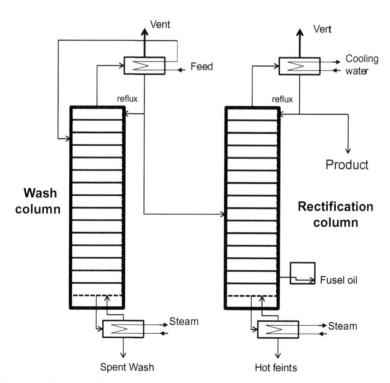

FIG. 15.8   Schematic of a simple double column distillation system.

Non-condensable gases are vented from the condenser. The overhead condensers, in practice, are normally installed in pairs to ensure complete condensation of the vapour. In most stills, the feed is preheated by using preheaters. The most energy-efficient stills use the feed as the cooling media in the overhead condensers and recover the heat into the feed. The spent wash is removed from the system in a manner similar to that of a Coffey still. The bottom product, hot feints from the rectification column, is pumped to the wash column to maintain balance. Some designs of this type of distillation include a heads column, which allows for concentration of the heads and bottom product from the different columns prior to re-entry into the system.

Vacuum column still

The third type of distillation, shown in Fig. 15.9, carried out in grain whisky production uses a vacuum still. This has a design very similar to that of the previous distillation type, but a vacuum is created in the overhead condenser at the top of the column by installing a vacuum pump on the condenser vent. This lowers the boiling point of the ethanol–water mixture, resulting in a significantly reduced energy requirement.

The design of column stills allows for numerous take-off points up the height of the column; therefore, a greater degree of congener-level control in the final product is possible in this type of still compared to a Coffey still. This is important if different cereals are used in the production of the fermented feed, as each, along with different fermentation conditions, gives rise to different levels of congeners in the feed. This will alter the quality of the spirit being produced, but it can be mitigated if good control is exercised during the predistillation process stages.

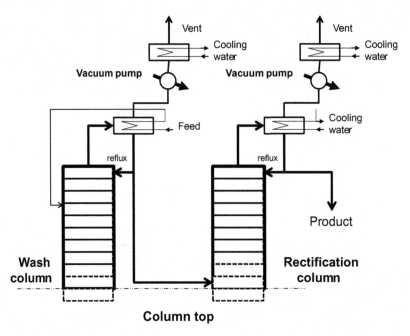

FIG. 15.9   Top section of a vacuum still illustrating the location of the vacuum pumps.

Operation

Upon starting the still, hot water is introduced to warm the columns, and steam is added to the base of the columns. Feed or feints from the previous still run are gradually introduced, and the build of ethanol begins. The bottom product is monitored, and when the liquid exiting the still shows signs of containing solid material, the drain valve is closed, and the flow is directed to the spent wash collection system. In all three still types, liquid produced at start-up will be of reduced ethanol concentration and not have the desired quality. This liquid, cold feints, is collected through the spirit take-off line but directed to a separate tank and then fed back into the process once the system is balanced. During this start-up process, the spirit off-take needs to be closely monitored. When the correct strength and quality have been achieved, the spirit off-take is diverted from the cold feints collection system to the product collection system.

When the still is required to be shut down for maintenance, cleaning, or lack of production requirements, the feed is cut off and water at the same temperature is introduced. The spirit take-off is monitored in the same way as on start-up, and, once the quality and strength fall below that desired, the off-take is directed to the cold feints collection system. The steam supply is then stopped and the contents of the rectification column collected to be used on the restart. The wash/analyser column continues to be fed water until the bottom product is free from solids, at which point the water feed is stopped and the liquid in the column is run to the drain.

Plate and tray design

The analyser and stripping columns have been designed to contain the required number of liquid trays to reflect the chosen reflux ratio. The design of the trays, like the stills, varies but the principles of design are the same. Each tray must achieve four functions (Fig. 15.10A and B):

1. Ensure mixing of the vapour with the falling liquid on the tray
2. Allow separation of the vapour created from the liquid
3. Provide a path for the liquid to fall to the tray below
4. Allow the vapour to rise to the next tray.

The predominant styles are perforated trays, where the tray is constructed of a plate with drilled holes to allow the vapour to rise through the holes, which mixes with liquid on the tray. The number and size of the holes are critical design features. Other designs, such as bubble caps, are also used in an effort to reduce the degree of plate fouling by providing better vapour distribution on the plate. More advanced trays, such as disc–doughnut, are designed to be self-cleaning. Sufficient space between the trays is needed to ensure that the froth created above the liquid level does not come into contact with the next tray, thereby allowing the vapour to separate from the liquid and pass through the tray above. The pressure of the rising vapour prevents the liquid from passing down through the perforations in the plate. A pipe or weir, called a downcomer, connects the trays and is designed to form

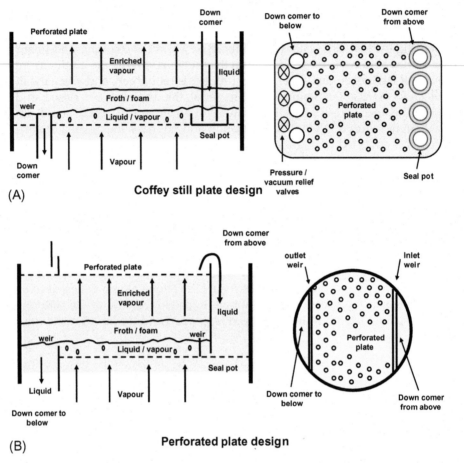

FIG. 15.10 (A) Schematic of typical internal design of plate in Coffey still and (B) schematic of typical internal design of plate in a column still. Source: Courtesy of Diageo PLC, London.

a dam, which maintains a liquid level on the tray. Liquid from one tray flows across the plate, overflows down the pipe, and is allowed to flow over the next lower tray. The bottom of the pipe sits in a cup that also acts as a dam that creates a vapour seal, thus the name *seal pot*. In this fashion, enhanced ethanol concentration vapour travels up the still, while ethanol-stripped liquid flows down the column. The design of the trays has advanced from the earliest design. The thickness has been increased so solid material can be handled, and the perforated sheets are in sections to allow easy replacement without major engineering. The manufacture from stainless steel, which lasts much longer than the traditional copper, is now common. This last development does give rise to some quality issues which are discussed later. In addition, thought has been put into safe operation, and built-in pressure and vacuum relief valves are included.

Thermocompressor operation

Spent wash from the base of the analyser at 100°C is passed into a vessel where steam is flashed off from the liquid by creating a partial vacuum, generated by high-pressure steam passing through a restriction in a pipe (Venturi) connected to the flash vessel. The reduced-pressure steam from the Venturi, along with the steam created in the flash vessel, is then injected into the base of the analyser. This reduces the temperature of the spent wash by about 10°C, but also reduces the steam required by about 25%.

Development of flavour

Congeners and distillation—Their role in flavour creation

The predistillation regimes of the various grain distilleries are different; these differences, along with differences in the cereal mix used, create varying levels of minor compounds (congeners) in the wash. These number in the hundreds and are derived from the cereal, yeast, and its metabolic products during fermentation (Nykanen and Suomalainen, 1983). Typical examples of congeners are methanol, propanol, butanol, ethyl acetate, and acetaldehyde. Each of these is measured in parts per million and, for some, parts per billion or trillion. Their different relative volatilities allow them to reach very high concentrations in specific areas within the still. They can be roughly categorised into compounds more volatile than ethanol, similar to ethanol, or less volatile than ethanol. Although the boiling points of several congeners, such as *n*-propanol and isobutanol, are different (97°C for *n*-propanol, 108°C for isobutanol), they have similar azeotropic boiling points (88°C for *n*-propanol, 90°C for isobutanol), and they behave similarly in a still and concentrate near the top of the rectifier. Similarly, all the other congeners concentrate at different areas within the still depending on their volatility. The aim is to recover these in the spirit off-take stream. The picture is complicated, as the volatility of these congeners is related not only to their boiling point but also to the concentration of ethanol. The higher volatility congeners (e.g. methanol) are more volatile than ethanol. At all concentrations, they rise up the column and if not condensed and returned would be lost in the vapours leaving the still. These are known collectively as *heads*. The congeners with lower volatility (e.g. pentanol) will concentrate in the area of the still where the ethanol concentration is low (i.e. the base of the rectifier). Congeners that have lower volatility compared to ethanol when the ethanol concentration is high and higher volatility at low ethanol concentrations collect in the column with the ethanol and are removed with the product stream. Fig. 15.11 shows the distribution of low-volatility congeners, *n*-propanol, isobutanol, ethanol, and isoamyl alcohol within the rectifier section of a still (see also Table 15.3).

As stated above, the high-volatility congeners will move up the still into the overhead vent, and, by careful operation of the condensers on this pipe, the desirable congeners will be returned to the still and build up near the spirit plate; they will then be removed from the still into the product stream. In some cases, a proportion of the condenser condensate is passed to a separate heads distillation column to allow even greater control of these congeners. This is similar to operating a demethanolisation column for producing grain-neutral spirit. The

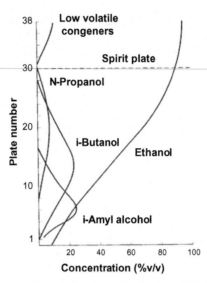

FIG. 15.11　Congener profile in the rectifier column. From Whitby, B.R., 1992. Traditional distillation in the whisky industry. Ferment 5 (4), 261–266.

TABLE 15.3　Comparison of typical levels of the main congeners in malt and three styles of grain new-make spirit (g/100 L absolute alcohol).

New make type	Malt	Light	Medium	Heavy
Acetaldehyde	7	0.5	2.4	12
Total aldehydes	8	1	1.5	2
Ethyl acetate	40	14	17	23
Total esters	110	35	50	75
Methanol	8	5	8	8
Propanol	40	58	75	92
Isobutanol	80	35	60	75
2-Methyl butanol	46	0.22	3	8
3-Methyl butanol	130	0.62	8.5	23
Total higher alcohols	400	100	200	250

Source: Data courtesy of Diageo PLC, London.

lower volatile congeners are either removed in the spent wash steam or by using a side stream to remove them at the base of the rectifier, as in fusel oil removal. All others will concentrate where their volatility and that of ethanol are equal. This concentration would continue to increase, but the upper tail of this concentration crosses over the spirit plate region of the still and is therefore drawn off along with the ethanol. When a continuous still has been in operation for a short time, all of these factors reach a steady state, and, provided the concentration of the congeners in the feed is stable, the level in the product will also be stable. The removal of the congeners in the spirit maintains the balance.

During the predistillation process, several compounds are created that are detrimental to quality. These may be concentrated during distillation and must be removed before they reach the spirit plate as they would then be present in the product. These include fusel oils, sulphur compounds, and ethyl carbamate.

Fusel oil removal

During fermentation, compounds such as higher alcohols are formed. The compound iso-amyl alcohol (pentanol) is desirable in malt whisky but not in grain whisky and therefore is removed. Fusel oil, as it is more commonly known, is more volatile in low ethanol/high water concentrations than ethanol and consequently concentrates very quickly, lower down the rectification column, than other higher alcohols such as butanol and propanol. This concentration effect allows the fusel oil to be removed from the system through a side stream. The peak concentration will change depending on the level of ethanol and pentanol in the feed, so operators must determine where the maximum concentration is to select the appropriate side stream. If the fusel oil is not removed, this undesirable compound will be detected as an off-note in the spirit and, if allowed to build up in the still, will eventually reach its solubility limit, causing foaming to occur in this area of the still and significantly reducing the ability of the still to function.

Unfortunately, the fusel oil withdrawn from the still contains a significant level of ethanol and flavour congeners such as isobutanol. This represents a loss of product and quality and could result in being above the legal limit imposed by Her Majesty's Revenue and Customs (HMRC). Traditionally, the ethanol level is brought below 8%, as required by HMRC, through mixing with water and then transferring to a tank where the immiscible fusel oil is decanted from the top. This method can be modified to use less water and allows for the aqueous liquid to be returned to the still for further distillation. In modern plants, the side stream is introduced into a small rectification column that does not utilise perforated plates. Instead, the column is packed with copper, stainless steel, or ceramic rings. These provide a large surface area with sufficient reflux capability. Steam is introduced at the base, usually through a re-boiler, and controlled to allow the ethanol and flavour compounds to be fractionated from the fusel oil. The almost pure stream of fusel oil produced collects at the base of the column and is removed. The ethanol/flavour stream from the top of the rectification column is returned to the still below the fusel oil take-off point.

Sulphur removal

The main sulphur-based compounds are dimethyl sulphide (DMS), dimethyl disulphide (DMDS), and dimethyl trisulphide (DTS), although many others are present at very low concentrations that have low odour threshold values. Their removal is achieved by exposing them to active copper, where they are held and removed from the vapour. Traditionally, the still itself would be constructed from copper, as in a Coffey still, which then would provide sufficient copper to remove the sulphur. After a period of operation, this copper will become exhausted, and the sulphur will break through into the product stream. At this point, distillation is stopped, the still doors are opened, and air is allowed into the interior of the still. This removes the sulphur compounds on the copper surface and rejuvenates the copper. In more modern designs of column stills, the copper is introduced either by constructing some of the trays from copper, which are located at either the top or base of the rectification column, or by providing sacrificial copper in the overhead condensers, the heads, or feints tanks, where the copper dissolves and then reacts with, and removes, the sulphur when introduced back into the column.

Ethyl carbamate

In addition to the undesirable compounds already discussed, detection of a compound of concern, ethyl carbamate (EC), in distilled beverages in the 1980s led to the Scotch whisky industry adopting a policy to reduce the level of this compound to as low a level as is practical. All new barley varieties used within the industry must have zero or near-zero potential to produce the precursors that lead to EC formation. This has had the desired impact on all spirit being produced. In relation to malt whisky, where all of the raw material is malted barley, control is achieved by careful barley selection and care during the malting process. In grain whisky, only a portion of the feed material is malted barley, so there is some potential to have a level of EC in the spirit. This is mitigated by the use of sacrificial copper, which converts the EC precursor into a non-volatile compound that does not distil into the spirit. The location of this copper is the same as for the control of sulphur aromas. Air, at a controlled flow, can also be bubbled into the vessel containing the sacrificial copper to ensure that adequate levels of copper ions are in solution.

Energy and utilities

As stated in the section on distillation equipment design, the overall energy requirement to distil the fermented wash is one of the main decision factors in modern still design. In comparison with single batch distillation, the energy savings are obvious, but the amount of energy used is still a major part of the overall production cost. Just as for many other parameters, determining the energy requirements for alcohol distillation is a compromise between capital costs (e.g. larger column with more plates requiring lower reflux) and running costs (e.g. smaller column with fewer plates requiring increased reflux). It is the level of reflux that determines steam consumption. With higher reflux, more liquid flows down the column from plate to plate and there is greater vapour flow up the column. The typical energy requirement for a feed with 8% ethanol is in the region of 4.9 MJ/LA. This is reduced to 4.4 MJ/LA energy when the feed strength is 10% ethanol and further reduced to about 3.3 MJ/LA at 15% ethanol. Further efficiencies, such as increased heat recovery from the spent wash, can reduce steam usage by about 10% (Blair, 2013). The use of steam direct injection, as in a Coffey still, gives little scope for reducing energy, providing a thermocompressor is fitted, as the design is not flexible. The focus is on preventing liquid streams such as feints being cooled, as that would require reheating in the still. To allow this, care has to be exercised when designing and specifying plant and pumps.

Modern stills use reboilers to utilise the energy in the spent wash and, when coupled with vacuum distillation, can achieve significant energy savings. However, despite its age, the energy requirement of a Coffey still is not that dissimilar from modern designs, but it can only operate at a fixed throughput. Modern stills can vary their throughput by over 30% with little effect on energy usage. This is a key differentiation point when considering which still design to install at a distillery.

Acknowledgements

Thanks to Dr. Iain Campbell, formerly of Heriot–Watt University, who wrote the first edition's chapter on grain distillation and whose layout and words have been used as the basis for the second edition and now the third edition. Thanks also go to my colleagues and friends within Diageo PLC, both past and present, principal among them being George Blair, who have over the years given freely of their knowledge and experience. Finally, thanks to my friends within the Scotch whisky world who have helped me gain valuable insights into operational practice in companies other than Diageo.

References

Blair, G., 2013. Diageo PLC Technical Learning Manual. Diageo PLC, London.
Campbell, I., 2003. Grain whisky distillation. In: Russell, I. (Ed.), Whisky Technology, Production and Marketing. Elsevier, Cambridge, UK, pp. 179–206. Chapter 6.
McCabe, W.L., Thiele, E.W., 1925. Graphical design of fractionating columns. Ind. Eng. Chem. 17 (6), 605–611.
Nykanen, L., Suomalainen, H., 1983. Beer, Wine and Distilled Beverages. Kluwer, Dordrecht.
Perry, R.H., Green, D.W. (Eds.), 1984. Perry's Chemical Engineer's Handbook, sixth ed. McGraw-Hill, New York.
Seader, J.D., Kurtyka, Z.M., 1984. Distillation. In: Perry, R.H., Green, D.W. (Eds.), Perry's Chemical Engineer's Handbook, sixth ed. McGraw-Hill, New York, pp. 13–21.

Further reading

Palmer, G.H. (Ed.), 1989. Cereal Science and Technology. Aberdeen University Press, Aberdeen, Scotland, p. 154, Table 24.
Whitby, B.R., 1992. Traditional distillation in the whisky industry. Ferment 5 (4), 261–266.

Maturation

John Conner
The Scotch Whisky Research Institute, Edinburgh, United Kingdom

Introduction

In the production of Scotch whisky, many factors are known to influence the final quality of the product, including water, barley variety, extent of peating, fermentation, still type, and distillation conditions. However, one of the most important contributors is the container in which maturation takes place. During this time, major changes occur in its sensory character. The pungent, feinty aromas of the new distillate are transformed into the typical mellow characteristics of a mature whisky. The colour of the spirit also changes, from virtually clear to golden brown.

Historically, the origins of maturation are obscure; it is likely that the majority of whisky distilled in the 18th and 19th centuries was drunk without being matured. However, the making of whisky was traditionally a seasonal process and would have involved a period of storage. This storage would inevitably have involved the use of wooden casks, the nature of which depended on the availability of suitable timber. In the United States, with its large reserves of white oak, storage in new casks was normal. In Scotland, suitable supplies were limited and second-hand containers such as old sherry casks or brandy barrels were used. With time, the benefits of this storage were realised, and a period of maturation became an accepted part of whisky production and was incorporated into the legal definitions of whisky throughout the world.

Outwardly, the maturation process appears simple. Casks are filled with spirit and set aside in a warehouse to mature. However, a wide range of variables can influence the quality of the matured whisky. The type of cask used, its method of manufacture, and the climatic conditions during storage all affect the maturation process and are described in detail in this chapter. When selecting and controlling maturation variables, the traditions and product expectations for each whisky have to be considered. American bourbon and Tennessee sour mash whiskies are matured in new charred oak casks, while whiskies produced in Scotland, Ireland, Japan, and Canada are matured in oak casks previously used for the maturation of bourbon or for the fermentation and shipment of sherry (Booth et al., 1989). Consequently, while maturation of a Scotch malt whisky in a new charred oak cask may produce a well-matured whisky, it may not be readily identifiable as Scotch (Clyne et al., 1993).

This chapter presents an overview of the current technical understanding of the maturation process and describes the main variables that influence product quality. Current control of the maturation process is achieved by the careful selection and sourcing of casks and their reuse. This wood policy varies from company to company, and even within a company for individual products, and is used to ensure a continued supply of quality and diverse whiskies for blending and bottling.

Finally, the author's expertise has been gained from working within the Scotch whisky industry, and for this reason, the chapter is based primarily on the maturation of Scotch whisky. The production of other whiskies around the world uses a similar range of cask types as those used in Scotland. The exception is American straight whiskies, most notably bourbon and Tennessee sour mash, which are matured in new charred oak casks.

Cooperage oak wood

The majority of whiskies around the world are matured in casks made from American white oak. Smaller numbers of casks made from oaks grown in Spain and Japan are also used. Furthermore, the increasing practice of conducting a second period of maturation (finishing) using wine or fortified wine casks requires specific casks from other countries, for example red wine casks from France and port pipes from Portugal.

Most American oak casks are initially made for the maturation of straight (bourbon) whiskies in the United States. The oak species used for tight cooperage in the United States is predominantly *Quercus alba*, but may include similar species such as *Q. bicolor*, *Q. muehlenbergii*, *Q. stellata*, *Q. macrocarpa*, *Q. lyrata*, and *Q. durandii*. The main areas of bourbon barrel production in the United States are Kentucky and Missouri, but the wood may be sourced from a much larger number of states in the central and eastern United States (Swan, 1994). After use for straight whiskies, the barrels are sold to whisky producers around the world. In Scotland, most former bourbon barrels are imported from the United States as standing barrels, referred to as American standard barrels (ASBs), and these have a capacity of 190 L. The practice of disassembling barrels and exporting them as staves is in decline. These stave 'shooks' were commonly rebuilt using a greater number of staves per cask and new plain oak ends to form larger casks called hogsheads with a capacity of 250 L.

American oak is also imported into Spain for the manufacture of sherry casks. Sherry producers favour the use of this oak, and any cask used in the production of sherry is likely to be American oak (González Gordon, 1990). The use (and reuse) of this type of cask is long established. The analysis of Scotch whisky bottled in the first decade of the 20th century showed that it had been matured in sherry or wine casks made from American oak (Pryde et al., 2011). However, sherry butts (500 L) may also be made from pedunculate (*Q. robur*) or sessile oaks (*Q. petraea*) grown in Spain. Harvested from the regions of Galicia, Asturias, Cantabria, and Pais Vasco in Northern Spain, these were used to transport sherry to the United Kingdom for bottling. When the definition of Spanish sherry changed to include a requirement for bottling in Spain, the source of this type of sherry cask disappeared and required whisky producers to deal directly with sawmills and sherry-producing bodegas for supplies of suitable timber (Ramsay, 2008). With increasing demands on Spanish oak forests, cooperages in Spain have been sourcing oak from throughout Europe including France, Romania, Hungary, Russia, and Ukraine (Owen et al., 2018). Studies of American, French, and Eastern European oaks (Prida and Puech, 2006) found considerable variation within species and by origin, with the chemical composition of some Eastern European oaks closer to American oak than the traditional oak grown in northern Spain for sherry casks. Infrared spectroscopic techniques may be used to confirm the origins of oak wood (Owen et al., 2018) or classify staves based on the tannin content (Michel et al., 2013).

A small number of casks made from Japanese oak (*Q. mongolica*) have been used for the maturation of whisky in Japan. Japanese oak grows in East Asia, including Japan, the Korean Peninsula, northeast China, and south Sakhalin. Japanese oak casks have a capacity of 500 L and are generally seasoned with Oloroso sherry wine for one year before use (Noguchi et al., 2010).

Many nations define whisky as being matured in casks made specifically of oak. However, some countries and customs unions (e.g. the European Union) refer simply to wooden casks (Council of Europe, 2008). Many types of wood have been used for the production of tight cooperage containers, and there has been renewed research into using species other than oak for the maturation of wines (De Rosso et al., 2009; Fernández de Simón et al., 2014) and sugar cane distillates (Bortoletto and Alcarde, 2013). Differences in the extractives and variations in the degradation of wood polymers and lipids during heat treatment result in aged products that can be differentiated from those matured in oak. Their use for prolonged storage of whiskies has not been tested, and the sustainability of their supply is unknown. Consequently, it remains likely that only a limited number of oak species will continue to be used to produce casks for maturation.

Wood structure

Certain oaks remain the woods of choice for cask construction due to a combination of their physical properties and the sensory characteristics of their extractives. Key factors are their ability to contain liquids while allowing some exchange with the surrounding environment, and their strength and durability with the ability to bend staves during construction The former is strongly influenced by oak anatomy through a combination of genetics and the environment of the living tree. The latter arises from the properties of lignin, one of the macromolecular constituents of oak wood.

The ability of some oaks to hold liquids is due to the formation of tyloses in the conducting vessels. Tyloses are outgrowths from parenchyma cells into an adjoining lumen which form when the living sapwood is converted to heartwood, sealing off the conducting vessel (De Micco et al. 2016). The process of wall formation within the tylosis is like other woody plant cells, and chemical analyses have confirmed the presence of lignin, cellulose, and hemicelluloses in broadly similar proportions (Sachs et al., 1970). Tylosis formation only occurs in approximately 20 species of oak (Schahinger and Rankin, 1992). The extent of formation varies from species to species and can influence how the wood is cut for barrel manufacture. In American white oak, the tyloses are highly effective at sealing the conducting vessels, meaning it is possible to saw-cut staves without loss of water tightness. However, in some European oak the tyloses do not provide an adequate barrier against liquid seepage (Chatonnet and Dubourdieu, 1998a) and these must always be split so that the medullary rays remain parallel to the inner side of the staves and maintain a barrier to the diffusion of liquid (Singleton, 1995). In most trees, the medullary rays are only one cell wide, but in oaks and some other hardwoods they can be several cells wide (multiseriate). In oak species, these unusually large rays represent between 19 and 32% of the wood volume and contribute significantly towards the strength and flexibility of the wood (Schahinger and Rankin, 1992).

The structure of the wood influences its permeability, with greater oxygen permeation through finer grained wood of both European and American oak (del Alamo-Sanza and Nevares, 2018). Oak grown in temperate climates is classed as a ring porous wood due to the contrast between the large earlywood vessels formed prior to bud break and the smaller vessels formed later in the year (Wiedenhoeft, 2010). The parameters such as density, vessel diameter, and growth ring width are influenced by growing conditions that include soil conditions, rainfall, shading, and competition from other trees (Haneca et al., 2009; Lebourgeois et al., 2004). These growth characteristics determine the grain of the wood. A higher proportion of large earlywood vessels create a coarse-grained wood, whereas fine-grained wood has a more even distribution of vessels with a smaller average diameter (Vivas, 1995).

Wood is composed of cellular and intercellular material. The cell wall structure is constructed from the macromolecular polymers cellulose, hemicellulose, and lignin, whereas the intercellular region consists mainly of lignin. In addition to the main polymeric fraction of the wood, free low-molecular-weight components (extractives) are present in smaller amounts but are not an integral part of the cell wall structure.

Cellulose is the most abundant component in wood, accounting for 38 to 46% of the dry weight (Le Floch et al., 2015). It is a linear polymer of glucose molecules linked by β-(1,4)-glycosidic linkages. Hydrogen bonding between hydroxyl groups on adjacent cellulose polymers results in the formation of fibrils, which further assemble to make the cell wall layers. This produces a superstructure for the remaining wood components to build upon (Fengel and Wegener, 1984).

In addition to cellulose, the cell walls contain other polysaccharides known as hemicelluloses that act as a matrix for the cellulose superstructure (Parham and Gray, 1984), binding cellulose microfibrils and strengthening the cell wall. In oak, hemicellulose comprises between 19 and 30% of the dry weight of wood (Le Floch et al., 2015). Its backbone consists of xylose molecules linked by α-(1,4)-glycosidic bonds. Approximately seven xylose units in every ten are substituted at the C2 or C3 positions with O-acetyl groups. Hydrolysis of oak hemicellulose yielded xylose as the main sugar (20% of the extractive-free wood weight) with mannose (3.7%) and galactose (3.1%) the second and third most abundant, respectively (Assor et al., 2009).

The third major component of wood is lignin, which has a highly branched three-dimensional structure of high molecular weight. Lignin serves as a binding agent for the wood cells and accounts for between 22% and 30% of the dry weight (Le Floch et al., 2015). Although lignin is found in intercellular regions, the majority, more than 70%, is in the cell walls (Parham and Gray, 1984). The complex structure of lignin is built from phenylpropane units that are substituted with hydroxyl and methoxyl groups. In hardwoods, such as oak, lignin is formed by the polymerisation of coniferyl and sinapyl alcohols. The reaction of these two groups produces a mixed polymer linked by a variety of ether and carbon–carbon bonds involving both aromatic rings and side chains (Monties, 1992). Lignin also covalently binds to hemicellulose and cellulose within hardwoods (Tarasov et al., 2018). The type of bonds within lignin is important for the structural properties of oak and is an important factor in raising a cask. Mild heat treatments break β-0-4 bonds in the lignin, producing a considerable decrease in wood rigidity and allowing the cask to be shaped. Increasing the temperature induces further changes in lignin and hemicellulose, leading to a substantial reorganisation within cell walls that almost fully restores rigidity. Higher temperatures after bending are used to fix the cask in shape (Assor et al., 2009).

Oak also contains low-molecular-weight components that do not contribute to the structure of the wood. These may be extracted by the maturing spirit and have a significant influence on flavour development during maturation. Making up to 12% of the dry weight of the wood, the composition is complex and includes polyphenolics and other secondary metabolites. These are formed during the conversion of sapwood to heartwood in which they are thought to act as chemical defences and increase durability (Taylor et al., 2002). The concentration of polyphenols varies depending on the species and origin of the tree (Mosedale et al., 1996). The components of interest in the maturation of whisky are hydrolysable tannins, lactones and volatile phenols (described in later sections) and organic acids. Hydrolysable tannins are partly responsible for the astringency of whisky, and a variety of structures based on gallic and ellagic acids have been identified. These degrade during heat treatment, and the major constituents extracted by maturing spirits are the free phenolic acids (Mosedale, 1995). The major organic acids are acetic and linolenic acids. Linolenic acid is a potential haze constituent in mature whiskies, and its degradation produces odorous aldehydes and alcohols (Chatonnet and Dubourdieu, 1998b). Oak lactone is one of the main volatile compounds in oak wood and plays an important role in the aroma of mature whiskies, particularly the cis isomer, which has a much lower perception threshold than the trans isomer. Free volatile phenols such as eugenol and vanillin are also present in small quantities, although these may be increased by the thermal degradation of oak lignin during cask manufacture.

Cask manufacture

While the basic shape and structure of a cask have not changed for a long time, production methods have progressed and modern cooperages make full use of fail-safe systems and robotics to ensure a safer working environment and improve consistency and productivity (Zimlich and Beach 2015; Crow and Carson, 2018).

Timber processing

Oak wood logs, destined for cask production, are initially cut into either stave or heading (for cask ends) lengths, depending on any defects that may be present in the log. These shorter lengths are then either split, ensuring medullary rays remain parallel to the inner side of the staves, or quarter sawn, with the cuts being made through the heart of the log and running along its radius (Fig. 16.1).

From the split or sawn wood, flat stave and heading timber can be cut to the appropriate thickness. For quarter sawn wood, the first saw cut is removed parallel to the radius of the tree, the quarter turned 90°, and the second cut removed from the other flat surface. This is

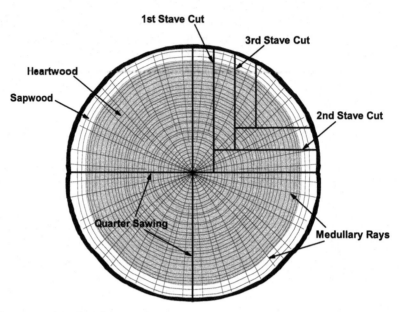

FIG. 16.1   Quarter sawing of oak logs.

repeated until the amount of timber left in the quarter is too small to be of any use. The sapwood and any defective heartwood are then removed from the sawn timber to produce the finished stave and heading blanks. Quarter sawing recovers approximately 50% useable timber from the log (Schahinger and Rankin, 1992) but the yield from split wood is much lower.

The next stage is the drying or seasoning of the timber. In the United States, the oak to produce bourbon barrels is dried in a kiln over a period of approximately one month. This reduces the moisture content of the wood to a workable level (approximately 12%). Throughout the drying process, the timber is held under specific temperature and humidity conditions to ensure that drying is efficient and that wood damage, such as the appearance of splits or cracks in the stave ends, is minimised. In Spain, the seasoning of timber to produce sherry casks is quite different. The stave and heading blanks are initially air dried in the growing regions of Northern Spain for about 9 months, which reduces the moisture content of the wood to approximately 20%. The timber is then shipped to the warmer sherry-producing regions in the south of Spain, where it is further air seasoned for a period of six to nine months, or until its moisture content is reduced to a workable level (approximately 14 to 16%).

Bourbon cask construction

In the manufacture of bourbon barrels, the dried, stave blanks are jointed or cut to stave shape. The jointed staves have smooth angled edges and are slightly wider in the middle compared to the ends. This is essential for forming the desired barrel shape. The introduction of computer-controlled cutters for jointing staves improves wood utilisation and produces more consistent staves, each with the correct angle on the edge of the stave. The result is a more uniform barrel with tighter joints that reduce or eliminate leaks (Zimlich and Beach, 2015). The assembly of the barrel begins by arranging the straight staves into a circular structure that is closed at one end. This assembly is steamed for a period of between 10 and 20 min at a temperature of approximately 95°C to soften the fibres of the wood and enable the staves to be bent. The staves are then drawn into the conventional barrel shape using a windlass, and a temporary iron hoop is used to hold them in place.

The next part of the process is to heat the inside of the cask shell. In terms of whisky maturation, this is probably the most important stage of barrel production as it defines the character the cask will confer to a maturing spirit. The cask shell, which is still wet from the steaming process, is initially heated to between 230°C and 260°C for a period of approximately 15 min. This drives off the surface water and sets the staves in the shape of a barrel (Hankerson, 1947). The shell is then charred by setting the inside on fire and allowing it to burn until the required degree of char has been obtained. The role during maturation of the char and toast layers formed is described in the following sections.

The heading pieces, which form the end of a cask, may either be pinned together using dowels or slotted together using a tongue and groove system. They are then cut into a circular head with a bevelled edge. These ends are then charred before being inserted into the cask shell. The bevelled edge fits into a groove, known as a croze, cut around the inside of each end of the shell. Finally, the hoops are driven on to create a tight seal between the staves, thus producing the finished barrel.

Sherry cask construction

As in the production of a bourbon barrel, the staves for a 500-L sherry butt are similarly tapered but are longer and thicker. Where the 190-L bourbon barrel requires around 30 staves for its construction, the larger butt will contain approximately 50 staves. These are initially raised into a circular structure, as in bourbon barrel production, which is held in shape using temporary hoops. This assembly is then placed over an open fire at approximately 200°C. This heating both toasts the wood, turning it brown in colour, and makes it more pliable, allowing it to be slowly pulled into the conventional butt shape using a windlass. No steaming is involved in the construction of a sherry cask, but water is applied to the outside of the staves during heating to prevent cracking. As in the production of a bourbon barrel, the ends are initially prepared by pinning together heading timber, but with the use of wire instead of wood dowels. Ends are then cut into the circular head and inserted into the cask shell to produce the finished sherry butt.

Control of heat treatment

Controlling the toasting and charring of casks has a major effect on the sensory properties of the matured spirits (Perry et al., 1990; Spillman et al., 2004). The charring of bourbon casks is controlled using high-temperature gas fires for a set time. More consistent results are achieved when timing is started after a predetermined temperature is reached or a specific change in combustion flame or exhaust characteristics is observed. For sherry casks, the intensity of toasting is also controlled by time and temperature and is classified as light, medium, or heavy (Mosedale and Puech, 1998). Toasting normally uses oak-chip fires maintained by individual coopers; consequently, there are large variations between coopers and cooperages. New methods are being investigated to provide a better classification of toasted casks. These are based on analysis of the volatiles produced during toasting using either chromatographic techniques or metal-oxide-based sensors (Chatonnet, 1999). More recently, Fourier transform infrared (FTIR) spectroscopy has been investigated for the quality control of charcoal (Labbé et al., 2006) and found to be a powerful tool for monitoring chemical changes in wood during heat treatment. In addition, alternative methods of heat treatment that are potentially more controllable have been developed, including infrared heating (Wickham, 2009) and hot air convection (Fantoni-Salas and Fernandez-Mesa, 2007).

Cask regeneration

Repeated reuse of a cask depletes the levels of available colour and extractives to the point that it will fail to produce a satisfactory maturation. These casks can be regenerated, and in Scotland, this is achieved by first removing the old char using a rotating brush or flail system before re-charring using a gas burner. When casks are re-charred, thermal degradation of lignin yields flavour compounds similar to those produced in a new charred cask. However, other constituents of oak are not regenerated, for example oak lactones and hydrolysable tannins, and consequently, the balance of extractives is different to new charred casks. Extraction and analysis of wood from different depths within an exhausted stave have shown that only the inner surface of the stave is depleted of extractives whereas deeper layers retain some of the oak lactones and tannins found in new wood (Conner et al., 2012). Char may be removed by a rotating brush or flail system but this does not remove sufficient exhausted wood for the maturing spirit to extract from these deeper layers. However, new cutting heads are being used that remove more of the exhausted wood and, combined with better control of the re-charring process, can increase the levels of cask extractives in the maturing spirit by an average of 80% when compared to traditional regeneration methods (Crow and Carson, 2018).

Chemistry and whisky maturation

Modern analytical techniques have been used to identify an increasing number of constituents in mature whiskies, often without any attempt to clarify their impact on sensory character. Studies directed at identifying aroma-active constituents paint a much simpler picture. Of the 45 most odour-active volatile constituents identified in bourbon whisky by gas chromatography–olfactometry (GCO), only 7 were constituents of oak extracted from the cask during maturation. It was found that cis-oak lactone, eugenol, and vanillin had the highest flavour dilution factors, with progressively lower factors for trans-oak lactone, sotolon (3-hydroxy-4,5-dimethyl-2(5H)-furanone), 4-ethyl-2-methoxyphenol, and 2-methoxyphenol (Poisson and Schieberle, 2008a). Aroma recombination studies showed that omission of cis-oak lactone and vanillin caused highly significant changes in flavour (Poisson and Schieberle, 2008b), but the addition of only these two compounds to grain whisky matured in a refill cask did not duplicate the mature aroma of the same spirit matured in new toasted oak casks (Conner et al., 2001). This suggests that other wood constituents, such as eugenol, 4-ethyl-2-methoxyphenol, and 2-methoxyphenol, must contribute to the aroma of mature spirits. Recombination studies also showed that omission of components of the original distillate (e.g. ethyl esters and 3-methylbutyl acetate) caused a highly significant change in aroma. Indeed, few of the aroma compounds present in new-make distillates are lost or degraded during maturation. Consequently, the large sensory change that occurs through maturation is brought about by the loss of a small number of new-make distillate aromas and the addition of a limited number of wood aromas. It is likely that the interaction of aromas from these two sources plays a significant role in the development of mature character.

Wood-derived aromas

Wood-derived aromas can originate from the unprocessed heartwood, the thermal degradation of wood polymers during cask manufacture, and the carry-over from previous use of the cask. The most important aroma compounds that are constituents of the unprocessed heartwood are the oak lactones; two of the four possible diastereoisomers are naturally occurring in oak wood species used for cooperage: cis (3S,4S) and trans (3S,4R) (Fig. 16.2). The ratio of these two isomers varies among species and even within species, depending on the source forest. The cis isomer predominates in whiskies matured in American oak (Waterhouse and Towey, 1994), whereas the trans isomer predominates in whiskies matured in Japanese oak (Noguchi et al., 2010). European oaks tend to give equal concentrations of each isomer, although the ratios can vary, with the cis isomer being predominant in wood from some locations (Masson et al., 1995). Glycosidic precursors have been identified in oak wood (Hayasaka et al., 2007) and may break down to liberate additional oak lactone during heat treatment and maturation (Wilkinson et al., 2013). However, the precursor is also lost during cask use and re-charring exhausted American and Japanese oak casks does not recreate the sensory character of a first-fill cask (Noguchi et al., 2012).

Vanillin is naturally present in oak wood, but concentrations are greatly increased by heat treatment, during which it is formed by the thermal degradation of lignin. In addition to vanillin, lignin degradation produces coniferaldehyde, sinapaldehyde, syringaldehyde, and vanillic and syringic acids (Nishimura et al., 1983; Reazin, 1983), but the levels of these do not exceed their odour threshold. The intensity of heat treatment affects the levels of aromatic aldehydes and acids. Studies using oak chips have shown that temperatures up to approximately 200°C increase the levels of these compounds (Nishimura et al., 1983), whereas higher temperatures and charring decrease levels due to volatilisation and carbonisation. In casks,

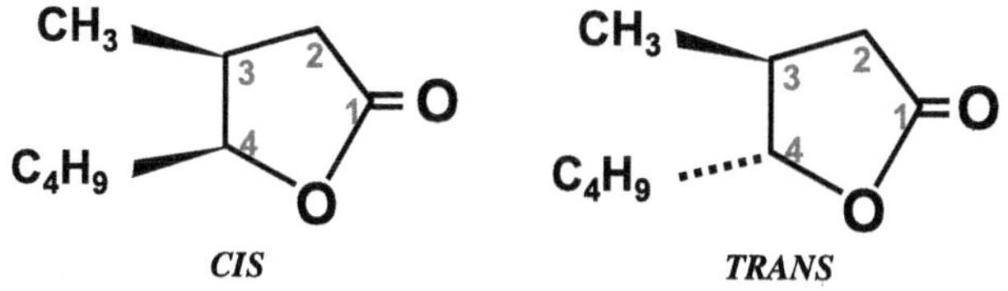

FIG. 16.2   Structures of *cis* (3S, 4S) and *trans* (3S, 4R) oak lactones.

charring increases the levels of lignin breakdown products extracted by the spirit. Although the char layer contains few aromatics, heat penetration to subsurface layers promotes thermal degradation reactions and increases aromatic aldehydes and acids to a depth of 6 mm (Perry et al., 1990). Although deeper in the stave, the char layer does not hinder their extraction because the disruption of the wood structure during charring increases the penetration of the maturing spirit. Although the majority of vanillin is formed during the initial heat treatment, this may be supplemented by the breakdown of a small portion of labile lignin by processes such as hydrolysis and oxidation during the life of the cask (Conner et al., 1993). This route of formation becomes increasingly important with cask reuse.

Eugenol is present in the heartwood of oak and is also generated during heat treatment. Concentrations in whiskies matured in first-fill casks appear to be more variable than vanillin (Conner et al., 2001; Poisson and Schieberle, 2008b). Unlike vanillin, eugenol is not regenerated by re-charring exhausted casks, which suggests that it is not formed by degradation of structural lignin in the cask wood (see Fig. 16.3). Glycosidic precursors have been identified for eugenol in *Quercus petraea* (Nonier et al., 2005). Consequently, variability in the amounts generated during heat treatment may be related to its concentration in the original oak and/or the amount of precursor that degrades during heat treatment. The origins and sensory impact of other phenolic constituents have not been fully characterised. Some of these (e.g. 4-ethyl-2-methoxyphenol and 2-methoxyphenol) are also present in new-make spirits, and their impact in the mature product will be a combination of both sources.

Breakdown products of sugars, hemicellulose, and cellulose are also detected in matured whiskies. Furaldehydes (furfural and 5-hydroxymethyl furfural) have little sensory impact, but their formation may be accompanied by that of other molecules with sweet, caramel, and toasted aromas. Maltol and 2-hydroxy-3-methyl-2-cyclopentenone were identified in toasted oak after heating (Nishimura et al., 1983), but comparison of the amounts present with their odour thresholds suggests that their sensory impact may be limited (Cutzach et al., 1997). Sotolon is frequently detected in whiskies and sherries by GCO, but analysis using liquid chromatography (without heat) rarely gives concentrations above its odour threshold.

Another group of constituents that have a sensory impact is the hydrolysable tannins, although these impact on taste rather than aroma. Oak wood contains monomeric and dimeric glycosidic ellagitannins, such as castalagin, vescalagin, grandinin, and roburins A–E, which undergo a series of transformations during heat treatment and maturation resulting in the polyphenols in whisky being different from the original oak wood tannins. Heat treatment of castalagin gives mainly dehydrocastalagin and ellagic acid with smaller amounts of castacrenin F (Fujieda et al., 2008). Vescalagin produces mainly deoxyvescalagin, and similar deoxy and dehydro compounds were observed for roburins A and D, respectively (Glabasnia and Hofmann, 2007). Both studies reported the formation of an uncharacterised

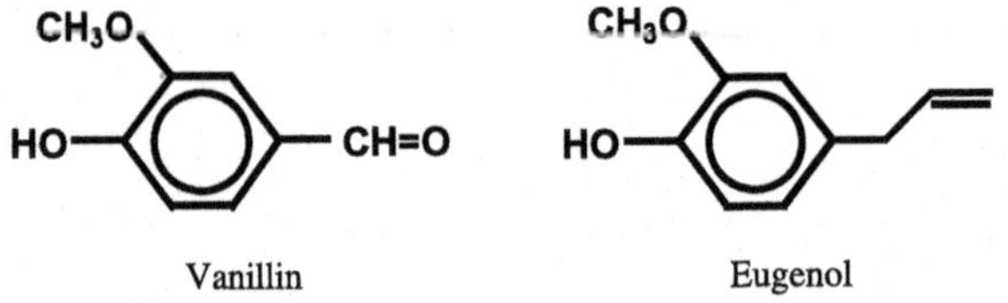

FIG. 16.3   Structures of vanillin and eugenol.

phenolic/melanoidin substance during heat treatment of ellagitannins, and when tested, this produced sensations of complexity, mouth fullness, and astringency (Glabasnia and Hofmann, 2007). Comparing concentrations of oak wood ellagitannins in bourbon whisky with their taste recognition threshold showed that all except ellagic acid were below threshold (Glabasnia and Hofmann, 2006), suggesting that the combined effects of heat treatment and maturation reduce the overall astringency of hydrolysable tannins.

Reactions affecting distillate components

The sensory character of the new-make spirit at the start of maturation may be modified by a number of interactions that have been shown to take place during maturation (Wanikawa, 2020). The formation of the char layer on the inner surface of the cask is the result of carbonisation of the polymeric constituents and contributes little in the way of colour or extractives to the maturing whisky. However, it plays an important role in the removal of immature character and has been shown to promote the oxidation of dimethyl sulphide (Fujii et al., 1992). Char may reduce the concentration of other sulphur compounds by a combination of adsorption and oxidation (Philp, 1986).

Evaporation of volatile compounds through the cask occurs during maturation. For a model whisky, the rate of evaporation ranges from 32% of the total present in the spirit for acetaldehyde to 5% for isoamyl alcohol, and 1% for ethyl hexanoate and acetic acid (Hasuo and Yoshizawa, 1986). Evaporation is thought to be the main route for the loss of dimethyl sulphide (Fujii et al., 1992) and dihydro-2-methyl-3(2H)-thiophene (Nishimura and Matsuyama, 1989). Evaporation will affect the concentration of all wood and distillate components during maturation. Most of these components have boiling points much greater than ethanol and water, and their concentrations increase as the spirit volume reduces (Baldwin and Andreasen, 1974).

Chemical reactions that alter distillate components include oxidation, esterification, and acetal formation, and this sequence of reactions is typified by the oxidation of ethanol to acetaldehyde and acetic acid and their subsequent transformation to 1,1-diethoxyethane and ethyl acetate (Reazin, 1981). The formation of ethyl acetate accounts for a greater part of the increase in ester levels observed during maturation. The factors that govern the rate of this reaction are still obscure. Nishimura et al. (1983) observed higher concentrations of acetaldehyde due to the presence of wood extractives in model solutions, and mechanisms have been proposed that involve an interaction among hydrolysable tannins, dissolved oxygen, and copper ions to produce active oxidants such as superoxide and peroxide (Philp, 1986). More recent studies suggest that the dynamics of this interaction change during maturation as copper, an important promoter, is adsorbed by the wood of the cask (Muller and McEwan, 1998). Also, there is now a considerable body of research that identifies cask extractives as antioxidants with the ability to adsorb free radicals and prevent oxidation (e.g. McPhail et al., 1999; Koga et al., 2007). Consequently, the role of these compounds and their impact on the rate of reactions during maturation need further clarification.

The presence of wood aromas and extractives may mask the immature character of a spirit. Direct sensory interactions may occur when the presence of strong wood aromas lessens the impact of sulphury or feinty characters. Less dominant wood aromas may also interact by enhancing the perception of positive distillate characters. However, the nature and extent of this type of interaction have not been studied due to difficulties in creating realistic models of whisky aroma. Chemical and physicochemical interactions have been suggested as ways in which the presence of wood extractives may alter the perception of immature characters in mature whiskies. The pH reduction during maturation, which is cask dependent, affects the ionisation state of weak bases and consequently their volatility (Delahunty et al., 1993). Physicochemical changes have been studied using differential scanning calorimetry (Nishimura et al., 1983), small-angle light scattering (Aishima et al., 1992), mass spectrometric analysis of liquid clusters (Furusawa et al., 1990), and proton nuclear magnetic resonance (Nose et al., 2004). The measurements suggest a change in the structuring of ethanol and water in mature whisky, mostly related to the presence of non-volatile constituents. The studies of mature whiskies using small-angle X-ray scattering and dynamic light scattering revealed the presence of two sizes of clusters. Small clusters ($r \sim 0.75$ nm) increased with maturation age and were thought to be aggregates of small numbers of aromatic extractives (Morishima et al., 2019). The impact of aggregation on aroma and taste is still poorly understood. When the deodorated, non-volatile residue of bourbon was included in aroma recombination studies, there was only a small improvement in similarity to the original sample compared with the same odorants in neutral alcohol (Poisson and Schieberle, 2008b). This suggests that for bourbon (and other whiskies matured in reused casks) physicochemical changes are not a crucial factor determining aroma. The levels of extractives required to induce such changes may only be achieved in first-fill sherry casks or after prolonged maturation.

Cask type

Maturation should not be thought of as a homogeneous process, with the same reactions occurring irrespective of the cask type. It is most likely a specific combination of one type of distillate with any one type of cask leading to the development of a flavour profile relative to time (Philp, 1986). The typical aromas and their development in the main types of cask used to mature whisky are summarised as follows:

- New charred casks—Typical characteristics are woody, vanilla, coconut, and resinous. Wood aromas are mostly heartwood constituents and thermal degradation products. Distillate character is modified by char-mediated adsorption or degradation. Strong wood aromas may mask some distillate characters.
- Ex-sherry casks—Typical characteristics are vanilla, fruity, and sweet. Wood aromas are mostly heartwood constituents and thermal degradation products, potentially modified and added to by the period of sherry contact. Distillate character may be masked by wood aromas and the high level of extractives in European oak casks. The lower level of extractives from American oak sherry casks may be responsible for their limited ability to reduce (mask) such characters.
- Ex-bourbon casks—Typical characteristics are dry, floral, scented, and vanilla. Wood aromas are still mostly heartwood constituents and thermal degradation products but possibly augmented by hydrolysis and oxidation of wood polymers. Distillate character is modified by char-mediated adsorption or degradation, although the activity of char

may be lower than first fill. The lower levels of wood aromas may enhance positive distillate characteristics, although prolonged maturation may give strong wood aromas that mask some distillate characters.

- Refill casks—Typical characteristics are smooth, vanilla, and sweet. Wood aromas come mostly from subsurface layers of the stave where they may be formed by hydrolysis and oxidation of wood polymers. The extraction of constituents from previous fills also occurs, and for grain spirits, this may have a significant impact on whisky character. The activity of the char layer is unknown, but almost certainly reduces with each fill. The loss of immature character/sulphur compounds by evaporation may be more important as the other routes of degradation are diminished.
- Regenerated casks—Typical characteristics are woody, vanilla, and sweet. Wood aromas are mostly thermal degradation products. Some constituents from previous fills can survive the regeneration process (e.g. peaty characters). Distillate character is modified by char-mediated adsorption or degradation.

In the first three types of cask, heartwood constituents are important contributors to the character of the mature spirit, and consequently, the origin of the oak may affect the mature character of the whisky. The impact of different oak species is most clearly seen when comparing sherry casks made from American and European oaks. European oak produces whiskies with high levels of colour and extractives and prominent vanilla, fruity, and sweet aromas. In contrast, American oak casks produce whiskies that are relatively light and floral and have a limited ability to reduce immature characters in new-make spirit. Consequently, American oak sherry casks are more suited to clean, more delicate distillates, whose character would be masked by the heavy extract from a European oak cask.

There are no casks in common use that offer a comparison between toasting and charring heat treatments. In laboratory studies, charring generates lower levels of colour and extractives than toasting, as both are lost by volatilisation and carbonisation at the higher temperatures of charring. Toasting is generally reserved for wine casks, and their use before whisky introduces other variables such as the transfer of wine components and the impact of wine contact on the cask wood.

Repeated use of casks results in decreased yields of wood compounds (Reazin, 1981). In tandem with this decrease in colour and extractives is a decrease in the development of mature characters, such as smooth, vanilla, and sweet, and less suppression of immature characters such as soapy, oily, and sulphury (Piggott et al., 1993). Oak wood aromas are present but are not dominant, and the role of the wood is to integrate the individual characteristics of the distillate and enhance product complexity (Swan, 1994). Comparison of first-fill and refill casks shows that most wood aromas present in the first-fill spirit can still be detected in spirit from refill casks but at much lower levels (Conner et al., 2001). However, the relative amounts may change, and the different balance can give rise to changes in mature character.

Other maturation variables

Although the cask type is the dominant factor in maturation, other variables can influence the mature qualities of the whisky. The most important of these is maturation time, but other factors such as fill strength and warehouse conditions also play a role.

Maturation time

Time is an important variable in the maturation of distilled spirits. Maturation periods of 10 to 20 years are not uncommon, and, although these produce high-quality mature spirits, the important reactions in generating these qualities have not been identified. One problem of identifying reactions over these time scales is that, when modelled in the laboratory, no appreciable activity may be observed within a practical experiment time.

The extraction of colour and wood constituents during maturation has been followed in a large number of studies. In first- and second-fill casks, there is a rapid initial extraction within the first 6 to 12 months. Thereafter, the extraction rate is reduced, although a steady increase in colour and cask constituents is maintained throughout the maturation period. The initial rapid extraction is attributed to the fast diffusion of free extractives from the cask surface. The steady increase thereafter is due to the release of further concentrations of hydrolysable tannins and lignin breakdown products through a combination of diffusion from deeper in the stave or formation by spirit hydrolysis and oxidation. The evaporation of spirit, which increases the concentration of all non-volatile congeners, may also contribute to this steady increase. In refill casks, most free extractives have been depleted and the initial rapid extraction is greatly reduced. Extractives generally increase linearly through the maturation period, reflecting the slower diffusion from deeper in the stave or formation by hydrolysis and oxidation. Consequently, increased levels of cask extractives accompany prolonged maturation, although this, to a large extent, will be dependent on cask type. The effect of prolonged maturation on the relative amounts of cask-derived congeners has not been investigated, but decreases observed in the ratio of sinapaldehyde to syringaldehyde and of coniferaldehyde to vanillin, along with an increase in the proportion of vanillic and syringic acids, suggest slow conversion to a more oxidised extract.

Colour and cask extractives provide easily measured markers for additive reactions during maturation. No such markers have been followed for subtractive reactions, with the result that their contribution to the sensory properties of old whiskies is not known. Evaporation will take place throughout the maturation period, and in Scotland, this is accompanied by a decrease in strength. On prolonged maturation, large reductions in spirit strength may occur, which could affect the solubility of wood and distillate components. Consequently, the concentration of long-chain ethyl esters, fats, and ethanol lignin may decrease and the concentration of sugars and hydrolysable tannins increase. Recent studies have shown that char-mediated reductions in dimethyl disulphide occur to a large extent within the first 18 months of maturation. Adsorption of copper by the cask will reduce the level of active oxidant present in older spirits. However, the development of rancio character in old brandies is attributed to the oxidation of fatty acids to ketones, so there may be other, as yet unidentified reactions that are promoted by prolonged maturation. Finally, the increase in the level of extractives and wood aromas with longer maturation could lead to both sensory and physicochemical masking of distillate aromas.

Fill strength

Distillates are generally filled into casks at a constant strength. Normally, malt whiskies are diluted to between 57 and 75% alcohol by volume (abv). The fill strength of grain whiskies may be higher but is generally less than 80% (abv). Spirit strength influences the extraction and formation of flavour congeners in a maturing whisky. Lower alcoholic strengths favour the extraction of water-soluble wood components such as hydrolysable tannins, glycerol, and sugars (Reazin, 1981) but higher strength distillates extract more ethanol-soluble congeners such as ellagic acid. Overall, increasing the fill strength has been found to reduce the levels of colour, solids, and volatile acids developed during maturation (Baldwin and Andreasen, 1974; Reazin, 1983). Some wood/distillate reactions may be dependent upon the presence of water. Therefore, as spirit strength increases, water concentration decreases and reaction rates will be slower, changing the levels of congeners in the final product. One notable exception is ester formation, which remains constant over the maturation period and is not affected by fill strength. To control the extraction of wood components, casks are filled with distillates at empirically derived strengths. This strength rarely exceeds 80% (abv), as strengths above this can lead to the extraction of excessive amounts of wood lipids and ethanol lignins that give filtration problems before bottling.

Warehousing

The maturation of whisky requires that companies have suitable storage facilities for the large cask inventory that this entails. Traditionally, maturing whisky was accommodated in stone-built, single- or multi-storey warehouses located beside the distillery. As production expanded, large centralised multi-storey warehouses were used to increase storage capacity. During maturation, the cask is not an impermeable container, so it allows the evaporation of spirit (both ethanol and water) and the ingress of air (oxygen). The loss of a small percentage of spirit has long been an accepted part of maturation (called the 'angels' share'). These losses have been found to vary with the environmental conditions in a warehouse (Reid and Ward, 1994). Local, regional, and national differences exist in warehouse temperature and environment. These can affect both evaporative losses and the rate and progress of maturation.

Warehouse types

The traditional maturation warehouse was a stone-built, single- or multi-storey building with a slate roof on timber sarking. The bottom storey of these warehouses had cinder floors, and additional levels had wooden floors. Warehouses generally were constructed without damp courses, and humidity often depended on the surrounding soil type and water table. Casks were stored in stows, usually two or three high, sitting on top of one another with wooden runners between each layer. Large, centralised multi-storey warehouses have a basic construction of brick walls, concrete floors, and insulated aluminium or asbestos roofs. Steel racking with wooden runners allows casks to be tightly packed on their sides in long parallel rows. Racking can extend to up to 12 rows high. More recently, warehouses have been steel-framed and used corrugated metal panels for both roof and walls. They have also dispensed with racking, and casks are stored on their ends on pallets and stacked up to eight high using forklift trucks. Palletised storage offers considerable benefits in terms of cask handling efficiency and warehouse capacity but can result in higher evaporative losses and lower levels of extractives.

Environmental conditions and evaporative losses

Differences in construction give rise to important variations in ventilation and insulation. The large roof area of warehouses and their generally poor insulation allow relatively high rates of solar heat transfer through the roof to upper levels. If no natural or forced air circulation is provided, then hot stagnant air builds up around the upper tiers and there is a sizeable temperature difference between the top and bottom of the warehouse. This effect is most marked in continental America during the summer, when temperatures of 50°C to 60°C can develop on the top floor while temperatures at the bottom are only 18°C to 21°C (USEPA, 1978). In Scotland, temperature variations in summer are less, with top tiers typically being 16°C to 20°C in summer and 10°C to 15°C on bottom tiers. There is no stratification of temperatures in Scottish warehouses in winter (Conner and Forrester, 2017). Traditional warehouses are generally lower with better insulation characteristics and so do not experience the internal temperature gradients of the multi-storey racked warehouses.

Warehouse temperature and humidity can influence evaporative losses. Within a warehouse, there is an inverse relationship between temperature and humidity. Under controlled climatic conditions, temperature and humidity have been shown to affect the relative rates at which ethanol and water are lost. Raising the temperature increases the evaporation losses of both ethanol and water. Humidity influences the relative rate at which ethanol and water are lost. At high humidity, more ethanol than water is lost, and strength is decreased; at low humidity, more water than ethanol is lost and strength increases (Philp, 1989). Application of these results to the warehouse environment is not straightforward, as conditions can vary on a seasonal, monthly, and even daily basis.

For large racked and palletised warehouses, there are marked differences between top and bottom tiers. The environment around bottom tiers is generally stable with a relatively high humidity. On top tiers, there are marked daily decreases in humidity as the temperature increases, and this translates to higher losses of water from the top tier (Reid and Ward, 1994). Consequently, for equivalent casks maturing the same spirit, the alcoholic strength can decrease when matured at the bottom of the warehouse but increase when matured at the top. However, total evaporative losses will be higher from the top tier. Humidity also explains international differences in strength changes. In the United States, the relatively hot and dry climate encourages preferential loss of water vapour relative to ethanol, so strength increases during maturation (Reazin, 1981). In Scotland, the cool humid environment favours the loss of ethanol over water, and strength decreases during maturation.

In a study of environmental factors (Conner and Forrester, 2017), temperature was the major determinant of losses. In contrast, trials comparing different air change rates showed no significant difference in losses, indicating that within normal ranges, ventilation did not directly impact on ethanol loss. The trials were conducted at temperatures typical of Scottish warehouses and suggest the rate-limiting step is the diffusion of ethanol to the cask surface. This may be increased around leak sites and at higher temperatures, and for these, air flow may influence the rate of loss. Management of warehouses is influenced by factors such as health and safety regulations and the labour involved in moving barrels and opening and closing vents. Often, in Scotland, warehouses do not have any active ventilation systems, and the exchange of air will depend on the prevailing weather conditions, how sheltered the warehouse is, and the number of times it is opened on a weekly basis.

Effects on quality

Current methods of warehouse operation have not been developed by design and calculation; rather, each distiller's operation is for the most part the result of tradition and experience. The chemical effects of temperature are straightforward. Higher temperatures increase the rates of extraction, reaction, and diffusion. Under controlled conditions, the non-volatile content extracted during maturation significantly increases with temperature (Philp, 1989; Conner and Forrester, 2017). In large warehouses, spirit matured on the top tier is generally darker and has a higher non-volatile content than spirit matured on the bottom tier.

Under controlled climatic conditions, no consistent differences were observed for Highland and Lowland malt spirits that could be related to temperature or humidity; however, for grain spirits, some differences were noticed. Those grain spirits matured at a higher temperature were described as sweeter but less clean, while smoother and more pleasant whiskies were produced at lower temperatures (Philp, 1989). This suggests that the influence of warehouse conditions is subtle and can be masked by distillate character and cask-to-cask variation. A more recent study showed that the physical and chemical reactions typical of maturation proceed at a greater rate in the warmest (top) tier of a warehouse. Substantial differences in character were noted for whisky from casks on the top and bottom tiers of Japanese warehouses (Nakajima and Fujii, 2012). Casks on the top tier gave higher levels of phenolic compounds, esters, and higher alcohols, and the whiskies were described as woody, vanilla, fruity/estery, and floral. Whisky from casks on the bottom tier was described as green/grassy, cereal, and feinty. Despite these differences, no optimal temperature has been determined for producing the desired product quality (Reazin, 1981), and the effects of different conditions are normally averaged out by cask selection prior to bottling.

Wood policy

Distillers strive to produce a range of unique spirits with recognisable but distinctive characteristics from their different distilleries. Consistency in the character of these individual whiskies is vital, not only for spirits that will be sold as single malts and grains, but also for whiskies used in blending. Because wood is one of the primary influences in whisky character and accounts for a considerable proportion of the production costs, it is vital that distillers establish a successful and cost-efficient wood policy to control overall product quality.

An important aspect of wood policy is the introduction of new casks. To maintain whisky quality, it is important to have a healthy influx of new wood into cask stocks. Therefore, on reaching the point of exhaustion, casks should be removed from the system and new casks introduced. A proportion of these exhausted casks may be sent for regeneration, but this does not recreate the original state of the casks, so the number of casks that can be recycled into the system is limited. Most new casks are former bourbon barrels from the United States, with a lesser number of sherry butts from Spain. In the case of bourbon barrels, cask construction and heat treatment are not under the control of Scotch whisky companies and the quality of the purchased cask is to a large extent unknown. Sherry casks can be made from both American and European oak woods. This variable, in conjunction with the heat treatment applied to the wood during cask construction, has a large impact on the maturation performance of the cask and the character of the spirit it will produce. Again, if casks are purchased directly from the sherry producer without any prior knowledge of their history, then the quality of cask performance will be unknown. To overcome this problem, some Scotch whisky companies have wood policies that extend to the first use of the cask, either through obtaining knowledge of the cask history or providing specifications for variables such as wood type, sherry type, and storage period.

An effective wood policy should ensure full utilisation of the maturation potential of a cask. This can be achieved by matching cask types and activities with spirit types. Cask management involves matching the flavour characteristics of a new-make spirit with casks that produce the quality of whisky required for the target products. Casks that begin life maturing one type of spirit may end life maturing another. For example, a cask displaying a high degree of maturation activity may be capable of maturing a malt whisky spirit that has heavy sensory characteristics. After several refillings, its activity will have dropped, and it may then be used to mature lighter spirit, such as a grain whisky. Other variables are also important in ensuring the maximum use of a cask. These include the number of times a cask is refilled, the period of spirit storage, and the warehouse environment in which the cask is stored (Roullier-Gall et al., 2020).

In summary, careful management of cask stocks and their uses is required to ensure a continued supply of quality and diverse whiskies for blending and bottling.

Acknowledgements

The author would like to thank Jim Beveridge of Diageo Scotland, Ltd., and George Espie of The Edrington Group for their help and advice during the preparation of the original version of this chapter.

References

Aishima, T., Matsushita, K., Nishikawa, K., 1992. Measurements of brandy ageing using O17 NMR and small angle x-ray scattering. In: Cantagrel, R. (Ed.), Élaboration et Connaissance des Spiritueux: Recherche de la Qualité, Tradition et Innovation. Lavoisier, Tec. & Doc, Paris, pp. 473–478.

Assor, C., Placet, V., Chabbert, B., Habrant, A., Lapierre, C., Pollet, B., Perre, P., 2009. Concomitant changes in viscoelastic properties and amorphous polymers during the hydrothermal treatment of hardwood and softwood. J. Agric. Food Chem. 57, 6830–6837.

Baldwin, S., Andreasen, A.A., 1974. Congener development in Bourbon whisky matured at various proofs for twelve years. J. AOAC 57 (4), 940–950.

Booth, M., Shaw, W., Morhalo, L., 1989. Blending and bottling. In: Piggott, J.R., Sharp, R., Duncan, R.E.B. (Eds.), The Science and Technology of Whiskies. Longman Scientific & Technical, Harlow, Essex, pp. 295–326.

Bortoletto, A.M., Alcarde, A.R., 2013. Congeners in sugar cane spirits aged in casks of different woods. Food Chem. 139 (1–4), 695–701.

Chatonnet, P., 1999. Discrimination and control of toasting intensity and quality of oak wood barrels. Am. J. Enol. Vitic. 50 (4), 479–494.

Chatonnet, P., Dubourdieu, D., 1998a. Comparative study of the characteristics of American white oak (*Quercus alba*) and European oak (*Quercus petraea* and *Q robur*) for production of barrels used in barrel aging of wines. Am. J. Enol. Vitic. 49, 79–85.

Chatonnet, P., Dubourdieu, D., 1998b. Identification of substances responsible for the "sawdust" aroma in oak wood. J. Sci. Food Agric. 76, 179–188.

Clyne, J., Conner, J.M., Piggott, J.R., Paterson, A., 1993. The effect of cask charring on Scotch whisky maturation. Int. J. Food Sci. Technol. 28, 69–81.

Conner, J., Forrester, A., 2017. Building for the future: innovative warehouse design for efficient maturation of Scotch whisky. In: Brewer and Distiller International, May 2017, pp. 28–33.

Conner, J.M., Patterson, M., Owen, C., Freeman, J., 2012. Reducing the need for new wood by regenerating and re-using casks. In: Walker, G., Fotheringham, R., Goodall, I., Murray, D. (Eds.), Distilled Spirits—Science and Sustainability. Nottingham University Press, Nottingham, UK, pp. 47–54.

Conner, J.M., Piggott, J.R., Paterson, A., 1993. Analysis of lignin from oak casks used for the maturation of Scotch whisky. J. Sci. Food Agric. 60, 349–353.

Conner, J.M., Reid, K., Richardson, G., 2001. SPME analysis of flavor components in the headspace of Scotch whiskey and their subsequent correlation with sensory perception. In: Leland, J., Schieberle, P., Buettner, A., Acree, T. (Eds.), Gas Chromatography—Olfactometry: The State of the Art. American Chemical Society, Washington, DC, pp. 113–122.

Council of Europe, 2008. Regulation (EC) No. 110/2008 of the European parliament and of the Council on the definition, description, presentation, labelling and the protection of geographical indications of spirit drinks. Off. J. Eur. Union L39, 16–54.

Crow, M., Carson, J., 2018. Ergonomic and flavour innovation in a modern Scottish cooperage. In: Jack, F., Dabrowska, D., Davies, S., Garden, M., Maskell, D., Murray, D. (Eds.), Distilled Spirits – Local Roots. Global Reach. Context Products Ltd., Packington, UK, pp. 159–164.

Cutzach, I., Chatonnet, P., Henry, R., Dubourdieu, D., 1997. Identification of volatile compounds with a "toasty" aroma in heated oak used in barrelmaking. J. Agric. Food Chem. 45, 2217–2224.

De Micco, V., Balzano, A., Wheeler, E.A., Baas, P., 2016. Tyloses and gums: a review of structure, function and occurrence of vessel occlusions. Int. Assoc. Wood Anatom. J. 37 (2), 186–205.

De Rosso, M., Panighel, A., Dalla Vedova, A., Stella, L., Flamini, R., 2009. Changes in chemical composition of a red wine aged in acacia, cherry, chestnut, mulberry, and oak wood barrels. J. Agric. Food Chem. 57, 1915–1920.

del Alamo-Sanza, M., Nevares, I., 2018. Oak wine barrel as an active vessel: a critical review of past and current knowledge. Crit. Rev. Food Sci. Nutr. 58 (16), 2711–2726.

Delahunty, C.M., Conner, J.M., Piggott, J.R., Paterson, A., 1993. Perception of heterocyclic nitrogen compounds in mature whisky. J. Inst. Brew. 99, 479–482.

Fantoni-Salas, A., Fernandez-Mesa, A., 2007. Process and Apparatus for Inner Wall Toasting of Casks for Wine Guard by Hot Air Convection. U.S. Patent No. US7179082 B2.

Fengel, D., Wegener, G., 1984. Wood: Chemistry, Ultrastructure, Reactions. Walter de Gruyter, New York, pp. 66–131.

Fernández de Simón, B., Martínez, J., Sanz, M., Cadahía, E., Esteruelas, E., Muñoz, Á.M., 2014. Volatile compounds and sensorial characterisation of red wine aged in cherry, chestnut, false acacia, ash and oak wood barrels. Food Chem. 147, 346–356.

Fujieda, M., Tanaka, T., Suwa, Y., Koshimizu, S., Kouno, I., 2008. Isolation and structure of whiskey polyphenols produced by oxidation of oak wood ellagitannins. J. Agric. Food Chem. 56, 7305–7310.

Fujii, T., Kurokawa, M., Saita, M., 1992. Studies of volatile compounds in whisky during ageing. In: Cantagrel, R. (Ed.), Élaboration et Connaissance des Spiritueux: Recherche de la Qualité, Tradition et Innovation. Lavoisier, Tec. & Doc, Paris, pp. 543–547.

Furusawa, T., Saita, M., Nishi, N., 1990. Analysis of ethanol water clusters in whisky. In: Campbell, I. (Ed.), Proceedings of the Third Aviemore Conference on Malting, Brewing and Distilling. Institute of Brewing, London, pp. 431–438.

Glabasnia, A., Hofmann, T., 2006. Sensory-directed identification of taste-active ellagitannins in American (Quercus alba L.) and European oak wood (Quercus robur L.) and quantitative analysis in bourbon whiskey and oak-matured red wines. J. Agric. Food Chem. 54, 3380–3390.

Glabasnia, A., Hofmann, T., 2007. Identification and sensory evaluation of dehydro- and deoxy-ellagitannins formed upon toasting of oak wood (Quercus alba L.). J. Agric. Food Chem. 55, 4109–4118.

González Gordon, M.M., 1990. In: Doxat, J. (Ed.), Sherry: The Noble Wine. Quiller Press, London.

Haneca, K., Čufar, K., Beeckman, H., 2009. Oaks, tree-rings and wooden cultural heritage: a review of the main characteristics and applications of oak dendrochronology in Europe. J. Archaeol. Sci. 36, 1–11.

Hankerson, F.P., 1947. The Cooperage Handbook. Chemical Publishing, New York, pp. 24–27.

Hasuo, T., Yoshizawa, K., 1986. Substance change and substance evaporation through the barrel during whisky ageing. In: Campbell, I., Priest, F.G. (Eds.), Proceedings of the Second Aviemore Conference on Malting, Brewing and Distilling. Institute of Brewing, London, pp. 404–408.

Hayasaka, Y., Wilkinson, K.L., Elsey, G.M., Raunkjær, M., Sefton, M.A., 2007. Identification of natural oak lactone precursors in extracts of American and French oak Woods by liquid chromatography–tandem mass spectrometry. J. Agric. Food Chem. 55, 9195–9201.

Koga, K., Taguchi, A., Koshimizu, S., Suwa, Y., Yamada, Y., Shirasaka, N., Yoshizumi, H., 2007. Reactive oxygen scavenging activity of matured whiskey and its active polyphenols. J. Food Sci. 72 (3), S212–S217.

Labbé, N., Harper, D., Rials, T., Elder, T., 2006. Chemical structure of wood charcoal by infrared spectroscopy and multivariate analysis. J. Agric. Food Chem. 54, 3492–3497.

Le Floch, A., Jourdes, M., Teissedre, P.-L., 2015. Polysaccharides and lignin from oak wood used in cooperage: composition, interest, assays: a review. Carbohydr. Res. 417, 94–102.

Lebourgeois, F., Cousseau, G., Ducos, Y., 2004. Climate-tree-growth relationships of *Quercus petraea* Mill. stand in the Forest of Berce´ ("Futaie des Clos", Sarthe, France). Ann. For. Sci. 61, 361–372.

Masson, G., Guichard, E., Fournier, N., Puech, J.L., 1995. Stereoisomers of beta-methyl-gamma-octalactone. II. Contents in the wood of French (*Quercus robur* and *Quercus petraea*) and American (*Quercus alba*) oaks. Am. J. Enol. Vitic. 46 (4), 424–428.

McPhail, D.B., Gardner, P.T., Duthies, G.G., Steele, G.M., Reid, K., 1999. Assessment of the antioxidant potential of Scotch whiskies by electron spin resonance spectroscopy: relationship to hydroxyl containing aromatic compounds. J. Agric. Food Chem. 47, 1937–1941.

Michel, J., Jourdes, M., Le Floch, A., Giordanengo, T., Mourey, N., Teissedre, P.-L., 2013. Influence of wood barrels classified by NIRS on the Ellagitannin content/composition and on the organoleptic properties of wine. J. Agric. Food Chem. 61 (46), 11109–11118.

Monties, B., 1992. Composition chimique des bois de chêne: composés phénoliques, relations avec quelques propriétés physiques et chimiques susceptibles d'influencer la qualité des vins et des eaux-de-vie. In: Le bois et la qualité des vins et eaux-de-vie. Vigne et Vin Publications Internationales, Martillac, pp. 59–72.

Morishima, K., Nakamura, N., Matsui, K., Tanaka, Y., Masunaga, H., Mori, S., Iwashita, T., Li, X., Shibayama, M., 2019. Formation of clusters in whiskies during the maturation process. J. Food Sci. 84 (1), 59–64.

Mosedale, J.R., 1995. Effects of oak wood on the maturation of alcoholic beverages with particular reference to whisky. Forestry 68 (3), 203–230.

Mosedale, J.R., Charrier, B., Janin, G., 1996. Genetic control of wood colour, density and heartwood ellagitannin concentration in European oak (*Quercus petraea* and *Q. robur*). Forestry 69 (2), 111–124.

Mosedale, J.R., Puech, J.-L., 1998. Wood maturation of distilled beverages. Trends Food Sci. Technol. 9, 95–101.

Muller, S., McEwan, A., 1998. Observations on the changes in copper concentration during maturation of malt spirit. In: Campbell, I. (Ed.), Proceedings of the Fifth Aviemore Conference on Malting, Brewing and Distilling. Institute of Brewing, London, pp. 318–321.

Nakajima, N., Fujii, T., 2012. Creating various whiskies utilizing differences in warehouse conditions in Japan. In: Walker, G., Fotheringham, R., Goodall, I., Murray, D. (Eds.), Distilled Spirits—Science and Sustainability. Nottingham University Press, Nottingham, pp. 199–204.

Nishimura, K., Matsuyama, R., 1989. Maturation and maturation chemistry. In: Piggott, J.R., Sharp, R., Duncan, R.E.B. (Eds.), The Science and Technology of Whiskies. Longman Scientific & Technical, Harlow, pp. 235–263.

Nishimura, K., Ohnishi, M., Masuda, M., Koga, K., Matsuyama, R., 1983. Reactions of wood components during maturation. In: Piggott, J.R. (Ed.), Flavour of Distilled Beverages: Origin and Development. Ellis Horwood, Chichester, pp. 225–240.

Noguchi, Y., Hughes, P.S., Priest, F.G., Conner, J.M., Jack, F., 2010. The influence of wood species of cask on matured whisky aroma—identification of a unique character imparted by casks of Japanese oak. In: Walker, G.M., Hughes, P.S. (Eds.), Distilled Spirits—New Horizons: Energy, Environment and Enlightenment. Nottingham University Press, Nottingham, pp. 243–251.

Noguchi, Y., Hughes, P.S., Priest, F.G., Conner, J.M., Jack, F., 2012. The behaviour of whisky lactone isomers in Japanese oak casks. In: Walker, G., Fotheringham, R., Goodall, I., Murray, D. (Eds.), Distilled Spirits—Science and Sustainability. Nottingham University Press, Nottingham, pp. 191–198.

Nonier, M.-F., Vivas de Gaulejac, N., Vivas, N., Vitry, C., 2005. Glycosidically bound flavour compounds in *Quercus petraea* Liebl. wood. Flavour Fragr. J. 20, 567–572.

Nose, A., Hojo, M., Suzuki, M., Ueda, T., 2004. Solute effects on the interaction between water and ethanol in aged whiskey. J. Agric. Food Chem. 52, 5359–5365.

Owen, C., MacDonald, S., Muller, S., Tang, L., Miller, G., 2018. The use of the Agilent 4100 ExoScan series FTIR device for determination of the origin of oak for construction of sherry casks. In: Jack, F., Dabrowska, D., Davies, S., Garden, M., Maskell, D., Murray, D. (Eds.), Distilled Spirits – Local Roots; Global Reach. Context Products Ltd., Packington, UK, pp. 169–172.

Parham, R.A., Gray, R.L., 1984. Formation and structure of wood. In: Rowell, R. (Ed.), The Chemistry of Solid Wood. American Chemical Society, Washington DC, pp. 3–56.

Perry, D., Ford, A., Burke, G., 1990. Cask rejuvenation. In: Campbell, I. (Ed.), Proceedings of the Third Aviemore Conference on Malting, Brewing and Distilling. Institute of Brewing, London, pp. 464–467.

Philp, J.M., 1986. Scotch whisky flavour development during maturation. In: Campbell, I., Priest, F.G. (Eds.), Proceedings of the Second Aviemore Conference on Malting, Brewing and Distilling. Institute of Brewing, London, pp. 148–163.

Philp, J.M., 1989. Cask quality and warehouse conditions. In: Piggott, J.R., Sharp, R., Duncan, R.E.B. (Eds.), The Science and Technology of Whiskies. Longman Scientific & Technical, Harlow, pp. 264–294.

Piggott, J.R., Conner, J.M., Paterson, A., Clyne, J., 1993. Effects on Scotch whisky composition and flavour of maturation in oak casks with varying histories. Int. J. Food Sci. Technol. 28, 303–318.

Poisson, L., Schieberle, P., 2008a. Characterization of the most odor-active compounds in an American bourbon whisky by application of aroma extract dilution analysis. J. Agric. Food Chem. 56, 5813–5819.

Poisson, L., Schieberle, P., 2008b. Characterization of the key aroma compounds in an American bourbon whisky by quantitative measurements, aroma recombination and omission studies. J. Agric. Food Chem. 56, 5820–5826.

Prida, A., Puech, J.-L., 2006. Influence of geographical origin and botanical species on the content of extractives in American, French, and east European oak Woods. J. Agric. Food Chem. 54 (21), 8115–8126.

Pryde, J., Conner, J., Jack, F., Lancaster, M., Meek, L., Owen, C., Paterson, R., Steele, G., Strang, F., Woods, J., 2011. Sensory and chemical analysis of "Shackleton's" Mackinlay Scotch whisky. J. Inst. Brew. 117 (2), 156–165.

Ramsay, J., 2008. Getting the right wood. Brewer Distiller Int. 4 (9), 46–47.

Reazin, G.H., 1981. Chemical mechanisms of whiskey maturation. Am. J. Enol. Vitic. 32 (4), 283–289.

Reazin, G.H., 1983. Chemical analysis of whisky maturation. In: Piggott, J.R. (Ed.), Flavour of Distilled Beverages: Origin and Development. Ellis Horwood, Chichester, pp. 225–240.

Reid, K.J., Ward, A., 1994. Evaporation losses from traditional and racked warehouses. In: Campbell, I., Priest, F.G. (Eds.), Proceedings of the Fourth Aviemore Conference on Malting, Brewing and Distilling. Institute of Brewing, London, pp. 319–322.

Roullier-Gall, C., Signoret, J., Coelho, C., Hemmler, D., Kajdan, M., Lucio, M., et al., 2020. Influence of regionality and maturation time on the chemical fingerprint of whisky. Food Chem. 323. https://doi.org/10.1016/j.foodchem.2020.126748, 126748.

Sachs, I., Kuntz, J., Ward, J., Nair, G., Schultz, N., 1970. Tyloses structure. Wood Fiber Sci. 2, 259–268.

Schahinger, G., Rankin, B., 1992. Cooperage for Winemakers. Ryan Publications, Adelaide, pp. 4–15.

Singleton, V.L., 1995. Maturation of wines and spirits: comparisons, facts, and hypotheses. Am. J. Enol. Vitic. 46 (1), 98–115.

Spillman, P.J., Sefton, M.A., Gawel, R., 2004. The effect of oak wood source, location of seasoning and coopering on the composition of volatile compounds in oak-matured wines. Aust. J. Grape Wine Res. 10 (3), 216–226.

Swan, J.S., 1994. Sourcing oak wood for the distilling industries. In: Campbell, I., Priest, F.G. (Eds.), Proceedings of the Fourth Aviemore Conference on Malting, Brewing and Distilling. Institute of Brewing, London, pp. 56–70.

Tarasov, D., Leitch, M., Fatehi, P., 2018. Lignin–carbohydrate complexes: properties, applications, analyses, and methods of extraction: a review. Biotechnol. Biofuels 11, 269.

Taylor, A.M., Gartner, B.L., Morrell, J.A., 2002. Heartwood formations and natural durability – a review. Wood Fiber Sci. 34 (4), 587–611.

USEPA, 1978. Whiskey warehousing and aging. In: Cost and Engineering Study—Control of Volatile Organic Emissions from Whiskey Warehousing. U.S. Environmental Protection Agency, Research Triangle Park, NC, pp. 1–16.

Vivas, N., 1995. The notion of grain in cooperage. J. Cooper. Sci. Tech. 1, 17–32.

Wanikawa, A., 2020. Flavors in malt whisky: a review. J. Am. Soc. Brew. Chem. 78 (4), 260–278.

Waterhouse, A.L., Towey, J.P., 1994. Oak lactone isomer ratio distinguishes between wines fermented in American and French oak barrels. J. Agric. Food Chem. 42, 1971–1974.

Wickham, N., 2009. American, French oak trials support Phoenix revolution. Aust. N.Z. Grapegrow. Winemak. 548, 112–114.

Wiedenhoeft, A., 2010. Structure and function of wood. In: Forest Products Laboratory, Wood Handbook—Wood as an Engineering Material. U.S. Department of Agriculture, Forest Service, Forest Products Laboratory, Madison, WI, USA, pp. 1–16. General Technical Report FPL-GTR-190.

Wilkinson, K.L., Prida, A., Hayasaka, Y., 2013. Role of glycoconjugates of 3-methyl-4-hydroxyoctanoic acid in the evolution of oak lactone in wine during oak maturation. J. Agric. Food Chem. 61 (18), 4411–4416.

Zimlich, J., Beach, M., 2015. Tale of a modern bourbon barrel cooperage. In: Goodall, I., Fotheringham, R., Murray, D., Speers, R.A., Walker, G.M. (Eds.), Distilled Spirits – Future Challenges, New Solutions. Context Products Ltd., Packington, UK, pp. 143–148.

Blending

Stephanie J. Macleod

John Dewar & Sons Ltd., Glasgow, United Kingdom

The purpose of blending is to create a Scotch whisky with a particular flavour profile, which can be recreated batch after batch with minimal variation. 'Blending' is more commonly associated with combining single-malt whiskies and single-grain whiskies together to form blended Scotch whisky. However, the principles of blending are also crucial to the consistency and spirit quality of other types of Scotch whisky such as single malts. The guiding principles of blending and associated processes are explored in detail in this chapter.

The role of the blender

The blender is generally considered to be the sensory authority of a whisky company, who creates the recipes for new blends or malt expressions, manages the inventory to ensure the sustainability of the existing and future brand portfolio, and assigns a sensory judgement to all stages of the Scotch whisky manufacturing process. The exact scope, nature, and indeed the title of the role can vary depending on the company, and the size and complexity of the company may dictate additional responsibilities for the blender. These may include new product development; scientific, technical, and regulatory matters; ambassadorial duties; and internal and external education. The main element is, of course, management of the inventory and development and maintenance of the wood policy to ensure that the spirit profiles of the whisky portfolio remain the same and that future blends are consistent with the 'house style' of the brand. The blender is, in essence, the 'Guardian' of the whisky portfolio and the brand and should pay particular attention to anything that could impact on flavour and appearance—in the distillery, the cask, the warehouse, the blend centre, or the bottling hall; during transit to the market; and, finally, off-trade and on-trade in the hands of the consumer. Thus, the knowledge and experience of the blender must be extensive.

In order to understand blending and indeed the role of the blender more fully, it is important to understand the spirit category of Scotch whisky and how it is defined in law. It is of critical importance that the blender is aware of, and understands, the law concerning Scotch whisky and operates strictly within its parameters (see Chapter 1).

Scotch whisky

Two distinct types of Scotch whisky are produced: Scotch malt whisky and Scotch grain whisky. Scotch malt whisky is made from malted barley using a batch distillation process in copper pot stills; it is typically full flavoured, and each distillery produces a spirit that possesses a unique flavour profile. Scotch grain whisky, on the other hand, is typically made from wheat (or maize), with a small amount of malted barley as an enzyme source, using a continuous distillation process that produces, on average, ten times the volume of a malt distillery. Therefore, the resulting spirit is much cleaner and lighter in character with a higher alcoholic strength. From these two types of Scotch whisky, five different categories of Scotch whisky can be identified, as depicted in Fig. 17.1. Beginning with single-malt Scotch whisky—and single-grain Scotch whisky, if two or more of either the single-malt Scotch whiskies or the single-grain Scotch whiskies are blended together—a blended malt Scotch whisky and blended grain Scotch whisky, respectively, are created. One or more single malts blended together with one or more single grains creates a blended Scotch whisky.

Blended Scotch whisky

Blending of Scotch whisky began commercially in the mid- to late-1800s when the continuous distillation process used to produce Scotch grain whisky was introduced (see Chapter 15). Before this time, the main purpose of blending was to make the then-bland Lowland whiskies more interesting and the then-fiery Highland whiskies more palatable. With little understanding of the flavour profiles of individual Scotch whiskies, the process control at the distilleries, or the effects of maturation, the resulting blends could be somewhat variable in quality.

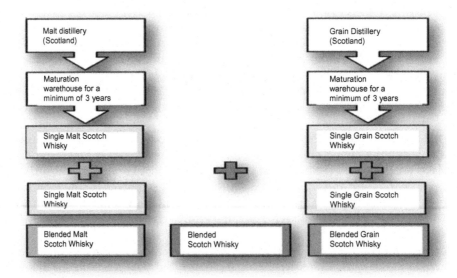

FIG. 17.1   Scotch whisky categories.

Due to the efforts of generations of distillers, blenders, and, laterally, scientists who had a great desire to understand and improve the quality of Scotch whisky, the industry now has achieved a much greater insight into the nature of the distinct characteristics of the spirits from our distilleries, as well as the way in which they interact with the wood when maturing and with each other when blended. Scotch malt whiskies and Scotch grain whiskies can be easily discerned from one another on the nose, with the grains having a much lighter, cleaner character than the malts. In fact, it is very easy to assert that malt whiskies have more flavour or more character than grain whiskies, but what flavour is this exactly? Each malt distillery in Scotland produces a spirit that is unique and distinct from all other malt whisky distilleries and can produce a vast array of different flavour profiles, and the same is true of grain distilleries—although the array of flavour is not as varied as malt whisky, each distillery has its own character. It is the task of the blender to harness these disparate flavour characteristics and meld them into balanced and well-rounded blends that have a consistent sensory profile from batch to batch. The manner in which this is achieved varies among whisky companies, but most companies follow similar processes and are interested in similar success criteria for their Scotch whiskies.

The blending process

Based on bottling production requirements and in accordance with the recipe (or, in more technical parlance, the formula), the required number and type of casks are retrieved from the maturation warehouses and assembled in a manner that is aligned with the configuration requirements of individual blend centres. Exploring this simple statement in more detail will reveal the complex process of blending and the way in which a number of different elements are required to come together on the day of blending.

Market forecasts and bottling demand

The bottling production requirements for a particular brand of Scotch whisky are driven by the demands of the market. Ideally, these demands are based on a forecast created a number of years prior to bottling. This forecast is important, as Scotch whisky must spend at least three years in oak casks in a maturation warehouse, and, of course, many Scotch whisky products are much older than this. Additionally, the single malts or blends may have an age statement on the label, and for Scotch whisky, the age statement represents the age of the youngest spirit—we refer to this as the AYS. Consequently, the creation of an accurate forecast is necessary to ensure that each whisky company makes or procures enough malt and grain spirit to satisfy the forecast and ultimately the demand, several years into the future. Before laying down these stocks, maturation losses must be factored into the calculations to ensure that, with losses, there will still be sufficient stocks to satisfy future demand. The volume of stock to be laid down is driven by the forecasts, but the individual makes of malts and grains that are required are driven by the recipe. Each Scotch whisky product has a recipe, which stipulates the age of the youngest spirit and contains the exact percentage of each malt and grain spirit that is required for a particular blend, as well as the types of cask from which the malts and grains should be disgorged. Therefore, it is important that the recipes used during the procurement process, for both spirit and casks, are accurate in order to reduce the risk of a surplus or shortage situation developing in the future.

Wood policy

Maturation is the process during which the clear and pungent new distillate is transformed into Scotch whisky, with its distinctive colour, aroma, and taste. Maturation must take place in oak casks and in bonded warehouses in Scotland for a minimum period of three years in order for the matured spirit to be categorised as Scotch whisky. There are a number of different types of cask that can be used in the maturation of Scotch whisky, and, provided that the casks are made of oak and hold a volume of less than 700 L, the blender has quite a variety at their disposal.

The types of casks used for the maturation of a whisky company's inventory are determined by the wood policy. The wood policy varies from company to company but, in general, it states the percentage of wood types that should be filled for each product, as well as the life cycle of the cask. The new-make distillate can be filled into casks that have previously held Scotch whisky, and they can also be filled into casks that have previously held bourbon, generally American standard barrels. The sherry industry also typically supplies the whisky industry with butts from Bodegas in Spain. Casks that have held wines and other spirit types can be used provided they have a long history of use in the whisky industry. However, in 2019, the Scotch Whisky Association (SWA) announced a change to the Scotch whisky Technical File (Spirit Drinks Verification Scheme, 2019). This broadened the range of casks permitted for use by the Scotch whisky industry and gave the industry more of a choice regarding the types of casks that could be used in the maturation and finishing of Scotch whisky. The Technical File now states that Scotch whisky can now be matured in new oak casks and/or oak casks that have only been used to mature wine (still or fortified) and/or beer/ale and/or spirits. However, there are exceptions and conditions, which are outlined in the Technical File, such as for the cask selected, the previous maturation is part of the traditional processes for those wines, beer/ales, or spirits, and the final Scotch whisky product still retains the colour, appearance, aroma, and taste that is typical of Scotch whisky.

The Scotch whisky industry endeavours to be as sustainable as possible, and, to that end, the casks are reused. Casks have a finite amount of maturation potential to impart to the spirit within its walls; consequently, after a prescribed number of fillings, the casks have to be removed from the maturation cycle. At this stage, the cask may be rejuvenated. For this process, the inside surface is partially removed to reveal fresh wood, and then, heat is applied to create a new char layer. Rejuvenation does not return the cask to its virgin state, but it does provide enough potential to enable at least another couple of productive maturation cycles. Sherry casks and bourbon barrels have traditionally been used in the whisky industry to mature its inventory, but it is not tradition alone that dictates their continued use. These types of casks impart congeners to the spirit that form part of the distinctive flavour characteristic of Scotch whisky. The spicy notes from the sherry casks and the toffee and vanilla notes that the bourbon casks impart to the whisky give it depth, complexity, and added interest. The percentage use of all these types of cask usually form part of the recipe, and it is important that this element is adhered to because too much or, indeed, too little of a particular cask type could result in a perceptible change in the flavour profile.

The wood policy is as equally applicable to single-malt Scotch whiskies as it is to blended Scotch whisky, as a single-malt Scotch whisky is still a blend of different casks, albeit from the same distillery, brought together to create a specific flavour profile.

Marrying and finishing

Marrying is a traditional term used in the industry to describe a process that is usually carried out immediately after blending has taken place. It is a specified period of maturation or resting, depending on the vessel in which this takes place, which allows the different flavour profiles of the whiskies to interact with one another. Marrying is not a legal requirement of Scotch whisky, but some companies, the author's company included, believe that marrying creates a more rounded and smoother blend. Expert and non-expert panels, when presented with married and non-married blends, can often distinguish between the two samples, giving more favourable scores to the married variant. However, this perception cannot currently be explained through chemical analysis.

The purpose of finishing is to create another dimension to the sensory characteristics of a particular Scotch whisky by using the flavour characteristics of a type of cask that is different from the casks the Scotch whisky was matured in originally. The casks used to finish a spirit must have a history of use in the Scotch whisky industry, and the resulting spirit must still have the characteristics identifiable as Scotch whisky. The finishing process should be carefully monitored in order to avoid the character of the finishing cask overwhelming the final Scotch whisky flavour profile.

Maturation warehouses

During the maturation period, the casks are held in maturation warehouses. These warehouses have two important functions. First and foremost, they provide a stable environment for the casks to mature and they protect the revenue until the time comes to disgorge the contents of the casks. Casks are available in different types and sizes, and the same is true for maturation warehouses. However, one feature that is common to all types of warehouses is that they must be situated in Scotland if the spirit contained in the casks within the walls of the warehouse is to be called Scotch whisky at the end of three years. There are three main types of warehousing used in Scotland today: dunnage, racked, and palletised. Palletised is the most recent warehousing development in relative terms and was introduced to the industry in the 1980s. Dunnage warehouses are low lying or may be two-storey buildings; the casks are balanced on rails and held in place with sturdy wooden wedges. Dunnage warehouses are constructed without a damp course, and the floor of the dunnage warehouse tends to be ash or soil, so a damp atmosphere is created. Racked warehousing is approximately the equivalent of three stories high; the casks are stowed in steel racking frames and are held in place on the rack by a series of metal stops. The third and most recent type of warehousing is palletised. These warehouses are also around three stories high, but instead of the casks being stowed in racks they are stowed on their ends on specially designed pallets and stacked to around six pallets high. Specialist forklifts and expert forklift drivers are required to ensure that the cask-laden pallets are stowed and then retrieved safely.

The blenders' challenge

Earlier in this chapter, it was stated that a blended Scotch whisky is a blend of one or more malt whiskies with one or more grain whiskies. Usually, the reality is a great many more malts and a variety of grains are used to make a blended Scotch whisky. The malt and grain profiles are distinctly different from one another, due to differences in the method of production. The grains tend to have a flavour profile that contribute light and clean characteristics to the blend. However, it would be folly to presume that different grains can be interchanged in a blend, as each grain make has its own character and interacts in a different way with the malt component; therefore, a wrong selection could result in an unacceptable difference in the blend. The malts, on the other hand, exhibit a variety of sometimes very different flavour characteristics and have higher levels of congeners. The flavour profiles of the resulting malt and grain blends are dependent not only on the malts and grains chosen but also on the exact percentage of each of the makes. Therefore, given that the main objective of the blender is to create a consistent blend from batch to batch, no matter the inventory situation, it is important to create a system of categories or classes in order to make the inventory more manageable. The first level of category is generally 'malt' and 'grain', and the percentage that each type contributes to the blend is usually the top line of the recipe. The next level of category is generally based on the flavour profile of the malts and the grains. The exact nature of these categories varies from company to company, and this is not the information that is readily shared in the industry. However, it generally involves grouping makes of Scotch whiskies together based on common organoleptic characteristics. The use of categories or classes, as they are sometimes known, means that, in practical terms, if a particular make is not available on the day of blending, then a substitute make can be selected from its category in order to maintain the flavour profile of a particular blend. These categories are again assigned a percentage based on their contribution to the recipe. The next level contains the individual makes of malts and grains and their percentage contributions to the particular blend. Another important aspect of the blend is, of course, the casks in which the individual components of the blend have been matured. The wood profile, governed by the wood policy, for individual blends is a key element of the recipe that must be adhered to, so the recipe not only must conform to the individual makes and their percentages but must also fall within the specifications for the individual wood categories. Bringing all of these elements together on the day of blending requires careful planning!

New product development and innovation

New product development (NPD) is the process through which ideas for new products are assessed and subsequently developed into physical products for the marketplace. The aim of the NPD process is always to launch a successful product onto the market in the shortest possible time. The NPD process varies from company to company.

But usually the process starts with a perceived need from an individual market, or a category of the global market, depending on the size and product range of the company. For the liquid part of the idea, a *liquid brief* is generated that is shared with the blender and other relevant parties. The information contained in the brief, as a minimum, should include the alcoholic strength and the age, as well as any claims or unique selling points that the market wants to claim on the label. The blender will then engage with the market representatives to ensure that all aspects of the brief are clearly understood by all parties. The blender will then check that the attributes fall within the Scotch Whisky Regulations of 2009 and that the inventory is sufficient to cover the anticipated demand. The next step is to proceed to create test samples that the blender considers matches the liquid brief. The test liquids are then assessed both internally and by the market through a series of blind sensory assessments; that is, panellists assess the liquid without prior knowledge of the samples presented. At this stage, a preferred liquid may have been identified, in which case the sample is officially approved by the market, rendering the liquid brief 'frozen', which means that no more changes to the brief are permitted without impacting on the launch date.

If the new liquid is also required to be prepared for bottling in a different manner or packaged in material that may be new to the brand, the blender will be involved in testing to ensure that the new conditions do not have a detrimental impact on the flavour profile of the product.

Ensuring that the legacy of the whisky brands is passed down to the next generation of blenders is crucial. Learning from and so developing the techniques and recipes will keep Scotch whisky relevant to generations of whisky appreciators to come.

Acknowledgements

The author would like to thank John Dewar & Sons, Ltd., and Bacardi for permitting the time required to write this chapter and would also like to thank Dr. John R. Piggott, without whom the author may not have realised a career in the Scotch whisky industry.

Reference

Spirit Drinks Verification Scheme, 2019. Technical Guidance Scotch Whisky Verification – Notice June 2019. Available from: https://assets.publishing.service.gov.uk/.

18

Sensory analysis

Frances Jack

The Scotch Whisky Research Institute, Edinburgh, United Kingdom

Whisky flavour and other sensory characteristics

Flavour is the key driver of consumer preference and brand loyalty in the alcoholic beverage sector. Consequently, whisky companies focus on controlling the sensory attributes of their products, differentiating them from other categories of spirit drinks and, on a finer level, from competitors' whisky brands. The sensory properties of whisky can be subdivided into aroma, taste, mouthfeel, colour, and appearance. Because no added flavourings are allowed in most whiskies, these attributes are due to the presence of naturally occurring compounds, often referred to in the industry as *congeners*, which are introduced with the raw materials or created during the production process.

Aroma

By far, the largest group of compounds are those responsible for aroma. These are odour-active, volatile compounds that evoke a sensory response through stimulation of the olfactory epithelium, located at the top of the nasal cavity. Research has been carried out using gas chromatography olfactometry to determine how many individual aroma compounds are present in a whisky. This instrument splits the whisky into its individual constituents, which are then evaluated to determine which ones are odour active. This has revealed that a typical malt whisky contains in excess of 80 such aroma congeners (Steele et al., 2004). The levels of these compounds are generally very low. Sensory detection thresholds are typically in the parts per million level, but can be as low as parts per trillion quantities (the equivalent of 1 drop in 20 Olympic-sized swimming pools). Due to the complexity of whisky aroma, it is not possible in this chapter to describe the origins of all odour-active compounds. Instead, examples from each of the four main production steps (raw materials, fermentation, distillation, and maturation) have been chosen to illustrate the diversity of the aromas present.

- *Raw materials*—The use of peated (smoked) malt provides a good example of how aromas associated with the raw materials can be carried through the production process into the final product. Use of this type of malt gives smoky, medicinal notes to the whisky, which are due to phenolic compounds originating from the smoke produced by the peat as it burns. Other aromas also come from the cereals, such as malty or biscuity notes, with different types of cereals giving different aromas due to variation in their composition.
- *Fermentation*—When yeast ferments sugars to ethanol, it also produces a number of secondary metabolites, such as esters, aldehydes, acids, and sulphur compounds. Many of these compounds are odour active, with the types of aromas produced depending on yeast strain, fermentation time and temperature, and other parameters. The presence of wild yeast or bacteria can contribute other congeners and hence give the spirit different flavour characters.
- *Distillation*—Not all of the aroma compounds originating from the raw materials and produced during fermentation end up in the spirit. Distillation not only separates water from ethanol but also selectively separates the other congeners. In broad terms, compounds with similar volatilities tend to end up in the spirit, whereas compounds with very high or lower volatilities do not. One exception to this is sulphur compounds. Although volatile, they react with copper in the still system, limiting the levels distilling into the spirit. Distillation does not just eliminate compounds; new aromas are also created. Heat is applied during distillation which causes aroma-forming reactions. Maillard reactions, for example, occur between amino acids and residual sugars, generating compounds with notes ranging from cereal, floral, and grassy aromas through to burnt and sulphury.
- *Maturation*—The aroma of the spirit alters considerably as it matures. The main sensory changes that occur are due to the extraction of odour-active compounds from the wood, with vanillin (vanilla aroma) and oak lactone (coconut aroma) being two of the most significant maturation-related congeners. Again, this is not the only change. Other aroma compounds are lost, evaporating from the cask during maturation as part of the 'angel's share'. Finally, reactions occurring between spirit components are also important. These tend to take place over a longer period of time, giving the flavour characters that can only be achieved with extended maturation.

As a result of their varied sources, the aroma compounds present in whisky are diverse in terms of their chemical properties and their sensory characteristics. To further complicate our understanding of whisky aroma, the relationship between levels of these congeners and sensory perception is not straightforward. Some aromas dominate, masking the perception of other compounds, while other congeners have synergistic effects, acting in combination to give an enhanced perception of a particular aroma note. The remaining sensory characteristics, other than aroma, are simpler at least in terms of the number and diversity of the congeners responsible.

Taste

Taste characteristics are detected by the taste buds on the tongue; the main attributes in this category are sweet, bitter, and sour. Sweet tastes are due to the presence of sugars that enter the spirit during maturation. For example, if a cask has previously held wine, low levels of residual sugars from the wine (e.g. fructose and glucose) can carry over into the spirit. Bitter tastes are also related to maturation, this time due to the presence of tannins that are extracted from the wood. Sour tastes are caused by the presence of acids. These are undesirable attributes in whisky, but, thankfully, the levels of acids are rarely high enough to have a detectable influence on taste.

Mouthfeel

Mouthfeel encompasses attributes such as mouthwarming, astringency, and mouthcoating. The perception of these attributes is associated with the stimulation of the mucous membranes, the soft tissues lining the inside of the nose, lips, mouth, and throat. Mouthwarming effects (namely mild burning or fiery sensations) are primarily due to the presence of alcohol. Astringency is related to compounds extracted from the cask. Again, tannins play an important role in this. Some whiskies give the perception of being more mouthcoating than others. As the compounds responsible for this effect are yet to be identified, more research is required in this area.

Visual characteristics

Colour is obviously an important visual attribute of whisky. The new spirit coming off the still is clear. The golden colour typically associated with whisky develops during maturation. Again, this is due to the extraction of compounds from the cask, although the exact nature of the colour-forming compounds is as yet unknown. The degree of colour development varies from one cask to another. Subsequently, most whisky legislation allows the addition of small amounts of caramel to adjust batch-to-batch colour differences. Other factors can also influence visual appearance. The main one of these is the presence of insoluble compounds, which have a detrimental impact on the clarity of a product, causing haze or cloudiness in the whisky. The compounds that give this haze (namely long-chain fatty acid esters) can be removed by chill filtration (discussed in Chapter 19).

Variations exist between whiskies in terms of all of the sensory characteristics outlined above. The previous chapters of this book have described the various types of whiskies and processes used in their production. Differences in flavour are created by altering the raw materials, fermentation and distillation practices, maturation parameters, and blending procedures described in Chapters 10–17. The result is a spirit category that exhibits an interesting range of diverse flavours and other sensory attributes.

Evaluation of sensory characteristics

This section describes the sensory evaluation of whisky. The focus is on the assessments carried out within the industry to manage sensory quality, although consumer perceptions are also considered.

Consumer assessments

When drinking a glass of whisky, a consumer will evaluate its sensory characteristics. Often these evaluations are subconscious, although sometimes whisky drinkers will discuss their perceptions, describing flavours in their own words. Although these assessments are often not clearly defined, they are of vital importance, as consumer perceptions ultimately determine whether a brand will be successful. Sensory descriptions are widely used in the marketing of whiskies, directing the consumer in terms of the types of flavours that they should expect to encounter. It is therefore critical that companies understand the sensory qualities that the consumer is looking for and carefully control these characteristics, making sure that individual brands are in line with the consumer's desires and expectations.

Measurement by instruments

In an ideal world, we would have an instrument (e.g. an electronic nose) that would be able to provide data on flavour development at each stage in whisky production, alerting the distiller to the emergence of any unusual or undesirable notes. Chapter 19 describes the compositional analysis of whisky. Such analyses have a wide range of useful applications but are limited when it comes to the measurement of flavour. We are, as yet, unable to measure some of the compounds responsible for aroma, taste, and mouthfeel effects, and we do not fully comprehend the interactions between the various sensory stimuli described previously. New techniques are being developed that, rather than examining composition, measure the response of biological receptors (biosensor-based technologies) to aromas. These are not currently sensitive enough to detect the subtle differences between whiskies. But, this is a rapidly progressing area of flavour research where we may expect to see developments in the future.

At present, although we can measure aspects of flavour, human perceptions are much more sensitive than any instruments currently available. As a result, the whisky industry relies on sensory analysis, the detailed evaluation of a product by trained assessors, to achieve the required control of flavour. Colour and clarity, on the other hand, can be determined analytically, and, in both cases, instruments have largely replaced traditional sensory assessments for the measurement of these attributes.

Sensory evaluation during whisky production and of the finished product

Because flavour develops during the production process, it is important to monitor the flavour changes on an ongoing basis. Generally, the first point at which a sensory assessment takes place is immediately after distillation. Each distillery will have its own target flavour characteristics, and evaluations are carried out to ensure that the newly distilled spirit exhibits the desired sensory attributes and has no off-notes.

Identification of problems at this stage allows remedial action to be taken, ensuring that further quantities of unsatisfactory spirit are not produced. Once approved, the spirit is filled into casks for maturation. Where the casks are being reused, the aroma of the empty cask may be assessed prior to filling. Growth of mould or bacteria in the cask between fills can result in off-notes, such as musty, sour, or solvently aromas, which can carry over into the spirit. Consequently, it is good practice to identify and eliminate any problem casks.

Maturation practices vary from company to company, with one option being to mature for a predetermined time interval. In this case, once the time period has elapsed, samples will be taken to ensure that the required flavour development has occurred. Alternatively, companies may opt to mature some or all of their stock for extended periods. Due to the longer maturation, these whiskies tend to be higher value, and as such, more detailed evaluation of flavour development is often undertaken. Samples may be taken at a number of time points and a full assessment of aroma and taste carried out. The results of these tests will be used as the basis to decide when the whisky has achieved a suitable quality for bottling. It is unusual for casks to be bottled individually, with blending, of some form, being common practice.

Sensory evaluation is again used to control the final quality of the product. To achieve this, the blender/blending team must have a good understanding of how the flavours of different whiskies interact when mixed. Further details of blending and the role of sensory evaluation in this process are given in Chapter 17. A final check of product quality is carried out in the bottling hall to ensure that there has been no cross-contamination with other products and to identify taints in this final stage of the process.

Many of the production-based tests described above are simple screening checks, which aim to determine whether a sample exhibits target sensory characteristics. The comprehensive evaluation of aroma, taste, and other sensory characteristics, though not routine, is also conducted. These types of sensory tests may be used to provide product information for marketing or promotional purposes. Thorough sensory assessments are also used for research purposes—for example, to determine the impact that a change in the raw materials or a production parameter has had on the sensory attributes of the whisky—or in product development where the aim is to develop a whisky with particular flavour attributes.

Type of assessment: Nosing versus tasting

Whisky contains relatively high levels of alcohol, which causes problems in the routine tasting of multiple samples, both in terms of alcohol intake and sensory fatigue. As a result, much of the sensory testing carried out in the whisky industry is based solely on the aroma of the product, a practice referred to as 'nosing'. The same organ, the olfactory epithelium, is involved in aroma detection regardless of whether the whisky is nosed or tasted. In nosing, the aroma compounds travel orthonasally, directly up the nose to the top of the nasal cavity where this organ is located. When tasting, the compounds reach the olfactory epithelium retronasally, travelling through the passage that links the back of the mouth with the nose, as shown in Fig. 18.1.

Although recent research has demonstrated that there may be difference in the intensity of perception between nosing and tasting, nosing provides the benefit of allowing all of the volatile flavour compounds to be evaluated without having to ingest any of the samples. Because these compounds cover by far the largest proportion of the overall sensory character of the product, encompassing the key flavours that differentiate one whisky from another and the compounds responsible for most taints and off-notes, nosing is widely used. As described previously, detection of taste and mouthfeel attributes requires consumption of the product. Taste tests are generally reserved for evaluation of the final product or other situations where these types of attributes are of particular importance.

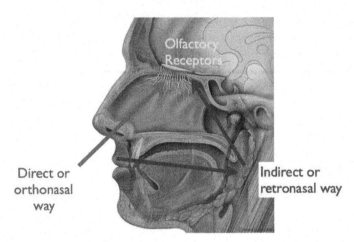

FIG. 18.1   Physiology of aroma perception.

Sensory assessors

Sensory evaluation is always carried out by trained assessors. Traditionally, within the whisky industry, each company has its own sensory expert, who has direct responsibility for the flavour quality of their products. The titles given to these experts vary from company to company, with Master Blender or Master Distiller being two of the more common designations. An individual requires years of experience in the sensory assessment of whisky to reach the required level of expertise, combined with in-depth knowledge of the process, an understanding of how flavours develop, and an appreciation for the sensory interactions that occur when different whiskies are combined. As companies have grown and whisky brands have become global commodities, flavour and, in particular, flavour consistency have often become too large a responsibility for just one person. Although the individual expert still plays an important role in most whisky companies, many have devolved full responsibility for flavour to a larger group of assessors: the sensory panel.

A sensory panel is a group of individuals who have been specifically trained to recognise, describe, and quantify the sensory characteristics of a product—in this case, whisky. The sensory panel provides data relating to aroma, taste, and other attributes and can be considered as the equivalent of an 'instrument' for measuring flavour. Again, this is a skill that cannot be learned overnight. After being screened for sensory aptitude, panellists require training in the differentiation of products and recognition of key sensory attributes. Once this is complete, they need to be exposed to a wide range of different samples and given time to further develop their skills. The level of experience required will depend on the type of sensory assessment being carried out. For example, this may be less for a panel giving a pass/fail grading, as opposed to a panel tasked with producing a detailed quantification of a full range of sensory attributes. Nonetheless, in both cases, skills need to be maintained, which requires frequent involvement in the sensory panel assessments.

Many companies regularly test the performance of their panellists. This can either be done by checking their ability to pick up off-note or unusual spirits, or by using samples that have been deliberately spiked with flavour compounds. Another option is to enrol the panel in a formal sensory proficiency scheme. Such schemes provide an independent assessment of individual panellists, as well as an evaluation of the overall performance of the panel in relation to the other panels participating in the scheme.

Within the normal range of sensitivity, individuals can vary significantly in their sensory perceptions. For example, an aroma that smells very strong to one person may only be weakly detected by another person and vice versa. Statistical analysis of sensory panel data can be carried out to produce a consensus opinion of the group's perceptions. This is one of the main benefits of using a panel, as an individual, due to low personal sensitivity, may on occasion overlook a particular attribute that is of importance to the consumer. Sensory panels also tend to have better availability than individual experts, where illness or other engagements can have an impact on the ability to give rapid turnaround on sample assessments. Sensory panels, however, do not have the same process knowledge as the expert. Consequently, a panel leader is required who is able to interpret and act on the data produced.

Sample preparation and presentation

The first stage in sensory evaluation is to prepare the sample. This is an important consideration, as the way that the sample is prepared and presented to the assessor has been shown to influence perceptions (Jack, 2003). This section describes best practice for sensory sample preparation, although it should be appreciated that this cannot always be fully achieved. As outlined previously, assessments can be carried out in very different environments, some of which are easier to control than others. Ideally, sensory tests are carried out where there are no distractions that prevent the assessor from focussing on the task in hand and there are no background interfering aromas. Panel assessments generally take place in dedicated sensory rooms, where temperature, lighting, and odour can be controlled. These rooms often have booths where each panellist can carry out his or her own independent assessments without conferring with or being influenced by other members of the group. However, some tests may have to be carried out under less ideal conditions, such as in the maturation warehouse or at the bottling plant.

A wide range of appropriate glasses are available on the market, some of which have been specifically designed for whisky evaluation. Tulip-shaped glasses, or similar, are best for the assessment of aroma, and it is advisable to cover the glass with a watch glass in order to trap the whisky vapours. When a whisky is being tasted, it may be more practical to use a straighter glass. Coloured glasses can also be used in situations where the visual attributes of the product may influence other perceptions; for example, a darker whisky may be perceived as having more mature flavours, such as vanilla notes, than a paler whisky. A defined quantity of whisky/spirit is transferred into the glass and a measured amount of water added (if required). The addition of water is particularly important when assessing high-strength samples, such as the new spirit collected immediately after distillation. Common industry practice is to dilute to around 20% alcohol by volume for nosing, because high levels of alcohol cause burning sensations that can interfere with the perception of aromas. Dilution has the added benefit of increasing the level of certain volatile compounds in the headspace, by forcing them out of solution (Conner et al., 1998). Thus, the addition of water can increase rather than reduce the aroma of the sample. Conversely, where taste or mouthfeel characteristics are being evaluated, it may be preferable not to add water, as doing so may reduce these particular sensations. On occasion, samples may be assessed both at an original strength and at a reduced strength to replicate how the product is typically consumed.

The temperature of the sample also influences flavour perception. Cold temperatures generally suppress the level of volatile aroma compounds in the headspace in the glass (Jack, 2003). Heat is naturally released in the exothermic reaction that occurs when water and alcohol are mixed, such as when a whisky is diluted (Costigan et al., 1980). When evaluating samples in cold environments (e.g. in maturation warehouses), it may be advisable to assess them directly after the addition of water rather than leaving them to cool.

Finally, the samples may be coded prior to being evaluated. This is typically carried out in the panel assessments, where the provision of sample information could potentially influence the responses obtained.

Sensory methodology

Sensory tests are time consuming; consequently, it is important to select a method that is fit for the purpose. In some cases, rapid feedback is required, such as when checking samples during different stages in production. In other cases, there will be advantages to be gained from taking time to carry out a comprehensive evaluation, such as when sensory information is being generated for promotional purposes. As described earlier, some of the most frequently used sensory tests in the whisky industry are simple screening tests that are used to determine if a sample has the desired or expected flavour characteristics. This will often involve the assessment of multiple samples in one session. An expert carrying out this type of test will have a clear idea of the types of flavours that are required. A sensory panel, on the other hand, may have less familiarity with the product. In this instance, panellists may be asked to compare each sample with a preselected reference sample that exhibits the desired flavour profile. These tests are rapid and simple to carry out and are often used in production situations where a quick turnaround is required.

Flavour consistency is fundamental to maintain brand identity. Difference tests, a particular type of sensory test, can be used not only to examine how consistent the final product is from batch to batch but also to determine whether or not there is any detectable sensory variation at other stages in production. This may be particularly important if a change has been made to either the raw materials or process parameters. There are a number of difference tests available, such as the difference from control test, tetrad test, or paired comparison test, but the one most widely used in the whisky industry is the triangle test (Lawless and Heymann, 1998). This involves presenting assessors with three coded samples, two of which are the same and one that is different, and asking them to identify the 'odd' sample. This is a forced-choice test, where the assessors are asked to pick one of the samples even if they cannot detect any differences. The samples are regarded as being different if the number of correct responses obtained is statistically higher than the 1-in-3 chance of selecting the odd sample. An International Standard (ISO, 2021) outlines the full details of how to carry out this type of test. If these procedures are followed, a triangle test will give an unbiased judgement of sample differences. Other difference tests are also available, such as the duo–trio test or the paired-comparison test. All of these tests involve multiple assessments of the samples by different assessors, as the results rely on a statistical interpretation of the response. So, they require the use of a sensory panel. The disadvantages of these types of test are that, although they can be used to determine if samples are different, they do not provide any information regarding the degree or nature of the difference. If this type of information is required, then further descriptive evaluations must be carried out.

The aim of descriptive sensory evaluation is to produce a comprehensive description of a whisky's characteristics that can be readily understood and interpreted by others. There are two approaches that can be used. The first is where an assessor describes the product using their own words. These descriptions can be very personal and as such may not be the best form of communication. The alternative is to use a predetermined and agreed-upon vocabulary. The need for such a vocabulary to define whisky was recognised and addressed back in the 1970s with the development of a whisky flavour wheel by industry experts and whisky researchers (Shortreed et al., 1979). This has evolved to become The Scotch Whisky Research Institute's Flavour Wheel (Fig. 18.2), which is now widely used in the industry.

The flavour wheel encompasses aroma, mouthfeel, and taste attributes and was designed to cover not only desirable sensory characteristics but also off-notes or undesirable flavours that can occasionally occur. Hence, it is a useful tool for both the detailed evaluation of final products and control of flavour quality during production. The use of a fixed vocabulary, such as the flavour wheel, aids communication. Flavours are grouped into categories on the wheel, making it easier to understand which types of flavours are similar or related. The wheel can be used in two ways. The first is to work from the inside of the wheel outwards. Looking at the central flavour category of 'Peaty', for example, assessors can work their way out to give a more detailed description of the peat-smoke-related flavours in the product, such as 'smoky', 'burnt', or 'medicinal'. Alternatively, assessors can work from the outside in to give a more general description of a very specific note. One example might be the detection of a 'fresh herbs' aroma, which would be categorised under 'Green/grassy'.

Even with a fixed vocabulary, it is still not always easy to arrive at agreement among assessors. This can be improved by using training samples that exhibit distinct flavour characteristics. Companies exist that produce flavour training standards that can be added directly to whiskies or other spirits. These standards tend to be individual flavour compounds, which are used at recognisable, but not overpowering, levels. Some sensory attributes are relatively easy to define using single compounds; for example, a solution of glucose can be used to define sweet tastes or phenolic compounds to represent peat-smoke-related aromas. Other attributes are less easy to represent, particularly where the congeners responsible are yet to be identified or the aroma is due to an interaction of several compounds. In this case, the production samples may provide a useful alternative. 'Feinty' aromas, for example, cannot be readily reproduced chemically, but samples from the late stages of a batch distillation run are good training aids. Unfortunately, it is not always easy to obtain production samples with single dominant aromas that are suitable for descriptive sensory training.

After training, assessors generally require future experience to be able to produce good sensory descriptions. Discussion of perceptions with others may be useful and is often carried out to obtain a consensus description of flavour. It is important to bear in mind the point made earlier in this chapter that not all people perceive flavours in the same way. Differences in the descriptions produced by two individuals may be due to different sensitivities to the congeners present, rather than one person being able to provide a more accurate description than the other.

FIG. 18.2   The Scotch Whisky Research Institute's Flavour Wheel.

Descriptive sensory evaluation can be used to produce detailed summaries of appearance, aroma, taste, and other characteristics. This type of information is widely used for sales and marketing purposes. There are situations where a simple description is not sufficient. In certain circumstances, it may be useful to obtain a more in-depth quantification of sensory attributes, showing the relative levels and balance of flavours in the product. In other instances, it may be useful to compare the relative intensities of sensory attributes between samples.

Quantifying the intensities of flavour attributes requires the use of a scaling system. There are a number of different scale types and methods that can be used. The most common of these is quantitative descriptive analysis (QDA) (Stone et al., 1974). This technique requires the use of a sensory panel and involves panellists individually scoring the intensities of a number of predetermined attributes. An example of a scoring sheet used for QDA is shown in Fig. 18.3. When all of the panellists have assessed the samples, an average score is calculated for each attribute. These average scores can be tabulated or presented in graphical format as a flavour profile (also known as a spider diagram), an example of which is shown in Fig. 18.4. Flavour profiles can be compared by plotting more than one sample on the same chart. Statistical analysis of the data is required in order to conclude whether there is a difference between the

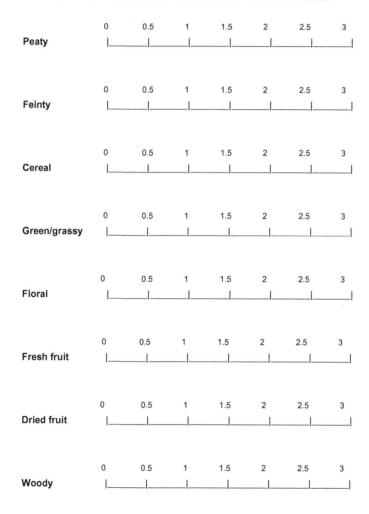

FIG. 18.3   Quantitative descriptive analysis scoresheet.

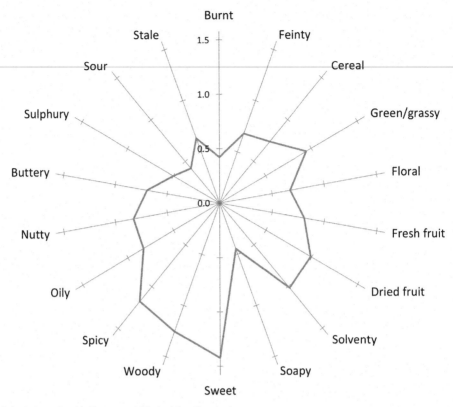

FIG. 18.4   Example of a flavour profile (spider diagram).

samples in terms of any of the attributes. A *t*-test is used to compare two samples, or analysis of variance for multiple samples. Where statistically significant differences are found, this can be followed up by a multiple comparison test, such as Tukey's HSD test, to determine which samples differ from one another.

If you have several samples, 6 or more, then you can use a multivariate statistical approach, such as principal component analysis, to summarise the sample differences. Using this analysis, you can 'map out' the relationships between your samples in relation to one another and identify the differences or similarities between samples in terms of flavour.

QDA and related methods are relatively time consuming, in terms of not only the time taken to evaluate the samples but also the time required for panel training. Rapid sensory profiling methods are available (Delarue et al., 2015). One such technique is Napping. This involves panellists arranging samples on a large piece of paper according to flavour, with similar samples placed near each other and dissimilar ones farthest away. Panellists are also asked to provide a description of each sample or group of samples. The placement of the samples is recorded by determining the XY coordinates of each on the paper. Multivariate statistical analysis, namely multiple factor analysis, is used to produce a consensus plot (map) of the samples over which the descriptions can be laid. Napping has the advantage of being much more rapid than QDA and is particularly useful when evaluating larger sample sets.

The other advantage is that it can be carried out by untrained assessors, as individuals use their own criteria to arrange the samples and their own words to describe them. The downside of this approach is that it does not give the flavour profiles (spider diagrams), which can only be generated using the more standard profiling techniques.

Benefits of sensory evaluation

Sensory evaluation continues to play an important role in the whisky industry. This is the main way in which flavour quality is controlled and by which companies ensure that their brands match consumer needs. It is crucial that whisky producers recognise its importance and commit the required resources to this function. This chapter has outlined the aspects that need to be considered. The whisky producer must identify the key points in the process that are important from a flavour control point of view and routinely monitor product quality at these stages. Suitably trained and experienced staff must be available to carry out these assessments and appropriate sensory methods used, with consideration being given to sample preparation, presentation, and the testing environment. Sensory evaluation can be time consuming but, when used correctly, it provides an understanding and control of the key attributes that ultimately govern the success of a whisky in the marketplace.

References

Conner, J.M., Birkmyre, L., Paterson, A., Piggott, J.M., 1998. Headspace concentrations of ethyl esters at different alcoholic strengths. J. Sci. Food Agric. 77, 121–126.

Costigan, M.J., Hodges, L.J., Marsh, K.N., Stokes, R.H., Tuxford, C.W., 1980. The isothermal displacement calorimeter: design modifications for measuring exothermic enthalpies of mixing. Aust. J. Chem. 33 (10), 2103–2119.

Delarue, J., Lawlor, B.J., Rogeaux, M., 2015. Rapid Sensory Profiling Techniques and Related Methods: Applications in New Product Development and Consumer Research. Woodhead Publishing, UK.

ISO, 2021. ISO 4120: Sensory Analysis—Methodology—Triangle Test. International Organization for Standardization, Geneva, Switzerland, Switzerland.

Jack, F.R., 2003. Development of guidelines for the preparation and handling of sensory samples in the Scotch whisky industry. J. Inst. Brew. 109 (2), 114–119.

Lawless, H.T., Heymann, H., 1998. Sensory Evaluation of Food. Kluwer Academic, New York.

Shortreed, G.W., Rickards, P., Swan, J., Burtles, S., 1979. The flavour terminology of Scotch whisky. Brewers Guardian, 2–6. November.

Steele, G.M., Fotheringham, R.N., Jack, F.R., 2004. Understanding flavour development in Scotch whisky. In: Bryce, J.H., Stewart, G.G. (Eds.), Distilled Spirits—Tradition and Innovation. Nottingham University Press, Nottingham, UK, pp. 161–167.

Stone, H., Sidel, J.L., Oliver, S., Woolsey, A., Singleton, R.C., 1974. Sensory evaluation by quantitative descriptive analysis. Food Technol. 28, 24–34.

19

Whisky analysis

Ross Aylott

Aylott Scientific, Dunblane, United Kingdom

Introduction

Our analytical knowledge of whiskies has expanded rapidly in recent decades, beginning in the 1960s with the introduction of gas chromatography (Duncan and Philp, 1966; Singer and Styles, 1965), gas chromatography–mass spectrometry (GC–MS) (Kahn, 1969), and then high-performance liquid chromatography (HPLC) (Lehtonen, 1983a) and ion-exchange chromatography. Hundreds of congeners are known to be in whiskies, and these include alcohols, acids, esters, carbonyl compounds, phenols, hydrocarbons, and nitrogen- and sulphur-containing compounds (Swan et al., 1981). Whiskies distilled on pot stills are richer in congeners than those made with greater rectification on continuous stills. Congener concentrations typically range from the high parts per million (mg/L) to low parts per billion (μg/L).

Congeners are natural constituents of the production process. Some clearly contribute sensory character and others do not. However, together all of the congeners help make each whisky unique. The presence of each congener is associated with specific parts of the process—for example, the cereals used and the fermentation, distillation, and maturation processes. These processes affect the concentration of each congener in the final product. Congeners such as higher alcohols formed in fermentation may be reduced in concentration and even eliminated by rectification during distillation. Ethyl esters formed in fermentation may increase in concentration during maturation. Wood-related congeners, not present at the end of distillation, are extracted from casks into the liquid during maturation.

Whiskies of the world and their regulatory definitions

Whisky in the distilled spirits sector

Whisky definitions are found in the laws of most countries. These laws, plus the many other regulations concerned with distilled spirits help protect the interests of the consumer, protect the business of the manufacturer, and protect the tax revenue of the state (Aylott, 2013; Goodall et al., 2018). The most important definitions are those of the countries where a particular whisky originates. Many other countries either refer to the definition of the home country in their own regulations or use words and phrases taken from the home definition. The following sections review definitions for whisky/whiskey and specifically whiskies in their countries of origin.

Whisky in the European Union

European Union Commission Regulation No. 110/2008 defines whisky/whiskey along with another 46 categories of spirit drinks and supersedes earlier Commission Regulation 1576/89/EEC (EEC, 1989). This regulation continues to apply to the UK post-BREXIT while intergovernmental negotiations continue during 2020 (European Union Withdrawal Act 2019). Regulation No. 110/2008 states:

(a) Whisky *or* whiskey *is a spirit drink produced exclusively by:*
  (i) *distillation of a mash made from malted cereals with or without whole grains of other cereals, which has been:*
    — *saccharified by the diastase of the malt contained therein, with or without other natural enzymes,*
    — *fermented by the action of yeast;*
  (ii) *one or more distillations at less than 94.8% vol., so that the distillate has an aroma and taste derived from the raw materials used;*
  (iii) *maturation of the final distillate for at least 3 years in wooden casks not exceeding 700L capacity.*

The final distillate, to which only water and plain caramel (for colouring) may be added, retains its colour, aroma, and taste derived from the production process referred to in points (i), (ii), and (iii).

(b) *The minimum alcoholic strength by volume of* whisky *or* whiskey *shall be 40%.*
(c) *No addition of alcohol as defined in Annex I(5), diluted or not, shall take place.*
(d) *Whisky or whiskey shall not be sweetened or flavoured, nor contain any additives other than plain caramel used for colouring.*

Annex III of Regulation 110/2008 lists five geographical designations for the major whiskies produced in the EU that cover 'Scotch Whisky', 'Irish Whiskey', 'Whisky Espanol', 'Whisky de Bretagne', and 'Whisky d'Alsace' (the latter two are both from France). National regulations described later refine the definitions for Scotch and Irish whiskies.

Whisky in the United States

The standards of identity for whisky, along with other distilled spirits, are documented in the U.S. Code of Federal Regulations (27 CFR 5.22—The standards of identity) (Code of Federal Regulations, 1998a):

Whisky is an alcoholic distillate from a fermented mash of grain produced at less than 190° proof in such a manner that the distillate possesses the taste, aroma, and characteristics generally attributed to whisky, stored in oak containers and bottled at not less than 80° proof.

Although the standard for whisky is straightforward, the standards go on to define the various types of whiskey produced in the United States, and these are discussed later.

The U.S. standards of identity also define Scotch, Irish, and Canadian whiskies by making reference to compliance with the specific laws in each of the producing countries; further details can be found in Chapter 7.

Scotch whisky

Scotch whisky is made in Scotland from cereals, yeast, and water only. There are two types of Scotch whisky: malt whisky and grain whisky. Malt Scotch whisky is fermented from malted barley and the resulting alcohol concentrated in a batch distillation in copper pot stills. Grain Scotch whisky is fermented from a mash containing other nonmalted cooked cereals (such as barley, wheat, or maize) and a smaller proportion of malted barley. The resulting alcohol is concentrated in a continuous distillation process, the most common design being the Coffey or patent still. Both malt and grain new-make spirits are then reduced with water to approximately 65%–70% alcohol by volume (abv) and matured in oak casks for at least 3 years. At the end of maturation, the whisky is either blended or taken as a single distillery product and reduced further with water for bottling at a minimum 40% abv.

Scotch whisky is defined under U.K. law in the Scotch Whisky Regulations 2009 (SWA, 2009). This recent regulation supersedes and sets tighter standards compared to the earlier Scotch Whisky Order 1990 made under the Scotch Whisky Act, 1988 (HMSO, 1990). In addition, the Scotch Whisky Technical File has been published (DEFRA, 2013). It includes much more information than the 2009 regulations and is used by U.K. Revenue and Customs (HMRC) to ensure compliance with the main requirements for Scotch Whisky and is essential for maintaining Scotch Whisky's Geographical Indication status. The technical file has been lodged with the European Commission by the U.K. Department for the Environment, Food, and Rural Affairs (DEFRA) and is now a law. A 2019 amendment to the file gave specific guidance on the range of casks that can be used to mature or finish Scotch Whisky.

Both the U.K. and European Union regulations define the process but not analytical parameters (with the exception of the maximum distillation strength and minimum bottling strength). The Scotch Whisky Regulations 2009 are read as follows:

In these Regulations, 'Scotch Whisky' means a whisky produced in Scotland—

(a) *that has been distilled at a distillery in Scotland from water and malted barley (to which only whole grains of other cereals may be added) all of which have been—*
  (i) *processed at that distillery into a mash;*
  (ii) *converted at that distillery into a fermentable substrate only by endogenous enzyme systems; and*
  (iii) *fermented at that distillery only by the addition of yeast;*
(b) *that has been distilled at an alcoholic strength by volume of less than 94.8% so that the distillate has an aroma and taste derived from the raw materials used in, and the method of, its production;*
(c) *that has been matured only in oak casks of a capacity not exceeding 700L;*
(d) *that has been matured only in Scotland;*
(e) *that has been matured for a period of not less than 3 years;*

(f) *that has been matured only in an excise warehouse or a permitted place;*

(g) *that retains the colour, aroma, and taste derived from the raw materials used in, and the method of, its production and maturation;*

(h) *to which no substance has been added, or to which no substance has been added except—*

    (i) *water;*

    (ii) *plain caramel colouring; or*

    (iii) *water and plain caramel colouring; and*

    (iv) *that has a minimum alcoholic strength by volume of 40%.*

    The regulations define five categories of Scotch whisky along with locality and region geographical indications:

The categories are

(a) *Single Malt Scotch Whisky;*

(b) *Single Grain Scotch Whisky;*

(c) *Blended Malt Scotch Whisky;*

(d) *Blended Grain Scotch Whisky; and*

(e) *Blended Scotch Whisky.*

The protected localities are 'Campbeltown' and 'Islay', and the protected regions are 'Highland', 'Lowland', and 'Speyside'.

Irish whiskey

Like Scotland, Ireland also makes malt and grain whiskies using mixtures of malted and unmalted cereals but, unlike Scotch whisky, exogenous amylolytic enzymes can be employed in mashing. Irish malt whiskies are triple distilled in three pot stills, and Irish grain whiskies are continuously distilled on three-column systems, making them lighter in character than many of their Scottish equivalents. Like Scotch whisky, the minimum maturation period for Irish whiskey is 3 years. Production and maturation may take place in the Irish State or Northern Ireland. While the industry had consolidated in Cork and near Dundalk (in Eire) and in Bushmills (in Northern Ireland), many new distilleries have begun production and are planned. Legal definitions may be found in the Irish Whiskey Act 1980, and its geographical designation is listed in Annex III of EU Regulation 110/2008. Further details can be found in Chapter 9.

American whiskies

The standards of identity for distilled spirits are documented in the U.S. Code of Federal Regulations as described earlier. In particular, bourbon whiskey is.

Whisky produced in the U.S. at not exceeding 80% alcohol by volume (160° proof) from a fermented mash of not less than 51% corn and stored at not more than 62.5% alcohol by volume (125° proof) in charred new oak containers.

Whiskies stored for a period of 2 or more years are further designated as 'straight'—for example, 'straight bourbon whiskey'. Tennessee whiskey is straight bourbon produced in the state of Tennessee. Further details are provided in Chapter 7.

Canadian whisky

Canadian whisky is defined in Canada under the Food and Drugs Act *(1993).*
Canadian Whisky, Canadian Rye Whisky and Rye Whisky

(a) *shall*
   (i) *be a potable alcoholic distillate, or a mixture of potable alcoholic distillates, obtained from a mash of cereal grain or cereal grain products saccharified by the diastase of malt, or by other enzymes and fermented by the action of yeast or a mixture of yeast and other microorganisms.*
   (ii) *be mashed, distilled, and aged in small wood for not less than 3 years in Canada.*
   (iii) *possess the taste, aroma, and character generally attributed to Canadian whisky, and*
   (iv) *be manufactured in accordance with the requirements of the Excise Act and the regulations made there under; and*
(b) *may contain caramel and flavouring.*

Other countries

Other countries follow a similar regulatory theme. Japan's regulations (see Chapter 2) for whisky contain requirements for cereals and minimum distillations strengths. China's standard (PRC, 2008) also requires cereals and defines whisky as 'a distilled spirit made from malt and grains through saccharifying, fermentation, distilling, aging and blending'. Australia and New Zealand also require whisky to be made from cereal grains. (See Chapter 6 for details on Indian whisky production and use of sugarcane.)

Composition and analysis of whisky

As already noted, whiskies contain many hundreds of congeners (Kahn, 1969; Maarse and Visscher, 1985a, 1985b; Swan et al., 1981) and are analysed for a wide range of purposes from research and process improvement to routine production quality assurance. Gas and liquid chromatographies are key to this knowledge and offer advantages in selectivity, sensitivity, and speed over many of the older colorimetric and titrimetric methods that they replaced. Most of the well-known chromatographic detectors have a role, from flame ionisation to mass spectrometric detection for gas chromatography and from ultraviolet to fluorescence detection for liquid chromatography. Sample preparation has been minimised, and direct injection of the whisky sample in the presence of an appropriate internal standard is used wherever possible.

However, more recently, spectroscopic techniques, such as UV/visible, near infrared, and Raman, are offering fast and precise analyses without the need for sample distillation and other time-consuming methods prior to analysis. These techniques find applications in alcoholic strength measurement, raw material, and authenticity analyses. This section reviews the current methods available, the context in which they are used, and typical results obtained.

Alcoholic strength measurement

Alcoholic strength measurement is the key quantitative measurement in whisky production, as it helps measure the efficiency of the overall carbohydrate conversion process, determines the excise tax paid to government, and ensures that consumers receive the alcohol declared on the product label. Alcoholic strength is normally defined as the concentration of ethyl alcohol in solution by volume at a specific temperature. Alcohol concentration is normally expressed as a percentage or as degrees proof, where pure alcohol at 100% by volume is equivalent to 200° U.S. proof. Methods for alcoholic strength determination are published in *Official Methods of Analysis of AOAC International* (Horwitz, 2000) and in the European Union Regulations No. 2870 (2000) on 'Community reference methods for the analysis of spirit drinks'. Two versions of the measurement are used: real and apparent alcoholic strength.

Real alcoholic strength is the percentage alcohol by volume in the distillate of a test sample, where all dissolved solids are removed. This gives the true result. Apparent alcoholic strength is the percentage alcohol by volume in a test liquid without prior distillation, and the true result is obscured by dissolved solids. The presence of dissolved solids increases the density of the sample and results in an apparent alcoholic strength measurement that is lower in value than the corresponding real alcoholic strength measurement. The difference between real and apparent alcoholic strength is known as the 'obscuration':

Obscuration (% vol.) = Real alcoholic strength−Apparent alcoholic strength

The amount of dissolved solids in most whiskies is low. Obscuration is normally <0.2% vol. in Scotch whisky; therefore, distillation is not normally required before taking a strength measurement for revenue or trading purposes. The U.S. Alcohol, Tobacco, and Firearms (ATF) regulations require sample distillation prior to alcohol strength measurement when dissolved solids exceed 600mg/100mL. When there are between 400 and 600mg dissolved solids per 100mL, the obscuration is added to the determined proof (Code of Federal Regulations, 1998b).

Alcoholic strength measurement is usually based on the density of the solution. Pure ethanol has a density of 789.24kg/m^3 at 20°C, whereas water has a density of 998.20kg/m^3 at 20°C. Ethanol/water mixtures will have a density between these two values; the lower the density, the higher the alcoholic strength. The three main techniques based on liquid density involve the use of hydrometers, pycnometers, and electronic densitometers.

Hydrometers have traditionally been the method of choice for alcoholic strength measurement in the production environment. The hydrometer is a calibrated weighted device that floats in the test liquid to a depth corresponding to the density of the liquid. The reading is taken where the liquid intersects the hydrometer scale. The measurement is normally precise to up to ±0.1% vol., depending on the hydrometer used and the volume of liquid under test.

More precise alcoholic strength measurements are obtained using pycnometers and electronic density meters. A pycnometer is a small (usually Pyrex glass) vessel, typically of 100-mL volume, which is used to determine the specific gravity or density of the test liquid. Alcoholic strength is then read from tables that give alcoholic strength by volume as a function of the density of the water/alcohol mixture. An international table has been adopted by the International Legal Metrology Organization (OIML) for this purpose. The procedure is highly skilled and slow, but is precise to at least ±0.1% vol. Accurate temperature control and measurement are required.

During the 1980s, electronic densitometry became the preferred method for alcoholic strength determination in laboratories, larger distilleries, and bottling plants. Measurement is based on the resonant frequency oscillation of the sample in an oscillation cell. The oscillation frequency of the cell is modified by the added mass of the test sample, and this can be used to determine its density and corresponding alcoholic strength. Small test volumes are required (<5mL), and precision up to $\pm0.1\%$ vol. is possible. Once calibrated, instruments are stable and can be automated. Commercial instruments are made by Anton Paar GmbH (Graz, Austria), Kyoto Electronics Manufacturing Co., Ltd. (Kyoto, Japan) and others.

Near-infrared spectroscopy is also applied to the real alcoholic strength measurement of obscured liquids. Commercial instruments include Infratec (Foss, Denmark) and Anton Paar GmbH (Graz, Austria). This approach eliminates the need for distillation prior to analysis and is particularly useful for alcoholic strength measurement in obscured liquids such as whisky-based liqueurs and lower strength ready-to-drink formulations.

Finally, gas chromatography may be used to separate the test liquid with a specific peak for ethanol, from which a value for real alcoholic strength can be calculated. Results are typically precise to $\pm1\%$ vol. The method can be automated but it is not precise enough for revenue purposes.

Major volatile congeners

After ethanol, the major congeners in whisky are the higher alcohols, namely *n*-propanol, isobutanol, and 2- and 3-methylbutanols. Considerable useful work on these congeners in various distilled spirits has been published (Duncan and Philp, 1966; Kahn, 1969; Shoeneman and Dyer, 1973). The gas chromatographic analysis for these congeners usually also includes separation of methanol, acetaldehyde, and ethyl acetate. All of these congeners are formed during fermentation, and their final concentrations are influenced by the distillation and blending processes. The major volatile congeners are analysed at the mg/L (ppm) level in one direct-injection gas chromatographic separation, using a polar stationary phase on a packed or capillary column (Martin et al., 1981). Common internal standards include 1- or 3-pentanol, and the method has been validated by comprehensive interlaboratory trials (Kelly et al., 1999).

A typical higher alcohol chromatogram for blended Scotch whisky is shown in Fig. 19.1 (using Carbowax 20M on Carbopak B as stationary phase), and the quantitative results are given in Table 19.1. The results show that Scotch grain whisky from the continuous Coffey still distillation contains very few congeners eluting after isobutanol, but the Scotch malt whisky from a double pot still distillation is much richer in the less volatile congeners that elute after isobutanol. The less volatile congeners, particularly 2- and 3-methylbutanol, are eliminated from grain spirit in the rectifier section of the Coffey still and are recovered as fusel oils. Blended Scotch and Irish whiskies have congener profiles representing the various malt and grain whiskies used in their blends. Bourbon whiskey is very rich in congeners, as there is little if any rectification in the bourbon distillation process beyond the beer still in the doubler. Canadian whiskies, being blends of grain neutral spirits and a bourbon-style distillate, have relatively low congener concentrations compared to the other

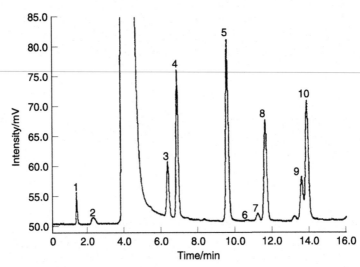

FIG. 19.1 Typical higher alcohol gas chromatogram for blended Scotch whisky. Peaks: 1, acetaldehyde; 2, methanol; 3, ethyl acetate; 4, propanol; 5, isobutanol; 6, butanol; 7, diethylacetal; 8, internal standard; 9, 2-methylbutanol; 10, 3-methylbutanol. Reproduced with permission of Royal Society of Chemistry from Aylott, R.I., Clyne, A.H., Fox, A.H., Walker, D.A., 1994. Analytical strategies to confirm the authenticity of Scotch whisky. Analyst 119, 1741–1746.

whiskies examined. Other trace congeners that may be detected in this analysis also include *n*-butanol and acetic acid.

The normal concentration unit used in whisky congener analysis is g/100L absolute alcohol (LAA). Thus, congener concentration is related to the volume of pure alcohol and not simply the volume of sample liquid. This concentration unit at first appears strange, but it proves very useful in that it can relate to a whisky that will occur at many different alcoholic strengths during its manufacture. For example, a Scotch grain spirit may be distilled at approximately 94.6%, matured at 65%, blended, diluted, and finally bottled at 40% vol.

The major volatile congener analysis is the most widely used gas chromatographic method applied to whisky and finds applications throughout the life cycle of the product. For example, it may be used to monitor the efficiency of rectification in the continuous stills used to distil Scotch grain spirit and Canadian grain neutral spirits (GNS), in competitor product analysis to determine the percentage malt whisky in a blended Scotch whisky, and in consumer protection activity to confirm brand authenticity. The spectrophotometric and titrimetric methods from the prechromatography era (Shoeneman and Dyer, 1973; Singer and Styles, 1965) are now all but redundant and only occasionally required for use in supporting dated regulatory requirements. When total esters are determined by these older methods, over 50% of the total ester content is ethyl acetate. Similarly, the major volatile acid is attributable to acetic acid. If required, methods for total aldehydes, esters, fusel oil, furfural, colour, and extract can be located in *Official Methods of Analysis of AOAC International* (Horwitz, 2000) and *Methods for the Analysis of Potable Spirits* (Research Committee on the Analysis of Potable Spirits, 1979).

TABLE 19.1 Major volatile congener concentrations (g/100 LAA) in single samples of Scotch malt, grain, and blended Scotch whiskies and Irish, American, and Canadian whiskies.

Whisky type	Acetaldehyde	Methanol	Ethyl acetate	n-Propanol	Isobutanol	2-Methylbutanol	3-Methylbutanol	2- and 3- Methylbutanol	Ratio 1	Ratio 2
Scotch malt whisky	17	6.3	45	41	84	46	130	176	2.2	2.8
Scotch grain whisky	12	8.5	18	72	68	6	17	23	0.3	2.8
Scotch blended whisky	5.4	8.9	23	55	62	19	53	72	1.2	2.8
Irish whiskey	4.1	10	13	28	15	13	36	49	3.1	2.8
Kentucky bourbon whiskey	15	17	89	28	160	104	281	385	2.4	2.7
Canadian whisky	3.3	7.9	7.1	6.2	6.9	5	11	16	2.3	2.2

Ratio 1 = 2- and 3-methylbutanol/isobutanol.
Ratio 2 = 3-methylbutanol/2-methylbutanol.

Trace congeners

The next major groups of congeners are longer chain alkyl alcohols, aldehydes, and esters at low and sub-mg/L levels. A few of these congeners are present at higher concentrations and can be determined by direct injection on a packed temperature programmed polar column (such as Carbowax® 20M on Chromosorb B). However, the best separations are achieved on polar capillary columns using an organic solvent extract of the test liquid. Chlorinated fluorocarbons (CFCs) were ideal extracting solvents until their withdrawal from use on environmental grounds. Pentane is also a good extracting solvent (with a 10 μL sample extracted into 1mL of pentane). A capillary column chromatogram of a typical blended Scotch whisky extract is shown in Fig. 19.2.

Capillary column systems can use both vaporising splitless and on-column injection. Solid-phase microextraction (SPME) offers a straightforward method of sample preparation prior to capillary column analysis (Lee et al., 2001). These high-resolution separations are often linked to mass spectrometric detection, with the mass spectrometer operating in an electron

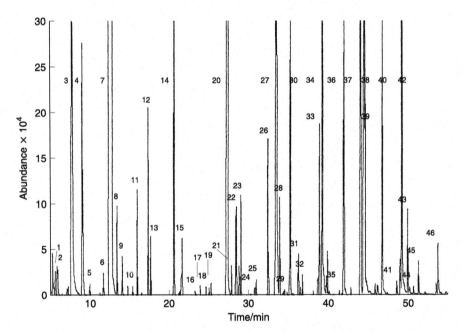

FIG. 19.2 A capillary column gas chromatogram of a typical blended Scotch whisky extract. Peaks: 1, ethyl butanoate; 2, propanol; 3, isobutanol; 4, isoamyl acetate; 5, butanol; 6, 3-ethoxypropanol; 7, isoamyl alcohol; 8, ethyl hexanoate; 9, diethoxypropane; 10, furfurylformate; 11, triethoxypropane; 12, ethyl lactate; 13, hexanol; 14, ethyl octanoate; 15, furfural; 16, acetylfuran; 17, benzaldehyde; 18, ethyl nonanoate; 19, ocatanol; 20, ethyl decanoate; 21, isoamyl octanoate; 22, diethyl succinate; 23, internal standard; 24, ethyl undecanoate; 25, decanol; 26, b-phenyl ethyl acetate; 27, ethyl dodecanoate; 28, isoamyl decanoate; 29, trans-oak lactone; 30, b-phenyl ethanol; 31, cis-oak lactone; 32, dodecanol; 33, ethyl hexadecenoate; 38, decanoic acid; 35, ethyl pentadecanoate; 36, tetradecanol; 37, ethyl hexadecenoate; 34, octanoic acid; 35, ethyl pentadecanoate; 36, tetradecanol; 41, ethyl octadecanoate; 42, dodecanoic acid; 43, ethyl octadecenoic acid; 44, vanillin; 45, b-phenyl ethyl butanoate; 46, tetradecanoic acid. Reproduced with permission of Royal Society of Chemistry from Aylott, R.I., Clyne, A.H., Fox, A.H., Walker, D.A., 1994. Analytical strategies to confirm the authenticity of Scotch whisky. Analyst 119, 1741–1746.

impact mode. The technique was applied to both Irish and Scotch whiskies (Fitzgerald et al., 2000). Fourier transform ion cyclotron resonance mass spectrometry (FT-ICR MS) was used to reveal the complexity of Scotch whisky by high resolution mass spectrometry (Kew et al., 2016). This technique offers a potential tool for investigating the chemistry of maturation processes.

Sensory analysis is covered in Chapter 18. Many trace congeners have been associated with specific flavour characteristics (Lee et al., 2001), and the methods described in this chapter are relevant to their analyses. Interestingly the impact of bromophenols in Islay Scotch whiskies at ng/L levels was reported (Bendig et al., 2014).

Maturation congeners

Whisky maturation at the molecular level involves processes of congener addition, congener reduction, and congener production (Philp, 1986). During these processes, the pungent and harsh characteristics of new-make spirit (new distillate) diminish, and the smoother more complex character of mature whisky develops.

Scotch and Irish whisk(e)y maturation casks are normally made from Spanish and American oak. Bourbon whiskey casks are made only from new charred American oak and are used only once. This results in a flourishing market for used American oak barrels, with the Canadian and Scotch whisky industries being major buyers. Maturation warehouses in Scotland and Ireland tend to be cool all year around, while those in the United States can become hot in the summer and cold in the winter. Some Bourbon whiskey companies heat their warehouses in the winter to accelerate their maturation process. As the character of the spirit develops, the volume in each cask decreases by 1%–2% per year due to evaporation (known by enthusiasts as 'the angels' share').

The process of congener addition is one of extraction from the oak cask into the liquid through interaction of ethanol with wood lignin (Reazin, 1979). The resulting congeners are known as lignin degradation products (LDPs). Polyphenolics are determined by high-performance liquid chromatography (HPLC) (Lehtonen, 1983a, b), and oak lactones are determined by capillary column gas chromatography. New-make whiskies are clear liquids prior to the start of maturation, but they acquire varying amounts of colour and particles of charred wood from the oak casks during the maturation process.

Direct-injection gradient elution reversed-phase HPLC with ultraviolet and/or fluorescence detection is the most commonly used type of analysis within the maturation process (Aylott et al., 1994). This analysis can be used to quantify the LDPs, which include gallic acid, vanillic acid, syringic acid, vanillin, syringaldehyde, coniferaldehyde, ellagic acid, and scopoletin, as well as furfural (found in Scotch malt spirit) and 5-hydroxymethylfurfural (which is mainly associated with spirit caramel). A typical chromatogram is shown in Fig. 19.3.

The congener reduction process involves volatilisation (such as the loss of dimethylsulphide), adsorption onto the charred wood, and oxidation of carbonyl compounds. The congener production process involves oxidation to form acetals, esterification to form more esters, and hydrolysis to form quinones.

Sulphur-containing congeners such as the volatile dimethylsulphide (DMS), dimethyldisulphide (DMDS), and less volatile dimethyltrisulphide (DMTS) may be determined by temperature-programmed capillary column gas chromatography with flame photometric

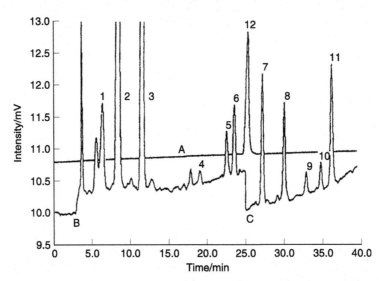

FIG. 19.3 A high-performance liquid chromatogram of cask-related and other congeners in blended Scotch whisky with (A) fluorescence detection (excitation at 345nm and emission at 450nm), and ultraviolet detection at (B) 280nm and (C) 340nm. Peaks: 1, gallic acid; 2, 5-hydroxymethylfurfural; 3, furfural; 4, vanillic acid; 5, syringic acid; 6, vanillin; 7, syringaldehyde; 8, internal standard; 9, coniferaldehyde; 10, synapaldehyde; 11, ellagic acid; 12, scopoletin. Reproduced with permission of Royal Society of Chemistry from Aylott, R.I., Clyne, A.H., Fox, A.H., Walker, D.A., 1994. Analytical strategies to confirm the authenticity of Scotch whisky. Analyst 119, 1741–1746.

detection (Beveridge, 1990). DMTS is an important flavour compound in Scotch malt whiskies. These ultra-trace compounds are concentrated from the test sample by a dynamic headspace concentration technique onto Chromosorb® 101 prior to thermal desorption, chromatographic separation, and detection at the low µg/L level.

Whisky age

The concentrations of oak-derived congeners in a given cask of whisky increase with maturation time. It is, therefore, possible to start to draw a graph plotting maturation congener concentrations against age. However, there are a great variety of barrel types in use from the new charred oak used in the bourbon whiskey process (which will produce relatively high concentrations of addition maturation congeners) to the second or third refill cask used for a Scotch grain whisky (which will result in less congener addition). Casks may also be rebuilt and rejuvenated. This means that age determination on the basis of maturation congener concentrations is imprecise. However, near-infrared reflectance (NIR) was used as a predictive tool for Canadian whisky ageing (Livermore, 2012). Furthermore, the chromatographic profiles for maturation congeners are consistent and can prove useful in checking false authenticity claims in spurious products where the use of wood flavour additives is suspected.

Natural ^{14}C in atmospheric carbon dioxide is absorbed by metabolism into all plants, including the cereals used for whisky manufacture. The natural levels of ^{14}C increased between 1945 and 1960 (corresponding to the start and end of atmospheric nuclear testing) (Baxter and

Walton, 1971) and have now fallen back to near the pre-1945 levels. Analysis of the ^{14}C levels in the ethanol concentrated from whisky samples has been used to estimate the year in which the cereal was grown and then relate this year to age. Analytical precision is limited and this obviously reduces the accuracy of the resulting data.

pH, residues, ash, anions, and cations

The natural pH of whisky at 40% ethanol volume strength is normally in the range of 4–4.5. This mild acidity is the result of trace organic acids, with acetic acid being the main component. Calibration solutions for the pH meter should be prepared in 40% ethanol. The pH of whisky can be made higher when reduced to bottling strength with softened rather than demineralised water.

Whisky residues are low (typically <0.2g/100mL) and generally represent nonsteam volatile materials derived from the cask during maturation. Samples are prepared for this analysis over a steam bath. Ash values are much lower (typically <0.02g/100mL) and represent the involatile inorganic compounds remaining when the test sample is put to dry in a furnace. The ash will include trace metals such as calcium, magnesium, sodium, and potassium (all typically at low mg/L levels), which are derived from the water and the casks used in the whisky process. These trace metals may be analysed in whisky directly by atomic absorption spectroscopy with flame atomisation or by ion chromatography with electrochemical detection.

Various trace sugars are also present and are derived as cask extracts and low-molecular-weight carbohydrates present in spirit caramel (when used). Sugar concentrations in Scotch whisky are typically <200mg/L. A 12-year-old deluxe blended Scotch whisky was found to contain 50mg/L glucose, 50mg/L fructose, and 20mg/L sucrose. Test samples may be analysed by direct-injection ion-exchange chromatography with pulsed amperometric detection or by gas chromatography (after taking the sample to dryness and separation of the sugars as their trimethylsilyl ether derivatives). Other sugars derived from carbohydrates extracted from oak wood may be detected at much lower relative concentrations.

Volatile phenolic congeners

Various volatile phenolic compounds are found in whiskies where peat has been used in the kiln-drying of the malted barley. These flavour congeners are present in many malt Scotch whiskies, particularly those malts from Islay, an island with nine distilleries in production off the west coast of Scotland. In addition, volatile phenols are found in many of the blended Scotch whiskies made with such malt whiskies.

Volatile phenol congeners are best determined by direct-injection, reversed-phase gradient elution HPLC, with fluorescence detection, in the presence of an internal standard (such as 2,3,5-trimethylphenol) (Aylott et al., 1994). This analysis detects phenol, guaiacol, isomers of cresol and xylenol, eugenol, and other phenol derivatives (Fig. 19.4). Alternatively, many of these compounds may be separated by capillary column gas chromatography and enhanced sensitivity and selectivity obtained with 2,4-dintrophenyl derivatives (Lehtonen, 1983b).

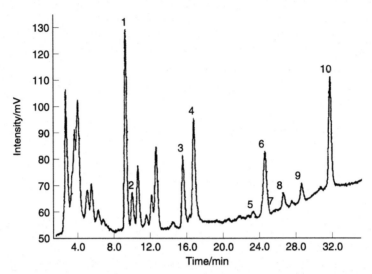

FIG. 19.4 A high-performance liquid chromatogram of volatile phenolic congeners in blended Scotch whisky with fluorescence detection (excitation at 272nm and emission at 298nm). Peaks: 1, phenol; 2, guaiacol; 3, *m*-, *p*-cresol; 4, *o*-cresol; 5, 3,5-xylenol; 6, 4-ethylphenol (2,5-xylenol); 7, 4-ethylguaiacol; 8, 2-ethylphenol; 9, eugenol; 10, internal standard. Reproduced with permission of Royal Society of Chemistry from Aylott, R.I., Clyne, A.H., Fox, A.H., Walker, D.A., 1994. Analytical strategies to confirm the authenticity of Scotch whisky. Analyst 119, 1741–1746.

Quality assurance and analysis in the whisky production process

Malting, fermentation, distillation, and maturation

Distillers require cereals with specific properties such as low nitrogen content, known moisture content, and appropriate malting characteristics. Malting barley varieties may require authentication. Potential distillery yield may be predicted by testing fermentation efficiencies in laboratory trials, and an array of techniques is available for distillery process troubleshooting (Hardy and Brown, 1989). Furthermore, routine process analyses during fermentation, distillation, and maturation are normally limited to sensory assessments and alcoholic strength measurements.

Blending and bottling

The main analysis (along with sensory assessment) during the blending and bottling process is alcoholic strength measurement, and modern facilities employ electronic density measurement. The quality of the water used for the final reduction to bottling strength is very important. The town water supply is normally used, but it is usually subjected to demineralisation and carbon and ultraviolet light treatment processes in order to reduce trace anions and cations to a minimum and to eliminate the risk of any off-odours. Conductivity measurement is used to monitor the water treatment process and the characteristics of bottled product in association with pH measurements.

Colour consistency is also an important quality parameter, particularly to those whiskies that include addition of trace spirit caramel for colour standardisation. Colour is normally monitored by visible spectrophotometry at a wavelength of around 500nm. The resulting tint is the absorbance at a chosen fixed wavelength × 100.

Whiskies are often filtered through both coarse and fine sheets or cartridge filters in order to ensure a clear bright product. Sometimes, chill filtration is used where the liquid is first chilled before filtration in order to achieve enhanced stability under prolonged cold environmental storage conditions. Clarity or turbidity is measured on a nephelometer using instruments made by Sigrist-Photometer (Ennetbürgen, Switzerland) and Hach® (Loveland, Colorado).

Whisky stability

In the bottle, whisky is a very stable product. Providing the closure gives a good seal on a glass bottle, alcoholic strength and major volatile congener concentrations will remain constant, and the product character will not change (Aylott and MacKenzie, 2010). Storage at warm temperatures can result in slightly elevated acetaldehyde and ethyl acetate concentrations. If the closure is loose, then ethyl acetate and secondly acetaldehyde concentrations will diminish through volatilisation and alcoholic strength will be lost. Stability testing with Scotch whisky in polyethylene terephthalate (PET) 5cL bottles over 2 years showed that alcoholic strength slightly increased due to preferential migration of water through the bottle wall. Ethylene terephthalate cyclic trimer was also detected by HPLC as a trace migrant from PET into spirit at the sub-μg/L level (Aylott and McLachlan, 1986).

Filtered whisky is normally a clear bright product, but two forms of flocculation may occasionally be experienced in Scotch whisky. The first form is known as 'reversible floc' and can form when whisky is stored for prolonged periods at very cold temperatures (as may be encountered in cold winter transit). The whisky develops a haze that disappears when the liquid is warmed and shaken. The main congeners detected in 'reversible floc' are the ethyl esters of long-chain fatty acids and larger alkyl esters, both detectable by capillary column GC–MS. As whiskies are produced for distribution around the world with a wide range of climatic conditions, it has been found that 'reversible floc' formation may be minimised by chill filtering whisky prior to bottling. This process reduces the concentrations of 'reversible floc' forming material and has no effect on product character.

The second floc is known as 'irreversible floc'. This shows itself as very small crystals of calcium oxalate, which slowly form and settle in the whisky when natural low mg/L concentrations of oxalic acid in the whisky react with similarly low concentrations of calcium ions. Oxalic acid may be determined by ion chromatography and calcium by flame atomic absorption spectrophotometry. 'Irreversible floc' formation is eliminated by ensuring that calcium concentrations are kept to a minimum by demineralising the water used for final reduction. Tennessee whiskies, whose process includes percolation of new distillate through maple charcoal prior to maturation, also pick up calcium from the wood, and levels may be reduced using ion-exchange treatment.

Off-odours as contaminants in whisky

Off-odours are occasionally reported in bottled products. The more common off-odour issues result from whisky being transported and stored under unsuitable environmental conditions. For example, transit and storage close to very smelly chemicals can result in odour ingress into the bottle. Bottles capped with roll-on pilfer-proof (ROPP) closures are more resistant to odour ingress. In each of the examples described below, GC-Sniff used in parallel with GC–MS helped to identify the offending contaminants.

Ingress of naphthalene odours (mothball smell) has been detected on many occasions as a contaminant in consumer complaint samples returned from distant markets where the product had been stored under unsuitable environmental conditions. The sensory threshold for naphthalene in whisky is very low (typically at µg/L levels in product), and its presence may be quantified by HPLC with fluorescence detection or GC–MS with selected ion monitoring. Musty-smelling compounds can also be encountered as contaminants in whisky. The first example is 1,3,5-trichloroanisole (TCA), a particularly smelly compound that has been reported on many occasions (Saxby, 1996). TCA contamination at the sub-µg/L level has been found when the good-quality bottled product was stored under hot, wet, and humid climatic conditions, such as those found in southern Asia. It is thought that trace trichlorophenols in the pulp used to make cartons and cases can undergo microbiological degradation to form TCAs, which then enter the product through the closure/bottle interface. The second musty-smelling example is that of geosmin contamination resulting from the microbial activity on cereals stored under damp conditions, prior to being taken for fermentation. Trace µg/L concentrations of this musty/earthy-smelling compound can be carried through fermentation, distillation, and maturation to contaminate the resulting whisky.

Consumer issues

Analysis can prove very useful in the examination of samples returned by consumers as complaints. Considering the enormous number of whisky bottles sold globally, the number of consumer complaints is usually very small. This is due to the inherent stability of whiskies and very high quality standards applied throughout the whisky manufacturing processes.

The first requirement of a quality laboratory on receipt of a consumer complaint is to confirm its validity and the likely cause. Most whisky producers now have 'track and trace' markings on their bottles that provide considerable detail on the supply chain used in the manufacture of their product. These markings, known as 'lot codes', are usually found near the neck of the bottle, on the side of the bottle near its base, or on the glass behind a label. Information such as batch number, bottling date, bottling line number, date of dispatch, and identity of the first customer can be found. Such information is very useful in the investigation of consumer and quality issues. Valid consumer complaints can be divided into two categories: those resulting from packaging issues and those from liquid issues. Complaints in the packaging category normally arise from supplier issues, quality issues on the production line, or damage caused during distribution. Packaging line issues may be minimised by employing appropriate standards and quality assurance procedures in the supply and acceptance of materials to the manufacturing process. Modern glass bottles are treated at the hot end of the glass-making process with tin oxide in order to toughen the glass and at the cold end of the process with surfactants and oleic acid in order to give the glass resistance to scuffing during handling. Occasionally, over-application of oleic acid can give rise to contamination by small oily globules of the acid in the product. Sometimes, a consumer may inadvertently contaminate their whisky with material that renders the whisky unstable. As an example, traces of milk result in a heavy precipitate of insoluble milk and whisky components. This contamination can be characterised through the detection of lactose (from milk) in the supernatant and dairy fats in the precipitate (analysed as their methyl esters after derivatisation with boron trifluoride/methanol).

In a similar way, low mg/L concentrations of iron react with maturation congeners and cause a green discolouration, which becomes virtually black at the 10mg/L iron level. Rust in any post-distillation process can be very damaging, and process engineers can minimise this risk by employing high-grade stainless steel in the construction of tanks and pipe work.

Whisky authenticity

Whisky authenticity deals with various forms of counterfeiting and falls into two categories: brand and generic authenticity. Brand authenticity addresses whether the liquid in the bottle purchased by the consumer corresponds to the brand name on the label. Generic whisky authenticity applies to whether the liquid in question is entitled to the broad description 'whisky'. Liquid analysis plays a major role in both cases by allowing the analytical fingerprint of the suspect sample to be compared with that of the genuine brand or whisky type. The investigation and occurrence of non-authentic distilled spirits are reviewed in Chapter 20. However, authenticity analyses appropriate to Scotch whisky are reported here.

Brand authenticity

The analysis of the major volatile Scotch whisky congeners (particularly the higher alcohols) allows normal congener concentration ranges to be set (Aylott, 2010; Aylott et al., 1994). As an example, Table 19.2 shows the normal analytical range data for three different brands of blended Scotch whisky. The three brands contain quite different blends of malt and grain whisky, as can be seen in their concentration ranges for 2- and 3-methylbutanol. As the less expensive brands tend to contain lower proportions of malt whisky, these are usually the type of product used to illegally refill the more expensive product bottles.

For a suspect sample, the analyst will first determine alcoholic strength and the major volatile congeners and then compare the results with the normal ranges for the genuine brand. If the suspect sample's results fall within the normal ranges, the sample's authenticity is normally accepted. If its results are outside the normal ranges, the sample is not authentic. Table 19.3 also contains the results for three suspect samples, numbered B1, B2, and B3, which were collected from bars by consumer protection officers and sold in bottles labelled as brand B. The result for sample B1 fell within the normal ranges for brand B and was, therefore, accepted as a genuine brand B. Sample B2 had congener concentrations within the normal ranges for the brand but an alcoholic strength of 30% vol. It was concluded that Sample B2 was the genuine brand but it had been adulterated and stretched by dilution with added water. Sample B3 had many congeners outside the normal ranges for brand B (particularly

TABLE 19.2 Major volatile congener concentration ranges (g/100 LAA) for samples of three different brands of Scotch whisky taken from 42, 54, and 42 production batches.

Sample	Declared alcoholic strength (%)	Acetaldehyde	Methanol	Ethyl acetate	n-Propanol	Isobutanol	2-Methylbutanol	3-Methylbutanol	2- and 3-Methylbutanol	Ratio 1	Ratio 2
Brand A	40	3.4–8.4	6.4–9.9	20–24	50–58	61–70	10–13.7	32–39	42–53	0.7–0.8	2.8–3.1
Brand B	40	3.3–8.6	6.2–10	21–27	51–66	61–72	16–21	46–56	62–76	0.9–1.1	2.7–3.0
Brand C	40	6.6–13	6.0–9.2	34–42	58–87	64–76	22–26	60–70	82–96	1.2–1.3	2.6–2.8

Ratio 1 = 2- and 3-methylbutanol/isobutanol.
Ratio 2 = 3-methylbutanol/2-methylbutanol.

TABLE 19.3 Major volatile congener concentrations (g/100 LAA) in Scotch whisky brand authenticity investigations.

Suspect Sample	Alcoholic strength (%)	Acetal-dehyde	Methanol	Ethyl acetate	n-Propanol	Isobutanol	2-Methyl-butanol	3-Methyl-butanol	2- and 3-Methyl-butanol	Ratio 1	Ratio 2	Authenticity	Conclusion
B1	40.0	7.2	8.5	25	60	70	17	48	65	0.93	2.82	Genuine	–
B2	30.2	5.0	6.0	22	55	70	20	54	68	0.97	2.7	Not genuine	Diluted with water
B3	40.0	9.0	5.0	15	39	75	13	35	48	0.64	2.7	Not genuine	Other Scotch whisky
C1	40.1	11.7	8.8	32	70	73	22	62	84	1.15	2.8	Genuine	–
C2	39.3	6.4	6.9	15	43	46	13	38	51	1.10	2.8	Not genuine	Other Scotch whisky
C3	47.2	1.7	1.7	4	7	9	5	15	21	2.18	2.9	Not genuine	Admixture
C4	38.0	8.0	4.0	16	30	35	11	31	42	1.15	2.8	Not genuine	Stretched

Ratio 1 = 2- and 3-methylbutanol/isobutanol.
Ratio 2 = 3-methylbutanol/2-methylbutanol.

isoamyl alcohol), and it was concluded that it was not a genuine brand B but another blended Scotch whisky with less malt whisky in its blend composition compared to brand B. This analytical approach regularly contributes to prosecution evidence against those who substitute popular brands in the on-trade with cheaper products.

In a similar way, samples C1 to C4 were suspect counterfeit samples collected in Asia and purporting to be deluxe Brand C. The results for suspect sample C1 fell within the normal ranges for Brand C; therefore, sample C1 was concluded to be genuine. Sample C2 results were outside the required ranges, particularly for 2- and 3-methylbutanol, which suggested that the liquid was not genuine but was another Scotch whisky with much less malt whisky in its blend composition compared to Brand C. Sample C3 was not genuine. Its higher alcohol concentrations were very low, but ratios 1 and 2 were characteristic of malt Scotch whisky. This suggested that Sample C3 was an admixture based on neutral alcohol flavoured with a smaller proportion of malt whisky. Sample C4 was also not genuine, with all congener concentrations at approximately one-half of their expected values for the genuine product. This suggested that sample C4 was based on genuine Brand C but had been adulterated (or stretched) by the addition of an equal volume of cheaper neutral alcohol.

Generic authenticity

Generic authenticity analysis is more complicated than brand authenticity analysis because the analytical data employed must encompass all of the whiskies in that category rather than the more narrow analytical fingerprints of specific brands (Aylott, 2010; Lisle et al., 1978; Simpkins, 1985). Generic authenticity is primarily a concern in the less regulated markets, where there are less stringent legal definitions for whisky and less stringent consumer protection laws.

Chromatographic techniques contribute qualitative and quantitative information for suspect samples to be compared with the known analytical ranges of the genuine product. It is also useful to check for the presence of components in the suspect product that are foreign to the generic whisky, or for the absence of required congeners from the suspect product. The presence of foreign components or the absence of key congeners will raise suspicions about generic authenticity.

The major volatile congener profile can be very useful in generic authenticity analyses. For example, an abnormally high methanol concentration may suggest the use of a non-cereal alcohol. The presence of 3-methylbutanol without any 2-methylbutanol may suggest the addition of 3-methylbutanol as a synthetic flavouring process. The maturation congener profile can be used in a similar way. The absence of maturation congeners may suggest that the suspect product has not been subjected to the required period of maturation, and the presence of a congener (such as vanillin) on its own without related maturation congeners may suggest its addition as part of a synthetic flavouring process. Analytical conclusions from such examples may be used to demonstrate that a suspect product purporting to be whisky has not been produced in compliance with the regulatory requirements. The analytical evidence can contribute to the disqualification of the suspect product as whisky.

Wide-ranging analytical ranges and congener ratios for grain Scotch whisky, malt Scotch whisky, and blended Scotch whisky were established (Aylott and MacKenzie, 2010). When coupled with the learning described above, it was possible to develop an experimental protocol for determining the generic authenticity of Scotch whiskies. The resulting analytical evidence can then assist in protecting the geographic indication of Scotch whisky and help to disqualify false products.

Isotopic techniques have been assessed for use in checking that the alcohols present in the suspect samples are derived only from cereal fermentation and not from a cheaper carbohydrate source such as cane or beet. One potentially stable isotope measurement is the $^2D/^1H$ ratio in different positions on the methyl and methylene hydrogens in ethanol, determined using site-specific natural isotope fractionation–nuclear magnetic resonance (SNIF–NMR). The methyl hydrogens are primarily influenced by the fermented substrate, and the methylene hydrogens are strongly influenced by the ratio in the fermentation water (Martin et al., 1995).

The $^{13}C/^{12}C$ stable isotope ratios are influenced by the photosynthetic pathway used by the fermentable substrate to assimilate CO_2 (Parker et al., 1998). Barley, wheat, and beet use the C_3 (Calvin) pathway, and maize (corn) and cane predominantly use the C_4 (Hatch–Slack) pathway. Unfortunately, this technique cannot help authenticate the many whiskies made from both barley and maize. Finally, synthetic alcohol is obviously not permitted in whisky, and a ^{14}C analysis by liquid scintillation counting can be used to discriminate between synthetic and natural ethyl alcohol in spirits and fortified wines (McWeeny and Bates, 1980).

Whiskies, along with all alcoholic beverages, normally contain trace and harmless concentrations of methanol. Counterfeiters have, on rare occasions, created health hazards by formulating their illicit products with denatured alcohol or methanol. On such occasions, the analyst may then have to interpret the significance of abnormally high results (Paine and Dayan, 2001). A rapid colorimetric method for detecting 1% or greater methanol in beverages has been developed (Graham et al., 2012).

New technologies in authenticity analysis

Various researchers have considered diverse techniques beyond gas chromatography in order to create novel ways of authenticating whisky brands. Although the authentication of Scotch whisky brands by gas chromatography has proven to be reliable and widely used, it requires specialist laboratories, and the analytical process is relatively expensive and time consuming. Counterfeit investigations benefit from rapid field tests, so a screening test based on the ultraviolet/visible spectra of specific whisky brands was developed that delivers preliminary results in a few minutes. The spectrum of a suspect sample is compared to those of genuine brands, using reference data stored in the memory of a hand-held instrument. Many samples can now be rapidly screened, and only suspect samples require confirmatory GC analysis back in the laboratory (MacKenzie and Aylott, 2004). Investigation can also be enhanced by conducting brand authenticity analyses without the need to open bottles and remove liquid samples for analysis either in a laboratory or in the field. Recent work is taking this approach forward using handheld Raman spectroscopy (Ellis et al., 2019).

The following summarises some other techniques that have been tested for whisky authentication. Pyrolysis–mass spectrometry, followed by multivariate analysis of the resulting mass spectra, enabled non-authentic samples to be discriminated from the authentic brand (Aylott et al., 1994). Trace copper and other metal analyses were used as indicators for the authenticity of Scotch whiskies (Adam et al., 2002). Carbon isotope ratios were used to detect the illegal addition of neutral alcohol to Scotch whisky, giving rise to a technique useful in both brand and generic authenticity analyses (Rhodes et al., 2009). The ^{2}H and ^{18}O stable isotope analyses of spirit and source water were demonstrated in brand authenticity analysis (Meier-Augenstein et al., 2012). Mid-infrared spectroscopy was used to detect counterfeit Scotch whisky (McIntyre et al., 2011), and near-infrared spectroscopic analysis on an optofluidic chip, followed by principal component analysis, was used to categorise malt Scotch whiskies by distillery, age, and cask type (Ashok et al., 2011). Straight American whiskies were authenticated by reference to the ratio of furfural to 5-hydroxymethyl-2-furfuraldehyde (Jaganathan and Dugar, 1999). Plain caramel (E150a) is the only additive permitted in Scotch whisky. Liquid chromatography–mass spectrometry (LC–MS) was used to detect and differentiate between caramels E150a, b, c, and d in whisky (Cubbon et al., 2012). Electrospray ionization mass spectrometry with chemometric data treatment was used to check origin and authenticity (Garcia et al., 2013; Møller et al., 2005). Most recently, desorption atmospheric pressure chemical ionisation mass spectrometry was applied to rapid Scotch whisky authentication (Smith et al., 2019). Many of these new applications use specialist equipment that is expensive and requires highly skilled operation, whereas existing authentication techniques based on chromatography are now available internationally. Hopefully, the new technologies will also become economic and more readily available.

There is also a need for analytical strategies to be established for checking the brand authenticities of Irish, Canadian, and bourbon whisky brands. Many brands, each with its own unique blend compositions, show overlapping analytical fingerprints, making the chromatographic methodology for Scotch whisky described above limited for these other whiskies.

Other industry issues

Nitrosamines

N-Nitrosodimethylamine (NDMA) was discovered in the late 1970s at low μg/L levels in certain beers and whiskies. NDMA is a suspected carcinogen; therefore, methods of analysis were required in order to determine its concentration and to understand and control its formation. The compound NDMA is analysed by gas chromatography with a thermal energy analyser (TEA) as a detector. This analysis requires no sample preparation for whisky samples, yet offers sensitivity below 1μg/L. Research, at that time, on Scotch whisky found that NDMA formation was influenced by the kilning conditions for malted barley. The presence of oxides of nitrogen (NO_x) in the kiln gases enhanced its formation. The levels of NDMA could be reduced by preventing NO_x from entering the system or by ensuring the presence of sulphur dioxide in the gas stream (Duncan, 1992). Sulphur dioxide levels were enhanced in those malt kilns fired by heavy fuel oil rather than natural gas. Now that the NDMA issue has passed, its regular analysis in Scotch whisky is all but over, with only occasional analyses required for aged samples.

Ethyl carbamate

Ethyl carbamate (urethane) was discovered as a trace component in a wide range of foods and alcoholic beverages in the mid-1980s. As ethyl carbamate (EC) is a carcinogen to laboratory animals, albeit at concentrations much higher than those found in alcoholic beverages, Health and Welfare, Canada set maximum concentrations in table wines ($10\mu g/L$), fortified wines and saké ($100\mu g/L$), distilled spirits ($150\mu g/L$), and fruit spirits, liqueurs, and grape brandy ($400\mu g/L$) (Connacher and Page, 1986). There followed an intense 5-year period in the European and North American whisky industries initially to determine natural concentrations of EC in whisky, with adequate sensitivity, selectivity, and precision and to understand its method of formation, so control measures could be implemented (Zimmerli and Schlatter, 1991). Various surveys have shown that most blended Scotch whiskies contain EC concentrations at $<100\mu g/L$ (Battaglia et al., 1990; Dennis et al., 1989; Food Standards Agency, 2000), although higher concentrations could be found in certain single malt Scotch whiskies. Similarly, certain bourbon whiskies exceeded the Canadian $150\mu g/L$ limit. U.S. distillers later agreed with the Bureau of Alcohol, Tobacco, and Firearms (BATF) on a $125\mu g/L$ EC target limit for new whiskey distillates. A $400\mu g/L$ EC limit has been introduced in Germany, due to the greater EC levels encountered in certain fruit spirits.

Methods for the determination of EC in whiskies and other distilled spirits require capillary column gas chromatography coupled with nitrogen specific or mass spectrometric detection (Aylott et al., 1987). Initially, EC was concentrated by solvent extraction, but subsequently sample preparation was minimised and the method automated, employing direct injection of whisky samples with added n-propylcarbamate as an internal standard. GC–MS on a polar capillary column with mass detection at m/z 62 enabled good selectivity and a limit of detection of $<5\mu g/L$ in whiskies. It was found that the chromatographic peak shape for EC was better when analysing distillation- and maturation-strength samples, compared to those at 40% vol. alcohol bottle strength. Therefore, bottle strength samples were first diluted with absolute alcohol (containing no EC) in order to bring the sample alcoholic strength up to 70% vol. prior to injection.

Intense research was initiated throughout the alcoholic beverage industry in order to understand EC formation. The breakthrough for whisky came when postdistillation EC formation from trace cyanide and cyanate precursors was discovered in the Scotch grain whisky process (Aylott et al., 1990; MacKenzie et al., 1990). Trace cyanide was found to come from thermal decomposition during distillation of the cyanohydrin of isobutyraldehyde (IBAC), which is present in fermented wash. The IBAC arises during fermentation by the hydrolytic action of yeast beta-glucosidase on a naturally occurring cyanogenic glycoside known as epiheterodendrin (EPH), which is located in the acrospires (growing shoots) of malted barley (Cook et al., 1990). Laboratory-based radiochemical studies verified this chemical pathway (McGill and Morley, 1990).

The knowledge that the relatively low EC levels in new-make grain spirit off a Coffey still could be increased by postdistillation EC formation from cyanide and cyanate precursors led to investigations into their formation, using ion chromatography with conductivity and pulsed amperometric detectors (MacKenzie et al., 1990). The precursors cyanide, copper cyanide complex anions, lactonitrile, and IBAC were collectively determined as 'measurable cyanide' (MC) along with cyanate and thiocyanate ions. A practical relationship was established between the MC concentrations in new-make spirit and the final EC concentrations after all the available MC precursor had converted during the first few weeks of maturation into EC (Aylott et al., 1990), giving the formula:

$$\text{Final EC} = 0.5\text{MC} + 15$$

Ion chromatographic or colorimetric MC analyses were introduced into the distillery process control. During this period of research, the sacrificial copper surfaces in Scotch grain whisky Coffey stills and the addition of sacrificial copper rings to stainless steel American bourbon beer stills were found to adsorb MC and thus help reduce postdistillation EC formation.

More importantly, it was found that different barley varieties have differing levels of EC precursors (Cook, 1990; Cook et al., 1990). First, this knowledge enabled maltsters and distillers to select low MC potential varieties, and the growers were able to develop new varieties of malting barley with low levels of epiheterodendrin precursor. Ethyl carbamate in all types of whisk(e)y can now be monitored and its formation minimised through distillation process control and the use of malting barley varieties with a low MC potential. Interestingly, EC is formed by different mechanisms in other distilled spirits and research to minimise its formation is ongoing.

Nutrient information for consumers

The labelling of whiskies, like all distilled spirits, follows a standard pattern that includes the product's brand name, the category name (such as Scotch whisky, straight bourbon whiskey, and so on), alcoholic strength, volume, name and address of the producer, and, in export markets, the local distributor. Labelling regulations are becoming increasingly complex, as certain markets require health warnings, unit alcohol labelling, maximum recommended daily intakes for men and women, drinking during pregnancy warnings, and packaging recycling information.

Although alcoholic beverages continue to be exempt from nutrition labelling, producers should be ready to answer consumer questions. For example, the energy value of whisky is derived from its alcohol component at 91kJ/222kCal/100mL; there is only a trace of carbohydrate and zero fat and protein (Food Standards Agency, 2002). In mid-2019, SpiritsEUROPE (a trade association) committed in the EU to provide the energy value on label and the list of ingredients off label.

One unit of alcohol in the United Kingdom is 10mL (8g) ethanol, so 28units of alcohol in a 70cL bottle of whisky result in 40% vol. alcoholic strength. However, elsewhere in the European Union and in Australia and New Zealand, one unit of alcohol is 10g ethanol. The United States requires nutrition facts to be quoted in terms of American serving volumes, whereas other countries base their values in terms of metric serving volumes and 'per 100 mL'.

Common labelling standards are obviously desirable in order to minimize the number of labels required across markets, to reduce costs, and to facilitate trade. It is imperative for producers to check local labelling regulations.

Allergen labelling questions

The European Union was among the first markets to introduce allergen labelling regulations for food ingredients such as cereals that contain gluten (Annex IIIa in Directive 2003/89/EC). While it was widely understood by industry, medical practitioners, and celiac support organisations, the fact that distilled spirits made from cereals do not contain allergenic material had to be demonstrated. As wheat and barley contain gluten, the industry undertook a program of study that showed the absence of allergenic materials in distillates that use wheat and barley as raw materials and presented their findings to the European Commission. This resulted in the Commission amending the original Directive to give cereals used for making distillates (among others) an exemption from allergen labelling (EU Commission Directive 2007/68/EC).

References

Adam, T., Duthie, E., Feldman, J., 2002. Investigations into the use of copper and other metals as indicators for the authenticity of Scotch whiskies. J. Inst. Brew. 108, 459–464.

Ashok, P.C., Praveen, B.B., Dholakia, K., 2011. Near infrared spectroscopic analysis of single malt Scotch whisky on an optofluidic chip. Opt. Express 19 (23), 22982–22992.

Aylott, R.I., 2010. Assuring Brand Integrity: Counterfeiting Issues in the Spirits Industry. Brewer & Distiller International, pp. 50–52. April.

Aylott, R.I., 2013. Analytical strategies supporting protected designations of origin for alcoholic beverages. In: Food Protected Designation of Origin Methodologies and Applications, Comprehensive Analytical Chemistry. vol. 60. Elsevier, pp. 409–438.

Aylott, R.I., MacKenzie, W.M., 2010. Analytical strategies to confirm the generic authenticity of Scotch whisky. J. Inst. Brew. 116, 215–229.

Aylott, R.I., McLachlan, I., 1986. Assessment of materials for contact with potable spirits. In: Campbell, I. (Ed.), Proceedings of the Third Aviemore Symposium on Malting, Brewing and Distilling. Institute of Brewing, London, pp. 425–429.

Aylott, R.I., McNeish, A.S., Walker, D.A., 1987. Determination of ethyl carbamate in distilled spirits using nitrogen specific and mass spectrometric detection. J. Inst. Brew. 93, 382–386.

Aylott, R.I., Cochrane, G.C., Leonard, M.J., MacDonald, L.S., MacKenzie, W.M., McNeish, A.S., Walker, D.A., 1990. Ethyl carbamate formation in grain based spirits. Part 1. Post-distillation ethyl carbamate formation in maturing grain whisky. J. Inst. Brew. 96, 213–221.

Aylott, R.I., Clyne, A.H., Fox, A.H., Walker, D.A., 1994. Analytical strategies to confirm the authenticity of Scotch whisky. Analyst 119, 1741–1746.

Battaglia, R., Connacher, H.B.S., Page, B.D., 1990. Ethyl carbamate (urethane) in alcoholic beverages and foods: a review. Food Addit. Contam. 7, 477–496.

Baxter, M.S., Walton, A., 1971. Fluctuations of atmospheric carbon-14 concentrations during the past century. Proc. R. Soc. Lond. A 321, 105–127.

Bendig, P., Lehnert, K., Vetter, W., 2014. Quantification of bromophenols in Islay whiskies. J. Agric. Food Chem. 62, 2767–2771.

Beveridge, J.L., 1990. Malt distillery flavour investigation. In: Campbell, I., Priest, F.G. (Eds.), Proceedings of the Third Aviemore Symposium on Malting, Brewing and Distilling. Institute of Brewing, London, pp. 449–452.

Code of Federal Regulations, 1998a. 27 CFR, Chapter 1. Part 5—Gauging Manual. Subpart C—Gauging Procedures. Section 5.22—The Standards of Identity.

Code of Federal Regulations, 1998b. 27 CFR, Chapter 1. Part 30—Gauging Manual. Subpart D—Gauging Procedures. Section 30.31—Determination of Proof.

Commission Directive 2003/89/EC of the European Parliament and of the Council, 2003. of 10 November, Amending Directive 2000/13/EC as Regards Indication of the Ingredients Present in Foodstuffs. (Annex IIIa).

Commission Directive 2007/68/EC, 2007 of 27 November, Amending Annex IIIa to Directive 2000/13/EC of the European Parliament and of the Council as Regards Certain Food Ingredients. (Annex).

Commission Regulation (EC) No. 110/2008 of the European Parliament and of the Council, 2008 of 15 January, On the Definition, Description, Presentation, Labelling and the Protection of Geographical Indications of Spirit Drinks and Repealing Council Regulation (EEC) No. 1576/89.

Commission Regulation (EC) No. 2870/2000, 2000 of 19 December, Laying Down Community Reference Methods for the Analysis of Spirit Drinks. Annex 1, III.2, L333/36-L333/46.

Connacher, H.B.S., Page, D.B., 1986. Ethyl carbamate in alcoholic beverages: a Canadian case history. In: Institute of Toxicology, Swiss Federal Institute of Technology, and University of Zurich (Ed.), Proceedings of Euro Food Tox II: Interdisciplinary Conference on Natural Toxicants in Food. Institute of Technology, Swiss Federal Institute of Technology, University of Zurich, pp. 237–242.

Cook, R., 1990. The formation of ethyl carbamate in Scotch whisky. In: Campbell, I. (Ed.), In: Proceedings of the Third Aviemore Symposium on Malting, Brewing and Distilling. Institute of Brewing, London, pp. 237–243.

Cook, R., McCaig, N., McMillan, J.M.B., Lumsden, W.B., 1990. Ethyl carbamate formation in grain based spirits. Part 3. The primary source. J. Inst. Brew. 96, 233–244.

Cubbon, S., McMillan, D., Owen, C., Goodall, I., 2012. Anticounterfeiting: using LC-MS to detect and differentiate between caramels E150a, b, c and d in whisky. In: Walker, G.M., Goodall, I., Fotheringham, F., Murray, D. (Eds.), Distilled Spirits IV: Science and Sustainability. Nottingham University Press, Nottingham, pp. 381–387.

Dennis, M.J., Howarth, N., Key, P.E., Pointer, M., Massey, R.C., 1989. Investigation of ethyl carbamate levels in some fermented foods and alcoholic beverages. Food Addit. Contam. 6, 383–389.

Department for the Environment, 2013. Food and Rural Affairs (DEFRA). https://www.gov.uk/government/publications/scotch-whisky-technical-file. downloaded 21/5/20.

Duncan, R.E.B., 1992. A History of Glenochil. United Distillers, Edinburgh, p. 41.

Duncan, R.E.B., Philp, J.M., 1966. Methods for the analysis of Scotch whisky. J. Sci. Food Agric. 17, 208–214.

EEC, 1989. EEC Council Regulation No. 1576/89 of 29 May 1989, laying down general rules on the definition, description and presentation of spirit drinks. Off. J. Eur. Communities 1, L160.

Ellis, D.I., Mumadali, H., Xu, Y., Eccles, R., Goodall, I., Goodacre, R., 2019. Rapid through-container detection of fake spirits and methanol quantification with hand-held Raman spectroscopy. Analyst 144, 324–330.

Fitzgerald, G., James, K.J., McNamara, K., Stack, M.A., 2000. Characterisation of whiskeys using solid-phase microextraction with gas chromatography-mass spectrometry. J. Chromatogr. A 896, 351–359.

Food and Drug Regulations, 1993. Sections B.02.010-023. Food and Drug Regulations, Ottawa, Canada. Sections B.02.020-021. Ottawa, Canada.

Food Standards Agency, 2000. Survey for Ethyl Carbamate in Whisky, Information Sheet 2/00. (May).

Food Standards Agency, 2002. McCance and Widdowson's the Composition of Foods, 6th Summary Edition. The Royal Society of Chemistry, Cambridge, p. 347.

Garcia, J.S., Vaz, B.G., Corilo, Y.E., Ramires, C.F., Saraiva, S.A., Sanvido, G.B., Schmidt, E.M., Maia, D.R.J., Cosso, R.G., Zacca, J.J., Eberlin, M.N., 2013. Whisky analysis by electrospray ionization–Fourier transform mass spectrometry. Food Res. Int. 51, 96–106.

Goodall, I., Harrison, S., Eccles, R., Cockburn, P., Tomaniova, M., 2018. Spirit drinks. In: Morin, J.-F., Lees, M. (Eds.), Food Integrity Handbook: A Guide to Food Authenticity Issues and Analytical Solutions, pp. 229–250. Available from: https://secure.fera.defra.gov.uk/foodintegrity/index.cfm?sectionid=83.

Graham, S., Holmes, S., McGhee, B., Tester, R., Kariagin, A., Steiner, B., Sarver, R., 2012. Alert® for methanol, a rapid diagnostic assay to detect methanol contamination in distilled spirits. In: Walker, G.M., Goodall, I., Fotheringham, F., Murray, D. (Eds.), Distilled Spirits IV: Science and Sustainability. Nottingham University Press, Nottingham, pp. 209–212.

Hardy, P.J., Brown, J.H., 1989. Process control. In: Piggott, J.R., Sharp, R., Duncan, R.E.B. (Eds.), The Science and Technology of Whiskies. Longman Scientific & Technical, Harlow, p. 182.

HMSO, 1990. The Scotch Whisky Order 1990 (Statutory Instruments: 1990). Her Majesty's Stationary Office, London, p. 1.

Horwitz, W. (Ed.), 2000. Official Methods of Analysis of AOAC International, seventeenth ed. vol. II. AOAC International, Gaithersburg, MD, pp. 26.1.07–26.1.11. Ch. 26.

Jaganathan, J., Dugar, S., 1999. Authentication of straight whiskey by determination of the ratio of furfural to 5-hydroxymethyl-2-furfuraldehyde. J. AOAC 82, 997–1001.

Kahn, J.H., 1969. Compounds identified in whisky, wine and beer. J. AOAC 52 (6), 1166–1178.

Kelly, J., Chapman, S., Brereton, P., Bertrand, A., Guillou, C., Wittkowski, R., 1999. Gas chromatographic determination of volatile congeners in spirit drinks: interlaboratory study. J. AOAC 82, 1375–1388.

Kew, W., Goodall, I., Clarke, D., Uhrín, D., 2016. Chemical diversity and complexity of Scotch whisky as revealed by high-resolution mass spectrometry. J. Am. Soc. Mass Spectrom. 28, 200–213.

Lee, K.-Y.M., Paterson, A., Birkmyre, L., Piggott, J.R., 2001. Headspace congeners of blended Scotch whiskies of different product categories from SPME analysis. J. Inst. Brew. 107, 315–332.

Lehtonen, M., 1983a. Gas–liquid chromatographic determination of volatile phenols in matured distilled alcoholic beverages. J. AOAC 66, 62–70.

Lehtonen, M., 1983b. High-performance liquid chromatographic determination of non-volatile phenolic compounds in matured distilled alcoholic beverages. J. AOAC 66, 71–78.

Lisle, D.B., Richards, C.P., Wardleworth, D.F., 1978. The identification of distilled alcoholic beverages. J. Inst. Brew. 84, 93–96.

Livermore, D., 2012. Near infrared reflectance (NIR) used as a predictive tool for Canadian whisky ageing. In: Walker, G.M., Goodall, I., Fotheringham, F., Murray, D. (Eds.), Distilled Spirits IV: Science and Sustainability. Nottingham University Press, Nottingham, pp. 35–45.

Maarse, H., Visscher, C.A., 1985a. Volatile Compounds in Food: Quantitative Data, Supplement 4. Zeist, The Netherlands, TNO-CIVO Food Analysis Institute, p. 66.1.

Maarse, H., Visscher, C.A., 1985b. Volatile Compounds in Food: Qualitative Data, Supplement 2. Zeist, The Netherlands, TNO-CIVO Food Analysis Institute, p. 66.14.

MacKenzie, W.M., Aylott, R.I., 2004. Analytical strategies to confirm Scotch whisky authenticity. Part II. Mobile brand authentication. Analyst 129, 607–612.

MacKenzie, W.M., Clyne, A.H., MacDonald, L.S., 1990. Ethyl carbamate formation in grain-based spirits. Part 2. The identification and determination of cyanide-related species involved in ethyl carbamate formation in Scotch grain whisky. J. Inst. Brew. 96, 223–232.

Martin, G.E., Burgraff, J.M., Dyer, R.H., Buscemi, P.C., 1981. Gas–liquid chromatographic determination of congeners in alcoholic products with confirmation by gas chromatography/ mass spectrometry. J. AOAC 64, 186–193.

Martin, G.G., Symonds, P., Lees, M., Martin, M.L., 1995. Authenticity of fermented beverages. In: Lea, A.G.H., Piggott, J.R. (Eds.), Fermented Beverage Production. Blackie Academic and Professional, Glasgow, pp. 386–412.

McGill, D.J., Morley, A.S., 1990. Ethyl carbamate formation in grain based spirits. Part 4. Radiochemical studies. J. Inst. Brew. 96, 245–246.

McIntyre, A.C., Bilyk, M.L., Nordon, A., Colquhoun, G., Littlejohn, D., 2011. Detection of counterfeit Scotch whisky samples using mid-infrared spectroscopy with an attenuated total reflectance probe incorporating polycrystalline silver halide fibres. Anal. Chim. Acta 690, 228–233.

McWeeny, D.J., Bates, M.L., 1980. Discrimination between synthetic and natural ethyl alcohol in spirits and fortified wines. J. Food Technol. 15, 407–412.

Meier-Augenstein, W., Kemp, H.F., Hardie, S.M.L., 2012. Detection of counterfeit Scotch whisky by ^{2}H and ^{18}O stable isotope analysis. Food Chem. 133, 1070–1074.

Møller, J.K.S., Catherino, R.R., Erberlin, M.N., 2005. Electrospray ionization mass spectrometry fingerprinting of whisky: immediate proof of origin and authenticity. Analyst 130, 890–897.

Paine, A.J., Dayan, A.D., 2001. Defining a tolerable concentration of methanol in alcoholic drinks. Hum. Exp. Toxicol. 20, 563–568.

Parker, I.G., Kelly, S.D., Sharman, M., Dennis, M.J., Howie, D., 1998. Investigation into the use of carbon isotope ratios (^{13}C/^{12}C) of Scotch whisky congeners to investigate brand authenticity using gas chromatography-combustion-isotope ratio mass spectrometry. Food Chem. 63, 423–428.

Philp, J.M., 1986. Scotch whisky flavour development during maturation. In: Campbell, I., Priest, F.G. (Eds.), Proceedings of the Second Aviemore Conference on Malting, Brewing and Distilling. Institute of Brewing, London, pp. 148–163.

PRC, 2008. GB/T11857—Whisky. National Standard of the People's Republic of China.

Reazin, G.H., 1979. Barrel chemistry and whisky maturation. Brew. Guard. 108 (6), 31–34.

Research Committee on the Analysis of Potable Spirits, 1979. Methods for the Analysis of Potable Spirits. LGC, Middlesex.

Rhodes, C.N., Heaton, K., Goodall, I., Brereton, P.A., 2009. Gas chromatography carbon isotope ratio mass spectrometry applied to the detection of neutral alcohol in Scotch whisky: an internal reference approach. Food Chem. 114, 697–701.

Saxby, M.J. (Ed.), 1996. Food Taints and Off Flavours. Blackie Academic and Professional, Glasgow.

Scotch Whisky Act, 1988. http://www.legislation.gov.uk/ukpga/1988/22/contents.

Shoeneman, R.L., Dyer, R.H., 1973. Analytical profile of Scotch whisky. J. AOAC 56, 1–10.

Simpkins, W.A., 1985. Congener profiles in the detection of illicit spirits. J. Sci. Food Agric. 36, 367–376.

Singer, D.D., Styles, J.W., 1965. The determination of higher alcohols in potable spirits: comparison of colorimetric and gas-chromatographic methods. Analyst 19, 290–296.

Smith, B.L., Hughes, D.M., Badu-Tawiah, A.K., Eccles, R., Goodall, I., 2019. Rapid Scotch whisky analysis and authentication using desorption atmospheric pressure chemical ionisation mass spectrometry. Sci. Rep. 9, 1–9.

SWA, 2009. The Scotch Whisky Regulations 2009: Guidance for Producers and Bottlers. Scotch Whisky Association, Edinburgh.

Swan, J.S., Howie, D., Burtles, S.M., Williams, A.A., Lewis, M.J., 1981. Sensory and instrumental studies of Scotch whisky flavour. In: Charalambous, G., Inglett, G. (Eds.), The Quality of Foods and Beverages. Academic Press, New York, pp. 201–223.

Zimmerli, B., Schlatter, J., 1991. Ethyl carbamate: analytical methodology, occurrence, formation, biological activity and risk assessment. Mutat. Res. 259, 325–350.

Investigation and occurrence of counterfeit distilled spirits

Ross Aylott and Irene Aylott

Aylott Scientific, Dunblane, United Kingdom

Introduction

Distilled spirits are alcoholic beverages intended for human consumption, manufactured, and sold worldwide. They are fermented from materials of agricultural origin and distilled to further increase their alcoholic strength. Additional ingredients, further processing, and maturation may then be used depending on the spirit drink category being manufactured.

Premium Western distilled spirits, such as whiskies, Cognac, gin, and vodka, form desirable high-value consumer products that are traded internationally. The economic impact of the industry is significant, despite recent tariffs on USA-EU trade, BREXIT, and the Covid-19 pandemic.

Spirits exports from the EU (2019)	€12.5 billion (Spirits Europe, 2021)
Scotch whisky exports from the UK (2020)	£3.8 billion (SWA, 2021)
Spirits exports from the USA (2020)	$1.4 billion (DISCUS, 2021)

Many distilled spirits are costly to make due to the materials used, the processes employed, and the time associated with their manufacture. Retail prices are further increased by various forms of taxation. However, because of their high prices, demand for distilled spirits such as Scotch whisky, Cognac, and vodka makes them a target of counterfeiters. Other spirits such as tequila (from Mexico), cachaça (from Brazil), rum and pisco (from Latin America), Indian whisky, soju (from Korea), shochu (from Japan), baijiu (from China), and local vodka brands (from Russia) are each particularly popular in their domestic markets and so may also be at local risk.

Illegal activities have long been associated with distilled spirits. Unlicensed distilleries have made spirits described as moonshine, poitin, country liquors, etc., whilst consumers have been sold and still can be sold as adulterated products. These practices were commonplace in Glasgow Public Houses in the late 19th century where whisky was diluted with water, stretched with other alcohols or adulterated with foreign ingredients such as wood alcohol (methanol) and turpentine (Burns, 2005).

Distilled spirits and their consumers have clearly required protection. First, the developing science of analytical chemistry in the late 19th century contributed to the formal characterisation of such products and helped start the control of illegal practices. Second, some of the earliest legal protection of distilled spirits was created with the maturation standards proposed by the 1908 UK Royal Commission on Whisky.

We now have laws and regulations that define each spirit category in most countries (Aylott, 2014; Goodall et al., 2019). The European Union Spirit Drink Regulation 110/2008 defines 46 categories made in EU countries, by reference to traditional processes and also by analytical standards concerning congener concentrations and the minimum limits for alcoholic strength (Commission Regulation (EC) No. 110/2008). Certain spirits, associated with specific countries and geographic locations, are also protected by geographical indications (GIs), protected designation of origin (PDO), and brand trademarks, together with international regulations. These are particularly useful when they reference compliance to the laws of the producing country; examples may be found in the US Code of Federal Regulations (27 CFR 5.22—The Standards of Identity) (Code of Federal Regulations (USA), 1998).

The counterfeit situation affecting distilled spirits is always changing from category to category, region to region, and brand to brand. The extent of counterfeit spirits is widely discussed within the industry, and the potential loss of business by brand owners is considerable. The reader can find many reports of the counterfeiting of spirits on the Internet. This article deliberately does not name brands; that being the business of the brand owner. It aims to help all distilled spirit producers and their consumers by presenting scientific strategies to help deal with both counterfeit liquids and counterfeit packaging.

Counterfeit spirit productions are investigated by first reviewing the key characteristics of genuine and counterfeit distilled spirits and their associated packaging. This chapter then describes the general occurrence of counterfeit; gives guidance on the detection of counterfeit and the practical aspects of counterfeit investigations involving Scotch whisky, vodka. and gin; and discusses threats and opportunities for brand owners and public authorities to further protect consumers.

Key characteristics of genuine and counterfeit distilled spirits

Production of genuine distilled spirits

Genuine distilled spirits are made from quality raw materials and processed under consistent conditions in compliance with relevant regulations. Materials such as cereals, grapes, molasses, and sugar cane are fermented by the action of yeast to produce alcohol. Distillation is then used to concentrate the alcohol into unique spirits. Some categories of newly distilled spirits are then matured in wooden barrels until the resulting spirits are ready for blending and consumption. Other spirit categories undergo further distillation, rectification, and/or processing with various additives and sweetening to produce gins, vodkas, liqueurs, and related products ready for immediate consumption.

Traditionally, distilled spirits are made from the raw materials available in the climatic conditions of their countries of origin using long-established processes—thus, cereals are used to make whisky, grapes are used to make brandy, and molasses and sugar cane are used to make rum and cachaça. Whilst the historical origins of whisky are in Scotland and Ireland, whiskies are now made in the United States, Canada, and Japan and an increasing number of other countries. Brandy is distilled from wine in many grape producing countries, the most renowned in this category being Cognac and Armagnac from south-western France. Tequila is fermented from agave in Mexico; rum is made worldwide from molasses, whilst cachaça is fermented from sugar cane juice in Brazil. Vodka is fermented primarily from cereals and potatoes, initially in Russia and Eastern Europe, but other substrates are frequently used for production worldwide.

Manufacturing plant and associated processes can be expensive and often require financing for many years before their products can be sold to the consumer. The resulting products have specific characteristics that highlight their authenticity compared to the fakes that are reported later. For example:

1. Genuine distilled spirit brands are consistent in their sensory characteristics (by nose, colour, viscosity, etc.) due to careful control of their production processes. The same liquid formulation is normally used in all markets for a specific brand.
2. Genuine distilled spirits have an alcoholic strength within tight limits as declared on their label, due to the careful production control when reducing high strength spirits with water to bottling strength.
3. Genuine distilled spirits contain trace compounds, known as congeners, which enable analytical fingerprints to be created and normal congener ranges established. The congeners and their concentrations in various spirit categories are associated with the raw materials used in fermentation, distillation conditions, maturation conditions (if applicable), and the presence of any additives, sweetening, and further processing.

Thus, genuine distilled spirits are the result of well-established processes, quality materials, and production expertise; these attributes are typically absent from counterfeiters' processes.

Production of counterfeit liquids

Counterfeit spirits are produced in numerous ways. Counterfeiters are driven by low costs, their limited knowledge and expertise, and their desire to bemuse their customers into believing that their products are authentic. Essentially, most counterfeiters mix ethanol and other ingredients using processes that are as simple, cheap, and fast as possible. Whilst counterfeiters may have to perform some operations similar to those of legitimate producers, such as bulk liquid storage, compounding, and strength reduction, their equipment is likely to be less sophisticated and more makeshift. If alcohol is not readily available, they may make it themselves. Common materials include

- Ethanol in various forms.
- Water.
- Additives, including flavourings, sweetening, and colourings.

Ethanol supply for counterfeiters

Counterfeiters like to procure a ready supply of ethanol for use in their fake products. Their products often contain alcohol previously manufactured for legitimate purposes elsewhere that they mix to produce their counterfeit liquid. Relatively small proportions of counterfeiters ferment and distil their own alcohol. Common ethanol supplies include

- Industrial alcohol.
- Neutral alcohol.
- Alcohol from illicit distillations.
- Genuine local brands of distilled spirits.
- Genuine cheaper brands within the spirits category being counterfeited.

Industrial alcohols can be made synthetically or by fermentation. Legitimate industrial alcohols are used in solvents, cleaning and sanitising materials, medicines, cosmetics, flavourings, and food products. Synthetic alcohol (as 100% ethanol/absolute alcohol) is made in the petrochemical industry from the reaction of ethane and steam. Fermented alcohol (as 95/96% ethanol) is made from inexpensive local sources of carbohydrates, dependent upon what is available in the country of production. These industrial alcohols are often rendered nonpotable by the addition of denaturants such as methanol (to produce methylated spirits), bitter-tasting materials, and colouring.

Neutral alcohol from fermentation is basically 95% ethanol with negligible congeners. It goes by many names such as neutral spirits, ethyl alcohol of agricultural origin (in EU regulations), extra neutral alcohol (ENA made from molasses in India), and grain neutral spirit (GNS for vodka and gin production). Whilst many countries have experience of illicit and crude small-scale pot still distillations, production of illicit higher strength neutral alcohol is less common as it requires both a more complex still and operator expertise.

Counterfeiters also acquire other sources of alcohol, such as local spirits and cheaper brands in the spirit category being counterfeited. When these products are mixed with other alcohols, they flavour the fake liquid and give it some of the required sensory characteristics.

Each of the liquids listed earlier generate unique analytical profiles that differ significantly from those of the genuine products being counterfeited (Table 20.1). For example, neutral alcohol contains negligible concentrations of major volatile congeners, whereas genuine products, such as whisky, are rich in such congeners. Liquid analyses are key to counterfeit investigation and to developing an understanding of the counterfeiter's process.

Counterfeiters may acquire their alcohol supply by the bottle, by the 200-L drum, or by the 1000-L palletised bulk container. Counterfeiters' supply of bulk neutral alcohol may be acquired covertly disguised as less pure alcohol (perhaps intended for use as windscreen wash or another cleaning solvent). A denatured alcohol may require purifying before being used in counterfeiting. An ignorant or negligent counterfeiter may use methylated spirits or indeed methanol instead of ethanol with the resulting toxicity issues.

Water

Counterfeiters require water as an ingredient in order to reduce the alcoholic strength of their fake liquids to normal bottling strengths. Their water supply is likely to be the local town supply rather than very high purity demineralised water as used by most genuine producers.

TABLE 20.1 Typical concentrations of major volatile congeners in genuine distilled spirits (upper) and three examples of counterfeit spirits (lower).

Alcohol category	Number of samples surveyed	Country of origin	Alcoholic strength	g/100L Absolute alcohol					Other congeners originating from:	Reference
			% Vol.	Methanol	n-Propanol	Isobutanol	2-Methyl-1-butanol	3-Methyl-1-butanol		
Synthetic alcohol/absolute alcohol		Various	100	ND	ND	ND	ND	ND	None	
Neutral alcohol (EAAOᵃ)		Various	>96	<30ᵃ	ND	<0.5		ND	None	Commission Regulation (EC) No. 110/2008
National whisky	16	India	40	1–10 (4.3)	0–39 (11)	1–15 (7)	0–15 (5)	2–49 (17)	Trace flavourings	Aylott and Mackenzie (2010)
National whisky	5	Brazil	40	3.6–8 (5.9)	2–41 (16)	7–43 (31)	21–29 (24)	62–76 (68)	Trace congeners from malt whisky	Aylott and Mackenzie (2010)
Scotch whisky (blended)	20	Scotland	40	5.2–14 (9.6)	36–121 (69)	56–94 (70)	7–38 (18)	19–107 (48)	Fermentation and maturation	Aylott and Mackenzie (2010)
Irish whisky	5	Ireland	40	7.7–10.7 (9)	28–46 (35)	16–34 (25)	14–26 (21)	35–77 (56)	Fermentation and maturation	Aylott and Mackenzie (2010)
Bourbon whisky	15	USA	40	12–20 (15)	18–38 (25)	57–255 (121)	74–178 (118)	205–420 (252)	Fermentation and maturation	Aylott and Mackenzie (2010)
Canadian whisky	6	Canada	40	5.4–12 (9)	1–4.3 (2.8)	3.4–9.9 (6.5)	4.3–9.0 (6.1)	9.4–19 (11.3)	Fermentation and maturation	Aylott and Mackenzie (2010)
Vodka	6	UK	37.5–40.0	2	ND	ND	ND	ND	Sugars/glycerol/propylene glycol	Aylott (1995)
Gin	7	UK	37.5	2	ND	ND	ND	ND	Juniper / other botanical congeners	Aylott (2003)
Cognac	81	France		14–140 (59)	23–350 (47)	8–145 (88)	33–71 (41)	74–385 (176)	Fermentation and maturation	Lisle et al. (1978)

Continued

TABLE 20.1 Typical concentrations of major volatile congeners in genuine distilled spirits (upper) and three examples of counterfeit spirits (lower)—cont'd

Alcohol category	Number of samples surveyed	Country of origin	Alcoholic strength	Methanol	n-Propanol	Isobutanol	2-Methyl-1-butanol	3-Methyl-1-butanol	Other congeners originating from:	Reference
			% Vol.	g/100L Absolute alcohol						
Tequila	30	Mexico		29–167 (57)	12–98 (38)	28–136 (66)	13–96 (35)	51–300 (169)	Fermentation and maturation	Simpkins (1985)
Suspect Whisky		Cyprus		125	<0.5	<0.5	<0.5	<0.5	Counterfeit	Aylott and Mackenzie (2010)
Suspect Whisky		Taiwan		1	0.6	33	0.6	151	Counterfeit	Aylott and Mackenzie (2010)
Suspect blended Scotch Whisky		Colombia		4	30	35	11	31	Counterfeit	Aylott (2014)

Average results are shown in brackets. *ND*, not detected.

[a] In reality methanol concentrations in ethyl alcohol of agricultural origin are normally <5g/100L absolute alcohol, whereas the standard continues to quote <30.

Town water contains various anions and cations depending on the locality. However, demineralised waters used by legitimate spirit producers are likely to contain fewer trace ions, so trace metal analyses can be useful in identifying counterfeit liquids.

Additives

Flavourings, sweetening (as common sugar), and colourings are easily obtainable locally by counterfeiters from legitimate specialist distributors and suppliers. Spirit caramel (E150a) is a commonly manufactured colouring that is used in many genuine products to give batch-to-batch colour consistency. Counterfeiters are known to make less consistent caramels by heating sugar until a brown colour develops. Analysis of mono- and disaccharides associated with caramels can therefore help identify fake formulations.

Poor quality control for counterfeit product

As suggested earlier, counterfeiters are generally less skilled than legitimate producers in blending, compounding, colouring, spirit reduction, and bottling. Therefore, batch-to-batch variations in liquid formulations, colouring, bottling strengths, and bottle fill heights are likely, and these can be useful markers in first identifying fake products.

Key characteristics of genuine and counterfeit distilled spirit packaging

Materials for spirit packaging involve five key components

- Bottles.
- Bottle closures.
- Bottle labels.
- Single bottle cartons.
- Shipping cases.

Normal packaging processes start with filling bottles with the correct volume of spirit from a bottling vat located just prior to the production line (Fig. 20.1). The correct volume is achieved by filling the bottle to a standard fill height. The spirit is at the correct alcoholic strength corresponding to that declared on the label. Closures are then applied to bottles, followed by labels (and lot codes). Bottles are then packed into single bottle cartons (if required) and then loaded into shipping cases ready for dispatch.

Global distilled spirit brands are normally presented with consistent packaging designs across all markets. There are a few exceptions about which investigators need to be aware. Bottle volumes may be market specific and text/language on labels can vary to reflect local labelling regulations and contact information for local distributors. Closure designs may vary to reflect packaging security requirements and the use of cartons may be market specific. The owners of genuine spirit brands are well aware of the need to market products in ways to give consumers confidence in their product's authenticity. Packaging security therefore plays a key role in the design, development, and procurement of the packaging materials supplied by many specialist producers. Although spirits packaging normally takes place in the country where the liquid was produced, some

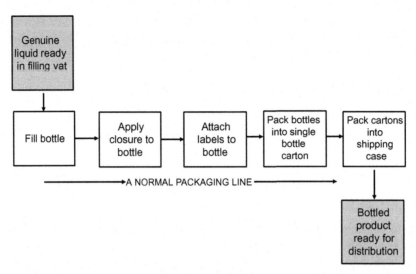

FIG. 20.1   Normal packaging process for genuine distilled spirits.

brand owners ship bulk liquids for packaging in a second country as this enables the use of locally produced packaging materials and labour that can reduce import taxes and other costs.

Although the same range of components is required for counterfeit production, the materials supply routes are very different to the legitimate process (Fig. 20.2). There are two distinct supply routes:

1. Supply of used genuine packaging materials.
2. Supply of counterfeited packaging materials.

The supply of already used genuine packaging components is usually separate from the manufacture of counterfeit liquids. Similarly, the supply of counterfeited packaging components is separate from the manufacture of counterfeit liquids but is part of the counterfeiters' supply chain. Those recycling genuine materials for reuse know that they are supplying counterfeiters and those making fake closures, labels, cartons, and cases invariably know that they are making counterfeit components. Packaging counterfeiters may sell their fake components to numerous spirit counterfeiters and their products can cross international borders for use by counterfeiters in another country. The manufacture of fake packaging not only damages genuine brand owners, it also damages genuine packaging manufacturers and contributes to the potential harm of consumers.

Most low-volume counterfeiters prefer to recover and reuse empty genuine packaging materials as this offers simpler and cheaper ways to procure packaging for their counterfeit product. Supply chains for these materials are established where previously used genuine bottles, single bottle cartons, and shipping cases are collected from bars, restaurants, and even homes and recycled for the subsequent use by counterfeiters. High-volume counterfeiters are likely to use increasing quantities of counterfeit packaging, starting with fake closures and then using fake labels and even fake bottles (Fig. 20.2).

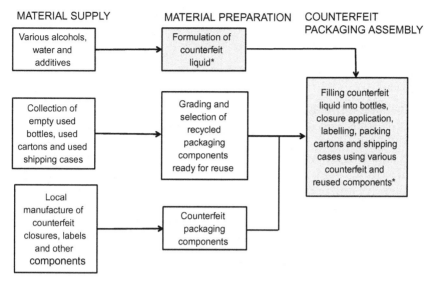

FIG. 20.2   Normal flow of product and packaging through the counterfeit process.
*Counterfeit liquid formulation and packaging are likely to be undertaken in the same premises.

Bottles

Genuine distilled spirits are packaged in bottles mainly made from glass and with much smaller proportions of bottles made from lead crystal glass and PET (polyethylene terephthalate) plastic. Bottle volumes are usually set according to the national regulations relevant to the market of their intended sale. For example, the EU has set volumes, known as prescribed quantities, for distilled spirits the most common being 70cl and 1 L. Other volumes (such as 75cl and nonmetric sizes) are encountered in markets outside the EU. Glass bottles are often designed with a unique shape for specific brands and are made for high-volume brands by major glass manufacturers in lightweight, thin walled glass and in very large quantities. PET plastic bottles are used primarily in the duty-free and 5cl airline trades. Flexible laminated containers may be encountered at the lower priced end of the distilled spirits market.

Most counterfeiters reuse genuine bottles, whereas only those making very large volumes of counterfeit acquire their own supply of new bottles. Empty used genuine bottles are regularly collected from bars and restaurants in Latin America and Asia, vodka bottles have been collected from pubs in Ireland, and in the poorest countries, all sorts of packaging are recovered from their refuse system and rubbish tips. Bar and domestic staff in some countries sell empty bottles, cartons, and cases to 'professional' bottle collectors, who then sell the materials to counterfeiters. Very little genuine packaging goes to waste!

When a supply of empty genuine branded bottles is not available, counterfeiters procure new bottles of similar design from the glass industry. Legitimate bottle manufacturers are often unaware of the end use of their bottles, especially when the bottle has a design common to many brands. An example is the 70cl or 75cl round bottle as used by numerous brands. However, brand owners can help to protect their brands against copying by using bottle designs specific to their brand and use only one bottle supplier. Bottle manufacturers will then know that their customer should only be the brand owner.

Bottle closures

Genuine distilled spirit manufacturers use bottle closures designed to protect their product. Most closures are made by specialist suppliers from either aluminium or plastics, whilst some specialised brands employ cork closures. The most common extruded aluminium closure is known as the 'roll-on pilfer-proof' (ROPP) cap. The normal ROPP closure manufacturing process starts with flat aluminium sheets, first printed and lacquered, and then cut and extruded into individual caps. ROPP caps are cut during their manufacture with a pilfer-proof 'break ring' whose metal bridges the consumer breaks on the first opening of the bottle. Caps normally receive their bottle threads as they are spun onto bottles after being filled with their liquid. Some pilfer-proof closures are made from plastics and some have plastic 'break rings', which are also destroyed as the bottle is opened for the first time. Specialised brands with cork closures are often covered with plastic shrink sleeves that are destroyed on the first opening. Some bottles are fitted with plastic inserts to act as pourers.

Brand owners, whose products are at risk from counterfeiting, use closures that incorporate one-way valves designed to prevent refilling. Consequently, spirits may be poured from bottles by consumers but may not be refilled with a fake liquid by a counterfeiter. These closures are commonly known as nonrefillable fitments (NRFs) and are made using combinations of materials that include: aluminium, plastics, glass, ceramics, and sealants. The many designs of NRF closures offer varying levels of protection against refilling and the closure industry continues to make efforts to optimise their effectiveness.

Brand security may be further enhanced by covering the closure and bottleneck with a seal made from a heat-shrink plastic or a metal foil. Some countries also require strip stamps made from paper with security printing and/or holographic images. These stamps can have two purposes, first to confirm that import/excise duties have been paid and secondly to enhance brand protection. Strip stamps are normally applied to bottles at the point of entry into a market, but some are applied during the production in the home country.

Counterfeiters both recycle and acquire counterfeit bottle closures. These are the most common components of spirits packaging to be counterfeited as it is an unopened closure that gives the consumer confidence that the product is authentic. Fake aluminium ROPP closures can be manufactured in many countries, especially where second-hand printing and extrusion machinery is often available to counterfeit packaging producers, who in turn supply brand counterfeiters.

Nonrefillable fitment closures are both reused and counterfeited. Counterfeiters develop ingenious methods to refill recycled bottles fitted with NRF closures in order to pass the fake liquid against the one-way valve into the bottle. They invent various methods and implements that render the valve ineffective. For example, insertion of a wire into the mechanism can disable some valves and enable fake liquid to fill the bottle. Use of a vacuum pump can create a vacuum within the bottle that enables fake liquid to pass against the valve into the bottle. However, brand owners and their closure suppliers are wise to these techniques and devise ingenious devices to make their NRF closures more effective.

Bottle labels

Genuine bottle labels come in many varieties, printed and cut to size by their specialist manufacturer. Print quality and complexity add to the security. The simplest and most common label is made from paper and is applied to the bottle after filling during liquid packaging. Labels are fixed to bottles by adhesives that come in two main forms, water-based glue that is applied to labels during the packaging operation or as an adhesive applied during label manufacture (creating a self-adhesive label).

The pallet that applies water-based glue to the back of the label usually leaves a distinct glue pattern when viewed from behind the label and this pattern can be useful when examining labels on suspect samples. When genuine labels have been applied using water-based glue, any damaged labels can be removed prior to relabelling, by immersing them in warm water until the label separates from the bottle. However, there is a general industry move from using water-based glue to self-adhesive labels.

Self-adhesive labels are received from their printers attached to a roll of backing tape that releases them onto filled bottles during liquid packaging. Some adhesives allow original labels to be peeled off whilst other adhesives create a strong bond between label and bottle and result in a damage to the label when attempts are made to remove it. As new technologies have been developed, some labels are now printed onto plastic sleeves that are shrunk onto bottles and some labelling is printed directly onto the glass surface. Bottles often contain multiple labels.

However, labels are also counterfeited. Their first use is on genuine bottles that are being reused by the counterfeiter, but the original label is damaged. Their second use is on counterfeit bottles.

Modern printing techniques (or even colour photocopiers) enable counterfeit printers to make reasonable copies that like genuine labels, come in two types—using water-based glue and self-adhesive. Self-adhesive labels are particularly helpful to counterfeiters as these may be easily applied to bottles by hand without the need for appropriate machinery. Inconsistencies of labels and their adhesives between genuine and suspect samples help to identify counterfeits.

Finally, some labels have involvement in taxation. National authorities apply various methods to protect their considerable tax revenues from distilled spirits. Some markets apply a tax stamp to the package to indicate that appropriate taxes have been paid and two examples are described. The first example involves a paper strip stamp as described under the closure section, whilst the second example is the UK tax stamp that is either printed within a product label or is applied as a separate small label on the bottle wall.

Cartons and shipping cases

Many genuine distilled spirit brands are sold in single bottle cartons usually made from a quality cardboard and printed with brand information corresponding to the bottle inside.

Most spirit bottles leave their packaging plant in shipping cases containing 12 or sometime 6 bottles, with or without single bottle cartons. Top-of-range products are often packaged individually.

Single bottle cartons and shipping cases are also counterfeited, especially when recycled genuine materials are not available. Like labels, counterfeit reproductions can be procured locally, although the materials used, the design by which they are cut, and print quality may differ from genuine originals. Fake shipping cases are usually hand assembled with random adhesive patterns compared to the linear adhesive patterns found on genuine machine-assembled cases.

Lot codes

Lastly, individual food and drink products are subject to various regulations within modern food law (such as EU Regulation 178/2002) that require them to have lot codes (or 'track and trace' markings) on their packaging. In the case of distilled spirit packages, these marks normally appear in the form of letters and numbers either inkjet printed on the bottle, typically behind a label, or laser etched onto the bottle wall. Typical examples from spirit bottles are

Example 1	Example 2	Example 3
LKGB12345678	L92540000	L728482001 12:11
	21947950	

Manufacturers can interpret these marks on their production databases to give information on what is in the bottle, where and when it was produced and the address of the first customer. Sometimes the markings are unique to each individual bottle, a characteristic used in counterfeit investigations. Specific information generated from the lot code may include

- Brand name of the product.
- Production batch number.
- 'best before' date (if applied).
- Production line number.
- Date and time of bottling.
- Date of dispatch.
- Name and address of first customer after dispatch from packaging operation, etc.

Occurrence of counterfeit distilled spirits

Brand counterfeits and generic counterfeits

The complex nature of distilled spirits counterfeiting is clarified by applying two basic definitions, brand counterfeits and generic counterfeits.

- **Brand counterfeits** are copies of genuine trademarked products that were originally produced by a legitimate manufacturer according to regulated processes. An example of a brand counterfeit could be a product with labels and packaging apparently corresponding to that of a genuine brand but produced without the authority of the brand owner. Brand counterfeits invariably involve fake liquids and varying degrees of fake packaging and/or reused genuine packaging.
- **Generic counterfeits** are where the liquid in the package does not correspond to the category description claimed on the product label. An example of a generic counterfeit could be a liquid labelled as 'Brand X' 'Blended Scotch Whisky', where 'Brand X' is not a trademarked brand and the liquid does not qualify as a blended Scotch whisky.

Counterfeit in the off-trade and on-trade

Genuine distilled spirit brands are sold to consumers in the following two key ways, off-trade and on-trade:

- **'Off-trade'** through distributors and wholesalers into shops, liquor stores, markets, and informal outlets. The products are normally sold in full-sealed bottles.
- **'On-trade'** in bars, pubs, hotels, and restaurants. Here spirits are normally sold by the glass in either standard volume or unregulated volume measures, with or without other mixer liquids such as tonic water or lemonade.

The entry of a counterfeit product into these trade routes is obviously illegal (Fig. 20.3) and will contravene excise and consumer protection laws. Brand owners and public authorities need to identify the appropriate national laws in order to take action against offenders.

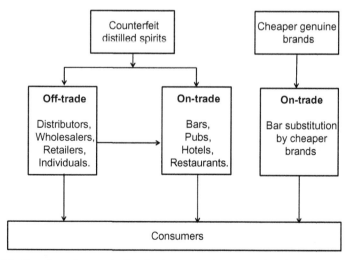

FIG. 20.3 Potential off-trade and on-trade distribution routes for counterfeit distilled spirits from counterfeiter to consumers.

Sale of counterfeit spirits in the off-trade

Here, we deal with the sale of full bottles of counterfeit spirits. These may be brand counterfeits or generic counterfeits.

Clearly, there are many routes to market starting with the sale of a single bottle of fake product from the counterfeiter directly to a consumer. This is most likely to be a fake product at the lower end of the counterfeit scale and to take place in unsophisticated markets, where a crude poor-quality counterfeit is sold at a low price. It is likely that the original packaging will be reused and refilled with a poor-quality fake liquid.

As one ascends the counterfeit scale, the fake product enters distribution networks and ends up in bars, hotels, and restaurants as well as shops and liquor stores. The fake goods may be packaged with some reused genuine packaging components and some fake components. The liquid inside the package is invariably counterfeit.

Distributors may or may not know that they are dealing with counterfeit goods. Fake product may also be exported from the country of its manufacture to another country, whilst customs officials believe that they are dealing with normal internationally traded goods. Counterfeit products and counterfeit packaging materials may also be smuggled from one country to another.

Sale of counterfeit spirits in the on-trade

When a customer asks for a specific brand in a bar, hotel, or restaurant that is what they expect to receive. Sometimes they may be served another brand, one that is invariably cheaper than the requested brand, resulting in someone in the supply chain making extra profit. Such **brand substitution**:

- may be the result of a counterfeit bottled product being supplied to the bar through its supply chain from a producer of counterfeit bottled product or
- may be the result of inappropriate brand substitution by bar staff.

On-trade brand substitution involves serving a different brand from that ordered by the customer in contravention to consumer protection law. The UK market is a good example where bottles of popular pouring brands of Scotch whisky, gin, vodka, and rum are refilled with cheaper and usually inferior products. This is a localised activity within a bar and corrective action may be taken by both brand owners and public authorities (Aylott, 1999).

In addition, the **on-trade** worldwide may knowingly or unknowingly be supplied with counterfeit bottled products. Sometimes bars buy bottles or full cases of popular spirit brands from a supplier offering lower prices. This product may be less expensive because it is counterfeit.

Consumer protection specialists (known as trading standards or environmental health officers) visit pubs and bars as a part of their normal duties in order to check that the alcoholic strength of spirits correspond to that quoted on the product label and that the volume dispensed at the bar meets the correct prescribed quantity. The officer may also take formal samples from the bottle for authenticity analysis, performed either in the field (if a test is available) or by their Public Analyst in a specialised laboratory. When a sample is shown not to correspond to the correct brand, the Public Analyst's report will form part of the evidence used in the prosecution of offenders.

Guidance on the detection of counterfeit spirits

The following guidance on the examination of suspect packaging applies to all distilled spirit packaging, whereas the analytical guidance is limited to those spirit categories being considered herein.

Sample collection in the off-trade

Sample collection in the **off-trade** involves the acquisition through purchase or seizure of full unopened bottles of product that may have been produced by the legitimate manufacturer or by a counterfeiter. Such samples should be immediately labelled and appropriate details recorded such as sample number, sampling location, time and date received, and collector's name. The closure on the bottle should be examined as soon as possible to see whether its pilfer-proof ring is intact or not. Such detail should be recorded.

During counterfeit investigations, enforcement agencies may enter counterfeiters' premises. Officers should not only be prepared to collect suspect filled bottles, but also be ready to seize other liquids, packaging materials, manufacturing equipment, and documentation (examples of which are listed in Table 20.2). Photographs of seized materials and equipment are valuable.

TABLE 20.2   Common materials used by spirit counterfeiters often found in their premises.

Packaging materials	Liquid materials	Manufacturing equipment
• Filled bottles	• Neutral alcohol	• Liquid vessels
• Empty bottles[a]	• Other alcohol supplies	• Mixing equipment
• Closures[a]	• Local distilled spirits	• Filtration equipment
• Labels[a]	• Cheaper imported brands	• Bottle filling machines
• Cartons[a]	• Industrial alcohols/methanol	• Capping machines
• Shipping cases[a]	• Water	• Labelling machines
• Used packaging	• Flavourings	• Adhesives
• Documentation	• Sweetening (sugar)	• Fermentation equipment
• Supplier details	• Glycerol	• Distillation equipment
• Customer details	• Colourings (caramel)	• Printing machinery
• Recipes/formulations	• Raw materials for fermentation	• Hydrometer
	• Yeast	• Measuring cylinder

a Counterfeit materials and genuine materials ready for reuse.

Sample collection in the on-trade

Informal sample collection can identify bars selling counterfeit spirits and give brand owners clarity on the extent of lost business. Informal collection from the on-trade is normally conducted covertly by investigators who purchase samples of the required brand, without any addition of water or mixer drink. These samples are also transferred into clean sample bottles, labelled and sent to a specialist laboratory for brand authenticity analysis.

As an example, one specific brand was covertly sampled from over 100 bars in an area of a large Latin American city. Interestingly, when the bars selling counterfeit product were identified on a street map, it could be seen that fakes were sold in a particular locality and were concentrated on particular streets. In addition, the analytical fingerprints of many fakes were similar showing that the supply of counterfeit was likely to have come from a single source. Such surveys can provide the enforcement agencies with useful information for their pursuit of on-trade offenders and their suppliers.

Formal sample collection of samples from the on-trade is conducted by consumer protection and fiscal officers as part of their normal duties. When undertaken in the United Kingdom, officers take liquid samples of spirits from specific branded bottles. Typically, three samples of the selected product (typically 50 mL) are collected into clean sample bottles. One sample goes to the Public Analyst, a second reference sample is retained by the bar, and the third sample is held in case of a later dispute. The Public Analyst will normally determine alcoholic strength and brand authenticity. When samples are determined to be nonauthentic, the owner of the offending bar may be subject to prosecution under an appropriate consumer protection law.

Packaging authenticity examination

Packaging authenticity examination is equally as critical as subsequent liquid analysis. The investigator examining the suspect package must obviously be knowledgeable about the corresponding genuine product sold in that market. Photographs make invaluable records.

1. The first step is to determine whether the closure is intact (Fig. 20.4); if it is intact and the liquid proves to be counterfeit, then the closure itself must be partially or totally counterfeit; if the break ring is intact, then the liquid may or may not prove to be a fake.
2. The second step is to determine whether the closure is an original or a counterfeit closure. The suspect sample is compared with a genuine example with reference to its dimensions, shapes, graphics, colours, and print quality.
3. Are specific marks, such as the closure manufacturer's trademark (if present), or any covert features present (known to the brand owner) that confirm authenticity?
4. The fourth step is to check the authenticity of the bottle (Fig. 20.5). The design, dimensions, and glass thickness are compared with a genuine reference example.
5. The punt (base) of the bottle is examined including any letters, numbers, and other marks, as these are likely to include a design number, a glass maker's mould number, and a code or symbol that can identify the name and location of the glassworks that made the bottle. The glasswork's code is known as the 'punt mark'

BOTTLE CLOSURE

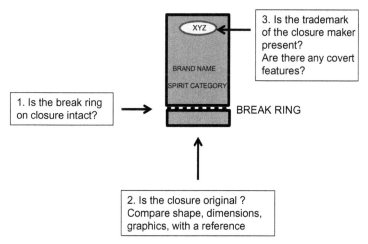

3. Is the trademark of the closure maker present?
Are there any covert features?

XYZ

BRAND NAME

SPIRIT CATEGORY

1. Is the break ring on closure intact?

BREAK RING

2. Is the closure original ?
Compare shape, dimensions, graphics, with a reference

FIG. 20.4 Examination of a suspect counterfeit bottle closure.

LOWER SECTION OF BOTTLE

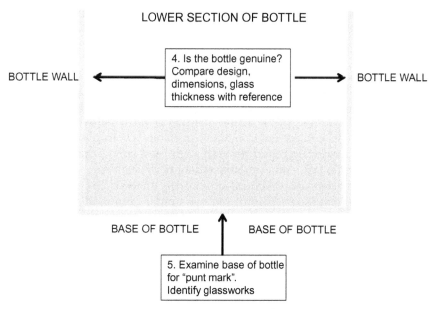

4. Is the bottle genuine?
Compare design, dimensions, glass thickness with reference

BOTTLE WALL

BOTTLE WALL

BASE OF BOTTLE

BASE OF BOTTLE

5. Examine base of bottle for "punt mark".
Identify glassworks

FIG. 20.5 Examination of a suspect counterfeit bottle.

and if different from that on genuine product, this can provide valuable information on the counterfeit bottle supply. A useful punt mark directory is published by Bucher Emhart Glass (Bucher Emhart Glass, 2012). Lastly, it must be remembered that it is possible for sophisticated counterfeiters to copy genuine punt marks and related details.

6. The sixth step is to check the authenticity of all labels including their precise position on the bottle, dimensions, paper quality, print quality, and method of adhesion to the bottle (Fig. 20.6). Any differences to the reference sample may suggest counterfeit. With reference to the genuine sample, whether the labels are self-adhesive or have been fixed using water-based glue is checked. Are counterfeit shrink sleeves and counterfeit tax stamps being used?

7. Step seven is to compare all texts against genuine reference labels. Brand and spirits category names, alcoholic strength and volume texts, producer and distributor names and addresses, etc., are checked. Is the text compliant with local labelling regulations?

8. The eighth step is to check for the presence of a lot code on the bottle noting its required position from the reference brand. Codes may be laser etched onto the glass, may be inkjet printed onto glass, and also be covered by a label, thus requiring reading from the opposite side of the bottle. Some genuine brands have unique codes for each bottle. Therefore, the repeated use of such codes on suspect samples suggests that the codes are counterfeited. The brand owner should be able to interpret the lot code and potentially be able to provide a reference liquid sample from their production that may be compared to the suspect liquid.

9/10. Steps nine and ten are to examine single bottle cartons and shipping cases focusing on dimensions, print, colours, cardboard quality, and cutting patterns. The presence of random glue patterns on the tuck under the sections of shipping cases suggests that they have been hand assembled and may be counterfeit.

Liquid authenticity analysis

Useful authenticity analyses are available for distilled spirit categories such as Scotch whisky, Cognac, gin, vodka, rum, and tequila. Liquid samples are withdrawn from their bottles followed by gas chromatographic analyses in a specialised laboratory. In addition, a field test is available for Scotch whisky using a small portable spectrophotometer and dipstick field tests are available for certain brands of vodka, gin, and rum. Counterfeit investigations are best served by accurate, fast, and economic authenticity analyses.

In the establishment of appropriate liquid authenticity analysis, the analyst should refer to the regulations that define the specific spirit category under examination. The analyst should also determine those congeners and their concentration ranges that can be used to characterise a specific genuine brand or specific spirit category. If the bottle lot code enables the analyst to access a sample of the original genuine liquid, then this can be very valuable in determining the authenticity of the suspect liquid.

Obviously, all distilled spirits should contain ethanol at the required concentration. Therefore, all suspect liquids should undergo alcoholic strength determination. An apparent alcoholic strength measurement is a good starting point but a real strength

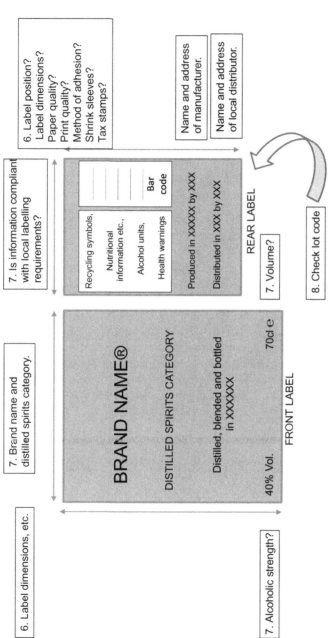

FIG. 20.6 Examination of labels on a suspect counterfeit bottle.

determination should be made if the liquid is sweetened or if other dissolved solids are present (Aylott, 2010).

Distilled spirits, made from neutral alcohol such as vodka, gin, and other flavoured spirits, normally contain negligible higher alcohols as these are removed during the rectification of the base alcohol. Whiskies, brandies, rums, cachaças, etc., are distilled at lower alcoholic strengths and therefore their major volatile congeners (that include acetaldehyde, methanol, ethyl acetate, *n*-propanol, isobutanol, and 2- and 3-methyl butanol) can exhibit profiles useful for authenticity analysis (Table 20.1). The methanol concentration in suspect liquids is particularly important in cases where potentially harmful liquids such as methanol (Paine and Dayan, 2001) and industrial solvents have been used by the counterfeiter. The profiles associated with different fermentation substrates and distillation conditions can give useful insights into the materials being used in a counterfeiter's formulation.

The following general approach to the examination of full suspect bottles is proposed, whilst the examination of subsamples should start at point 2.

1. Check bottle fill height against a genuine reference sample. Break pilfer-proof ring and open bottle. Check the sensory properties (colour and smell) of the suspect liquid against the reference sample—DO NOT TASTE.
2. Determine alcoholic strength (real and apparent alcoholic strength as appropriate). Alcoholic strength values more than ±0.2% vol. different from label alcoholic strength should be considered suspicious.
3. Determine the major volatile congeners by gas chromatography. Check for abnormal congener concentrations, particularly methanol. Determine brand or generic authenticity if sufficient analytical information is available.
4. Determine other congeners and additives as appropriate for the spirit category. This may involve the trace analysis of additional fermentation congeners, maturation congeners (for whisky and brandy), and botanical congeners (for flavoured spirits).

Practical aspects of counterfeit investigation

Scotch whisky

The production of Scotch whisky must follow the UK Scotch Whisky Regulations (SWA, 2009) and follow established processes that require its production in Scotland, from cereals, distillation under specific conditions, and with a minimum of 3 years of maturation.

Regulation and industry practice in Scotland help create malt Scotch whiskies and grain Scotch whiskies with characteristic major volatile congener profiles. The resulting brands of blended Scotch whiskies exhibit consistent congener concentration ranges that are used as the basis of brand authenticity analyses (Aylott et al., 1994; Aylott, 2010). Likewise, using similar data generated from samples representing the full spectrum of malt and grain Scotch whisky distilleries allow congener ranges to be established that describe Scotch whisky as a whole and thus enable the generic protection of Scotch whisky (Aylott and MacKenzie, 2010).

Other internationally important whiskies (such as Irish, bourbon and Canadian) also have distinct major volatile congeners, all of which can help in their protection.

It is important to consider potential liquids that Scotch whisky counterfeiters may use in their fake formulations. For example, many national whiskies produced in Asia and Latin America often containing locally made neutral spirits flavoured with a small proportion of malt Scotch whisky or another whisk(e)y and have distinct sets of congener profiles. When neutral alcohols (containing negligible congeners) are mixed with national or even genuine whiskies for flavour, other sets of congener profiles are created (Table 20.1). These liquids then end up as brand and generic counterfeits and are sold in both the off- and on-trades. Interpretation of congener concentrations found in fake samples and in seized liquids from counterfeiters' premises can be very useful in investigations by helping to explain how the counterfeiter formulated the fake liquids. Finally, when the analyst finds many examples of fake whiskies from different locations, all with similar congener profiles, this can indicate a common counterfeiting source whose product is widely distributed.

Vodka

Vodka is based upon neutral alcohol fermented and distilled from substrates such as potatoes, cereals, grapes, molasses, etc., which may then undergo further processing and finally be reduced to the required bottling strength. Neutral alcohol is essentially pure ethanol at approximately 96% vol. and contains negligible congeners. It is analytically defined in Annex I of the EU Spirit Drink Regulations 110/2008 (Commission Regulation (EC) No. 110/2008). However, some vodka brands contain trace processing components that may include: trace sugars, propylene glycol, glycerol, or fusel oil (Aylott, 2003).

Vodka counterfeiters essentially have two options for their alcohol supply. A small number of counterfeiters ferment and distil their own spirits. For example, common sugar (sucrose) can be fermented and distilled using relatively basic equipment. More often, counterfeiters obtain a neutral or an industrial alcohol and reduce the alcoholic strength with water and bottle it. Sometimes, they may purify an industrial alcohol by redistillation or carbon filtration and sometimes they may undertake further processing with the addition of sweetening, etc. Such liquids may be used to counterfeit established brands and may also be used to make a generic product with a made-up name, both of which can be sold in the on- and off-trades.

The absence of congeneric material means that there is very little on which to develop authenticity analyses. However, it is critical to always measure the alcoholic strength and major volatile congener profile of all suspect samples. As mentioned already, some genuine brands contain trace processing components that can act as useful authenticity markers (Aylott, 2003). In addition, some brand owners have developed simple field tests based upon a dipstick test that responds to trace sugars (Aylott, 2008). This enables investigators to undertake authenticity tests in the field prior to suspect samples being returned to the laboratory for confirmatory analyses.

Gin and other flavoured spirits

There are three categories of gin and many other flavoured spirits described within EU Spirit Drink Regulations 110/2008 (Commission Regulation (EC) No. 110/2008). Gin must be made from neutral alcohol, flavoured with juniper berries, and may also contain congeners from many other botanical materials including coriander seeds and angelica root (Aylott, 2003). These materials are distilled to create 'London Gin' and 'Distilled Gin' and compounded to create 'Gin'. Premium brands of gin are most at risk of on-trade bar substitution by cheaper products.

The botanical ingredients are responsible for the unique gin character of each brand and their resulting gas chromatographic botanical congener profiles enable brand authenticity analysis (Aylott, 1995). However, like vodka, some gin brand owners have developed similar dipstick tests that enable investigators to undertake authenticity tests in the field followed by confirmation in the laboratory (Aylott, 2008). Other flavoured spirits have their own unique processes, characteristic botanical ingredients, and regulatory standards, all of which offer new authenticity opportunities.

Threats and opportunities

The international success of many distilled spirit categories has resulted in them being subject to counterfeiting in many countries. As a result of these illegal actions:

1. Consumers are defrauded, consumers are sometimes subject to health risks.
2. Governments lose tax revenues, and local laws are broken.
3. Brand owners lose business, risk harm to their brand's public image, and there is a loss of asset value.

Many brand owners and spirit distributors take the threat of counterfeiting seriously and support actions to protect their own brands against counterfeit threats. In addition, this work is supported by their trade associations, particularly with respect to generic counterfeiting and misuse of category definitions. Major trade associations:

Scotch whisky	The Scotch Whisky Association (SWA), Edinburgh
Cognac	Bureau Nationale Interprofessionel du Cognac (BNIC), Cognac
Vodka and Gin	Wine and Spirits Trade Association (WSTA), London
Spirits Producers	International Federation of Spirits Producers (IFSP)

The following are specific actions that can assist efforts against counterfeiting.

1. Good regulation is required to define each spirit category both in its country of manufacture and in its country of sale. Trade associations are active in this area.
2. Brand owners and their distributors should always be alert to the risks of counterfeits in any market. Their brand protection activity should involve monitoring their market for counterfeit products, and liaison and training with government regulatory and enforcement agencies (such as police, customs, excise, and consumer protection, as appropriate). Brand owners and their associations may need to engage investigators and establish local authentication laboratories.

Such laboratories can be in the government, brand owner, or private sectors, the key requirements being their analytical accuracy and trustworthiness.

3. Brand owners should protect their products with appropriate packaging security to prevent tampering and refilling. Covert features within their packaging can support the detection of fakes. They should create a pipeline of developing technologies for use against counterfeiters' own developing skills.

4. Guidance on authenticity analysis and brand analytical range data should be made available to trusted third-party laboratories in order to support local investigations and provide evidence for use in the prosecution of offenders. Reference samples should be made available and interlaboratory analyses conducted to ensure analytical accuracy.

5. Brand owners need to understand not only the formulations of their products but also their congener analysis. They require the capability to authenticate their liquids through analytical chemistry and other appropriate technologies.

6. The protection of distilled spirits requires a pipeline of new authentication technologies and, in particular, brand authenticity analyses that can be conducted in the field and without the necessity to remove sample liquids from their bottle (Ellis et al., 2019). There is also a need for innovative methodologies for many other spirit categories. However, new methodology needs to be accessible, available, easily understood, usable as part of prosecution evidence, and relatively low cost.

References

Aylott, R.I., 1995. Analytical strategies to confirm gin authenticity. J. Assoc. Publ. Analysts 31, 179–192.

Aylott, R.I., 1999. That's the spirit: product authenticity, in the Scotch Whisky Industry. Chem. Rev. 3, 2–8.

Aylott, R.I., 2003. Vodka, gin and other flavored spirits. In: Lea, A.G.H., Piggott, J.R. (Eds.), Fermented Beverage Production, second ed. Kluwer Academic/Plenum Publishers, New York, pp. 289–308.

Aylott, R.I., 2008. Authenticity indicators—enhancing consumer and brand protection. In: Bryce, J.H., Piggott, J.R., Stewart, G.G. (Eds.), Distilled Spirits, Production, Technology and Innovation. Nottingham University Press, Nottingham, pp. 281–287.

Aylott, R.I., 2010. Assuring brand integrity: counterfeiting issues in the spirits industry. Brewer and Distiller International (April), 50–52.

Aylott, R.I., 2014. Whisky analysis. In: Russell, I., Stewart, G.G. (Eds.), Whisky, Technology, Production and Marketing, second ed. Academic Press, Oxford, pp. 243–270.

Aylott, R.I., MacKenzie, W.M., 2010. Analytical strategies to confirm the generic authenticity of Scotch whisky. J. Inst. Brew. 116, 215–229.

Aylott, R.I., Clyne, A.H., Fox, A.H., Walker, D.A., 1994. Analytical strategies to confirm the authenticity of Scotch whisky. Analyst 119, 1741–1746.

Bucher Emhart Glass, 2012. Punt Marks Guide. https://www.emhartglass.com/system/files/download_center/BR0068%20-%20BEG%20Punt%20Marks%20Guide.pdf. (Accessed December 2020).

Burns, E., 2005. Bad Whisky. Balvag Books, Glasgow.

Code of Federal Regulations (USA), 1998. 27 CFR, Chapter 1. Part 5—Gauging Manual. Subpart C—Gauging Procedures. Section 5.22—The Standards of Identity.

Commission Regulation (EC), n.d. Commission Regulation (EC) No. 110/2008 of the European Parliament and of the Council of 15 January 2008 on the definition, description, presentation, labelling and the protection of geographical indications of spirit drinks and repealing Council Regulation (EEC) No. 1576/89.

DISCUS (Distilled Spirits Council of the United States), 2021. https://www.distilledspirits.org/international-trade-exports/. (Accessed August 2021).

Ellis, D.I., Mumadali, H., Xu, Y., Eccles, R., Goodall, I., Goodacre, R., 2019. Rapid through-container detection of fake spirits and methanol quantification with hand-held Raman spectroscopy. Analyst 144, 324–330.

Goodall, I., Harrison, S., Eccles, R., Cockburn, P., Tamaniova, M., 2019. Spirit drinks. In: Morin, J.F., Lees, M. (Eds.), FI Handbook on Food Authenticity Issues and Related Analytical Techniques. Eurofins Analytics, France, pp. 229–250. (Chapter 14). Available from https://doi.org/10.32741/fihb.

Lisle, D.B., Richards, C.P., Wardleworth, D.F., 1978. The identification of distilled alcoholic beverages. J. Inst. Brew. 84, 93–96.

MacKenzie, W.M., Aylott, R.I., 2004. Analytical strategies to confirm Scotch whisky authenticity. Part II. Mobile brand authentication. Analyst 129, 607–612.

Paine, A.J., Dayan, A.D., 2001. Defining a tolerable concentration of methanol in alcoholic drinks. Hum. Exp. Toxicol. 20, 563–568.

Simpkins, W.A., 1985. Congener profiles in the detection of illicit spirits. J. Sci. Food Agric. 36, 367–376.

Spirits Europe, 2021. https://spirits.eu/issues/external-trade/key-data. (Accessed August 2021).

SWA, 2009. The Scotch Whisky Regulations 2009: Guidance for Producers and Bottlers. https://www.scotch-whisky.org.uk/insights/protecting-scotch-whisky/legal-protection-in-the-uk/scotch-whisky-regulations-2009-guidance-for-producers-and-bottlers/.

SWA (Scotch Whisky Association), 2021. https://www.scotch-whisky.org.uk/insights/facts-figures/. (Accessed August 2021).

Co-products

Duncan McNab Stewart

Cameronbridge Distillery (Retired), Fife, United Kingdom

Introduction

Most industrial processes and operations produce some form of co-products that have economic value. Within distilling, this is mainly cereal residues after mashing or distillation. From the earliest development of the process of distilling, these co-products have traditionally been sent to farms as animal feed. The first written account of this association was at the Kilbagie Distillery in Clackmannanshire in 1788.

Alfred Barnard quickly established this relationship when he wrote in 1887 about the first-mentioned distillery, Port Dundas, in his seminal tome on distilling:

> After the fine worts are drained off, the grains are pumped into a larger Draff Tun, and when properly exhausted the draff is dropped into carts, which come from various parts of the neighbourhood. For feeding cattle and more especially dairy cows, the draff from wort is unsurpassed.

From Barnard (2003), it can be gleaned that animal feed plays an important role in the distillery operation. Firstly, removing the co-product from the site allows the distillery to continue producing spirit. This may seem an obvious statement; however, it is not uncommon for a distiller to hope the truck will arrive soon to remove spent grains, before the storage area is full and the distillery has to shut down. Secondly, this is a revenue stream that on a regular basis brings cash into the business through its value to farmers.

After processing through the distillery, the cereal residues still have significant amounts of fibre, protein, and oils. For malt distilling draff, on a dry basis, the analysis range is as follows: fibre, 60% to 65%; protein, 20% to 23%; and oil, 9% to 13%. Draff is normally sold with a dry matter range of 20% to 26%. Even though draff has the least processing from an animal feeds perspective, it is still nutritionally rich and well suited to ruminant animals, in particular dairy and beef cattle. Although secondary to the distiller's main role, the potential added value from animal feeds has driven technological investment to improve the storage properties, nutrient content, and hence resale value of this co-product. To a large extent, this has been achieved by the deceptively simple task of removing water from the draff and liquid residues after distillation. The following summarises these technologies and challenges and introduces the new challenge of bioenergy.

Animal feeds

Moist co-products

The moist co-product of malt distilling is draff (termed *spent grains* in brewing), the malted barley residue at the end of mashing. This is produced in an infusion mash tun, lauter tun, or filter press. Removal of the draff from each of these mashing vessels is achieved by different methods.

Traditional mash tun

Within the mash tun, there normally is a pair of sweep arms that remove the draff by pushing the draff into a hole or slot that is uncovered at the end of mashing. The draff either drops directly into a buffer tank or is conveyed to a silo. Alternatively, the draff can be pneumatically conveyed to the draff silo, using the technique of dense phase blowing. A helical screw pushes the draff into the conveying pipeline past a non-return flap valve. Compressed air is introduced after the non-return valve; when the pressure builds up, the non-return valve closes and pushes the draff along the pipeline. As the draff moves forward, the pressure in the system reduces and the screw pushes open the non-return valve, allowing more draff to enter the pipeline, thus commencing another cycle of the operation. For best performance, the design of the system should have the pipeline rising vertically, within a few metres of the non-return valve, thus forming a 'slug' of draff that is of manageable length.

Lauter tun

Within the lauter tun, the draff is removed by reversing the lauter gear. The geometry of the lauter gear knives is designed to efficiently push the draff into the discharge.

Mash filter

Although these are not common in distilling (discussed further in Chapter 10), it is worth mentioning them with regard to draff production, as the unit operation of the filter press is also used within grain distilling and in bioenergy plants, referred to later. A filter press (Fig. 21.1) is constructed of a frame that supports a series of plates covered with porous cloths. The number and size of the plates are governed by the volume of mash or slurry to be treated. The frames are closed using a hydraulic ram, which is required to withstand the high pump pressure used in this process. The liquid slurry, mash in this instance, is pumped into the press, passing through the feed holes in each plate to evenly distribute the mash. The liquid passes through the cloths and is channelled out of the unit. As this is happening, the draff builds up on the cloths until the mash is complete. As each frame has two filtering cloths, the cake will build up on both surfaces until the space between the plates is full.

This should happen to all of the plates at the same time, assuming there is even distribution across the whole unit. In most modern filter presses, air can be injected behind the filter cloths to squeeze dry the draff, which is then followed by a sparge through the wort outlets.

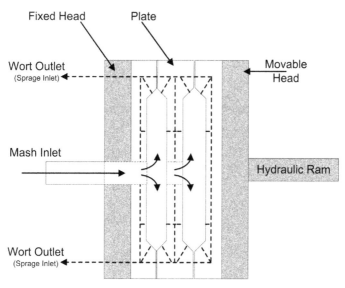

FIG. 21.1 Filter press operation.

The unit then opens and the draff falls into a hopper under the filter press and is conveyed to a silo as described above.

Draff is picked up by local farmers or feed merchants or is further processed into *dark grains*. The annual trend in the price of draff and other moist feeds is a reduction in the summer months, due to the availability of grass. During this period, draff is *ensiled*; that is, air is excluded from the draff by pitting and covering it with an airtight cover. This allows the farmer to manage the use of the feeds and to store them during the malt distilling silent season, although this now appears to be a practice of the past, as Scotch whisky and Irish whiskey sales are resurgent in global markets.

The above processes are applicable to grain distilling; however, for an all-grains process, there is no separation of cereal solids until after distillation. The spent wash from distillation can be dewatered by a number of different types of units, but centrifugation and filter pressing, described above, are readily suited to this duty.

There are many types of centrifuges. For large-scale operation, a horizontal decanter centrifuge is generally preferred. A decanter centrifuge (Fig. 21.2) operates by pumping the spent wash into the bowl assembly, which is rotating at high speed (1500 to 2500rpm). The internal scroll rotates in the same direction as the bowl but not at the same speed, referred to as the *differential speed*. This is important, as it is the differential speed combined with the depth of liquid (pond depth) adhering to the inner blow surface by centrifugal force and controlled by a dam ring on the end of the bowl that drives the efficiency of the centrifuge.

The dry matter, generally 24% to 26% (w/w) of the centrifuge cake, is governed by the properties of the spent wash. This is the percentage of total and suspended solids. Increasing the feed flow will require the differential speed to increase. However, a drier cake is generally produced by reducing the differential speed; therefore, to achieve a constant dry matter cake, these two control elements must be balanced.

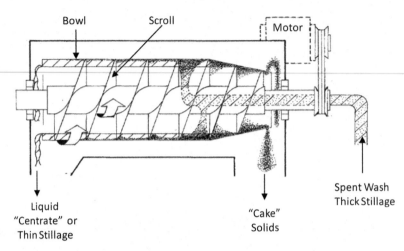

Bowl   Scroll   Motor

Liquid
"Centrate" or
Thin Stillage

"Cake"
Solids

Spent Wash
Thick Stillage

FIG. 21.2   Horizontal decanter centrifuge.

After the cake, solids are scrolled up the beach of the bowl and leave the machine at the opposite end from the liquid. The liquid exiting the centrifuge is referred to as the *centrate* or *thin stillage*. Centrate is normally about 3.0 to 3.5% total solids (w/w), with about 0.7 to 1.2% (w/w) suspended solids.

The cake produced from an all-grains process tends to command a higher price than draff because it contains autolysed yeast cells, and the protein content of the cake is higher than that of draff by about 8 to 10%. This product can be sold directly to local farms or, more normally, by a merchant who will undertake the marketing and selling of the cake on behalf of the distillery. The cake can be treated with a preservative to improve its keeping properties (e.g. propionic acid dosed at about 0.50% in the winter). A dose of 0.75% in the summer is required for ensiling the cake. These cereal solids have passed through distillation with a significant uptake of copper, thus feeding them to sheep is not recommended, as some breeds are susceptible to copper toxicity.

Dry co-products

Within malt distilling, distiller's dried grains with solubles (DDGS) are produced combining draff and pot ale to create a dried product. For an all-grain distilling process, DDGS are produced directly from the spent wash. If the separation of solids for a grain distillery takes place before fermentation, then the process is similar to malt distilling. Malt distilling dark grains plants are sized to service several malt distilleries, due to the scale of investment required for this type of plant and the significant amounts of energy required. A single grain distillery will generally have a purpose-built DDGS plant, generally referred to as dark grains plant. If the distillery is close to the sea or an estuary, it can utilise a long sea outfall for its centrate. There are several permutations of the dark grains process; the following is a generic description of this type of plant based on spent wash. Three key steps are used in the production of dark grains: dewatering, evaporation, and drying.

Dewatering

As indicated above, the use of decanter centrifuges to remove a solids cake is a proven and reliable method. However, to improve the dry matter, the centrifuged cake can be further dewatered using a screw press (Fig. 21.3). A screw press in this application will give a good-quality cake with an approximate dry matter of 29% to 34%. In malt distilling, the mash tun draff would be processed directly through a screw press. The screw press works by mechanically squeezing the cake or draff using two tapering screw conveyors that are enclosed within a rigid mesh screen. As the solids are compressed, the liquid, referred to as *expressate*, passes through the mesh. As a general rule, the slower the speed of the screw conveyor, the better the dewatering. Expressate at 4% to 6% (w/w) solids is generally returned to the spent wash buffer tank to be recycled back to the centrifuges for further solids recovery.

Evaporation

The process of evaporation is designed to concentrate centrate or pot ale by boiling off water. This is normally achieved by using steam as the heat source; therefore, as an operation, evaporation is energy-intensive. There are different types of evaporators, the specification of which is driven by the efficiency of the operation. The most common types of evaporators that have been and are currently being used for the production of DDGS are forced circulation and falling film. These can often be seen within the same evaporation plant, where multiple-effect falling film evaporators are followed by a forced circulation evaporator as the finisher. This arrangement achieves a concentration of the final syrup of > 45% (w/w). Fig. 21.4 illustrates these two designs in the form of a single-effect evaporator. To create a multiple-effect evaporator, the vapour off one evaporator feeds the steam inlet to the calandria of the next effect; the product of the first stage becomes the feed to the second, and so on. This process can have

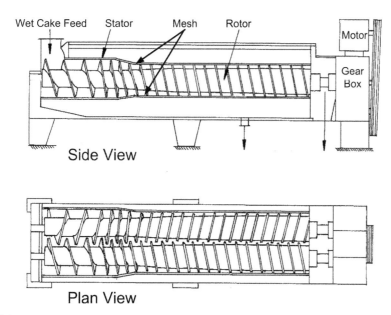

FIG. 21.3 Twin screw press.

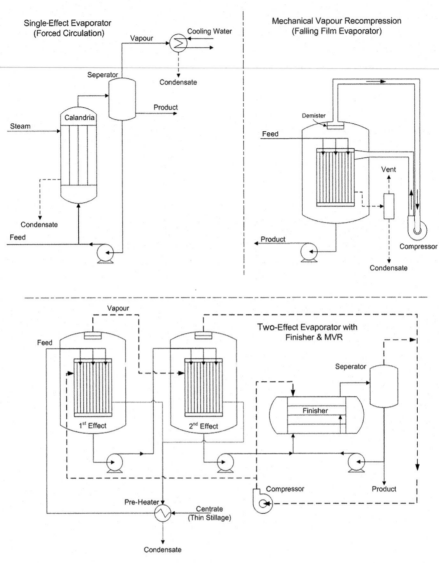

FIG. 21.4   Single-effect evaporator, two-effect evaporator, and mechanical vapour recompression (MVR) evaporator.

two to six effects, depending on the feed stock and the final desired concentration of the product. A two-effect falling fill evaporator, with a forced circulation finisher, is shown in Fig. 21.4.

The liquid from the centrifuges is pumped to a pre-heater to raise the temperature of the centrate as close to the operating temperature of the first effect as possible, based on using the condensate of the evaporators as heat-transfer fluids in the heat exchanger. The product from the first effect is fed to the second effect; it continues to lose water in each effect, thus increasing the solids concentration. With a starting solids concentration of 3% (w/w) for wheat centrate, a final syrup concentration of 45% to 50% (w/w) solids would be expected.

Mechanical vapour recompression (MVR)

Most modern evaporators use mechanical vapour recompression (MVR), which significantly improves the energy efficiency of the evaporation process. This technique utilises a compressor or axial pump to recompress the steam, which has been reduced in pressure after the final effects, thereby using significantly less steam than an evaporator using live steam drawn by a vacuum system. The relative energy efficiencies of a single-effect, a four-effect (without MVR), and an MVR evaporator are as follows:

- Single-effect evaporator—1kg of steam evaporates 1kg of vapour, thus requiring 2260kJ/kg evaporated.
- Four-effect evaporator—1kg of steam evaporates 4kg of vapour, thus requiring 565kJ/kg evaporated.
- MVR evaporator—To evaporate 1kg of vapour requires 70kJ/kg.

The difference in configuration of these systems is shown in Fig. 21.4. The MVR system is equivalent to a 30-fold increase in energy efficiency, when compared with a single-effect unit supplied with live steam. In addition to the increase in efficiency, the capital cost of the plant is smaller due to the requirement of fewer evaporation vessels. Operationally, the plant on start-up must be warmed through the use of live steam to commence the evaporation process. When a steady state has been achieved, a small top-up of steam is required to maintain efficient operation, as the vast majority of the energy input to the steam will be electrical, from the MVR. Cleaning the evaporator is very important. Production periods should not be overly extended, as this will result in longer periods of more difficult cleaning, which may result in the need for high-pressure jetting, which is time-consuming and expensive. The main source of fouling is scale formation from calcium oxalate ('beerstone'), calcium phosphate, and magnesium salts. These are precipitated by heat and evaporation and are inherent in this application. Spent wash will typically contain the following inorganics:

Calcium 50–100mg/L
Magnesium 100–200mg/L
Phosphate 900–1200mg/L as PO_4

As with 'beerstone' formation in wort coolers, the heat transfer will be affected across the evaporator tubes, significantly reducing their efficiency of operation. A general cleaning regime would include the use of caustic to clean organic deposits from pipework, vessels, and heat-exchange surfaces, with acid cleaning if there is a build-up of inorganic salts. Extensive fouling, if not removed, can lead to the carryover of scale and organics into the vapour pipework. Demisters are fitted on evaporators to prevent this, but if they are not maintained and kept clean, they can become ineffective, leading to steam/vapour-side fouling, which is difficult and time-consuming to clean.

Excessive fouling in the vapour line can lead to the MVR going into the surge. This occurs when the compressor is running at a higher rate than the available vapour to the suction side. This is a dangerous condition and can severely damage the MVR. Other reasons for surge are a build-up of condensate or over-thickening. When an MVR is in the surge, it makes a very distinctive noise that is never forgotten!

Fouling can be minimised by maintaining high flow rates though the evaporator. This is assisted by the use of beta-glucanase to reduce the viscosity. The addition of beta-glucanase to the spent wash buffer tank allows time for the enzyme to take effect, before being pumped to the evaporator.

Viscosity

Wheat syrup is highly thixotropic. If left standing without agitation, the syrup will quickly thicken with a jelly-like consistency; however, applying a shear force, such as vigorous mixing, will cause the syrup to become thin (i.e. less viscous and more fluid). The effect of shear force and temperature can be seen in Fig. 21.5. While there is very little impact on the viscosity with regard to temperature, the shear force has a significant effect on viscosity reduction. The addition of beta-glucanase reduces the viscosity of the syrup, so that in a storage tank it does not take on a jelly-like consistency and is still able to flow. The beta-glucanase operates as a catalytic enzyme in cleaving the polysaccharide glucans within the wheat spent wash.

Drying

Drying involves the thermal vaporisation of water from a material to reduce its moisture content. In this instance, it is the drying of the dewatered cake and the syrup from evaporation. Although the technical function of the dryer is the removal of moisture, the true purpose is to increase the retail value of the co-product by increasing the protein and nutrients on a w/w basis. Drying also significantly increases the shelf life of the finished product, which means its markets can cover a larger geographical area, allowing for export beyond the country of origin. Dryers come in various designs and configurations (e.g. rotary disc,

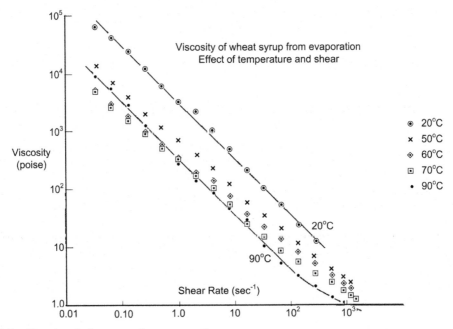

FIG. 21.5   Viscosity of wheat syrup from evaporation.

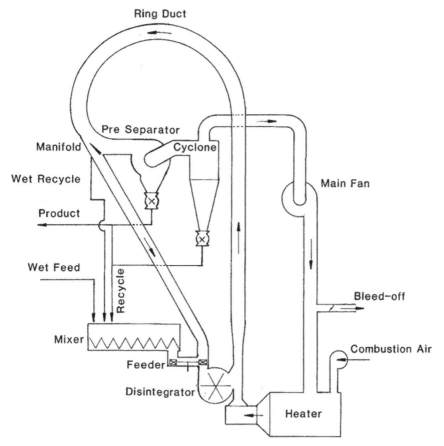

FIG. 21.6   Ring dryer.

pneumatic tube, ring) and are relatively standard with regard to the treatment of distillery moist co-products.

Using the example of a ring dryer (Fig. 21.6), the operation involves mixing the cake, which has been dewatered, with syrup from the evaporation of the centrate (or pot ale). Both are added to a paddle mixer to be introduced into the dryer. This also includes the addition of dry material from the dryer, as a recycle, which helps with the consistency of the feed material to the dryer and its friability, allowing the cake material to be broken up by the disintegrator. The disintegrator slings the solids into the ducting, where it is pneumatically conveyed upward into the ring of the dryer. Any foreign bodies, such as metal or unbroken lumps of material, fall to the bottom of the ducting, where they are collected in a trash box. This must be regularly cleaned due to the ever-present risk of fire in this type of dryer. The lift of the solids through the ducting is provided by the main or induced draft (ID) fan. This is a key item of plant equipment and must be kept clean, as any accumulation of dried material onto the fan blades can lead to an unbalanced fan, which can cause severe vibration and damage to the fan bearing. The heater provides the heat into the dryer by combustion of fuel oil or natural gas. Natural gas is the preferred option because it is a cleaner burning fuel. This style of dryer can readily be connected to the exhaust gas from a combined heat and power plant. In addition to being a good use of waste heat, it has the added advantage of not having a flame in close proximity to combustible solids.

By the time the solids have arced over the top of the dryer ring they are, on the whole, in a dry condition. The heavier wetter solids are flung to the outside edge of the dryer and return to the wet cake mixer to pass through the system again by means of an adjustable damper blade. The finer dry solids are sucked through the pre-separator to remove the solids as a product for onward processing.

The DDGS/dark grains meal can be sold as a product to merchants and feed compounders or, alternatively, it can be pelletised for ease of handling and safe storage. Care must be taken in the design of the storage systems, as large volumes of dark grains, if left undisturbed for a long period of time, are prone to self-heating.

After the pre-separator, the hot vapour-laden gases pass through a series of cyclones to remove any further dark grain fines, which are returned to the mixer. The cyclones protect the ID fan and minimise the amount of particulates that will exit the dryer at the bleed off. The bleed off is required to keep the dryer pressure in balance with the incoming air for combustion and the lean-phase pneumatic conveying of the solids around the dryer. The bleed will contain water vapour and small amounts of solids that have not been trapped by the cyclones. This will require abatement equipment to further treat the exhaust gasses (e.g. an electrostatic precipitator).

Steam drying is now being introduced into the industry due to the efficiency savings offered by utilising excess steam from the evaporator and to the significantly reduced risk of fire within the dryer.

Energy from co-products

With ever-increasing fuel prices and pressure from corporate stakeholders and environmental agencies, there has been a shift in the Scotch whisky industry to invest in renewable energy to improve their carbon emissions. Additionally, this shift presents an opportunity to stabilise distillery energy costs and to improve the overall green credentials of the business. Although the animal feed business is a sustainable practice, this does not in itself provide any direct benefit to the distiller with regard to improving carbon reduction. The Scotch Whisky Association (SWA) in 2014 stated a commitment for reducing Scotch whisky's reliance on fossil fuel, of 20% by 2020, with an ambition of 80% reduction by 2050 (SWA, 2014, 2018). From the SWA Environmental Strategy Report 2018, 'The industry has surpassed its 2020 target and achieved 21% of primary energy use for non-fossil fuels sources' (Performance for 2009–2016).

Bioenergy is one way the industry can achieve its longer-term fossil fuel reduction, although the investment in bioenergy has varied and has faced some significant challenges in optimising the technologies. The main technologies, with regard to the utilisation of co-products, are anaerobic digestion and biomass combustion. These can be stand-alone or used in combination.

Anaerobic digestion

Anaerobic digesters and reactors come in various formats, from agricultural to high-throughput reactors. Anaerobic treatment technology has been developing since the first example in Denmark in 1929. Distillers Company Limited (DCL) conducted pilot trials at St. Magdalene's Distillery in the early 1980s. The basic science of each type of anaerobic digesters is similar with regard to the production of biogas. In essence, anaerobic digestion follows the illustrated biological processes (Fig. 21.7).

- Hydrolysis of complex organic compounds in the spent wash or pot ale, using extracellular enzymes of the bacteria to break the long-chain molecules to produce fatty and amino acids and glucose
- Acidification of these acids by acid-forming bacteria to produce volatile fatty acids (VFA)
- Acetogenesis to form acetate, hydrogen, and carbon dioxide from the VFA
- Methanogenesis to convert the acetate to methane.

The methane concentration is highly dependent on the substrate used, but in the example of a wheat-based spent wash, 66% to 70% methane is expected, with the remainder of the biogas being composed of carbon dioxide and trace amounts of hydrogen sulphide. The efficiency of methane production generally follows the relationship of $0.34m^3$ methane per 1kg of soluble chemical oxygen demand (sCOD). Therefore, $1m^3$ of wheat-based spent wash with an sCOD of 40,000mg/L will yield approximately $14m^3$ of methane and approximately $20m^3$ of biogas at 70% CH_4. This equates to 544MJ of energy, assuming that the gross calorific value of methane is $40MJ/m^3$; therefore, converting to kWh gives $155kWh/m^3$ of spent wash. Assuming the energy used to produce 1L of alcohol is 3.5kWh/LA and a yield of 385 LA/tonne of cereal 'as-is' and a fermentation strength of 7% (v/v) ethanol, then the percentage recovery of energy for each litre of pure alcohol from anaerobic digestion would be as follows.

Spent wash generated from the above process equates to 385 LA at 7%=5500L of fermented wash. Based on live steam distillation, a further 5% of water is added, giving a total volume of spent wash, minus spirit, of 5366L/tonne of cereal processed. Hence, for every litre of pure alcohol, approximately 14L of spent wash would be produced, based on the above anaerobic digestion example. This would give 2.18kW of potential energy. As no engine is 100% efficient, for a small combined heat and power (CHP) system based on a spark engine, about 30% of the usable energy will be electrical and about 40% of the usable energy will be heat.

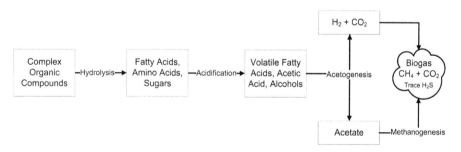

FIG. 21.7 Anaerobic digestion mechanisms.

Therefore, the overall efficiency would give 1.52kW of useful energy for every litre of pure alcohol produced. Thus, 43% of renewable energy can be returned to the process from the soluble component of the spent wash. For large engines, there is an improvement in overall efficiency in that further energy is available with regard to the spent grains, which traditionally are sold as animal feed.

Biomass combustion

In the above example of wheat-based spent wash from grain distilling, the solids from the spent wash can be separated by decanter centrifuge, belt, filter, or screw press. The suitability of these technologies is dependent on experiences with each type of machine. In general terms, they have different operational dependences as outlined in Table 21.1.

In relation to malt distilling draff, the centrifuge would be excluded from the above table. Considering the above units, the filter press and the screw press would operationally be the most effective.

Once the initial dewatering has been completed, the boiler selection is highly dependent on the biomass dry matter. Reducing the water content may require a secondary process, the use

TABLE 21.1 Dewatering technology assessment.

Unit	Decanter centrifuge	Belt press	Filter press	Screw press
Energy consumption	High	Low	Low/moderate	Low
Complexity of installation	Low	High	High	Low
Operation	Continuous	Continuous	Batch	Continuous
Unit cost	High/moderate	Low	High	Moderate
Throughput of each unit	High	Low	Moderate	High/moderate
Complexity of operation for spent wash	Low	High	Moderate	Low/moderate
	Complexity is low when the pond depth and differential speed range have been established.	Belt cleaning regime is difficult to establish.	Cloth selection and freeing of cake from cloth can present issues.	Polymer chemistry is required to obtain good flocculation.
	Operation is robust.	Operation requires a significant level of operator monitoring.	Once established, operation is robust.	Operation is difficult to establish.
	Careful monitoring of vibration and inspection of scroll tiles are required.		A hopper is necessary to smooth out operation to boiler.	Flocculated solids shearing under the force of the screw can be an issue.

of either a dryer or another mechanical dewatering unit. Currently in Scotland, the main type of biomass boiler used in malt distilling is the stepped grate boiler. This is best fed with a dry matter of between 45% and 50%, but certainly should never fall below 40%. This can be achieved by using a belt press and dryer combination. Only part of the draff flow is dried to ~90%, which is then recombined with biomass straight from the belt press to achieve a constant dry matter to the boiler. Alternatively, this has been achieved in one step using a filter press.

As a general rule of thumb, 1 tonne of draff per hour requires 1MW of installed capacity, producing 1.3 tonnes of steam at 7 barg. An example of this type of application is shown in Fig. 21.8.

Within a grain distillery, the above technology is applicable, but multiple units may be required. One strategy that has been pursued is the use of a bubbling fluidised bed (BFB) boiler. This is only applicable to large grain distilleries, due to the scaling of this technology. BFB boilers are mainly used in power station applications and do not lend themselves to the scale of malt distilling operations. The advantage of this technology is that the boiler is capable of burning high-moisture fuels, assisted by another fuel source. This is where the combination of technologies has an advantage within the scale of grain distilling by utilising both anaerobic digestion and biomass combustion together, either by utilising the biogas to assist the combustion in the boiler or by using the biogas to partly dry the biomass before entering the boiler. An example of a plant utilising both anaerobic digestion and biomass combustion is shown in Fig. 21.9.

In the above process, dewatering is achieved by a belt press. The use of a centrifuge could also be equally effective. The pH of the spent wash is increased by using lime and caustic. This aids the downstream anaerobic process, where the optimal conditions would be a pH

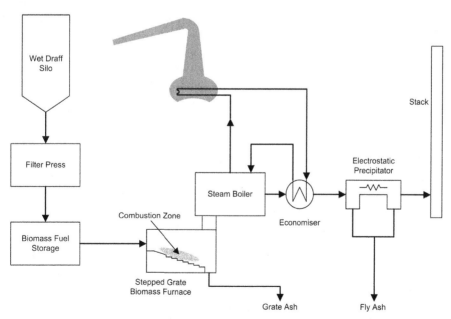

FIG. 21.8   Combustion of malt distilling draff with stepped-grate boiler.

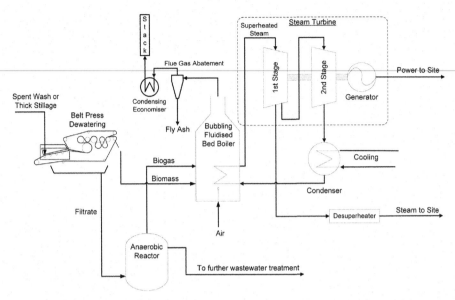

FIG. 21.9   Large-scale bioenergy plant utilising anaerobic digestion and biomass combustion.

of about 7.0 to 7.5. Additionally, the elevation in pH helps the performance of a flocculating agent to coagulate the solids within the spent wash. An anionic polyacrylamide flocculent, added between 0.1ppm to 10ppm, will aid the filtration of the spent grains and help to reduce the amount of suspended solids passing to the anaerobic digester—in this instance, a high-rate anaerobic reactor. Further clarification of the filtrate from the belt presses can be achieved by using a dissolved air flotation unit. The filtrate is pumped to the anaerobic reactor, producing biogas by the mechanism shown in Fig. 21.7, with the liquid passing to further water treatment and, ultimately, water recovery. The biomass, which at this point will be about 30% dry matter, is fed to the BFB boiler. The boiler will have been heated with biogas prior to the introduction of the biomass. The optimum bed temperature for the combustion of spent grain under these conditions is 810±50°C.

A BFB boiler works by fluidising a bed of sand in the bottom of the boiler by injecting air from the plenum chamber at the base of the boiler. The sand bed takes on the characteristics of a liquid, due to the air passing through the sand. The air steam has been preheated by combusting biogas in the air before it enters the sand bed. The bed material does not contain just sand but also ash and other non-combustibles. When new fuel is added, as its mass is significantly less than the mass within the bed, the heating and drying of the spent grains is rapid. Biogas over-burners positioned above the bed maintain the combustion of the spent grains.

As a biomass fuel, spent grains have a caloric value of about 6.5MJ/kg on an 'as-is' basis, which is 21.7MJ/kg on a dry basis, which is similar to wood. Wheat spent grains contain alkali salts, in particular K_2O ($\sim$18% ash basis) and Na_2O ($\sim$0.5% ash basis), which can react with silica to form an eutectic mixture of silicate. The melting point of this mixture will be dependent on the proportion of alkali salts in the spent grains. The formation of a eutectic mixture will begin to affect the bed by disturbing the fluidisation, resulting in hot spots that can lead to the agglomeration of the bed. This would result in a complete shutdown of the boiler to allow the bed to be dug out. Agglomeration of the bed can be prevented by the addition of limestone (which also controls SO_2 emission from the fuel stack) and by reduction of the bed temperature and careful selection of the bed material.

The heat from combustion is transferred to the boiler water through the tube walls, tube banks, and a superheater. In this type of boiler, the boiler tube's overall length is measured in kilometres! Superheated steam is collected in the steam drum at >60 barg, then fed to a steam turbine to generate electricity and supply steam to the site through a pass-out on the turbine at a reduced pressure. The remainder of the steam drives the second stage of the turbine, and then it is fully condensed to return to the boiler deaerator to be fed back into the boiler. In a large grain distillery, this combination of steam and electricity will supply 95% of the distillery's energy needs and power the bioenergy plant.

The burning of spent grains produces its own co-product—the ash from combustion. The incoming cereals to the distillation process contain the compounds of fertilisation used to aid their growth in the field. Therefore, some of these elements and compounds are present in the ash. In particular, there are high levels of phosphate (>20% as P_2O_5), which can be returned to the agricultural industry as fertiliser.

Other co-products

Carbon dioxide

The number of carbon dioxide plants has declined within the Scotch whisky industry. Currently, there is only one operational plant at the North British Distillery in Edinburgh. The decline in the collection and liquefaction of fermentation CO_2 has been driven purely by economics, due to more economic supplies of liquid CO_2 becoming available from the generation of CO_2 from natural gas. However, with ever-increasing natural gas prices in Europe and the United Kingdom, this is now less attractive. As the economics of CO_2 production fluctuates with the cost of natural gas and the 'green' credentials of distillery fermentation CO_2, there may be more of an interest in this source.

The main processes involved in the production of liquid CO_2 are removal of the impurities and liquefaction of the carbon dioxide gas. This is accomplished using the following operations. As the warm, moist, raw CO_2 gas vents from the washback (fermentation vessel), cereal husks are entrained in the gas steam. The gas scrubber has two parts: a wet bottom section and a dry section on the top. The bottom section washes out the cereal husks and any water-soluble compounds. The dry section contains packing (e.g. plastic saddles), which facilitates the removal of water droplets in the gas. At this point, the gas is ready to be compressed and cooled, typically completed in two cycles. The gas is first compressed to about 4bar through the process of compression. Heat is passed to the gas, typically raising the temperature to 120°C; hence, it must be cooled before the second compression to <30°C. The second compressor increases the temperature to 120°C. With further cooling, the gas leaves these stages at 20bar and <30°C.

The gas is then passed through activated carbon to remove odorous and sulphur compounds. This is followed by passing the CO_2 gas through a palladium-catalysed column with hydrogen injection for the removal of nitrous gas compounds. These must be removed, as NO_X gases are corrosive to stainless steel. The products of this catalysed reaction are water and nitrogen.

The final purification step is a silica gel tower, which desiccates the carbon dioxide gas by passing it though an activated alumina section before the silica. Finally, the gas is liquefied through the use of a refrigeration cycle and then condensation, expansion, evaporation, and compression. The now-purified liquid CO_2 is stored in cryogenic tanks at $-25°C$. One of the main customers for carbon dioxide is the food and drinks industry.

Fusel oil

Fusel oil is a co-product of continuous distillation in grain distilling; the production of fusel oil is covered in Chapter 15 on continuous distillation. As a co-product, fusel oil is predominantly amyl alcohol, which is used in the perfume and food industries. Through the esterification of amyl alcohol, amyl acetate is formed, which produces compounds that have a pleasant aroma. Specifically, the Fischer esterification of isoamyl alcohol with acetic acid produces isoamyl acetate, which has the aroma of banana or pear. Although there is a resale value for the fusel oil, this does not match its worth as a fuel. Fusel oil has been successfully burned in fuel oil boilers to raise steam or, alternatively, burned within a biomass boiler to assist with the combustion of spent grains.

Conclusions

The changes over recent years with regard to the processing of draff, pot ale, and spent wash have moved toward utilising these co-products as onsite fuel sources. Will this continue? In part, there will be a move toward the use of anaerobic digestion for the treatment of the liquid stream from distillation, as doing so offers both energy and environmental benefits. With regard to draff and spent grains, the situation is less clear because animal feed prices have been increasing in the United Kingdom. The economic argument for using them as fuel is now not as clear as previously. In the early 2000s, the potential growth of wheat-based bioethanol in the United Kingdom was predicted to flood the animal feed market with spent grains, thus reducing the price, but this has not happened. The fuel-versus-food debate is ongoing as far as determining what is truly sustainable.

This firmly places the decision to use potential animal feeds as a fuel on a site-by-site basis. The assessment needs to be based on location, scale of the distillery operation, current fuel sources, and the corporate environmental policy. In addition, there is a wider impact on other company operations or joint ventures, such as a distiller wanting to invest in anaerobic digestion for pot ale when currently pot ale and draff go to a central DDGS plant. Consideration needs to be given to the impact on the DDGS plant of removal of the pot ale. In essence, there is no one right answer.

One thing is clear regarding any investment in bioenergy—there is a need for good-quality data about the feed material to any bioenergy plant. This means detailed analyses of draff, spent grains, pot ale, and spent wash on an elemental basis, not just dry matter and suspended solids (in particular, alkali earth metals such as calcium and magnesium), due to potential scale formation on heat exchange and evaporative processes. Variability in mashing, fermentation, and distilling processes ripples through to variation in the co-products and the bioenergy process. Another great variable is the barley and wheat harvest. An analysis and assessment of these risks must be carefully considered before making the leap from feeds to energy.

Carbon dioxide may have an interesting future, depending on fuel cost and sustainability policies. A simple online search of YouTube reveals dozens of short videos on algae being utilised as a mechanism to produce liquid biofuels. Algae growth requires CO_2 in abundance, which is available in all distilleries. However, location will play a part, as sunlight is the other key ingredient; therefore, this approach may be more suited to distillers closer to the equator (e.g. United States, Brazil, and India), eliminating the seasonal variation at higher latitudes.

Irrespective of the changes that distillers make in the processing of co-products, there will still be a strong relationship with agriculture. Whether it is animal feed or fertiliser, these co-products close the loop on the ancient partnership between the farmer and the distiller.

References

Barnard, A., 2003. The Whisky Distilleries of the United Kingdom (Originally Published 1887). Birlinn, Edinburgh.

SWA, 2014. Environmental Strategy Reports. http://www.scotch-whisky.org.uk/what-we-do/environmental-strategy/.

SWA, 2018. Environmental Strategy. https://www.scotch-whisky.org.uk/newsroom/environmental-strategy-2018/.

Reading list

Akunna, J.C., Walker, G.M., 2017. Co-products from malt whisky production and their utilisation. In: The Alcohol Textbook. Ethanol Technology Institute, pp. 529–537.

Basu, P., 2006. Combustion and Gasification in Fluidized Beds. CRC, Boca Raton, FL.

Burrows, A., Holman, J., Parsons, A., Pilling, G., Price, G., 2009. Chemistry: Introducing Inorganic, Organic and Physical Chemistry. Oxford University Press, Oxford.

Crawshaw, R., 2001. Co-Product Feeds: Animal Feeds from the Food and Drinks Industries. Nottingham University Press, Nottingham.

Davis, M.L., Cornwell, D.A., 1998. Introduction to Environmental Engineering, third ed. McGraw Hill, New York.

Edwards, L., 2011. Utilisation of distillery by-products into energy and reclaimed water. In: Paper Presented to Chartered Institution of Water and Environmental Management (CIWEM).

Gunes, B., Stokes, J., Davis, P., Connolly, C., Lawler, J., 2019. Pre-treatments to enhance biogas yield and quality from anaerobic digestion of whiskey distillery and brewery wastes: a review. Renew. Sust. Energ. Rev. 113, 109281.

Hammond, J.R.M., van Waesberghe, J.W.M., Wheeler, R.E., 2007. European Brewery Convention Manual of Good Practice. Mashing and Mash Separation, Vol. 13 Getränke-Fachverlag Hans Carl, Germany.

Heald, C., Smith, A.C.K., 1974. Applied Physical Chemistry. Macmillan, New York.

Henze, P., Harremoes, J., Jansen, C., Arvin, E., 2002. Wastewater Treatment: Biological and Chemical Processes, third ed. Springer-Verlag, Berlin.

Leinonen, I., MacLeod, M., Bell, J., 2018. Effects of alternative uses of distillery by-products on the greenhouse gas emissions of Scottish malt whisky production: a system expansion approach. Sustainability 10 (5), 1473.

Maclean, R., 1987. Energy saving techniques in evaporation and drying of distillery effluent. In: The Chemical Engineer.

McCabe, W.L., Smith, J.C., Harriott, P., 1985. Unit Operations of Chemical Engineering, fourth ed. New York, McGraw-Hill.

Palmer, I., 2012. Building a brand new grain whisky distillery: a look at Glen Turner in Bathgate. Brew. Distiller Int. 8 (12), 18–23.

United Distillers, 1995. Grain Distilling: Animal Feeds. Open Learning Series - United Distillers, Edinburgh.

United Distillers, 1995. Grain Distilling: Carbon Dioxide. Open Learning Series - United Distillers, Edinburgh.

White, J.S., Stewart, K.L., Maskell, D.L., Diallo, A., Traub-Modinger, J.E., Willoughby, N.A., 2020. Characterization of pot ale from a Scottish malt whisky distillery and potential applications. ACS Omega 5 (12), 6429–6440.

White, J.S., Traub, J.E., Maskell, D.L., Hughes, P.S., Harper, A.J., Willoughby, N.A., 2016. Recovery and applications of proteins from distillery by-products. In: Protein Byproducts. Academic Press, pp. 235–253.

Designing for and maintaining cleanliness in the distillery

James W. Larson

Chemical Engineer (Retired), Lake Tapps, WA, United States

Introduction

Cleaning in distilleries is important for many reasons, which can differ depending on the area and facilities in the distillery. Likewise, the type of cleaning required can also differ depending on the various areas. Cleaning in process plants is important for:

- Preventing infestation by insects and rodents
- Maintaining product purity
- Controlling microbiological infection
- Maintaining process efficiency
- Safety
- Establishing a quality mindset amongst the employees
- Creating a good impression for visitors.

Cleaning basics

Cleaning can be divided into two broad areas in process plants. One is the general area of housekeeping, which includes the design of facilities such that effective housekeeping is possible. Housekeeping is important for worker safety and is crucial in any food manufacturing facility in order to ensure product safety and quality. In addition to providing a clean and safe workplace, proper housekeeping helps determine the mindset and priorities of the company. Cleaning and cleanability are components of good manufacturing practices (GMPs), a set of standards that exists in many countries around the world. In the United States, for example, relevant GMPs are regulated by 21 CFR 117 (21 CFR 117, 2020). Internationally, a set of voluntary food processing standards known as the Codex Alimentarius includes food plant design and cleaning standards, many of which are similar to those in the GMP standards (Codex Alimentarius, 2003).

The second area of cleaning addresses the processing equipment including tanks, pipes, pumps, and heat exchangers. We differentiate between cleaning and sanitising the process equipment. Cleaning is removing soils, whereas sanitising is removing or killing the remaining microorganisms to an acceptable level after the soil has been removed. It is important to recognise the difference between sanitising and disinfecting. For hard, nonporous surfaces, sanitising kills 99.9% of germs while disinfecting kills 99.999% of germs and viruses (UCSF Institute for Health and Aging, 2013). Sanitising in food plants normally refers to reducing viable germs to a level that is acceptable in the process. The goal is to achieve a level of cleaning and sanitising that ensures product integrity and process efficiency (Larson and Power, 2003). An effective cleaning program first removes the soil and then sanitises. The acceptable levels of cleaning and sanitising vary from process to process, and large variations of acceptable levels of cleaning and sanitising can be found throughout the distillery processes.

Acceptable cleaning and sanitising can be achieved by finding an effective combination of the four cleaning parameters: time, temperature, chemistry, and turbulence (Sinner, 1959). Cleaning deficiencies often arise because of inadequate turbulence or scrubbing action. It is usually possible to compensate for such a deficiency, in any of these parameters by increasing the intensity of one or more of the other parameters (Kretsch, 1994; Larson and Power, 2003).

Cleaning in place (CIP) is ideal. It occurs where permanently installed pumps, sprayheads, and hard piping systems are used to clean the entire process without having to disassemble equipment and to clean some parts by hand. CIP may be conducted manually or with automatic controls. In the real world, some parts of the process may also require manual disassembly and cleaning by hand to achieve acceptable levels of cleaning.

Cleaning in whisky production areas

Grain receiving, storage, and processing

Cleanliness and adherence to good manufacturing practices is important in grain receiving and all storage areas. Unattended spilled grains attract rodents and insects. Storage of grain under damp conditions supports microbial growth resulting in the production of geosmin, an earthy, musty compound that carries over from grain storage all the way to the packaged product (Aylott, 2014). Accumulated grain dust around mills and conveyors is also a fire and dust explosion hazard. Many distilleries completely empty and thoroughly clean all of the grain storage vessels as well as conveyors, mills, and scales on a set schedule.

Wort preparation

Cleaning the wort preparation equipment is extremely important in preventing infection as well as maintaining process efficiency. Mashing and sparging temperatures are not high enough to kill all the infecting bacteria. The build-up of soil can harbour bacteria, which then inoculate batch after batch. In batch mashing systems, it is important to rinse out the residual mash with water before the next batch. If the mash tub remains empty for several hours before refilling, steam is used to sanitise the vessel before starting the next batch. Mashing vessels, transfer lines, and heat exchangers are cleaned with hot caustic solution and sanitised with chemical sanitisers, hot water, and/or steam on a regular schedule.

Continuous cooking systems are operated initially at temperatures in the slurry tank that can allow for the growth of microbes, which may survive the subsequent higher cooking temperatures. Burn-on, fouling, and scaling of heating surfaces in the cook section can be a significant problem requiring cleaning with both acidic and alkaline cleaners.

Wort cooling, transfer to the fermenter, and yeast pitching are operations that require thorough cleaning and sanitising in order to prevent infection. The lower temperatures and abundance of nutrients in these areas are ideal for the growth of bacteria. Wort coolers are also prone to fouling both on the process side and on the cooling water side. Organic build-ups, carbohydrates, and proteinaceous materials on the process side can harbour bacteria and reduce heat transfer. These soils are best removed by frequent alkaline cleaning. Scale on the cooling water side can be removed with acidic cleaners and may only be required on an annual basis.

Fermentation

Fermentation with distiller's yeasts is often accompanied by an acceptable—and in some cases desired—degree of microbiological infection that exists along with the distiller's yeast culture (see Chapter 13). Effective cleaning of the fermenting tanks, accessories, and transfer lines must be conducted to maintain consistent fermentations with the controlled production of fermentation by-products. Along with flavour considerations, the acceptable level of infection in a distillery fermenter is also influenced by the fact that bacterial contamination is a major reason for reduced alcohol yield (Narendranath, 2003).

Hot alkaline solutions (sodium hydroxide is often the choice) are used for cleaning fermenters, and a chemical sanitiser or water (hot water or steam) is used for sanitising. Rotary sprayheads are recommended over fixed sprayballs for cleaning fermenters and other tanks (Resenhoeft, 2013). Wooden washbacks are cleaned by rinsing with water followed by steam or hot water for sanitising.

Cleaning the walls of the fermenting room and the exterior of tanks is effectively conducted with foam cleaning. The foam can reach all of the nooks and crannies of the surface, and additional manual scrubbing is normally not needed if foam cleaning is carried out regularly. Floor cleaning is usually conducted with scrub brushes, water hoses, and cleaning chemicals such as trisodium phosphate.

Distillation

Distillation of the fermenter wash, whether or not grain from mashing is present, is a cleaning challenge that can have a marked effect on heat transfer efficiency in the wash still, wash column coil, or heat exchanger. Frequent alkaline cleaning is necessary to remove organic soils and burn-on from the heating surfaces, especially with unfiltered wort. Descaling with an acid cleaner is also required to remove beerstone and other scales that build up in distillation columns, in heat exchangers, and on the downstream heating surfaces used in processing distillery by-products (Foote, 2013; Tegels and Mundell, 2013). These scales can significantly reduce heat transfer and increase resistance to fluid flow if they are not controlled.

Proofing

After distillation, the alcohol concentration in the product is so high that it is essentially sterile, and infection is no longer possible. On the other hand, the water used for proofing is usually pretreated by dechlorination and demineralisation or reverse osmosis. If the proofing water is to be treated with a reverse osmosis system, then the chlorine must be removed. This being the case, the proofing water storage tanks and transfer lines may be equipped with ultraviolet lights in order to maintain sterility. In addition, this equipment requires periodic cleaning and sanitising, especially if it is not continuously used and contains stagnant water for extended time periods.

Packaging

Because of the alcohol concentrations involved in traditional whisky bottling, microbiological infection is normally not a large concern during packaging. If there are large flavour changes with product changeovers, the equipment may be rinsed with water to prevent flavour carryover. Flavoured products, especially those containing sweeteners or dairy products, require more intense cleaning between brands. Microbiological infections can be a concern, including in the storage and delivery equipment for these materials.

Pipelines and the filler internal parts are cleaned and sanitised with circulating CIP solutions but external surfaces, e.g. filler tubes, cappers, and conveyors, also require sanitation. This is effectively accomplished using foam cleaning. A foam or gel is applied to the surfaces and allowed to rest for 10 to 40min, and then, the surface is rinsed with water to remove the foam and any loosened soil (Rench, 2019). Permanently installed external filler cleaning (EFC) systems are also used to clean and sanitise external filler and closure parts as well as conveyors.

Packaging operations also require strict attention to GMPs. The area around the filling operation and the packaging material storage area must be clean and completely free of any insect presence. Foreign materials, insects, and soils cannot be present in any form around or in the packaged products.

Warehouse and storage areas

Warehouse and storage areas require housekeeping and cleaning to meet the standards of Good Manufacturing Practices 21 CFR and the Codex Alimentarius. Basic housekeeping practices are maintaining floors and walkways clean and clear for access. A vacuum system or portable vacuums with filtered exhaust are essential. Ventilation must be sufficient to maintain dry conditions, in order to prevent mould growth. Rodent and pest control is critical in these areas as well as in the entire plant.

Sanitary design

Maintaining cleanliness in the workplace and in process equipment is much easier if the principles of sanitary design are applied at the very beginning of the planning process. It is important to design and construct buildings, structures, and equipment that are not dirt catchers and that can be cleaned effectively and efficiently with normal cleaning equipment.

Facilities' design

The cleanability of workspaces and product storage areas is improved when attention is devoted to the guidelines of sanitary design and good manufacturing practices. Critical areas such as exposed food areas and packaging lines are kept separate and isolated from less critical areas such as chemical storage, waste disposal and its transport, and maintenance areas. Accessibility for cleaning must be part of building and equipment layout decisions (Baking Industry Sanitation Standards Committee, 2004). In warehouse storage areas, it is recommended to have 0.5m (18in.) of clearance between the walls and the items being stored (Fig. 22.1). Machinery must be located so that maintenance and cleaning can be easily performed. Ideally, it will be possible to disassemble the machines without tools. Pipes and conduits on pipe racks should be spaced so that soils will not be trapped between them (Fig. 22.2). There should be no spaces behind control panels or other boxes mounted

FIG. 22.1   Storage of goods should be 0.5 m (18 in.) from the walls to allow access for cleaning.

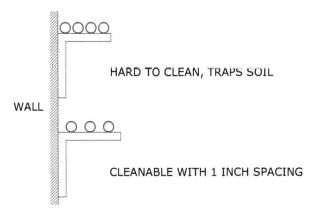

FIG. 22.2   Pipes and electrical conduits should be spaced at least 25 mm (1 in.) apart on pipe racks to avoid creating difficult-to-clean dirt traps.

on walls (Fig. 22.3). The design of structural components is important for cleanability and housekeeping. Flat horizontal surfaces, for example, can collect dust and dirt more readily than sloping surfaces. Frequently, it is possible to choose a sloping structural member rather than a flat horizontal one (Fig. 22.4). The ends of tubular members should always be capped to prevent liquids or other soils from collecting inside and becoming sources of infection. Floors in wet processing areas should have a minimum 2% slope to drain specification. All floor drains must have traps to prevent a backflow of sewer gas. Many more suggestions for the design of cleanable facilities are given in 21 CFR 117 and the Codex Alimentarius standards.

Floors are very important for maintaining cleanliness in the plant. Floor specifications must be matched to the operations that will be taking place in their areas. Conditions to be included are exposure to moisture, chemicals, heat, and physical abuse. Cleanability, drainage, slip resistance, and physical appearance are also important.

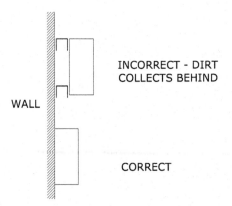

FIG. 22.3   Electrical panels and other boxes should be installed flush against the wall so that soil cannot collect behind them.

HORIZONTAL SURFACES COLLECT MORE SOIL

CLEANER   CLEANEST

FIG. 22.4   Structural members with horizontal surfaces will collect soils. Sloping or round pieces should be used whenever possible.

Walls and ceilings should be painted with a high-quality mould-resistant paint. Building ventilation must exhaust stale air from the building and provide fresh makeup air for worker safety and to prevent growth of mould and mildew.

Vacuum cleaners or central vacuum systems are recommended for cleaning these areas. Exhaust filters designed to trap the particles being removed are of prime importance for both indoor exhaust and outdoor exhaust. High-pressure water cleaning is not recommended due to the generation of aerosols that can redistribute and spread contamination throughout the plant.

Process design

One of the big challenges in designing cleanable process equipment is the ability to reach every surface that the product contacts and to supply sufficient solution scrubbing action to remove soils. Pipeline cleaning requires a minimum velocity of 1.5m/sec (5ft./sec) for the cleaning solution. This velocity gives the fully turbulent flow necessary for a sufficient scrubbing action (Singh and Fisher, 2003). A practical upper velocity limit is 3m/sec (10ft./sec). Flow rates required for different pipe sizes are shown in Table 22.1. Cleaning problems arise when a pipe route has different sizes of pipes connected in series. For example, if the pipe diameter changes from 75mm to 100mm (3in. to 4in.), the range of flows to achieve a velocity of 1.5 to 3m/sec in both is 689 to 770L/min (152 to 170 Imp gpm). If the diameter changes from 75mm to 150mm (3in. to 6in.), it is not possible to have a velocity between 1.5 and 3m/sec in both pipes.

TABLE 22.1   Volumetric flow rates to achieve minimum cleaning velocity of 1.5 m/sec (5 ft./sec) in pipes of different diameters.

Pipe diameter		Minimum flow for cleaning		
Millimetres	Inches	L/min	Imp gpm	US gpm
50	2	172	38	45
75	3	386	85	102
100	4	689	152	182
150	6	1552	342	410
200	8	2775	611	733
300	12	6130	1349	1620

Velocities or flow rates for cleaning process equipment are generally specified by the manufacturers of the equipment. This is often the case for plate heat exchangers. If there is no flow rate recommendation, a rule of thumb is that the cleaning flow rate should be 50% more than the normal process flow rate. It is also very effective to clean equipment, such as heat exchangers, in both the forward and reverse directions (Fig. 22.5).

Dead legs and shadow areas should always be designed out of the process. A dead leg can occur whenever there is a tee (T) in the piping. If the valve on the branch line is more than one pipe diameter away from the main line, there is a good chance that the branch line will not be cleaned. Soil will accumulate, and the dirty branch line will become a source of infection. Fig. 22.6 shows an acceptable branch line and an unacceptable branch line. Dead legs sometimes occur when a bypass line is built around the piece of equipment. Fig. 22.7 shows an unsanitary bypass with dead legs and a correctly installed sanitary bypass.

The orientation of tees is also important. Tees for branch lines should always have a horizontal orientation. If the branch is oriented upwards, it creates an air trap that does not become fully wetted during CIP; thus, it can be a source of infection. If oriented downwards, it can be an undrainable sump of stagnant liquid that promotes microbial growth (Fig. 22.8).

Proper gasketing of pipe joints is also an important consideration in sanitary design. The inside diameter of gaskets should be the same as the inside diameter of the pipes. The edge of the

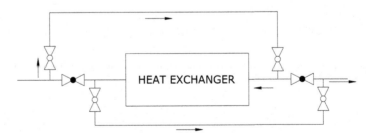

FIG. 22.5   Cleaning is improved by flow in both the forward and reverse directions.

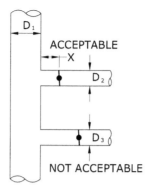

FIG. 22.6 When installing a tee, the valve on the branch line should be no more than one pipe diameter from the main line: $X \leq D_2$.

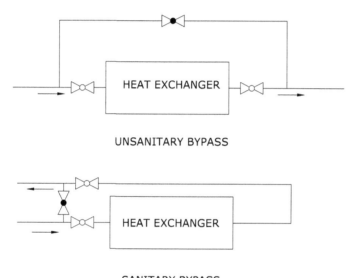

FIG. 22.7 Bypass lines can create difficult-to-clean dead legs. Use extra piping to ensure that all valves on tees are no more than one pipe diameter from the main line.

gasket will then become a smooth part of the pipe wall. Fig. 22.9 shows a sanitary gasket that is the correct size. Also shown are two gaskets, one with an ID that is too large and one with an ID that is too small. Both create unsanitary situations by providing places where soil can accumulate.

The construction and components of sanitary piping systems should comply with the standards outlined by the American Welding Society (American Welding Society, 2020), the European Hygienic Engineering Group (EHEDG Document No. 35, 2006), and 3-A Sanitary Standards, Inc. (3-A Sanitary Standards Inc., 2017). Sanitary welding of stainless pipes and tubing uses TIG welding, with inert gas purge and proper alignment, to ensure that the weld surfaces are inert and as hygienic as the pipe wall.

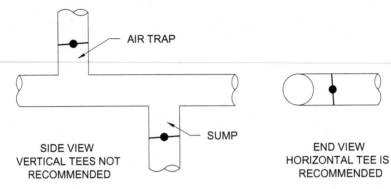

FIG. 22.8 The sanitary installation of tees is to have the branch line horizontal as shown in the end view *(right)*. When installed vertically upwards, an air trap can be created, and that branch does not become wet during cleaning. If the tee is installed vertically downwards, a sump is created and it may not be efficiently drained.

FIG. 22.9 For gaskets to be sanitary and cleanable, the inside diameter of the gasket must match the inside diameter of the pipe.

Turbulence and scrubbing action inside of tanks is achieved with a CIP sprayhead (or sprayheads). A CIP sprayhead, along with the correct flow and pressure for the cleaning solution, must be matched to the tank diameter and height. Low-pressure sprayheads operating up to 2bar (30psi) pressure are used in tanks up to 4m (13ft) in diameter (Barnes, 2018). Rotating sprayheads are preferred over fixed sprayballs because they achieve a more complete coverage of tank interiors. High-pressure sprayheads that operate at pressures around 7bar (105psi) are used in tanks up to 14m (46ft) in diameter (Resenhoeft, 2013). These are effective because in one cycle, usually lasting less than 20min, the entire inside surface of the tank receives the direct impact of the high-pressure stream of cleaning solution.

Shadow areas are common in tanks that are cleaned with a central sprayhead. When internal cooling coils are used, the tank wall behind the coil as well as the back side of the coil never receives a direct impact from the sprayhead—they remain dirty! Shadow areas can also exist around the tank's manway rim, behind the tank baffles, and on the back sides of the agitator blades. Cleaning of the shadow areas is achieved by adding more sprayheads or by employing manual cleaning. Fig. 22.10 illustrates two cleaning challenges in one vessel: a

FIG. 22.10 A wooden fermenter with internal cooling coils presents two cleaning challenges. The wooden walls and the floor have a porosity that makes cleaning and sanitising very difficult. The cooling coil has inherent shadow areas on the back side of the coil and on the tank wall behind the coil. Manual cleaning is often required.

wooden fermenter with internal cooling coils, where the porosity of wood harbours bacteria and the coils have shadow areas.

Cleaning chemistry

Cleaning processing equipment, as well as the building's walls and floors, requires cleaning materials and methods matched to the types of soil that must be removed. To select the right cleaning materials and determine the best way to use them, it is important to know the type of soil that is present on the dirty surface.

Types of soil

Carbohydrates are present in the grains and therefore present during mashing, fermentation, distillation, and the processing of by-products. Simple soluble sugars, including some non-fermentable sugars, may be left as a soil on surfaces but these are relatively easy to remove. Larger more complex carbohydrates, such as insoluble gums or retrograded starch, are a greater cleaning challenge, especially when they are deposited on hot surfaces such as heating jackets, mash coolers, and distillation heating surfaces.

Proteinaceous soils originate from soluble and insoluble proteins in the grains. Similar to carbohydrates, they can accumulate on surfaces and they sometimes form deposits that are difficult to remove. The fats and oils present in grains are insoluble in water but normally do not cause significant cleaning problems.

Carbohydrates, proteins, and fats are organic compounds, but scales are usually inorganic deposits, often calcium-based. Examples are calcium oxalate (beerstone) scales in fermenters and calcium carbonate scales in heat exchangers and heated tanks. Calcium, along with magnesium and iron, is present in the water as hardness.

Biofilms are combinations of several soils and contaminants, and they can develop into serious sources of infection. Biofilms become major cleaning challenges if they are not removed early. They start to form when microorganisms attach to surfaces and then combine with polysaccharide products (that they produce) and with other soils. A biofilm develops into a matrix of organic and inorganic soils with living microorganisms inside it. The last stage of biofilm development is the dispersion stage when it spreads to other surfaces and grows. The microorganisms contained in biofilms may have increased resistance to antimicrobials.

Types of cleaners and sanitisers

Water has been called the universal solvent, and by itself, it is an effective cleaner for some soils. Soluble sugars and proteins are easy to remove with water, and warming the water increases the solubility of many soils. Hard water (i.e. water with a high calcium content) is less effective for cleaning than soft water. When heated, it can produce a scale of calcium carbonate. For this reason, hard water should be softened when used for cleaning.

Water also has a relatively high surface tension, which keeps it from flowing freely and wetting the entire dirty surface, including the tiny cracks and crevices where soil can collect. For this reason, a wetting agent may be added to ensure that the water wets the entire surface.

Cleaners and sanitisers can be grouped into four categories (Kretsch, 1994):

- Alkaline cleaners—good for organic soils
- Acidic cleaners—good for beerstone and other scales
- Detergent additives—improve the performance of alkaline and acidic cleaners
- Sanitisers—kill organisms left by cleaners.

Alkaline cleaners have a pH above 7. Sodium hydroxide (NaOH), also termed caustic soda, is the most common alkaline cleaner. It is relatively inexpensive and is strongly alkaline and a good cleaner. A disadvantage is that it has poor rinsing properties, so rinsing agents (detergent additives) may need to be added to it. Other alkaline cleaners are more expensive than caustic soda but have some advantages. For example, potassium hydroxide (KOH) has better rinsability and is less corrosive to copper than sodium hydroxide. Trisodium phosphate (TSP, Na_3PO_4) has less alkalinity than caustic soda but is safer to use. It helps to soften hard water and is more effective when sodium hypochlorite is added. Sodium carbonate (Na_2CO_3) is also less alkaline than caustic soda, and it helps soften hard water. When a sodium hydroxide solution is used to wash a tank with a CO_2 atmosphere, the reaction (CO_2+NaOH) produces water and sodium carbonate, thereby decreasing the cleaning effectiveness of the original NaOH solution. Also, when the gaseous CO_2 reacts to form carbonate, the pressure inside the tank will decrease, sometimes rapidly enough to cause a tank implosion.

These alkaline cleaners are effective at breaking carbohydrate and protein bonds to solubilise organic soils. Fats and oils will be saponified by the alkaline cleaners to form free fatty acids and glycerol. However, if the water is hard, the free fatty acids (which become soap) will form a soap scum, making cleaning more difficult.

Acid cleaners are used to remove inorganic scales that are not removed by alkaline cleaners. Often an acid wash follows the alkaline wash that has already removed the organic soils. Phosphoric (H_3PO_4) and nitric (HNO_3) acid blends are typically used. Hydrochloric acid (HCl) and sulphuric acid (H_2SO_4) are used to remove rust and scale. These are much stronger acids, which has implications for worker safety and corrosion. Sulfamic acid (H_3NSO_3) is also used for inorganic soils because it has good descaling properties and is safer for workers than the stronger hydrochloric and sulphuric acids. Since acidic cleaning solutions are more corrosive than alkaline cleaners, care must be taken to use safe concentrations and temperatures. Acid cleaning is usually effective at 20°C to 40°C (68 to 104°F) whereas caustic detergents are used at 60°C to 80°C (140 to 176°F) (Rench, 2019).

Detergent additives are added to cleaning chemical solutions to improve their performance. Chelating agents are chemicals that combine with metal ions and solubilise them. In cleaning, chelators such as EDTA ($C_{10}H_{16}N_2O_8$) and sodium gluconate ($C_6H_{11}NaO_7$) are added to the base cleaning solution to prevent scale formation associated with calcium and magnesium (hardness) in the water. They can also be used to remove scales on surfaces; however, these chemicals are expensive and large amounts are required.

Surfactants improve the wetting ability of water by reducing its surface tension. This enables it to penetrate more places and clean more effectively. In CIP systems, nonionic surfactants are preferred over anionic surfactants because anionic surfactants create large amounts of foam.

Bleaches such as sodium hypochlorite (NaClO) or stabilised peroxides are blended with cleaners because they oxidise long-chain protein and polysaccharide molecules to smaller pieces that are more soluble. Chlorinated caustic and chlorinated TSP, usually at 200ppm (mg/L) chlorine, are effective in breaking down protein soils and removing beerstone.

Chemical sanitisers are used to kill the microorganisms that remain after cleaning; however, thorough cleaning must be conducted before the sanitising step. Sanitising dirt will not work in the long run. Sodium hypochlorite and iodophors are two widely used sanitisers. Iodophors have iodine complexed with a surfactant, and they release free iodine when dissolved in water. They are both inexpensive and have a wide range of microbiological kill. However, they have no residual activity and the hypochlorite is used up if any soil remains on the surface. Chlorine dioxide (ClO_2) is an effective sanitiser, and it can remove and prevent biofilms (Rench, 2019). Chlorine dioxide has a short half-life and must be generated in the plant daily. Peracetic acid (CH_3CO_3H), a combination of acetic acid and hydrogen peroxide, is expensive and corrosive. It has a broad range of kill and acid cleaning effects are sometimes claimed. Quaternary ammonium compounds (termed quats, positively charged polyatomic ions of the structure NR^+_4, where R is an alkyl group or aryl group) are expensive but have a fair range of kill and good residual activity. They are not always recommended for fermenters because their residual activity could kill yeast in the fermenter. Ozone, dissolved in water, is a powerful and effective sanitiser and disinfectant. Aqueous ozone kills bacteria and viruses by oxidation and also cleans the build-up of fat, fungi, and biofilms. In CIP systems, gaseous ozone is generated and dissolved in cold water at concentrations between 1.5 and 5.0mg/L (Hamil, 2017). Aqueous ozone has a short shelf life and cannot be stored.

CIP systems

Cleaning-in-place systems can be as simple as a portable pump connected with hoses to a tank that is being washed (Fig. 22.11). The tank is the solution reservoir, and the solution circulates from the tank outlet to the pump, then to the tank sprayhead, then down to the outlet. A circuit of additional product piping and hoses may be included en route to the sprayhead. A shortcoming of this arrangement is that the bottom of the fermenter is covered by the CIP solution and does not receive a direct impact from the sprayhead. Another disadvantage is that at the end of the cleaning cycle, the solutions are discarded into the sewer.

Other CIP systems have a series of tanks that may include fresh water, recycled final rinse water, caustic, acid, and sanitiser (McCrorie, 2012). These systems can be manually controlled but automatic computer control is preferred. Variables that should be controlled are chemical strength, solution flow rate, pressures, temperature, and time of circulation of each step. Different programs are used for the various cleaning routes and types of equipment in the plant.

Fig. 22.12 shows the same fermenter being washed using a CIP system that has only one solution tank. Depending on elevations, this may require a CIP return pump to move the solution back to the solution tank. A big advantage of this system is that the bottom of the fermenter is continually being drained by the return pump, allowing it to receive direct impact by the cleaning solution from the sprayhead. Another advantage is that detergent solutions can be saved and reused. The final water rinse in a CIP system is often saved in a recycled water tank and then used as a pre-rinse in the next cleaning cycle.

Fig. 22.13 shows an effective cleaning loop that includes the mash cooler, mash piping, and the fermenter. Developing cleaning loops like this makes it possible to clean and sanitise the

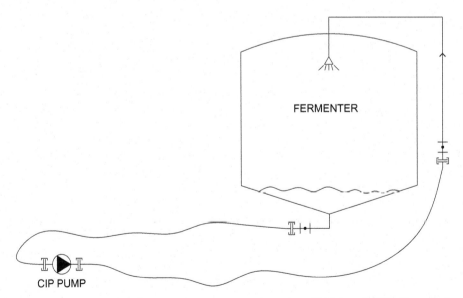

FIG. 22.11 Cleaning a tank where the solution reservoir is the tank itself. The pool of liquid on the floor of the tank prevents direct impact from the sprayhead.

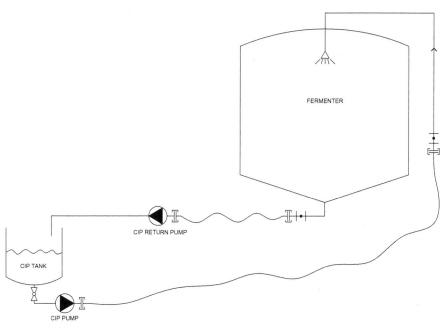

FIG. 22.12 Illustration of cleaning a tank where the solution reservoir is an external CIP tank. A CIP return pump may be required but the floor of the tank is drained, and it receives direct impingement from the sprayhead.

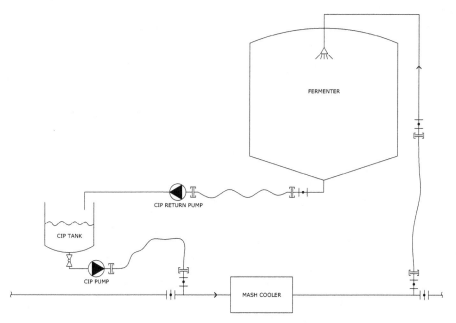

FIG. 22.13 Illustration of cleaning a tank where the solution reservoir is an external CIP tank. In this example, the mash cooler is included in the cleaning loop. In this way, the entire fermenter filling route can be cleaned and sanitised as a unit.

entire path that the mash must follow, from the time it leaves the mashing vessel when it is hot, through the mash cooler, and into the fermenter with the yeast.

Cleaning validation

After a cleaning process has been performed, the logical question is 'Is it clean enough?' This implies that there is a predetermined specification of cleanliness. Validation of the effectiveness of cleaning is a two-part process. First is verification that the process was performed correctly, meeting requirements for cycle times, temperatures, flows, pressures, and concentrations. These parameters should be recorded for every cleaning cycle. The second part is measuring the results of the cleaning to see if it was successful. There are several approaches to making these measurements.

Our senses

Although it is a subjective method, physical inspection of the cleaned equipment is an easy and economical option and, in many cases, an effective way to validate that the cleaning was performed properly. Can you see soil inside the tank that was washed (use of a flashlight is important)? Can you see if the CIP sprayhead or sprayballs are partially plugged? Does the vessel or pipe smell clean? Were the cleaning times, temperatures, flow rates, pressures, and concentrations correct and recorded? Is a heat exchanger heating as fast as it should? Tasting is often appropriate—for example, with water storage tanks. These are all easy and quick checks for validating the effectiveness of the cleaning process.

Wort stability

A simple, inexpensive, and tell-tale test for cleanliness is the wort stability test. A sample of cooled wort, aerated but not pitched, is collected in a small sterile jar, loosely capped, and then held at room temperature for 72h. If the sample remains clear with no bubbles or gas generation and no off-odours, the system cleaning was satisfactory. If the sample becomes cloudy, generates gas, or develops bad odours, cleaning needs to be improved. The faster any of the changes occur, the more urgent the need to improve cleaning.

Microbiological plating

Samples from pipelines and tanks can be inoculated onto selective agar plate media and incubated under controlled temperatures to test for microbes. A cleaned tank is opened, the wall or hard-to-clean area is swabbed, and the sample is streaked onto an agar plate or inoculated into liquid media. If microbes are present in the supposed sterile sample, colonies will appear on the agar plate, usually in 2 to 5days. The waiting time required for these results is not convenient, but microbiological plating can help identify the infecting bacteria if selective media are used.

Bioluminescence analyser

The bioluminescence analyser uses a swab method that gives a numerical cleanliness reading in less than one minute after sampling (Russell and Stewart, 2003). Adenosine tri-phosphate (ATP) is an indicator of biological contamination that can be detected on the swab in the presence of the enzyme luciferase. Light is emitted, and its intensity is measured with a luminometer and related to the concentration of ATP. The ATP concentration is used as a measure of the amount of soil on the surface that was swabbed. Based on experience, a 'go/no go' value can be established and used to decide if the cleaned equipment can be used or if it must be cleaned again. The identity of the soil or organism cannot be determined with this method.

An effective validation sample for vessels for all of these tests is the final rinse water from a freshly washed tank or pipe circuit. It can be difficult to obtain representative samples from large tanks or from pipelines in areas that are not accessible. A sample of this draining rinse water can checked visually and chemically for complete rinsing, it can be plated for microbi-ological identification of contaminants, and a bioluminescence test can be used as a go/no go test to decide if the tank or pipe circuit can be used or if it must be rewashed.

Epilogue

As the Third Edition of The Whisky Book is being published, the world is still learning how to control and defeat the COVID-19 pandemic. Providing clean, safe conditions in the workplace for employees and visitors has become paramount along with continuing to man-ufacture products that are safe and of the highest quality. Personal behaviour and practices have had to change to include social distancing, wearing face masks, and frequent washing or sanitising of hands. Changes to the plant itself have included increased physical distances between people within the plant and upgraded ventilation systems.

Prof. Michael Lewis has suggested that our routine battles against beer spoilage bacteria may be a metaphor for the immense battles being fought today against the COVID-19 coro-navirus. 'Unfortunately, science does not have a magic wand. Science takes time and circum-spection to do the job and it is up to all of us, as a community, to provide those necessary commodities' (Lewis, 2020).

The rules are constantly changing as more and more is being learned about how to control this virus and its variants and how to stay safe in our everyday activities. New methods and practices for keeping manufacturing plants and personal environments safe are continually being developed. As more is learned, it will be the responsibility of individuals, businesses, and governments to trust and adapt as this knowledge is expanded.

Acknowledgements

The author is grateful for the discussions and advice from distillers Roy Court, Conor O'Driscoll, and Jerry Summers, and especially for the insight that they provided in regard to the cleaning requirements and practices in distilleries.

References

21 CFR 117, 2020. Title 21 of the Code of Federal Regulations, Current Good Manufacturing Practice, Hazard Analysis, and Risk-Based Preventive Controls for Human Food. Available from https://www.ecfr.gov/cgi-bin/text-idx?SID=3f7125891958b3eb9380856f26065b5c&mc–true&node=pt21.2.117&rgn–div5.

3-A Sanitary Standards Inc, 2017. 21-A Sanitary Standard for General Requirements 00-01. 3-A Sanitary Standards Inc, McLean, VA.

American Welding Society, Miami, FL, 2020. D18.1/D18.1M:2020 Specification for Welding of Austenitic Stainless Steel Tube and Pipe Systems in Sanitary (Hygienic) Applications, third ed. D18_1_D18_1M_2020_PV https://pubs.aws.org/.

Aylott, R., 2014. Whisky analysis. In: Russell, I., Stewart, G.G. (Eds.), Whisky Technology, Production and Marketing, second ed. Academic Press, London, pp. 243–270.

Baking Industry Sanitation Standards Committee, 2004. Design Handbook for Easily Clean able Equipment, third ed. Available from http://www.bissc.org/designtoc.html.

Barnes, Z.T., 2018. Cleaning in place (CIP). In: Stewart, G.G., Russell, I., Anstruther, A. (Eds.), Handbook of Brewing, third ed. Boca Raton, FL, USA, CRC Press, pp. 415–432.

Codex Alimentarius International Food Standards Committee, 2003. General Principles of Food Hygiene, CAC/RCP 1-1969. World Health Organization, Food and Agriculture Organization of the United Nations, Geneva, Switzerland. Available from www.codexalimentarius.org/standards/list-of-standards.

EHEDG Document No. 35, 2006. Hygienic welding of stainless steel tubing in the food processing industry. Available from https://www.ehedg.org/guidelines/.

Foote, G., 2013. Efficiency gains possible through high heat transfer. Ethanol Producer Mag. 19 (12), 52–55.

Hamil, B., 2017. Ozone Sanitation – A Sustainable and Efficacious Approach to Food Safety. Available from https://www.researchgate.net/publication/320237844_Ozone_Sanitation-A_Sustainable_and_Efficacious_Approach_to_Food_Safety.

Kretsch, J., 1994. Practical considerations for brewery sanitation. Tech. Quart. Master Brew. Am. 31 (4), 124–128.

Larson, J., Power, J., 2003. Managing the four Ts of cleaning and sanitation: time, temperature, titration and turbulence. In: Jacques, K.A., Lyons, T.P., Kelsall, D.R. (Eds.), The Alcohol Textbook, fourth ed. Nottingham University Press, Nottingham, UK, pp. 299–318.

Lewis, M., 2020. Point of Brew: Beating microbes takes time and patience. Enterprise. May 26. Available from https://www.davisenterprise.com/features/food-and-drink/point-of-brew-beating-microbes-takes-time-and-patience/.

McCrorie, C., 2012. New developments in CIP technology. Brew. Distiller Int. 8 (7), 12–19.

Narendranath, N.V., 2003. Bacterial contamination and control in ethanol production. In: Jacques, K.A., Lyons, T.P., Kelsall, D.R. (Eds.), The Ethanol Textbook, fourth ed. Nottingham University Press, Nottingham, UK, pp. 287–298.

Rench, R.J., 2019. Brewery Cleaning: Equipment, Procedures, and Troubleshooting. Master Brewers Association of the Americas, St. Paul, Minnesota, USA.

Resenhoeft, J.W., 2013. Tank cleaning with rotary impingement technology. Ethanol Producer Mag. September 19. Available from http://ethanolproducer.com/articles/10252/tank-cleaning-with-rotary-impingement-technology.

Russell, I., Stewart, R., 2003. Rapid detection of microbial spoilage. In: Priest, F.G., Campbell, I. (Eds.), Brewing Microbiology, third ed. Kluwer Academic, New York, pp. 278–282.

Singh, M., Fisher, J., 2003. Cleaning and disinfection in the brewing industry. In: Priest, F.G., Campbell, I. (Eds.), Brewing Microbiology, third ed. Kluwer Academic, New York, pp. 337–366.

Sinner, H., 1959. Uber das Waschen mit Haushaltwashmaschinen. Haus/Heim, Verlag, Germany.

Tegels, J., Mundell, K., 2013. Absolute energy tackles evaporator fouling. Ethanol Producer Mag. 19 (6), 84–86.

UCSF Institute for Health & Aging, 2013. Green Cleaning, Sanitizing, and Disinfecting: A Curriculum for Early Care and Education. Available from https://www.epa.gov/sites/production/files/documents/ece_curriculumfinal.pdf.

Gin

Kirsty Black

Arbikie Highland Estate Distillery, Arbroath, United Kingdom

Medicinal origins

As with many of the spirits we drink today, the history and origin of gin are full of diverse and vibrant tales. Although we will leave these legends and myths to books dedicated to the subject, we can be reasonably sure that gin's origin was medicinal. For example, one of the first documented uses of juniper is from Egyptian times, around 1500BCE, for curing ails such as headaches or toothaches (Coates, 2012). From these medicinal beginnings, its path to the current day has, much like whisky, been shaped by parliamentary acts, spirit taxation, and improvements in equipment capabilities.

It is believed that the addition of plant materials to white spirit first arose as a method of improving the harsh and fiery character of newly made distillate; it was an alternative to ageing, delivering a relatively potable spirit in a far shorter time (Hastings, 1984; Simpson, 1977). By the thirteenth century in Flanders, one such drink had begun to evolve—a sweet, malt-flavoured spirit, whose less desirable character was tempered via the addition of juniper (Solmonson, 2012). This spirit was known as Jenever, after the Dutch word for juniper, and is believed to be the origin of the drink we know today as gin. Although it was not until the sixteenth and seventeenth centuries that this plant-flavoured distillate started to appear in the UK, it immediately became intertwined with the country's social and political history (Aylott, 2003; Solmonson, 2012). Over the years, gin has been accused of such crimes as causing the spread of poverty, promiscuity, high infant death rates and the crippling of the City of London as a whole, yet, it still became known as the quintessence of sophistication (Coates, 2012; Solmonson, 2012). British soldiers fighting in the Low Countries returned to the UK with a taste for Jenever that William of Orange, on becoming King of England in 1688, was only too happy to encourage. As an enemy of France, the import of French drinks was banned whilst home grain distillation was positively promoted (Aylott, 2003; Coates, 2012; Simpson, 1977; Solmonson, 2012). This led, ultimately, to an increase in gin production from just half a million imperial gallons across all of England in 1690 to London alone producing 20 million gallons of gin in 1743 (Aylott, 2003; Solmonson,

2012). By this point, the Dutch production techniques had been deemed too demanding and low-quality grain spirits diluted with substances such as oil of turpentine, vitriol, or alum were being used, thus increasing the need for masking, or flavouring agents more than ever (Solmonson, 2012). The gin craze had well and truly taken hold in London—the poor quality and high potency of the spirit led to malnutrition and crime within the lower classes. Parliament reacted by introducing new acts to try and reinstate some level of control, but the opposite resulted—illegal production and the sale of gin continued to grow and flourish (Simpson, 1977; Solmonson, 2012). Eventually, in 1751, the final Gin Act was passed introducing the concept of licensed sellers versus the previous approach of spirit taxation and order began to be restored (Dillon, 2003; Solmonson, 2012). The success of this Act was reinforced via changing alcohol preferences of the time, an overall decrease in wages and poor grain harvests (Solmonson, 2012). The latter Act briefly put a stop to gin production in the late 1750s, and on its return, a higher quality spirit appeared on the market, associated with the invention of new distillation equipment, removing the need for cutting or masking agents (Solmonson, 2012). Despite this, the use of botanicals and sweetening with sugar continued even though they were no longer required as flavouring aids—two of the distinct gin styles we know today had arrived! The sweetened version met the taste and preference of the gin drinkers of this time and is in the style of Old Tom. The unsweetened version become known as London Dry due to the absence of sugar and was seen as a healthier alternative favoured by the wealthier classes (Solmonson, 2012). Throughout the 1800s, London Dry became established as a quality and respected product. In parallel to this, the consumption of tonic water was also increasing but not due to its favourable flavour profile. As both preventative and curative against malaria, the bitter cinchona-based drink had become an essential medicine within the British colonies where mixing with gin, water, sugar, and citrus was deemed necessary to make it more palatable. The gin and tonic was born. Gin has continued to cycle in and out of favour throughout the last two centuries with its most recent peak in popularity, prior to today, occurring in the 1950s and 60s. It had once again become the stylish drink associated with the fame and glamour of Hollywood (Solmonson, 2012) but even this was short lived as the original, pure, white spirit—vodka—grew in favour. Although, today, vodka still dominates the spirit marketplace, gin's popularity is increasing. Despite gin representing only 5% of the 153 brands listed in the 2019 'Millionaires' Club', a list of the world's million case spirit brands published by Drinks International, totalling less than a third of the vodka volume (50.6 million versus 155.6 million 9-L cases), we are clearly in the midst of a boom in global gin consumption (Fig. 23.1).

Gin definitions

Gin, in the simplest of terms, can be defined as a juniper-flavoured spirit. Details of production methods and limitations vary by country with some examples summarised in Table 23.1. Below these general definitions are subclasses of gin, which, depending on the country, may be legally defined or used informally as descriptive styles. The most common of these are as follows:

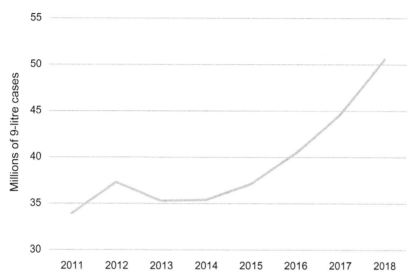

FIG. 23.1 Volume of 9-L cases sold by gin brands featured in the Drinks International "Millionaires' Club" (Drinks International, 2019).

1. **London Dry**—a style of distilled gin rather than an indicator of its geographic origin. With a classic dry (versus sweet) gin flavour profile with restrictions on base spirit quality, production methods, alcoholic strength, and additions as defined within the EU regulations (Reg, 2019/787):
   (a) *London gin* is distilled *gin* which meets the following requirements:
   (i) it is produced exclusively from ethyl alcohol of agricultural origin, with a maximum methanol content of 5g per hectolitre of 100% vol. alcohol, the flavour of which is imparted exclusively through the distillation of ethyl alcohol of agricultural origin in the presence of all the natural plant materials used;
   (ii) the resulting distillate contains at least 70% alcohol by vol.;
   (iii) any further ethyl alcohol of agricultural origin that is added shall comply with the requirements laid down in Article 5 but with a maximum methanol content of 5g per hectolitre of 100% vol. alcohol;
   (iv) it is not coloured;
   (v) it is not sweetened in excess of 0.1g of sweetening products per litre of the final product, expressed as invert sugar;
   (vi) it does not contain any other ingredients than the ingredients referred to in points (i), (iii) and (v), and water.
   (b) The minimum alcoholic strength by volume of *London gin* shall be 37.5%.
   (c) The term '*London gin*' may be supplemented by or incorporate the term '*dry*'.
2. **Old Tom**—a lighter, sweeter style of gin compared to London Dry with the sweetness being introduced through the botanical selection, for example, by using a higher ratio of liquorice root, or the addition of sugar. As with many aspects of the history of gin, the name is surrounded with tales and fables, but one of the most common is the result of

TABLE 23.1 Global regulatory or legal gin definitions.

Canada
(Canadian Food and Drug
Regulations, 2020)
	C.R.C., c. 870, B.02.041 Gin
	(a) shall be a potable alcoholic beverage obtained
	    (i) by the redistillation of alcohol from food sources with or over juniper berries, or by a mixture of the products of more than one such redistillation, or
	    (ii) by the blending of alcohol from food sources, redistilled with or over juniper berries, with alcohol from food sources or by a mixture of the products of more than one such blending;
	(b) may contain
	    (i) other aromatic botanical substances, added during the redistillation process,
	    (ii) a sweetening agent, and
	    (iii) a flavouring preparation for the purpose of maintaining a uniform flavour profile; and

EU (EU Council Regulation
No. 2019/787, 2019)
	Regulation (EU) 2019/787 Annex 1
	20. Gin
	(a) Gin is a juniper-flavoured spirit drink produced by flavouring ethyl alcohol of agricultural origin with juniper berries (*Juniperus communis* L.).
	(b) The minimum alcoholic strength by volume of gin shall be 37.5%.
	(c) Only flavouring substances or flavouring preparations or both shall be used for the production of gin so that the taste is predominantly that of juniper.
	(d) The term 'gin' may be supplemented by the term 'dry' if it does not contain added sweetening exceeding 0.1g of sweetening products per litre of the final product, expressed as invert sugar.
	21. Distilled gin
	(a) Distilled gin is one of the following:
	    (i) a juniper-flavoured spirit drink produced exclusively by distilling ethyl alcohol of agricultural origin with an initial alcoholic strength of at least 96% vol. in the presence of juniper berries (*Juniperus communis* L.) and of other natural botanicals, provided that the juniper taste is predominant;
	    (ii) the combination of the product of such distillation and ethyl alcohol of agricultural origin with the same composition, purity and alcoholic strength; flavouring substances or flavouring preparations as specified in point (c) of category 20 or both may also be used to flavour distilled gin.
	(b) The minimum alcoholic strength by volume of distilled gin shall be 37.5%.
	(c) Gin produced simply by adding essences or flavourings to ethyl alcohol of agricultural origin shall not be considered distilled gin.
	(d) The term 'distilled gin' may be supplemented by or incorporate the term 'dry' if it does not contain added sweetening exceeding 0.1g of sweetening products per litre of the final product, expressed as invert sugar.

USA (US Code of Federal Regulations, 2012)	TTB BAM Chapter 4/27 CFR 5.22 "Gin" is a product obtained by original distillation from mash, or by redistillation of distilled spirits, or by mixing neutral spirits, with or over juniper berries and other aromatics, or with or over extracts derived from infusions, percolations, or maceration of such materials, and includes mixtures of gin and neutral spirits. It shall derive its main characteristic flavour from juniper berries and be bottled at not less than 80° proof. Gin produced exclusively by original distillation or by redistillation may be further designated as "distilled".

the UK Gin Act of 1736, which aimed through the implementation of taxes to reduce gin consumption. Consequently, it forced people to become stealthier in their daily intake of gin, with one entrepreneurial developer devising a tom cat shaped gin dispenser, which on depositing a coin would provide the customer with a shot of gin.
3. **Navy Strength**—a higher strength gin with an alcohol content typically 57% alcohol by volume (A.B.V). Although no signature botanical flavour profile exists, it was historically thought to have an earthier flavour profile. Purported to be named as such due to the British Navy's love of gin and the above alcoholic strength that if mixed with gunpowder would still allow it to burn.

Gin production

Despite such a simple definition, with just two components: spirit and botanicals, the options for gin production are almost endless.

Base spirit

The base for any high-quality gin is a high-quality starting spirit. Such a spirit is produced traditionally from grain via a continuous distillation process, although with the increase in global gin production, alternative base raw materials are becoming more common including molasses, apple, grape, and potato, to name but a few. This process results in highly rectified ethanol with few flavour or aroma compounds (congeners) present. The purity of the neutral spirit is key to enable the flavours and aromas of the added botanicals to shine through and be the prime flavour source (Aylott, 2003).

Botanicals

The term 'Botanical' encapsulates any plant matter that can be used to flavour gin, from the flowers, foliage, bark, and roots of the plant through to its fruit, seeds, and berries. Each flavouring has the potential to add unique and varying flavours even when originating from the same plant source. It is their volatile essential oils that are key, primarily consisting of terpenes, principally mono-(C10) and sesquiterpene (C15) hydrocarbons.

As detailed in the regulations listed in Table 23.1, the main characteristic flavour of gin must be of juniper, making it the common ingredient for all white spirits labelled as gin. There are no other restrictions on the specific botanicals that must be employed, yet there is a set of commonly used botanicals, which can be referred to as 'classical botanicals'. The popularity of these botanicals is a result of the aromatic, volatile compounds that they contain, and the characteristic flavour of gin can be attributed to these botanicals. Not for the first time, distillers have taken advantage of a plant's inherent qualities. Occurring naturally, these compounds have a variety of roles, but the majority of these are produced to attract pollinators or repel attacks rather than be appealing to distillers (Brahmkshatriya and Brahmkshatriya, 2013; Kempinski et al., 2015)!

The oil content of the plant material used is impacted by its genetic structure, geographic location, environmental factors (altitude, rainfall, temperature, photoperiod), growing conditions (nutrients, fertilisers), the time of harvest (season, time of day), the age of the individual plant, and the time in storage (Letchamo et al., 1995; Carrubba et al., 2002; Chizzola, 2013; Telci et al., 2006; Ramezani et al., 2009; Ayala et al., 2017).

Juniper (Juniperus communis L. Cupressaceae family)

Over 50 species are found in the genus *Juniperus*, part of the Cupressaceae, or cyprus family and, as such, the botanical we refer to as berries are in fact cones where the scales have fused tightly together. These resinous shrubs have an expansive habitat, being extensively found across the northern hemisphere in both cold and warm climates. The most common source of berries is from *Juniperus communis* L., the common juniper, which is named specifically within the EU gin regulations (Table 23.1), but other berries such as *J. californica* and *J. occidentalis* are also used. However, not all of the species are safe for the production of gin and care must be taken in selecting the source and suppliers of the botanicals. For example, all parts of the savin juniper (*Juniperus sabina*) contain terpenes that are poisonous to humans if consumed (Wexler et al., 2015). The juniper plant is dioecious, meaning that it is not self-fertile; instead, male and female flowers are found on separate plants with pollination happening by the wind. This, combined with the fact that it takes many years for a juniper plant to reach flowering maturity, and once a cone has formed it takes 18–24 months to ripen, means that established juniper plants are of great value. This is true now more so than ever due to a fungus-like pathogen called *Phytophthora austrocedrae* that is currently posing a significant threat to wild UK juniper populations (Green et al., 2015; Donald et al., 2020).

Nearly all juniper berries are harvested in the wild rather than from cultivated plants. The berries are picked by hand, prior to cleaning and sorting by both size and colour. The ripe berry contains more than 100 volatile compounds with the exact composition varying with season and geographical location, but α-pinene remains the dominant component followed by β-myrcene. Both are monoterpenes with pine and woody qualities that are critical to the quintessential gin aroma. Other compounds include α-cadinol, α-terpineol, terpinen-4-ol, δ-cadinene, limonene, bornyl acetate, β-pinene, and β-phellandrene (Koukos and Papadopoulou, 1997; Butkienë et al., 2005). Table 23.2 lists the flavour and aroma compounds associated with these compounds.

Although it is the ripe, dark blue, purple, black berries we associate with gin, some producers are turning to use green berries and needles in flavouring their spirits. Although, again, the volatile compounds vary greatly depending on their geographic origin, generally unripe berries are higher in α-pinene and lower in β-myrcene than ripe berries. The needles of the plant are still dominated by α-pinene but higher levels of β-phellandrene and δ-cadinene occur (Butkiene et al., 2006; Labokas and Ložiene, 2013; Kazemi et al., 2017).

Coriander (Coriandrum sativum L., Apiaceae family)

Coriander is an annual plant cultivated across the globe as a herb, spice, and as a source of essential oil. Its earliest mention is in a Babylonian clay tablet as a stew flavouring. Historically referred to as 'pungent', its smell earned it the title of 'koris', meaning bug in Ancient Greek,

TABLE 23.2 Common botanical essential oil sources and their associated flavours (Brenna et al., 2004; Burdock, 2016).

Compound	Descriptors	Aroma threshold	Taste threshold
Bornyl acetate	Fresh, woody, piney, mentholic	75μg/L–1.38mg/L	10mg/L
δ-Cadinene	Herbal, woody	–	–
α-Cadinol	Herbal, woody		
Camphene	Camphoraceous, cooling, piney, woody, minty, citrus	10%	50–100mg/L
Camphor	Warm, minty, medicinal	1–1.29mg/L	20mg/L
δ-3-Carene	Sweet, pungent, turpentine	–	–
Citronellal	Lemon, rose	31–100μg/L	10mg/L
Geraniol	Rose, sweet, citrus	4–75μg/L	10mg/L
Geranyl acetate	Floral, green, fruity	9–460μg/L	20mg/L
α-, β-, γ-Irone	Violet, floral, warm, fruity, sweet floral-woody	–	10mg/L
Limonene	Sweet, citrus, lemon, orange	4–229μg/L	30mg/L
Linalool	Green, apple and pear, citrus, floral	4–10μg/L	5mg/L
Linalyl acetate	Floral, green, waxy, citrus, herbal	1mg/L	5mg/L
Myrcene	Woody, balsamic, citrus, floral, tropical	10%	5–100mg/L
β-Phellandrene	Fresh, citrus, peppery, minty, herbaceous	40–200μg/L	20mg/L
α-Pinene	Woody, piney, minty	2.5–62μg/L	10mg/L
β-Pinene	Fresh, piney, woody, minty	140μg/L	15–100mg/L
Sabinene	Black pepper	–	–
Terpinen-4-ol	Grapefruit, lime, herbaceous, pepper	–	–
γ-Terpinene	Citrus, herbaceous	–	–
α-Terpineol	Lilac, pine, woody, citrus	280–350μg/L	2–25mg/L

which developed into the Latin *Coriandrum sativum* (literally meaning 'bug-cultivated') and ultimately the shortened term, coriander, as we know it today (Charles, 2012b).

Although both the leaves and seeds can be used in gin production, with very different flavour profiles associated with each flavouring, it is the seeds that are most commonly used, and they form the second grouping of 'classical' botanicals. Although referred to as seeds, they are in fact fruits, with each consisting of two fused pericarps both containing a single seed inside. The odour and flavour of these 'seeds' are described as fresh, sweet, and aromatic, with citrus, warm, spicy, mint tones (Krist, 2020; Charles, 2012b). The oil content is affected by the aforementioned conditions, and, although the overall oil yield increases with water and nutrient availability, linalool, the main oil constituent, can be seen to increase in drier growing conditions suggesting it is a plant stress reaction in response to drought and higher temperatures (Telci et al., 2006; Nadjafi et al., 2009; Hani et al., 2015; İzgı et al., 2017; Milica et al., 2016). The pleasant, fresh aroma of the seeds can mainly be attributed to linalool, frequently accounting for more than 70% of the oils present. Other components include γ-terpinene, camphor, α-pinene, geraniol, limonene, and geranyl acetate with further details of each presented in Table 23.2 (Bandoni et al., 1998; Carrubba et al., 2002; Telci et al., 2006; Padalia et al., 2011; İzgı et al., 2017).

Angelica (Angelica archangelica L., Apiaceae family)

Both the Latin name and the common name, Holy Ghost, would suggest angelica's past is thought to be intertwined with angels and powerful curative properties. It has a long history of being used for medicinal purposes (Charles, 2012a). Unsurprisingly, as with the many botanical extracts used today, it has found its way into a variety of common herbal-flavoured drinks such as Chartreuse, Strega, Fernet, and, of course, gin.

Angelica is cultivated in northern, central, and western Europe. All parts of the angelica plant have been traditionally used but, as commonly seen in members of the Apiaceae family, the highest oil content is found in the roots, rhizomes, and fruits (Roslon et al., 2016). It is the roots and rhizomes that are usually used in gin production. Angelica imparts spicy, herbaceous, earthy, and woody flavours (Charles, 2012a), which can be attributed to the main components: α-pinene, δ-3-carene, myrcene, limonene, and β-phellandrene. The dominant compound varies with the origin and growing conditions (Bernard and Clair, 1997; Chalchat and Garry, 1997; Paroul et al., 2002; Nivinskienė et al., 2005; Forycka and Buchwald, 2019). More details of each can be found in Table 23.2.

Citrus (Citrus spp, Rutaceae family)

True citrus is thought to have originated in Southeast Asia millions of years ago in the form of *Citrus medica* (citrons), *Citrus maxima* (pomelos), *Citrus micrantha* (papedas), and *Citrus reticulata* (mandarins). Through their ease of cross fertilising and hybridisation, the many thousands of cultivated species currently available have all resulted from just these four aforementioned species (Ghada et al., 2019). The rinds of orange (sweet, *C. sinensis* L., and bitter, *C. aurantium* L.), lemon (*C. limon* L.), and lime (*C. aurantiifolia* L.) are most commonly used in gin production; however, the flowers, leaves, and whole fruits can also be used. Although different species and cultivars result in congener yield and compositional differences, some generalisations can be made. Citrus peels are dominated by D-limonene (approx. 90%), whereas the leaves and flowers are mainly linalool (Ayala et al., 2017). Other peel components include γ-terpinene, myrcene, sabinene, camphene, α-pinene, and β-pinene, and the leaves can also contain limonene, linalyl acetate, citronellal, and β-pinene (Sato et al., 1990; Boussaada et al., 2007; Viuda-Martos et al., 2009; Kirbaslar et al., 2009; Hosni et al., 2010; Ayala et al., 2017). Therefore, despite their similarities, the citrus family can introduce a variety of citrus notes to gin, from fresh and crisp to herbal and rich.

Liquorice (Glycyrrhiza glabra L., Fabaceae family)

Liquorice is cultivated across Europe, the Middle East, and Asia. The Latin name translates to sweet ('glykys') root ('rhiza'), and this characteristic of the liquorice plant has been well documented, appearing in a Roman Empire papyrus, in Assyrian tablets, and in one of the first Chinese herbal books (Charles, 2012c; Hosseini et al., 2018). The sweetness is due to the presence of glycyrrhizin, a triterpene saponin 50 times sweeter than sucrose (Omar et al., 2012), which, in the production of gin, adds an element of sweetness along with volatile compounds such as geraniol, terpinen-4-ol, and α-terpineol (Pastorino et al., 2018).

Orris (Iris germanica L., I. pallida L., I florentina L., Iridaceae family)

Orris root, or 'root of iris' (Skeat, 1907), is reputed to be worth more than gold, making it, once hydrodistilled to 'orris butter', the most expensive ingredient used in the fragrance industry (Masson et al., 2014). Obtained from the rhizomes of the bearded iris, an extensive, low-yielding process over many years must be completed to produce this sought-after product. The rhizomes must be at least two years old for harvesting, after which they are subsequently cleaned, peeled, and dried over multiple years in a dry, aerated environment. During this curing period, oxidative degradation of iridals to irones occurs (Brenna et al., 2003; Schütz et al., 2011). It is these irones (α-irone, β-irone, and γ-irone) that are at the heart of the distinctive violet-like smell that has made it, along with its natural fixative properties, so popular in perfumery and also to the distiller (Krick et al., 1983). The different isomers vary in aroma as does the quantity and ratio of the irones present, which vary with species and origin, with the Italian orris obtained from *Iris pallida* L. thought to be of the highest quality (Brenna et al., 2003, 2004).

This is just a small selection of the botanicals used. Most gins on the market will contain at least some of these (in addition to juniper). As there are an estimated 391,000 vascular plant species in the world, with on average over 2000 new species identified annually, there is no shortage of options for an imaginative distiller (Kew, 2016, 2017) (Box 23.1).

Box 23.1   Botanical selection and sourcing

In all cases, the botanicals chosen must be non-toxic and safe for human consumption. Certain countries have regulations in place detailing those plants that can be used in the production of food and drink. For example, the USA FDA list of substances is classified as GRAS (Generally Recognised As Safe).

Botanicals should be sourced from a reputable supplier with procedures and processes in place to assure that the delivered material is of the correct species and free of contamination (physical, chemical, biological). Storage should be in a cool, dry environment to maintain their integrity and prevent future contamination.

How to get botanical flavours into the spirit

Fundamentally, gin is produced by introducing juniper and other botanical flavourings to an ethanol-water mixture. There are, however, two main production methods, and therefore, subclasses of the spirit (see Table 23.1) are recognised legally *per* country-specific regulations—non-distilled gin and distilled gin.

Non-distilled gin

As the name would suggest, non-distilled gin does not involve a distillation step. Instead, flavouring substances are mixed with spirit of a suitable quality as defined by a country's specific regulations. For example, in the EU, the 'ethyl alcohol of agricultural origin' to be used must be of the grade detailed in Article 5 of Regulation 2019/787.

The botanical flavourings used can be essences, extracts, flavours, or plant materials, depending on what a country's specific regulations allow. Therefore, in some countries, a non-distilled gin can be produced by merely mixing neutral spirit with juniper essence.

Where plant materials are used as the flavouring source, discussed in more detail below, decisions must be made regarding the strength of the alcohol used, the duration of botanical maceration, the quantity, number, and preparation of botanicals used, and, following removal of the botanicals via filtration, whether the spirit is to be sweetened or not. Colour, as well as flavour, is extracted when plant material is used. Therefore, non-distilled gins can have a pale yellow hue unless the distiller chooses to remove this colouration by filtration.

Non-distilled gins are commonly referred to as compounded gins or bathtub gins. The latter referring to tales of prohibition gin made by taking spirit and essences and, of course, mixing it all up in a bathtub!

Distilled gin

Again, the name distilled gin is reasonably self-explanatory—botanicals and spirit are distilled together to result in a botanically flavoured spirit. The method, equipment, and approach to distilling with botanicals can vary greatly. In general terms, diluted neutral spirit and botanicals are loaded into the still and heat applied. At this point, the still charge is at its highest A.B.V., and as the temperature rises within the still, the most volatile congeners begin to evaporate. These congeners are mainly monoterpenes although some sesquiterpenes can be present from the previous distillation procedure. On condensing, these are collected and classified as the 'heads' fraction of the run. The 'heads' may include unfavourable flavours and can cause quality issues in the final gin. Once the A.B.V. of the distillate decreases to approximately 80%, desirable congeners begin to appear, such as terpineols and sesquiterpenes, and the level of monoterpenes present decreases (Clutton and Evans, 1978). These are collected and on dilution form the final gin. As alcohol content in the still continues to drop, the water-to-ethanol ratio adjusts and less desirable compounds become volatile—once these start to appear in the distillate, the final cut is made, and after this point, the distillate is collected separately and classified as 'tails'.

When producing a distilled gin, the same considerations as when producing a non-distilled gin must be made along with some additional factors relating to the distillation technique. This includes the strength of alcohol used both for maceration and pot charge, the duration of botanical maceration, the quantity, number, preparation, and still position of the botanicals used, the still type, and, following distillation, whether the spirit is to be sweetened or not.

In the European Union definition (EU Regulation 2019/787), gin is produced by flavouring ethyl alcohol of agricultural origin and it may be called 'distilled gin' if the alcohol is redistilled in the presence of the botanicals, or 'compound gin' where essences or flavourings are added to the alcohol without further distillation. The definition of 'distilled gin' in the US Code of Federal Regulations Title 27 Part 5 includes the EU definition and also includes the option of the original distillation of the mash and juniper berries together (see Table 23.1).

Production consideration details

The following factors should be considered when developing a gin recipe and the production methodology. Most are dependent on the equipment available and the distiller's personal preference, although it is worth noting that once a recipe has been finalised, changing any of the other factors is likely to modify the flavour profile of the end spirit.
• Botanical recipe

Although a large number of gin producers advertise and promote their botanical usage, very few are as open when it comes to the weights and quantities used. Table 23.3 shows a recipe for a classical gin developed using published theoretical recipes, where the weight of juniper is given as an absolute value and all other botanicals are calculated as a percentage of this. Despite all other botanicals seemingly having such a low value, this is a reflection of their relative oil content and will all still feature in a final balanced gin recipe.

TABLE 23.3 Theoretical recipe for a classical gin.

Botanical	Formula	Dose (g/L 50% A.B.V. charge)
Juniper Berries	x	4.00
Coriander Seeds	$x/2$	2.00
Cinnamon Bark	$x/10$	0.40
Angelica Root	$x/10$	0.40
Liquorice Root*	$x/10$	0.40
Lemon Peel	$x/100$	0.04
Orange Peel*	$x/100$	0.04
Cardamom	$x/100$	0.04
Orris Root*	$x/100$	0.04

Data calculated based on a hypothetical gin formula published by Willkie et al. (1937). Those botanicals identified with an asterisk (*) were not included in the original formula.

- Botanical preparation

Botanicals can be used in their whole state or processed further by crushing or milling. For example, botanicals such as orris root can be purchased in both kibbled and powdered form. Whilst such processing techniques increase the surface area by weight, allowing greater ethanol contact and accelerated flavour extraction, it can also increase the extraction of undesirable compounds, and therefore, maceration times and botanical quantities may need to be adjusted accordingly. The processing conditions themselves can also result in a decrease in the essential oil content. For example, Saxena et al. (2014) found both a significant loss and a significant increase in coriander volatile oil content depending on the grinding technique used when compared to whole seeds.

- Botanical positioning

The most commonly used gin production method is steep infusion, extracting the compounds through direct contact with the ethanol solution. There are conflicting views on this topic with one side believing that steeping, or macerating, allows for fuller extraction whereas the opposite view claims that it causes stewing and flavour loss (Coates, 2012). As a result, the use of macerating and its duration varies greatly between distilleries. Alternatively, extraction can be achieved using ethanol vapours, known as the vapour infusion method. With this method, the botanicals are positioned above the surface level of the ethanol within the pot—the exact position and container can vary from a mesh bag suspended inside the pot, to trays within the head of the still or separate in the lyne arm, external from the pot but before the condenser. Anecdotally, vapour-infused gins are typically considered to be lighter in flavour. However, Hodel et al. (2019) when comparing an identical recipe through steep and infusion found the vapour infusion method to be richer in 9 out of the 10 terpene compounds studied. This is thought to be the result of the higher ethanol concentration in the vapour, yet conflicting with the belief that a long extraction time via maceration should result in a higher extraction rate.

Some distillers choose to use a combination of these methods, with certain botanicals steeped in the ethanol and with others suspended above the surface for extraction via the vapour. Furthermore, some choose to handle each botanical separately, varying maceration times or distillation methods to tailor the extraction process to each botanical's specific requirements. Ultimately, the individual botanical distillates are blended together to form the final gin.

- Maceration conditions

In any extraction process, the solvent used, its concentration, and the exposure time are critical. As the goal in gin production is to produce an alcoholic beverage that is safe for consumption, the choice of solvent, ethanol, is very appropriate. The ethanol concentration and therefore the polarity of the water-ethanol mixture can be adjusted to influence the relative volatility and which compounds are extracted. Unsurprisingly, there is not a 'one size fits all' strength to extract the target desirable compounds with various studies demonstrating how differing ethanol concentrations influence extraction rates for different groups of compounds (Hodel et al., 2020; Jacotet-Navarro et al., 2018). However, a starting maceration strength in the range of 45%–65% is usually used. Extraction time and temperature are the decision of the individual distiller and the recipe, although in general it can be said that the longer and warmer the maceration procedure, the greater the extraction obtained. It is worth noting that excessive extraction can result in off flavours in the final spirit. Consequently, maceration times are typically 12–24h at temperatures from ambient up to 40°C in order to prevent breakdown of the botanicals and off-flavour development.

- Ratio of botanicals to spirit

Depending on the category of distilled gin being produced, either a one-shot or multi-shot method can be followed. With the one-shot method, the still is charged with the neutral spirit, water, and botanicals. This mixture is distilled and then diluted to bottling strength only with water. Alternatively, with the multi-shot method, a concentrated distillate is made by charging the still with neutral spirit, water, and an increased volume of botanicals. Following distillation, dilution is performed by adding additional spirit and then water, in order to achieve bottling strength. The quantity of botanicals used can be multiplied as many times as deemed appropriate by the distiller resulting in time, efficiency, and energy savings. As the botanical ratio increases, the extraction rate does not, however, necessarily correlate directly. For example, Hodel et al. (2020) found that doubling the botanicals used only resulted in a 1.5 increase in flavour compound extraction.

- Distillation strength

Although unusual, following botanical maceration and ahead of distilling, there is the option of adjusting the alcohol strength by adding more spirit to increase it or water to reduce it. The extracted compounds consist of a mixture of compounds, some of which are more soluble in ethanol and others that are more soluble in water. An adjustment in the water-to-ethanol ratio within the pot of the still will impact the relative volatilities of the compounds and will spread out the flavour compounds as they come across the still.

When selecting a starting distillation strength, it must be ensured that it is high enough to produce a distillate with sufficient alcohol content to meet legal or regulatory requirements.

- Distillation cut points

As with most batch-produced spirits, a gin run has a head, a heart, and a tails portion. The head, the first spirit across the still, contains the most volatile compounds but also cleans the still from the oily tails left behind from the previous run. The heads make up 0.5%–1.0% of the total charge alcohol and, when diluted with water, will turn milky. This is the same misting test as would occur when making whisky. The white appearance is due to oils that are soluble in high alcohol spirit coming out of solution at lower alcohol strengths. It is important not to include these in the heart of the gin as there is the risk, when dilution to bottling strength occurs, that the spirit will become cloudy. Once cleared and the flavour is acceptable, one can start collecting the heart of the gin—this is what is bottled. Note: when producing a gin using the multi-shot method, a clearing of the distillate will not necessarily be seen due to the higher botanical loading, and consideration should be made of the ethanol and water to be added post-distillation when determining the heart of the distillation. The heart cut is typically when the distillate has an ethanol content of 80%, but this will be starting alcohol strength, recipe, and still design dependent. Collection will continue until the flavour is no longer desirable—this will be purely dependent on the recipe but ultimately the flavour and/or aroma will change, and one can cut to collecting tails. The quantity of tails collected will be dependent on the flavour, whether they are to be recycled, and the balance of alcohol recovery with energy use. The recycling of heads and tails into the subsequent gin distillation depends on the recipe and the distiller's preference.

As with the distillation starting ethanol concentration, when making the distillation cuts, it must be ensured that the final distillate has a sufficient alcohol content to meet legal or regulatory requirements.

- Sweetening

The term 'dry' when referring to gin indicates that the spirit will have minimal sweetness, whereas other styles, such as Old Tom, traditionally will have had a sweeter palate. Within the legal and regulatory definitions (Table 23.1), the allowable sweetener level may be defined. For example, in the EU, a maximum of 0.1g of sweetening products per litre of final product is allowed to still refer to the spirit as 'dry'.

- Post-flavour extraction activities

After distillation, the resulting 'heart' middle cut is diluted to bottling strength—the minimum bottling strength in the E.U. is 37.5% A.B.V whereas in the USA. and Canada, it must be at least 40% A.B.V (EU Council Regulation No. 2019/787, 2019; Canadian Food and Drug Regulations, 2020; US Code, 2012). The water used for proofing must be of a potable standard and usually treated by demineralisation or reverse osmosis, to reduce the likelihood of haze or deposits forming within the bottle.

- Gin still features

Gin stills come in a variety of shapes and sizes, each influencing the final gin's flavour profile. They are most commonly made out of copper or stainless steel and heated indirectly via a steam jacket or water bath. Smaller scale distilleries may use glass stills, including when vacuum distillation is used, allowing flavour extraction to occur at lower temperatures and with gentler extraction of the botanicals.

In all instances, the still should have easy access to allow the botanicals to be added to the pot or vapour chamber, and just as importantly, an easy method for removing them after distillations such as a large discharge pipe. A CIP system is advantageous for flushing out botanicals but also in cleaning any oil build-up from the still's internal surfaces.

Sensory

The techniques and methods used for the assessment of a gin's sensory characteristics are the same as for other spirits and are discussed in a subsequent chapter within this book. Although the methods are the same, a distinct set of vocabulary is required to describe the aroma, mouthfeel, and taste attributes of gin. The flavour wheel in Fig. 23.2 is an example of one such set of terminology, developed by the Scotch Whisky Research Institute to aid in describing desirable sensory characteristics as well as the less desirable off notes.

The future

With the ever-expanding gin market, producers are continually striving to develop unique products. Botanicals, in some instances, are becoming more bizarre and exotic. Conversely, some distillers are focusing on the 'story' behind their raw ingredients with the rise in consumer interest in both sustainability and environmental concerns. Local, more transparent

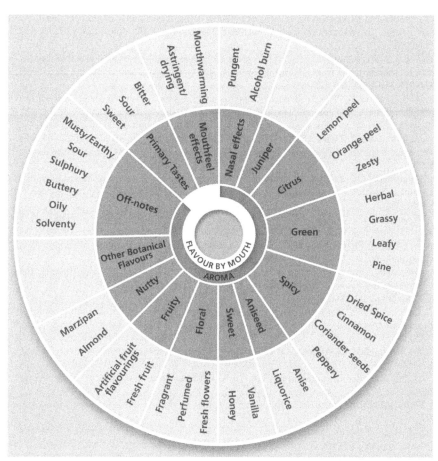

FIG. 23.2 The Scotch Whisky Research Institute's gin flavour wheel. Reproduced with permission from the Scotch Whisky Research Institute.

sourcing of both spirit and botanicals, accompanied with carbon accounting, may well become the new normal for the gin industry.

References

Ayala, J.R., Montero, G., Campbell, H.E., García, C., Coronado, M.A., León, J.A., Sagaste, C.A., Pérez, L.J., 2017. Extraction and characterization of orange peel essential oil from Mexico and United States of America. J. Essent. Oil Bear. Plants 20 (4), 897–914. https://doi.org/10.1080/0972060X.2017.1364173.

Aylott, R.I., 2003. Gin: the product and its manufacture. In: Caballero, B., Trugo, L., Finglas, P.M. (Eds.), Encyclopaedia of Food Sciences and Nutrition. Academic Press, London, pp. 2889–2893.

Bandoni, A.L., Mizrahi, I., Juárez, M.A., 1998. Composition and quality of essential oil of coriander (*Coriandrum sativum* L.) from Argentina. J. Essent. Oil Res. 10 (5), 581–584. https://doi.org/10.1080/10412905.1998.9700977.

Bernard, C., Clair, G., 1997. Essential oils of three *Angelica* L. species growing in France. Part I. root oils. J. Essent. Oil Res. 9 (3), 289–294. https://doi.org/10.1080/10412905.1997.10554246.

Boussaada, O., Skoula, M., Kokkalou, E., Chemli, R., 2007. Chemical variability of flowers, leaves, and peels oils of four sour orange provenances. J. Essent. Oil Bear. Plants 10 (6), 453–464. https://doi.org/10.1080/09720 60X.2007.10643579.

Brahmkshatriya, P.P., Brahmkshatriya, P.S., 2013. Terpenes: chemistry, biological role, and therapeutic applications. In: Ramawat, K., Mérillon, J.M. (Eds.), Natural Products. Springer, Berlin, Heidelberg, pp. 2665–2691, https://doi.org/10.1007/978-3-642-22144-6_120.

Brenna, E., Fuganti, C., Serra, S., 2003. From commercial racemic fragrances to odour active enantiopure compounds: the ten isomers of irone. C. R. Chim. 6 (5–6), 529–546. https://doi.org/10.1016/S1631-0748(03)00087-0.

Brenna, E., Fuganti, C., Serra, S., 2004. Changing the odor properties of commercial mixtures of α-irones by simple chemical transformations. J. Essent. Oil Res. 16 (4), 339–341. https://doi.org/10.1080/10412905.2004.9698736.

Burdock, G.A., 2016. Fenaroli's Handbook of Flavor Ingredients, Sixth ed. CRC Press, Boca Raton, USA, ISBN: 9780429150838.

Butkienė, R., Nivinskienė, O., Mockutė, D., 2005. Volatile compounds of ripe berries (black) of *Juniperus communis* L. growing wild in North-East Lithuania. J. Essent. Oil Bear. Plants 8 (2), 140–147. https://doi.org/10.1080/0972 060X.2005.10643434.

Butkiene, R., Nivinskiene, O., Mockute, D., 2006. Leaf (needle) essential oils of *Juniperus communis* L. growing wild in Eastern Lithuania. J. Essent. Oil Bear. Plants 9 (2), 144–151. https://doi.org/10.1080/0972060X.2006.10643486.

Canadian Food and Drug Regulations, 2020. (C.R.C., c. 870) Part B Foods. Available at: https://laws-lois.justice.gc.ca/eng/regulations/c.r.c.%2C_c._870/index.html.

Carrubba, A., la Torre, R., Prima, A.D., Saiano, F., Alonzo, G., 2002. Statistical analyses on the essential oil of Italian coriander (*Coriandrum sativum* L.) fruits of different ages and origins. J. Essent. Oil Res. 14 (6), 389–396. https://doi.org/10.1080/10412905.2002.9699899.

Chalchat, J.C., Garry, R.P., 1997. Essential oil of Angelica roots (*Angelica archangelica* L.): optimization of distillation, location in plant and chemical composition. J. Essent. Oil Res. 9 (3), 311–319. https://doi.org/10.1080/10412905.1997.10554250.

Charles, D.J., 2012a. Angelica. In: Antioxidant Properties of Spices, Herbs and Other Sources. Springer, New York, NY, pp. 151–157.

Charles, D.J., 2012b. Coriander. In: Antioxidant Properties of Spices, Herbs and Other Sources. Springer, New York, NY, pp. 255–263.

Charles, D.J., 2012c. Licorice. In: Antioxidant Properties of Spices, Herbs and Other Sources. Springer, New York, NY, pp. 385–392.

Chizzola, R., 2013. Regular monoterpenes and sesquiterpenes (essential oils). In: Ramawat, K., Mérillon, J.M. (Eds.), Natural Products. Springer, Berlin/Heidelberg, pp. 978–983, https://doi.org/10.1007/978-3-642-22144-6_130.

Clutton, D.W., Evans, M.B., 1978. The flavour constituents of gin. J. Chromatogr. A 167, 409–419. https://doi.org/10.1016/S0021-9673(00)91173-7.

Coates, G., 2012. The Mixellany Guide to Gin. Mixellany Limited, London, UK, ISBN: 1907434283.

Dillon, P., 2003. Gin: The Much Lamented Death of Madam Geneva: The Eighteenth-Century Gin Craze. Justin, Charles & Co, London. ISBN: 0747235694, 9780747235699.

Donald, F., Green, S., Searle, K., Cunniffe, N.J., Purse, B.V., 2020. Small scale variability in soil moisture drives infection of vulnerable juniper populations by invasive forest pathogen. For. Ecol. Manage. 473. https://doi.org/10.1016/j.foreco.2020.118324, 118324.

Drinks International, 2019. The Millionaires' Club. Available: https://drinksint.com. (Accessed May 2020).

EU Council Regulation No. 2019/787, 2019. Regulation (EU) N° 2019/787 of the European Parliament and of the council of 17 April 2019. Available at https://eur-lex.europa.eu/legal-content/EN/TXT/?uri=CELEX%3A32019R0787.

Forycka, A., Buchwald, W., 2019. Variability of composition of essential oil and coumarin compounds of *Angelica archangelica* L. Herba Polonica 65 (4), 62–75. https://doi.org/10.2478/hepo-2019-0027.

Ghada, B., Amel, O., Aymen, M., Aymen, A., Amel, S.H., 2019. Phylogenetic patterns and molecular evolution among 'True citrus fruit trees' group (Rutaceae family and Aurantioideae subfamily). Sci. Hortic. 253, 87–98. https://doi.org/10.1016/j.scienta.2019.04.011.

Green, S., Elliot, M., Armstrong, A., Hendry, S.J., 2015. *Phytophthora austrocedrae* emerges as a serious threat to juniper (*Juniperus communis*) in Britain. Plant Pathol. 64, 2,456–466. https://doi.org/10.1111/ppa.12253.

Hani, M.M., Hussein, S.A.A.H., Mursy, M.H., Ngezimana, W., Mudau, F.N., 2015. Yield and essential oil response in coriander to water stress and phosphorus fertilizer application. J. Essent. Oil Bear. Plants 18, 1,82–92. https://doi.org/10.1080/0972060X.2014.974080.

Hastings, D., 1984. Spirits and Liqueurs of the World. Chartwell Books, Secaucus, NJ, ISBN: 0890097143.

Hodel, J., Burke, M., Hill, A.E., 2020. Influence of distillation parameters on the extraction of *Juniperus communis* L. in vapour infused gin. J. Inst. Brew. 126, 184–193. https://doi.org/10.1002/jib.607.

Hodel, J., Pauley, M., Gorseling, M.C., Hill, A.E., 2019. Quantitative comparison of volatiles in vapor infused gin versus steep infused gin distillates. J. Am. Soc. Brew. Chem. 77, 149–156. https://doi.org/10.1080/03610470.201 9.1629263.

Hosni, K., Zahed, N., Chrif, R., Abid, I., Medfei, W., Kallel, M., Brahim, N.B., Sebei, H., 2010. Composition of peel essential oils from four selected Tunisian citrus species: evidence for the genotypic influence. Food Chem. 123 (4), 1098–1104. https://doi.org/10.1016/j.foodchem.2010.05.068.

Hosseini, M.S., Samsampour, D., Ebrahimi, M., Abadía, J., Khanahmadi, M., 2018. Effect of drought stress on growth parameters, osmolyte contents, antioxidant enzymes and glycyrrhizin synthesis in licorice (*Glycyrrhiza glabra* L.) grown in the field. Phytochemistry 156, 124–134. https://doi.org/10.1016/j.phytochem.2018.08.018.

İzgı, M.N., Telci, İ., Elmastaş, M., 2017. Variation in essential oil composition of coriander (*Coriandrum sativum* L.) varieties cultivated in two different ecologies. J. Essent. Oil Res. 29 (6), 494–498. https://doi.org/10.1080/104129 05.2017.1363090.

Jacotet-Navarro, M., Laguerre, M., Fabiano-Tixier, A.S., Tenon, M., Feuillère, N., Bily, A., Chemat, F., 2018. What is the best ethanol-water ratio for the extraction of antioxidants from rosemary? Impact of the solvent on yield, composition, and activity of the extracts. Electrophoresis 39, 1946–1956. https://doi.org/10.1002/elps.201700397.

Kazemi, S.Y., Nabavi, J., Zali, H., Ghorbani, J., 2017. Effect of altitude and soil on the essential oils composition of *Juniperus communis*. J. Essent. Oil Bear. Plants 20, 5,1380–1390. https://doi.org/10.1080/0972060X.2017.1387080.

Kempinski, C., Jiang, Z., Bell, S., Chappell, J., 2015. Metabolic engineering of higher plants and algae for isoprenoid production. Adv. Biochem. Eng. Biotechnol. 148, 161–199. https://doi.org/10.1007/10_2014_290.

Kew, 2016. State of the World's Plants. https://stateoftheworldsplants.org/2016/report/sotwp_2016.pdf.

Kew, 2017. State of the World's Plants. https://stateoftheworldsplants.org/2017/report/SOTWP_2017.pdf.

Kirbaslar, F.G., Kirbaslar, S.I., Pozan, G., Boz, I., 2009. Volatile constituents of Turkish orange (*Citrus sinensis* (L.) Osbeck) peel oils. J. Essent. Oil Bear. Plants 12 (5), 586–604. https://doi.org/10.1080/0972060X.2009.10643762.

Koukos, P.K., Papadopoulou, K.I., 1997. Essential oil of *Juniperus communis* L. grown in Northern Greece: variation of fruit oil yield and composition. J. Essent. Oil Res. 9 (1), 35–39. https://doi.org/10.1080/10412905.1997.9700711.

Krick, W., Marner, F.J., Jaenicke, L., 1983. Isolation and structure determination of the precursors of a-and y-irone and homologous compounds from *Iris pallida* and *Iris florentina*. Zeitschrift für Naturforschung C 38 (3–4), 179–184. https://doi.org/10.1515/znc-1983-3-404.

Krist, S., 2020. Coriander seed oil. In: Vegetable Fats and Oils. Springer, Cham, pp. 175–178, https://doi.org/10.1007/978-3-030-30314-3_26.

Labokas, J., Ložienė, K., 2013. Variation of essential oil yield and relative amounts of enantiomers of α-pinene in leaves and unripe cones of *Juniperus communis* L. growing wild in Lithuania. J. Essent. Oil Res. 25 (4), 244–250. https://doi.org/10.1080/10412905.2013.775678.

Letchamo, W., Gosselin, A., Hölzl, J., 1995. Growth and essential oil content of *Angelica archangelica* as influenced by light intensity and growing media. J. Essent. Oil Res. 7 (5), 497–504. https://doi.org/10.1080/10412905.1995.9698573.

Masson, J., Liberto, E., Brevard, H., Bicchi, C., Rubiolo, P., 2014. A metabolomic approach to quality determination and authentication of raw plant material in the fragrance field. Iris rhizomes: a case study. J. Chromatogr. A 1368, 143–154. https://doi.org/10.1016/j.chroma.2014.09.076.

Milica, A., Mirjana, C., Jovana, S., 2016. Effect of weather conditions, location and fertilization on coriander fruit essential oil quality. J. Essent. Oil Bear. Plants 19 (5), 1208–1215. https://doi.org/10.1080/0972060X.2015.1119068.

Nadjafi, F., Damghani, A.M., Ebrahimi, S.N., 2009. Effect of irrigation regimes on yield, yield components, content and composition of the essential oil of four Iranian land races of coriander (*Coriandrum sativum*). J. Essent. Oil Bear. Plants 12 (3), 300–309. https://doi.org/10.1080/0972060X.2009.10643724.

Nivinskienė, O., Butkienė, R., Mockutė, D., 2005. The chemical composition of the essential oil of *Angelica archangelica* L. roots growing wild in Lithuania. J. Essent. Oil Res. 17 (4), 373–377. https://doi.org/10.1080/10412905.2005.96 98934.

Omar, H.R., Komarova, I., El-Ghonemi, M., Fathy, A., Rashad, R., Abdelmalak, H.D., Yerramadha, M.R., Ali, Y., Helal, E., Camporesi, E.M., 2012. Licorice abuse: time to send a warning message. Ther. Adv. Endocrinol. Metab. 3 (4), 125–138. https://doi.org/10.1177/2042018812454322.

Padalia, R.C., Karki, N., Sah, A.N., Verma, R.S., 2011. Volatile constituents of leaf and seed essential oil of *Coriandrum sativum* L. J. Essent. Oil Bear. Plants 14 (5), 610–616. https://doi.org/10.1080/0972060X.2011.10643979.

Paroul, N., Rota, L., Frizzo, C., Atti dos Santos, A.C., Moyna, P., Gower, A.E., Serafini, L.A., Cassel, E., 2002. Chemical composition of the volatiles of Angelica root obtained by hydrodistillation and supercritical CO_2 extraction. J. Essent. Oil Res. 14 (4), 282–285. https://doi.org/10.1080/10412905.2002.9699855.

Pastorino, G., Cornara, L., Soares, S., Rodrigues, F., Oliveira, M.B.P., 2018. Liquorice (*Glycyrrhiza glabra*): a phytochemical and pharmacological review. Phytother. Res. 32 (12), 2323–2339. https://doi.org/10.1002/ptr.6178.

Ramezani, S., Rasouli, F., Solaimani, B., 2009. Changes in essential oil content of coriander (*Coriandrum sativum* L.) aerial parts during four phonological stages in Iran. J. Essent. Oil Bear. Plants 12 (6), 683–689. https://doi.org/10.1080/0972060X.2009.10643775.

Roslon, W., Wajs-Bonikowska, A., Geszprych, A., Osinska, E., 2016. Characteristics of essential oil from young shoots of garden Angelica (*Angelica archangelica* L.). J. Essent. Oil Bear. Plants 19 (6), 1462–1470. https://doi.org/10.1080/0972060X.2016.1238322.

Sato, A., Asano, K., Sato, T., 1990. The chemical composition of *Citrus hystrix* DC (Swangi). J. Essent. Oil Res. 2 (4), 179–183. https://doi.org/10.1080/10412905.1990.9697857.

Saxena, S.N., Rathore, S.S., Saxena, R., Barnwal, P., Sharma, L.K., Singh, B., 2014. Effect of cryogenic grinding on essential oil constituents of coriander (*Coriandrum sativum* L.) genotypes. J. Essent. Oil Bear. Plants 17 (3), 385–392. https://doi.org/10.1080/0972060X.2014.895197.

Schütz, C., Quitschau, M., Hamburger, M., Potterat, O., 2011. Profiling of isoflavonoids in *Iris germanica* rhizome extracts by microprobe NMR and HPLC–PDA–MS analysis. Fitoterapia 82 (7), 1021–1026. https://doi.org/10.1016/j.fitote.2011.06.005.

Simpson, A.C., 1977. Gin and vodka. In: Rose, A.H. (Ed.), Economic Microbiology: Alcoholic Beverages, first ed. Academic Press, London, ISBN: 0125965508, pp. 537–593.

Skeat, W.W., 1907. Orris-root. Notes Queries 10-VIII (196), 247.

Solmonson, L.J., 2012. Gin: A Global History. Reaktion Books Ltd, London, ISBN: 9781861899248.

Telci, I., Toncer, O.G., Sahbaz, N., 2006. Yield, essential oil content and composition of *Coriandrum sativum* varieties (var. *vulgare* Alef and var. *microcarpum* DC.) grown in two different locations. J. Essent. Oil Res. 18 (2), 189–193. https://doi.org/10.1080/10412905.2006.9699063.

US Code of Federal Regulations, 2012. (CFR) 27 Part 5 Alcohol, Tobacco Products and Firearms: Labelling and advertising of distilled spirits. Subpart C Standards of Identity for Distilled Spirits. Available at https://www.govinfo.gov/app/details/CFR-2012-title27-vol1/CFR-2012-title27-vol1-part5.

Viuda-Martos, M., Ruiz-Navajas, Y., Fernández-López, J., Pérez-Álvarez, J.A., 2009. Chemical composition of mandarin (*C. reticulata* L.), grapefruit (*C. paradisi* L.), lemon (*C. limon* L.) and Orange (*C. sinensis* L.) essential oils. J. Essent. Oil Bear. Plants 12 (2), 236–243. https://doi.org/10.1080/0972060X.2009.10643716.

Wexler, P., Fonger, G.C., White, J., Weinstein, S., 2015. Toxinology: taxonomy, interpretation, and information resources. Sci. Technol. Libr. 34 (1), 67–90. https://doi.org/10.1080/0194262X.2014.993788.

Willkie, H.F., Boruff, C.S., Althausen, D., 1937. Controlling gin flavor. Ind. Eng. Chem. 29 (1), 78–84.

CHAPTER

24

Vodka

Douglas Murray

Institute of Brewing & Distilling, London, United Kingdom

Introduction

This chapter considers a short history and the main production steps involved in the manufacture of vodka. Vodka is traditionally an unaged spirit produced from a variety of substrates and then distilled and filtered through charcoal. The detailed design of equipment and its use is complex and varied. This chapter will allow the reader to identify the key elements and their role in the production of vodka. Sections of the chapter are applicable to a range of high-strength distilled beverages available to the consumer. In addition, some information in Chapter 15 on grain whisky production is relevant and is referenced rather than repeated in this section. To the consumer, vodka is not just the traditional 'clear' spirit but comes in many different flavoured varieties. This aspect of vodka production will be discussed but the detailed requirements of all these styles are beyond the scope of this chapter.

Short history

The origin of vodka has been the topic of discussion for centuries and has resulted in bitter legal disputes. These have not provided a definitive history or a single country who can claim ownership. Vodka production in the twelfth- or thirteenth-century origins are probably located in North-Eastern Europe and Scandinavia, with Poland and Russia both producing it early on. In 2019, the ten largest exporters of vodka were Sweden, France, Russia, Poland, the Netherlands, the USA, Italy, Latvia, Germany, and Finland (http://www.worldstopexports.com/vodka-exporters/).

The range of raw materials was originally confined to potatoes and cereals, (wheat, barley, rye, oats), but now includes other substrates such as molasses from sugar cane, sugar beet, and their molasses from sugar refining. Also, there are now vodkas derived from grapes and rice.

Whisky and Other Spirits
https://doi.org/10.1016/B978-0-12-822076-4.00015-2

441

Historically, vodka production was on a small farm scale using local raw materials, predominantly rye and potatoes, and batch distillation using pot stills. This required times over eight separate distillations to obtain the strength up to 96% v/v with low levels of fusel oils and methanol. When continuous distillation was pioneered in the 1870s, this method of vodka production was embraced and now, by far, the greater percentage of production uses multiple column procedures. Today, production comes in several forms, started by small collective groups and then rectified further at large centralised plants, produced on site from the raw material for use as a single product or produced in a large distillery and sold to a finishing company for final vodka production.

Vodka definitions

The definition of vodka, flavoured vodka, and what constitutes ethyl alcohol is specific to each country or geographic zone. Before commencing operations, the distiller must establish that the proposed production method and the category regulations are met for the geographic place of production and also the geographical sales location.

Below is the European Union definition of vodka and also the United States equivalent in 2019.

REGULATION (EU) 2019/787 OF THE EUROPEAN PARLIAMENT AND OF THE COUNCIL

of 17 April 2019

on the definition, description, presentation and labelling of spirit drinks, the use of the names of spirit drinks in the presentation and labelling of other foodstuffs, the protection of geographical indications for spirit drinks, the use of ethyl alcohol and distillates of agricultural origin in alcoholic beverages, and repealing Regulation (EC) No 110/2008

15. Vodka

(a) *Vodka is a spirit drink produced from ethyl alcohol of agricultural origin obtained following fermentation with yeast of either:*
- *potatoes or cereals or both,*
- *other agricultural raw materials,*
  distilled so that the organoleptic characteristics of the raw materials used and by-products formed in fermentation are selectively reduced.

This may be followed by additional distillation or treatment with appropriate processing aids or both, including treatment with activated charcoal, to give it special organoleptic characteristics.

Maximum levels of residue for the ethyl alcohol of agricultural origin used to produce vodka shall meet those levels set out in point (d) of Article 5, except that the methanol content shall not exceed 10 grams per hectolitre of 100% vol. alcohol.

(b) *The minimum alcoholic strength by volume of vodka shall be 37.5%.*

(c) *The only flavourings which may be added are natural flavouring substances or flavouring preparations that are present in distillate obtained from the fermented raw materials. In addition, the product may be given special organoleptic characteristics, other than a predominant flavour.*

(d) *Vodka shall not be coloured.*

(e) *Vodka may be sweetened in order to round off the final taste. However, the final product may not contain more than 8 grams of sweetening products per litre, expressed as invert sugar.*

(f) *The description, presentation, or the labelling of vodka not produced exclusively from potatoes or cereals or both shall prominently bear the indication 'produced from ...', supplemented by the name of the raw materials used to produce the ethyl alcohol of agricultural origin. This indication shall appear in the same visual field as the legal name.*

(g) *The legal name may be 'vodka' in any Member State.*

US Regulation in 2019

Subchapter A. ALCOHOL, Part 5. LABELING AND ADVERTISING OF DISTILLED SPIRITS Subpart C. Standards of Identity for Distilled Spirits

(a) ***Class 1; neutral spirits or alcohol****. 'Neutral spirits' or 'alcohol' are distilled spirits produced from any material at or above 190° proof, and, if bottled, bottled at not less than 80° proof.*

    (1) *'Vodka' is neutral spirit which may be treated with up to two grams per liter of sugar and up to one gram per liter of citric acid. Products to be labelled as vodka may not be aged or stored in wood barrels at any time except when stored in paraffin-lined wood barrels and labelled as bottled in bond pursuant to § 5.42(b)(3). Vodka treated and filtered with not less than one ounce of activated carbon or activated charcoal per 100 wine gallons of spirits may be labelled as 'charcoal filtered'.*

In addition to the legal definition of what constitutes a Vodka, there are limits on what constituents make up the chemical compounds, or groups. Again, these are market specific and must be considered when producing a product. An example is Article 5 in the definition of ethyl alcohol of agricultural origin in the European Community.

REGULATION (EU) 2019/787 OF THE EUROPEAN PARLIAMENT AND OF THE COUNCIL

of 17 April 2019

Article 5 **Definition of and requirements for ethyl alcohol of agricultural origin**

For the purposes of this Regulation, ethyl alcohol of agricultural origin is a liquid which complies with the following requirements:

(a) *it has been obtained exclusively from products listed in Annex I to the Treaty;*

(b) *it has no detectable taste other than that of the raw materials used in its production;*

(c) *its minimum alcoholic strength by volume is 96.0%;*

(d) *its maximum levels of residues do not exceed the following:*

    (i) *total acidity (expressed in acetic acid): 1.5g per hectolitre of 100% vol. alcohol;*

    (ii) *esters (expressed in ethyl acetate): 1.3g per hectolitre of 100% vol. alcohol;*

    (iii) *aldehydes (expressed in acetaldehyde): 0.5g per hectolitre of 100% vol. alcohol;*

    (iv) *higher alcohols (expressed in 2-methyl-1-propanol): 0.5g per hectolitre of 100% vol. alcohol;*

    (v) *methanol: 30g per hectolitre of 100% vol. alcohol;*

    (vi) *dry extract: 1.5g per hectolitre of 100% vol. alcohol;*

    (vii) *volatile bases containing nitrogen (expressed in nitrogen): 0.1g per hectolitre of 100% vol. alcohol;*

    (viii) *furfural: not detectable.*

These regulations are subject to regular review, and the original document must be consulted to ensure compliance. As an example, (viii) furfural is under review as methods of analysis have been developed where the limit of detection now allows furfural to be detected.

The reader may have noted that several different units are used to express the quantity of ethanol there is in a liquid. Fig. 24.1 is a short explanation of the difference and how one is converted to the other.

Production technology

For a vodka producer to meet the regulatory requirements to be able to call the product a vodka and also to meet the quality expectation, the first step is to produce a spirit ethanol mixture with the correct type and quantity of flavour congeners. This is normally accomplished by producing or purchasing a raw spirit of ethyl alcohol of agricultural origin (EAAO) that complies with the specification discussed above for the geographical area where the producer is located. To achieve this high-strength spirit (96% v/v) and with relatively low levels of congeners, as well the removal of undesirable odour and flavour compounds (mainly sulphur derived), additional equipment is required.

The make-up of the feed to the stills is determined by the original raw materials and the method of fermentation. These have an impact on the level of congeners in the feed and on the final product. This in turn determines the equipment needed to achieve the level of congener removal desired. The initial feed distillation and the final rectification are similar to those

Quantity of Ethanol in a liquid

Alcohol by volume = the number of millilitres (mL) of pure **ethanol** present in 100 mL of liquid. Commonly expressed as ABV

Alcohol by weight (ABW) = sometimes used as this is easily measured by weighing the liquid and determining the density difference from readily available tables. The relationship is not linear but a rough estimate can be determined from the formula.

$$ABV = ABW \times 1.267$$

US proof = introduced in 1848 this is a simple factor where proof double the amount of ethanol by volume

US proof = ABV/2   E.g.   100 proof US = 50% ABV

Temperature,
Density changes with temperature then Ethanol content is converted to a standard temperature, normally 20°C. However care needs to be taken as other temperatures such as 15°C are used. To further complicate matters the Fahrenheit (F) scale rather than Celsius (C) can be used.

FIG. 24.1   Ethanol concentration definitions.

used in Scotch grain whisky distillation in that it utilises continuous stills. This is discussed in Chapter 15 and requires a two-stage distillation process and normally a side column on the rectification column to remove the fusel oils. This two-stage distillation is mainly conducted to reduce the height of the column but some distillers use a single column. To produce the quality required, EAAO producers normally add two further columns, a heads column, and a hydro-selection column, sometimes called an extractive column, to remove volatile congeners such as aldehydes. The operation of these columns is discussed below. In addition, a de-methanolisation column may be required if the feedstock is high in methanol. The number of these additional columns depends on the final product specification and incurs a financial penalty due to increased capital expenditure, increased energy usage, and decreased ethanol recovery through the inclusion of a residual level in the side stream products.

Fig. 24.2 is simplified and only shows the key liquid streams. Modern plants have integrated energy recovery and multiple heads recovery systems. To understand how these columns remove certain congeners, one needs to discuss a chemical engineering term, relative volatility. When discussing the process of distillation, it is normally assumed that the liquid being distilled is a binary mixture, which is a mixture of only ethanol and water. In reality, the liquid derived from fermentation is a complex mixture of many compounds. When this liquid is distilled, the ease at which a compound will distil compared to others is termed its relative volatility. To make things even more complicated, this volatility is dependent on the ethanol/water mix you start with. This is demonstrated in Fig. 24.3, where the relative volatility of some of the more well-known congeners is shown as ethanol levels from 0% to 100%.

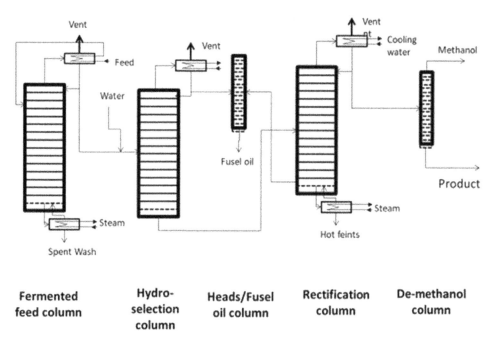

Schematic of continuous neutral spirit still

FIG. 24.2   Schematic of a continuous neutral spirit still.

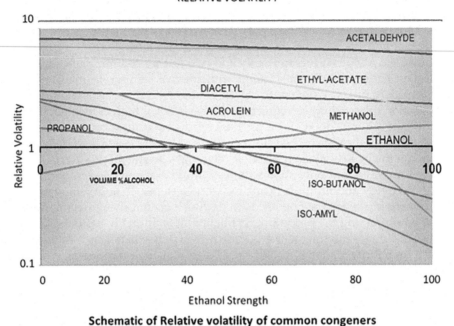

FIG. 24.3 Schematic of relative volatility of common congeners. Reproduced with permission of R. Piggot, Lallemand Biofuels & Distilled Spirits.

There are three groups of compounds:

1. Always above the ethanol line;
2. Start above, then cross under the ethanol line;
3. Start below and cross above the ethanol line.

When aiming to distil at a strength of about 96% ethanol, from the above compounds in group 1 and group 3, such as acetaldehyde or methanol, respectively, these will have a peak concentration further up the column than ethanol. Compounds in group 2, such as propanol and isobutanol, will have their peak concentration below the ethanol peak.

The typical pattern of congeners in a column is shown in Fig. 24.4.

In the Fig. 24.4 example, the product is removed at plate 30. It needs to be noted that in this 'simple' one-column setup, where there is no side stream removal, all of the congeners in the feed will build up in the column until the tail or leading edge of the peak crosses plate level 30 and are removed as a part of the product. In grain whisky distillation, this is the case except for the very volatile compounds that are not condensed and are allowed to vent from the top of the column, while isoamyl alcohol (fusel oil) is removed from the liquid using a side column. To remove the other compounds requires additional columns with specific designs and operational techniques. These columns cost money to purchase and operate, and they therefore increase costs. In addition, the

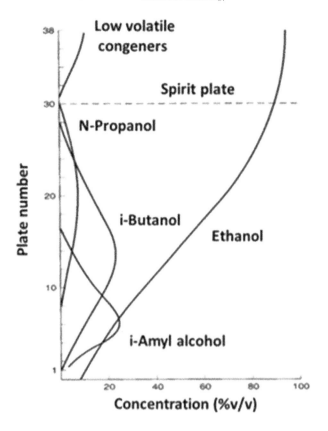

Congener concentration at different still plate levels

FIG. 24.4 Congener concentrations at different still plate levels.

removal of these undesirable compounds results in some loss of ethanol, as separation is never 100%, therefore reducing distillation efficiency and again increasing the cost of production.

Fermented feed distillation

The first stage requires taking the feed and concentrating the ethanol and the congeners by putting energy into the system at the base, which will drive vapour up the column. This also separates any solids, which are 'flushed' down the column to exit with water at its base. The strength of the spirit produced depends on the initial feed strength and the number of plates in the column. Typical values are 45%–60% v/v ethanol. As there are no side streams, all the congeners will be carried over in the ethanol. The operation of this column is explained in greater detail in Chapter 15.

Hydro-selection

This is where the distiller uses the relative volatility to assist the removal of some of the congeners. The feed to this column is diluted to 20%–30% v/v with water. This can be purified water or the bottom product from the next stage, the rectifier column, and is fed into the middle of the column. The energy input is controlled. The relatively low ethanol concentration brings several of the congeners above the ethanol line (see Fig. 24.3) allowing separation into the vapour phase. Thus, the vapour in the column is relatively strong in the more volatile congeners, which rise up the column and are condensed as heads. The bulk of the ethanol and methanol remains in the liquid cascading down the column and exits the still at the bottom. The liquid leaving this point is lower in ethanol than the feed, as in any distillation process (methanol column is an exception—see later). This results in a 'product' stream at 5%–10% v/v. This purified ethanol then needs to be re-rectified to a higher strength. This is classed as a distillation stage as it does separate the components.

Rectification

The low-strength spirit is then distilled in a conventional rectification column that contains more plates than what is used in grain whisky distilling, to bring the ethanol strength back to around 96% v/v.

Fusel oil removal

Some fusel oils are removed in the hydro-selection column but in most cases not all, therefore a separate fusel oil removal column is used. This is normally positioned towards the bottom of the rectification column. The fusel oils are removed by having a side stream feeding a fusel oil column. The still is designed to recover the ethanol while increasing the concentration of fusel oils. Note that in whisky production, one would also recover some higher alcohols that are considered to be part of the product but in vodka production, the only wish is to return the ethanol.

Heads column

The vapour from several columns is collected and this comes mainly from the hydro-selection column, but to maximise ethanol yield this stream is combined with the heads from the other columns. The design and operation of this column is similar to the fusel oil column, where careful control of the feed, energy, and temperature allows the non-ethanol components to be removed. The ethanol is returned to the system. This column is not needed to produce a high grade of ethanol but does impact on the overall efficiency of ethanol recovery and therefore the cost.

Methanol column

Where the feedstock produces a high level of methanol, this is then removed using a specific column. Looking back at the relative volatility graph (Fig. 24.3), it can be seen that maximum separation will occur at high ethanol strengths. For this reason, the column is placed high up the final rectification column. In practice, the column is placed in the line from the top of the column back to the still, in the reflux system. Methanol is removed from the top of the column and ethanol from the base.

At the culmination of all of this distilling activity, the spirit produced is very low in congeners and is at high strength. This spirit is ideal for vodka production but is also often used as the base for other beverages such as Gin.

Finishing

The 'neutral spirit' is ready then for the final part of the vodka production process. This is where the 'vodka is created' and where some of the harshness of the spirit is removed and some chemistry is allowed to occur to give the vodka a 'drinkability' feel. This is achieved mainly by taking the spirit and passing it through charcoal.

Each producer has their own way of achieving this and this helps to build up brand differentiation. The main factors to consider are:

- Type of charcoal—how it is produced;
- Raw material—wood variety, nutshells, coconut husk etc.;
- Physical size of the charcoal—granular or charred;
- Retention/contact time—the length of time spirit is in contact;
- Spirit strength of feed to filtration equipment.

Most charcoals used in the beverage industry are classed as 'common charcoal'. These are produced through the slow pyrolysis of the source material, through heating in the absence of air. Large-scale producers have specially designed kilns for this process. This converts the organic material into a carbon structure where the water content is removed. This process also removes many of the volatile components. This charcoal is then further processed in a heating tube, where introducing a steam at 900°C drives the final water and volatiles out of the carbon. This produces 'activated charcoal', where the internal structure of the carbon is pitted with multiple pores that increase its adsorption capacity. The typical yield of activated charcoal is between 30 and 45% of the original material. Activated charcoal can be reused after suitable reconditioning but the production plant is expensive to purchase and operate. Therefore, most vodka producers do not re-use the charcoal but rather use fresh charcoal. Spent charcoal can be used as a fuel to generate energy for the production facility.

The raw material has an impact on the adsorption capacity along with the level of ash in the charcoal. Providing that the charcoal has no residual compounds of concern contained in the source material, then any organic produce can be used. In practice, wood, husk (rice or coconut), and nut shells are the most used. Each one impacts in a different way the liquid being filtered through the finished charcoal. Several well-known vodkas state the type of wood used.

Once produced, the charcoal is sieved into different size groupings and may even be ground into a fine powder (Fig. 24.5). This increases the external surface area and helps the adsorption capacity, but it should be remembered, that the internal area of the charcoal is heavily pored

Activated Charcoal
Rough mill and fine mill

FIG. 24.5   Activated charcoal—rough mill and fine mill. Reproduced with the permission of A. Wardlaw, Diageo Plc.

and even in large flakes, it has a huge surface area. The distiller needs to take into consideration the separation of the product from the filter, and a compromise in size is usually obtained. Part of the specification when purchasing charcoal is how porous is the material.

Once the type of charcoal has been selected, the next stage is how long does one wish the spirit to be in contact with the charcoal. This can range from minutes to hours. The impact on additional chemistry and mouthfeel is specific to each brand and can only be determined by trial and error. The longer the ethanol is in contact with the charcoal, the greater will be the amount of ethanol-derived organic molecules such as acetaldehyde. The above parameters are, in a sense, fixed by design, but retention times can easily be allowed to be increased or be reduced. This will impact the final quality of the vodka.

Lastly, the distiller needs to consider the spirit strength passing through the filter. Once again there is a compromise. The favourable reactions occur faster at lower strength but this means a greater cost in purchasing the filter housing as it increases in size. One also needs to ensure that the slight decrease in strength through the filter does not drop the alcohol level below that declared on the bottle label.

Design of filter

The decisions discussed above have an impact on the design and operation of the charcoal filter. The overall scale of the production facility also has an impact. For a small producer, the filter can be as simple as a stainless-steel tank where charcoal is added, mixed, and then there is filtration through a fine cloth to separate the product from the charcoal (Fig. 24.6). Another alternative is to use charcoal-impregnated cartridge filters (Fig. 24.7). These have the benefit of low maintenance and ease of operation.

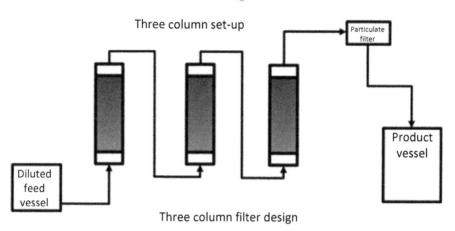

FIG. 24.6   Three-column filter design.

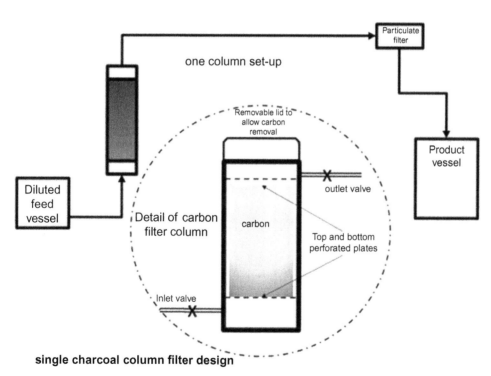

FIG. 24.7   Single charcoal column filter design.

Larger producers require an in-line facility with a range of equipment that allows for the following:

- High-strength storage tank;
- Dilution tank;
- Charcoal housing;
- Post-charcoal filter;
- Low-strength 'vodka' storage tank.

Some of these operations can be carried out in the same vessel, but in-line continuous design is highlighted below in order to illustrate the overall concept.

To allow a filtration plant to operate without the need to stop and replace the carbon, modern designs have an additional column. An example is the Smirnoff vodka process, where filtration is through ten columns, allowing for a long contact time. Fig. 24.8 shows a design using four columns, where three columns are in the product stream, and a fourth can be taken off-line. This allows the charcoal to be removed and refilled with a fresh batch. These setups have sufficient valves and pipework to allow each of the four columns to be isolated and replenished in rotation. Not shown on the simplified Fig. 24.8, a water and ethanol flush pipework is also installed to allow the ethanol in the column to be removed before the charcoal is emptied and also to flush the fresh charcoal before reconnecting into the process.

In some vodka production plants, activated charcoal that has been impregnated with rare metals, such as silver or gold, is employed. The chemistry behind their use is still not fully understood, but it does have an impact on the quality. The process tends to 'clean up' the spirit and remove the last traces of some organoleptic taints, such as sulphur derived compounds.

After carbon filtration, the liquid is put through a fine mesh filter to remove any carbon remaining in the liquid.

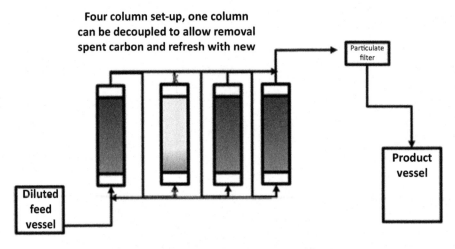

FIG. 24.8   Three-column filter design multipass to allow continuous operation.

Flavoured vodka

When considering producing a flavoured vodka, the distiller must once again consider the regulatory requirements of production and sales countries. They are extremely variable and two examples are shown below. These two were accurate at the time of publication but the distiller always needs to check the up-to-date versions before commencing production.

REGULATION (EU) 2019/787 OF THE EUROPEAN PARLIAMENT AND OF THE COUNCIL

of 17 April 2019

31. Flavoured vodka

(a) *Flavoured vodka is vodka which has been given a predominant flavour other than that of the raw materials used to produce the vodka.*

(b) *The minimum alcoholic strength by volume of flavoured vodka shall be 37.5%.*

(c) *Flavoured vodka may be sweetened, blended, flavoured, matured or coloured.*

(d) *When flavoured vodka is sweetened, the final product shall contain less than 100g of sweetening products per litre, expressed as invert sugar.*

(e) *The legal name of flavoured vodka may also be the name of any predominant flavour combined with the word 'vodka'. The term 'vodka' in any official Union language may be replaced by 'vodka'.*

US Regulation in 2019

Title 27. Alcohol, Tobacco Products and Firearms, Chapter I. ALCOHOL AND TOBACCO TAX AND TRADE BUREAU, DEPARTMENT OF THE TREASURY

Subchapter A. ALCOHOL, Part 5. LABELING AND ADVERTISING OF DISTILLED SPIRITS Subpart C. Standards of Identity for Distilled Spirits

(2) Vodka flavoured with natural flavouring materials, with or without the addition of sugar, bottled at not less than 30% alcohol by volume (60 proof) · The name of the predominant flavour shall appear as part of the class and type designation, e.g. 'Orange Flavoured Vodka' · Wine may be added but if the addition exceeds 2½% by volume of the finished product, the classes and/or types and percentages (by volume) of wine must be stated as part of the class and type designation.

There are three main ways to produce a flavoured vodka:

1. Essence addition;
2. Maceration;
3. Re-distillation with flavoured ingredients.

These techniques are very similar to what is used in the gin industry and are considered in Chapter 23 of this book.

Which one of the three methods used is mainly a brand-driven decision, however, some general rules apply. Care must be taken to ensure no compounds of concern are created by breakdown after contact with ethanol, either during distillation/filtration or in the bottle, and that the flavour or colour does not alter over time, primarily before the 'use by' period. Changes may be due to chemical reactions or the influence of sunlight through the glass bottle. Some markets and indeed consumers do not allow or like flocculation.

The easiest way to produce a flavoured vodka is by adding an essence. Providing that a safe food essence is used, it is a straightforward addition of the essence. Consideration needs to be taken of mixability and potential flocculation over time. The producer has the ability to select from natural essences, extracted natural essences, or flavour substitutes.

Maceration can also be derived in several ways, and the simplest way is to soak the botanic in ethanol to extract the desired flavour. Different flavours can be derived by altering the concentration of the botanic, the ethanol, and the maceration time. Nuances, such as grinding and crushing, can also play a part in the final flavour. After completion of this process, the liquid can be filtered to remove the solid material or be used as the feedstock for the third category of product.

Redistillation, in this context, is similar to that discussed in the gin Chapter 23, where dilution, followed by distillation, is used to produce a flavour derived from the botanics used.

Summary

The art of producing a vodka has a long history in which the range of styles has evolved and embraced innovation and modernisation of the way the vodka is produced. It can be seen that production is not straight forward and that it requires every bit as much knowledge of the different stages and attention to detail as any other distilled beverage. How the distiller operates is based on several key steps. What is the quality in terms of pureness that is required and what is the level of physical removal of congeners against the influence of the materials used to produce the base spirit? Distillation equipment selection is from 'simple' batch processing to complex multi-still regimes. Even once the spirit has been produced, it still needs to be made into vodka and this requires the selection and operation of the filtration medium. Finally, one must always remember the geographical regulatory requirements. The legislative framework in which the distiller is required to operate within does seem daunting. However, the number and range of operational variables discussed in this chapter has hopefully shown that the distiller continues to have great flexibility in order to create new and unique liquids.

Acknowledgements

Thanks to my friends and colleagues within the distilling community, too numerous to mention individually, from whom I have drawn inspiration and gained the knowledge set out in this chapter. A special mention goes to George Blair, who has contributed to the technical aspects covered and checked the work from a Chemical Engineering perspective, and to Alan Wardlaw who has reviewed the text. Thanks to Diageo Plc, for allowing me access to their extensive library of knowledge and for allowing me to author this chapter. The author would also like to thank Robert Piggot for granting permission to replicate Fig. 24.3 and Alan Wardlaw for Fig. 24.5.

Further reading

Lea, A.G.H., Piggott, J.R. (Eds.), 2003. Fermented Beverage Production, second Edition. Publishers, Kluwer Academic/Plenum. ISBN 0-306-4725-9.

McCabe, W.L., Thiele, E.W., 1925. Graphical design of fractionating columns. Ind. Eng. Chem. 17 (6), 605–611.

Nykänen, L., Suomalainen, H., 1983. Aroma of Beer, Wine and Distilled Alcoholic Beverages. Vol. 3 Springer Science & Business Media.

Piggott, J.R., 2003. Flavour of Distilled Beverages (Origin and Development). Ellis Horwood Publishers, Chichester, West Sussex, ISBN: 0-85312-546-5.

Seader, J.D., Kurtyka, Z.M., 1984. Distillation. In: Perry, R.H., Green, D.W. (Eds.), In Perry's Chemical Engineer's Handbook, sixth ed. McGraw Hill Book Co, Singapore.

Shinskey, F.G., 1977. Distillation Control for Productivity and Energy Conservation. McGraw-Hill Higher Education Publishers, ISBN: 0-07-056893-6.

Walker, G.M., Abbas, C., Ingledew, W.M., Pilgrim, C. (Eds.), 2017. The Alcohol Textbook, sixth edition. Ethanol Technology Institute, Duluth, Georgia, ISBN: 978-0-692-93088-5.

A short history of rum

Graham G. Stewart

GGStewart Associates, Cardiff, Wales, United Kingdom

Introduction

Rum is defined by Murtagh (1999) as 'an alcoholic liquor distilled from fermented sugar cane, molasses, etc.' There are a plethora of rum types available and it is produced in numerous countries and regions including the Caribbean and American countries (as well as Canada). Also, there are many other rum-producing countries, such as the Philippines, India, and Australia. In Newfoundland, there is a drink known as Swish, which is the remains of Jamaican rum from a cask rinsed with hot water when the rum arrives in Newfoundland. Screech rum is made from a blend of Jamaican rum, local Newfoundland water, caramel colouring, and flavouring. This East Coast spirit has been enjoyed by the locals in Newfoundland for centuries!

Rum is produced in a plethora of locations. 'Light' rums are commonly used in cocktails, whereas 'golden' and 'dark' rums have been typically consumed straight, neat, iced or used for cooking, but now they are commonly consumed with mixers (rum with cola is popular). Premium rums are produced to be consumed either straight or with ice.

Cachaça is a distilled potable spirit closely related to rum. The major difference between the two products is that rum is usually produced with molasses as the major raw material, whereas cachaça is made with fresh sugar cane juice, fermented, and distilled. Some rums, particularly from the French Caribbean, are also produced with sugar cane juice. Cachaça is also known as Brazilian rum. In the United States, cachaça is recognised as a type of rum and a distinctive product following an agreement, signed in 2013, between the United States and Brazil. In this agreement, the United States will cease using the term 'Brazilian rum.'

Christopher Columbus

An Italian-born explorer, who sailed under the royal flag of Queen Isabella of Spain, was responsible for introducing sugar cane to the New World and thereby, this changed history forever! When Columbus landed on the tropical island of Hispaniola, in 1493, his crew planted sugar cane. Columbus was very surprised how easily sugar cane grew in the local volcanic soil and under the humid warm soil conditions that prevailed there.

Columbus's Spanish successors, including the conquistadores, soon developed sugar cane in more Caribbean Islands, including Cuba, Puerto Rico, Jamaica, and Barbados. Around the same time, Dutch explorers landed in northern South America and established sugar cane plantations in the Guyana, Martinique, Trinidad, and Haiti. At this time, the emphasis was on sugar refining, not rum production. Rum production, mostly from molasses, was a profitable afterthought!

During the next two centuries, the Portuguese, French, Dutch, Spanish, and British explorers established a sugar cane industry that became the foremost agricultural crop in the Caribbean region with rum production being a very profitable 'spin-off'.

It is important to emphasise that rum is deeply associated with the Caribbean because of the tremendous quantities of quality sugar cane that are grown on these tropical islands and coastal regions. Sugar cane is an integral component of all varieties of rum. From the cane, the distiller extracts pure, sweet juice and the complex syrup known as molasses.

The first pot still in the New World was located on Staten Island in the Dutch colony that was known as New Amsterdam (now New York) circa the 1640s. Rum distilling became a major industry in New England in the late 1600s, especially in Massachusetts and Rhode Island colonies. Between 1700 and 1750, Massachusetts boasted 63 rum distilleries and Rhode Island 30, with 22 in Newport alone! Medford, Massachusetts served as the colonial rum capital due to the high quality of its 'Old Medford' rum.

All these products were molasses-based rum and there was a significant amount of molasses available. The enormous success of New England's rum distilleries also gave rise to other key regional industries, such as shipbuilding, logging, and ironworks (for ships) and cooperage (to produce, mature, store, and ship rum, brandy, and whisky).

After WWI (1914–1918), United States Prohibition (1916–1923), the Great Depression (1929–1938), and WWII (1938–1945), rum made a significant comeback in the early 1970s and 1980s. Rum became very popular in Australian bars and the proliferation of cocktails such as Mai Tai, Daiquiri, and Piña Colada re-established rum as a major spirit drink!

The Royal Navy and Nelson

The close association of rum with the Royal Navy began in 1655 when the British fleet captured Jamaica. With the availability of domestically produced rum, the British modified the daily ration of liquor provided to seamen from French brandy to rum.

Navy rum was originally a blend mixed with rum produced in the West Indies. It was originally at a strength of 100% proof (50% alcohol by volume [ABV]) and it was the only strength that could be tested (the gunpowder test) before the invention of the hydrometer. The term 'Navy strength' is still used in modern Britain to specify spirits bottled at 57% ABV.

The Royal Navy continued to give sailors a daily ration of rum. Known as a tot (totty) of rum, this is still issued on special occasions, on an order given by the Queen, members of the Royal Family, or on certain occasions, such as the Admiralty Board, in the UK, with similar restrictions, in other British Commonwealth Navies. In the days of daily rum rations, an order to 'splice the main brace' meant that double rum rations would be issued.

A legend regarding a rum in Britain involves Admiral Lord Horatio Nelson following his victory and death during the Battle of Trafalgar in 1805. The legend states that his body was preserved in a cask of rum en route back to England for burial in St. Paul's Cathedral in London (he did not wish to be buried at sea).

When the cask was opened, it was found to be empty of rum. Nelson's 'pickled' body was subsequently removed, and it was thought that the sailors on the Victory had drilled holes in the cask's bottom and had drunk of the rum! Hence, the term 'Nelson's blood' was used to describe the rum. However, this story is disputed by many! Some historians claim that the cask contained French brandy, not rum. Others claim instead that the term originated from the toast to Admiral Nelson. Variations of this anecdote of note have been circulated for many years!

It is worthy of note that the New Zealand Royal Navy was the last naval force to allow sailors a free daily tot of rum. The Royal Canadian Navy still gives sailors a rum ration on special occasions.

Types of rum

Dividing rum into meaningful groupings is complex because no single standard exists for a definition of this distilled beverage. Rum is defined by the laws and reputations of nations producing this potable spirit. Definitions include issues such as spirit proof, minimum ageing period, and even naming standards.

Mexico requires that rum is aged for a minimum of eight months, whereas rums from the Dominican Republic, Panama, and Venezuela require ageing for at least two years. Also, naming standards vary. Argentina defines rums as white, gold, light, and extra light. Granada uses the terms white, overproof, and matured. The United States defines rum liquor as flavoured rum. In Australia, rum is divided into dark, red, and white types.

Within the Caribbean, each island or production area has a unique style. Puerto Rican rum has a great influence and, as a consequence of this, most rum consumed in the United States is of the 'Spanish-speaking' style.

English-speaking styles are known for their darker rums with a fuller taste because they retain a greater amount of the underlying molasses flavour. Rums from the Bahamas, Antigua, Trinidad and Tobago, Grenada, Barbados, Saint Lucia, Saint Vincent and the Grenadines, Belize, Bermuda, Saint Kitts, and Jamaica are typical of this style. In the term denoting homemade, strong rum has been available since at least the early 19th century.

French-speaking areas are best known for their agricultural rums. These rums, produced extensively from sugar cane juice, retain a greater amount of the original flavour of the sugar cane and are typically more expensive than molasses-based rums. Rums from Haiti, Guadeloupe, and Martinique are typical of this style.

Areas that had formerly been part of the Spanish Empire traditionally produce añejo rums with a fairly smooth taste. Rum from: Colombia, Cuba, the Dominican Republic, Guatemala, Nicaragua, Panama, the Philippines, Puerto Rico, and Venezuela are typical of this style. Rum from the U.S. Virgin Islands is also of this style. The Canary Islands produce a honey-based rum known as Ronmiel de Canarias, which carries a protected geographical designation.

Rum production—Fermentation

In the production of rums, yeast strains of *Saccharomyces cerevisiae* and *Schizosaccharomyces pombe* are primarily used. Details of the use of *S. cerevisiae* have been discussed in Chapter 12. *S. pombe*, a 'fission yeast' yeast species, is a unicellular eukaryote, whose cells are rod shaped and that divide by division (Fig. 25.1), not budding (Fig. 25.2) (Dhamija et al., 1996). *S. pombe* cells typically measure 3 to 4μm in diameter and are 7 to 14μm in length. The genome of *S. pombe* is approximately 14.1 million base pairs and it also consists of 3 pairs of chromosomes compared to *S. cerevisiae*, which contains 16 chromosome pairs. In addition to its importance in the production of rum, *S. pombe* has evolved to become an important yeast species to study the cellular responses to DNA damage and the process of DNA replication (Fantes and Hoffman, 2016). Paul Nurse, a fission yeast researcher, successfully merged the fission yeast genetics and cell cycle research groups together with Lee Hartwell and Tim Hunt and was awarded the 2001 Nobel in Physiology or Medicine for the studies on cell regulation.

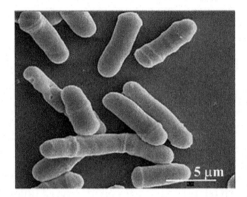

FIG. 25.1   Dividing yeast (*Schizosaccharomyces pombe*). Photograph by David O. Morgan.

FIG. 25.2   Budding yeast (*Saccharomyces cerevisiae*). Photograph by M. Das Murtey and P. Ramasamy.

Approximately 160 natural strains of *S. pombe* have been isolated and many of them have been characterised. These strains have been isolated from a variety of locations including Europe, North and South America, and Asia. The majority of these strains have been isolated from cultured fruits such as apples, grapes, and molasses and various alcoholic beverages, such as Brazilian cachaça and the fermented tea, kombucha (Teoh et al., 2004).

Among the flavour compounds found in many rums, it is mainly the result of the molasses fermentation with *S. pombe* that produces more of these compounds than *S. cerevisiae*. Under similar conditions of fermentation and distillation, *S. pombe* strains result in double production of ethyl acetate compared to *S. cerevisiae*. *S. pombe* strains generally produce lower levels of higher alcohols and there is a prominence of n-propanol. The metabolic route for propanol biosynthesis is specific with α-ketobutyric acid as an intermediate, which is also an intermediate in the formation of higher alcohols.

In conclusion, fermentation studies have demonstrated that a different compositional criterion for rums can easily be achieved with *S. pombe* as a fermentation agent rather than *S. cerevisiae*. The selection of appropriate yeast strains has been made possible by studies on *S. pombe*'s microbiological and physiological properties (Davey, 1998; Nurse 2004). The use of *S. pombe* in media based on molasses and sugar cane juice presents a series of biochemical, technological, and organoleptic challenges, where the solution exists in a better knowledge of its overall metabolism. Preliminary studies have made it possible to determine the focus that must be the subject of future research and development on this species.

Rum production—Distillation, ageing, and blending

As with other aspects of rum production, no standard procedure is used for the distillation procedure. Some rum producers operate in batches using pot stills but most rum production employs column still distillation. Pot still output contains more flavour congeners than the output from column stills and, as a consequence, it produces fuller tasting rums (Coulombe, 2004).

It has already been discussed that the rum ageing period varies between a plethora of producing countries. This period is usually performed in used bourbon casks but may also use different types of wood casks or stainless steel tanks. The ageing period is a primary factor in determining the rum's colour. When aged in oak casks, it becomes dark, whereas when rum is aged in stainless steel tanks, it remains virtually colourless.

Due to the typical tropical climate that exists in most rum-producing climates, rum matures at a much faster rate than whisky or brandy. An indication of this rapid maturation rate is the 'angel's share' or the amount of product lost by evaporation. When a product is being aged in France or Scotland, there is about a 2% loss of product per year, whereas tropical rums lose as much as 10% of the product during the maturation period.

Following ageing, rum is normally blended to ensure a consistent flavour, and this is the final step in the rum-making process (Smith, 2005). During blending, light rums are usually filtered to remove the colour gained during storage. For dark rums, caramel may be added for colour embellishment.

Rum in drinks and food

As well as rum being used as a traditional cocktail base, it is also employed in a number of cooked dishes as a flavouring agent in such products as rum balls, rum cakes, and sherries. It is commonly used as a marinade in some Caribbean dishes and in the propagation of rum-topf, bananas Foster, and other dishes. It can be mixed with ice cake and a variety of other sweet dishes. In addition, a number of non-alcoholic diluents are being used to prepare mixed drinks based on rum. Rum sales have been on the increase in 2020 and premium rums have relatively recently become a very popular drink served as a neat alcohol spirit.

References

Coulombe, C.A., 2004. Rum: The Epic Story of the Drink that Changed Conquered the World. Citadel Press/Kensington Publishing, New York.

Davey, J., 1998. Fusion of a fission yeast. Yeast 14 (16), 1529–1566.

Dhamija, S.S., Bhaskar, D., Gera, R., 1996. Thermotolerance and ethanol production are at variance in mutants of *Schizosaccharomyces pombe*. Biotechnol. Lett. 18 (11), 1341–1344.

Fantes, P.A., Hoffman, C.S., 2016. A brief history of *Schizosaccharomyces pombe* research: a perspective over the past 70 years. Genetics 203 (2), 621–629.

Murtagh, J.E., 1999. Feedstocks, fermentation and distillation for production of heavy and light rums. In: Jacques, K.A., Lyons, T.P., Kelsall, D.R. (Eds.), The Alcohol Textbook, third ed. Nottingham University Press, UK, pp. 233–255.

Nurse, P., 2004. Wee beasties. Nature 432, 557. https://doi.org/10.1038/432557a.

Smith, F.H., 2005. Caribbean Rum: A Social and Economic History. University Press of Florida, Gainesville, Fl.

Teoh, A.L., Heard, G., Cox, J., 2004. Yeast ecology of Kombucha fermentation. Int. J. Food Microbiol. 95(2, 119–126.

Spirits—Global packaging developments

Donna Abdelrazik[a] and Mike Mitchell[b]

[a]School of Graphic Communications Management, Ryerson University, Toronto, ON, Canada
[b]Diageo (Retired), London, United Kingdom

Introduction

Hidden somewhere between the mystery and craft of the distillers and blenders and consumers pouring a dram of their choice into their glass is the process of packaging. Consumers who are interested in whisk(e)y and white spirit production have no shortage of opportunities to indulge in distillery tours. These tours give a fantastic insight into the process from the field and stream to cask. Perhaps less well-understood, and indeed less well toured, is the process of converting matured and blended spirits into packaged products.

The primary objective of packaging is to provide a safe and effective delivery method of the product to its consumers. Packaging has several secondary goals that support this primary aim. These marketing and technical elements are considered further in Chapters 27–29. It is the technical elements from a global lens that is the focus of this chapter. Key technical objectives of any packaging process involving spirits include the following:

- Protect the health and safety of employees involved in the packaging process.
- Protect the end consumer's health and safety by adopting a food safety methodology during the packaging process (e.g. Hazard Analysis and Critical Control Points, or HACCP).
- Ensure that statutory and regulatory requirements are met in the packaging process.
- Protect spirits quality during packaging. The packaging process cannot improve spirits quality, but it can be adversely affected.
- Protect the brand's equity and reputation by providing a finished pack that is correctly filled (quantity), formed, labelled, and undamaged.
- Design the finished pack so that it can be warehoused and distributed without adversely affecting any of the above objectives.
- Keep the costs of packaging to a minimum.
- Design an agile packaging operation for competitive advantage and growth.

- Consider the Triple Bottom Line sustainability pillars throughout the packaging life cycle to consciously make decisions that will not negatively impact our imprint on the world.

Packaging brings together many components and materials to create the finished article, the actual spirit itself being the most important of these. The nature and characteristics of the spirit being packaged are determined by the distillation, maturation, and blending processes. It is important to re-emphasise that packaging cannot improve the quality of the spirit itself. Still, if it is carried out with insufficient care and attention to detail, it can adversely affect its quality.

When spirits arrive at the packaging plant, most are not ready for packaging and require some 'finishing'. They are usually at an alcoholic content higher than at which they are sold and contain various proteinaceous and wood materials from the distillation and maturation processes. Traditional and emerging preparation processes of spirits for packaging and the management of packaging materials are considered in the following sections.

Global spirits packaging process technologies

Preparation of traditional whisky for packaging

Consumers have come to expect that spirits will have a bright, brilliant finish. A cloudy whisky is assumed to be of inferior quality; many markets reject hazy whisky. Although yeast and other suspended solids from mashing remain behind in distillation, the aged whisky will have picked up wood fragments (so-called 'barrel char') that are now suspended in the whisky. When the casks are disgorged, this suspended material is contained within the whisky, and it is not uncommon for the whisky to have a cast or haze at this point, which can also occur if it has been stored at low temperatures in the cask warehouse. These materials must be removed by filtration.

Blended whisky arrives at packaging sites differently, depending on the proximity of the bottling hall to the maturation/blending warehouse. If they are located on the same site, whisky may be transferred directly by pipeline; however, in the vast majority of cases, packaging sites are remote from the maturation/blending warehouse. The whisky must be transported in bulk. It is common for the bulk whisky to be trap filtered at intake (via a basket filter) before subsequent sheet filtration (e.g. on a plate-and-frame filter). If maturation, blending, and packaging occur on the same site, then the trap filters may be sited between bulk storage vessels and the filtration operation buffer tanks.

Before filtration, it is typical for whisky strength reduction (through the addition of treated water) and colouring (with caramel) to take place. To ensure that those correct additions are made to bring the whisky to final packaging specifications, the parameters being affected (alcohol by volume and colour) will already be known from the bulk whisky vat.

When the whisky has been reduced and coloured correctly, it is filtered. A trim chiller is often used before the filtration operation to ensure that haze compounds remain suspended and do not re-dissolve in the whisky; the filter removes these particles. Unlike beer filtration, it is unusual for any filter powder to be used—the whisky is presented directly to the filter. This would quickly block the filter in a beer operation, as rough beer contains much more solid material than rough whisky. The filter (Fig. 26.1) is commonly a plate-and-frame unit with liners (paper filters).

Plate and frame filter

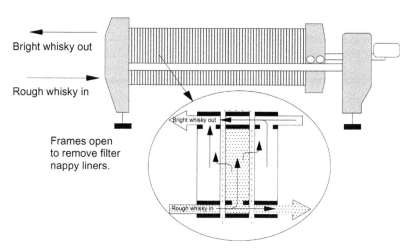

Bright whisky out

Rough whisky in

Frames open
to remove filter
nappy liners.

FIG. 26.1   Standard whisky plate-and-frame filtration equipment.

These paper filters have a finite lifespan. When they are blocked (as indicated by rising differential pressure across the filter), the filter must be separated, and the liners removed and replaced before further use. The filter works on the depth filtration principle. The liners themselves consist primarily of a cellulose fibre base and are recycled rather than discarded.

When a batch of filtered whisky is complete and ready for packaging, the most commonly analysed parameters are alcohol by volume (abv), colour, pH, and turbidity. Sensory analysis may also be performed at this stage. The whisky is stored in a bright whisky vat, from which the samples for this analysis are drawn. The length of time that the whisky stays in the bright whisky vat depends on several factors:

- The packaging program may be critical to supply customers, and a plentiful supply of whisky ready for packaging is essential.
- After filling, the contents of the bright whisky vat must be stabilised before a sample is taken for a quality check.
- Quality checks on the whisky take time, and the whisky must be stored until it is confirmed that it meets the required packaging specifications.

The whisky's quality will generally not change while stored in the bright whisky vat. Due to the high alcohol content, whiskies are much more stable than beers and do not deteriorate due to microbiological spoilage because of the elevated alcohol concentration. It is preferable to maintain a steady product throughout storage and handling. Still, if a tank of spirit has to stand for an extended period (e.g. through a packaging line failure), it generally is not an issue.

Global developments in spirits packaging processes

Traditional spirits preparations for packaging are well-understood, but are there nuances that need to be considered to support the expanding product development offerings in this category? Global trends in packaging are being driven by three primary trends: an experience-based economy, premiumisation, and sustainability.

The experience-based economy trend, a concept first introduced by Pine and Gilmore (1999), emphasises businesses' need to orchestrate memorable events for their customers. The memory itself becomes the product, the 'experience'. While the trend itself is not new, it has been heightened alongside the millennial generation entering the legal drinking age (LDA) population.

Ready-to-drink (RTD) mixed spirits cocktails and seltzers are taking the market by storm with their convenience and diverse flavour offerings. More recently, with the safety and comfort of off-premise occasion-based drinking experiences, with the influence of the COVID-19 pandemic, Nielsen reported a 91.8% increase from May to June 2020 in RTD consumer purchase behaviour (Nielsen, 2020). These blended beverages introduce carbonation to the spirits packaging process traditionally filled still, plus ingredients that can present special handling requirements and shelf-life constraints. Adapting CO_2 into the packaging lines has blurred the industry boundary and has introduced the new term 'brewstillery', whereby breweries and distilleries merge their technologies to create innovative products. An example is the recent collaboration by beer brewer Molson Coors with whisky distiller J.P. Wiser. The new product, Molson Common Bond Lager, is a 6.1% alcohol by volume lager infused with whisky-aged hops (CBN, 2021).

The second trend, premiumisation, is a global movement the spirits industry has been experiencing for over two decades and shows little slowing. According to IWSR (2020), the volume of premium (defined as spirits retailing for $22.50+ per bottle) spirits sold in the U.S. grew by 6.3% in 2019. Creating a memorable consumer experience is very connected to the premiumisation trend. It is 'table stakes' for brands that command premium prices. Regarding packaging process impacts, customisation, enhancements, and unique differentiation in packaging design strategies can challenge automation and packaging line speed efficiencies and negatively impact operational costs. However, as growth continues to validate the category, innovative methods to recover production costs from other areas across the packaging value chain, such as inventory automation, can offset the impacts.

Lastly, spirits companies are increasingly investing in sustainability-driven innovation and aligning themselves publicly to global sustainability goals important to their consumers, such as the UN Sustainable Development Goals. Specific sustainable packaging initiatives are covered more in-depth at the end of this chapter and in Chapter 27.

Packaging materials management and specification

Packaging materials are vital parts of the packaging process, and an understanding of them is crucial. Finished pack quality and packaging line performance (management of which is essential to cost control) can be seriously affected if the materials used have not been appropriately specified for the product/packaging line or if the material properties have changed in storage before use on the packaging line. Management of materials in a packaging plant is a critical element to ensure product quality and safety and allows the packaging line to operate at as high a level of efficiency as possible. Materials are expensive, and correct handling and storage are essential. Economic Order Quantity (EOQ) and First In First Out (FIFO) stock rotation are standard methods of inventory management. When brands are redesigned or refreshed in appearance, the warehouse's remaining materials relating to the old brand image must be exhausted as far as possible before the new design is produced.

Each brand and variant size of spirit will have its own specific bill of materials (BOM), listing several different components that must be brought to and staged at the packaging line for production to be successful. Adding in the complexity of servicing other world markets, which have their own legislative, labelling, and language requirements, it is easy to see that the number of different components being held in materials storage areas can be very high and complex! Materials are classified as primary, secondary, or tertiary. All three components make up a unit load or packaging system.

Product packaging components and considerations

For this chapter, the packaging component considerations that support the global spirits packaging manufacturing processes are outlined. A more in-depth comparison of spirits packaging containers, closures, and labelling enhancements is captured in Chapter 27.

Primary materials

The product cannot be sold without these materials. They contain the product itself and meet legislative and regulatory requirements regarding the vessel, closure, and label. The label will outline the net quantity statement, alcohol/volume, product description, list of ingredients, scannable barcode, and a manufacturing lot or batch code. The batch code supports quality control measures to source sensory analysis, processing dates, and overall product traceability for internal standard manufacturing practices and consumer enquiry. A 'best before date' may be applicable for blends or ready-to-drink mixes containing ingredients with shelf-life considerations; however, this does *not* apply to straight whisky.

The traditional global primary vessel material for the spirits category is glass. Fig. 26.2 illustrates typical glass terminology. Spirits can also be packaged into PET (plastic) bottles, aluminum cans, and most recently, a paper-based sustainable bottle innovation (Smith, 2020). These substrate alternatives to the classic glass portal offer their own unique advantages. PET is ideal for single-shot miniatures and for premises where breakage is a risk factor, such as outdoor occasions. The aluminum can is advantageous for the mixed ready-to-drink (RTD) offering, whereby the business strategy focuses on customer conversion, convenience, and variant flavour trialling. The food-safe wood pulp bottle alternative is a response to the global sustainability strategies that distillers and consumer product goods companies are spearheading to minimise their environmental footprint. As new container alternatives become available, the following questions should be reviewed:

- Is there a functional benefit to this new container?
- Is it more sustainable?
- Will it meet the same or better product containment and barrier property qualities?
- Will it support ease across the manufacturing and distribution processes?
- Will it increase the consumer experience and engagement?
- Will it have positive impacts on health and safety?
- Will it increase sales and revenue?

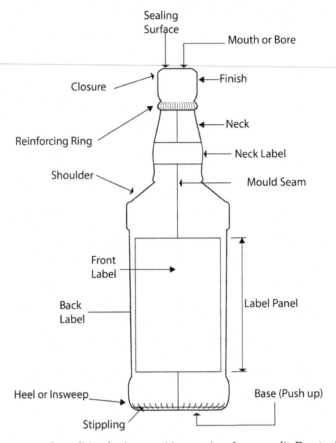

FIG. 26.2 The components of a traditional primary spirits container. Image credit: Dorotea Bajic.

Beyond the material choice, the shape of the vessel can lend inherent strength, accessibility, and brand awareness to the package. Global spirits brands are packaged into an array of different bottle shapes specifically designed to add to the brand appeal and to maintain the protective properties. Square, rectangular, oval, round, tall, and short bottles are all used, and these have particular handling characteristics. This is a key consideration when planning what to produce. Lightweighting (where the weight of the glass used is reduced) is a structural packaging consideration that many brands will opt to implement in order to save on material and distribution costs, but these bottles must be handled with additional care as they are more prone to breakage. A cost–benefit analysis exercise may be initiated to evaluate the quantitative savings and sustainability messaging against the increase in quality risks or damage potential. There are additional qualitative challenges to lightweighting that may only be measured once the new format is in circulation. Increased breakage may increase consumer complaint feedback, impacting quality perceptions of the product and ultimately the brand, in turn affecting consumer loyalty and the loss of repeat purchase intent.

A bottle's design must take into account its life cycle—from raw material sourcing, manufacturing processes, filling, labelling and packaging, storage, distribution, consumer usage, and end of life. Bottles are produced in bulk with a minimum order quantity (MOQ) by the glass supplier and are delivered to the spirits packaging site on pallets. Each pallet has several layers of bottles, separated by layer pads, and the entire pallet is wrapped in protective plastic film to prevent any ingress of foreign bodies into the bottles. The packaging site must have sufficient storage for empty vessels. They are bulky items and require sizeable warehousing capacity. Ideally, the glass supplier delivers bottles on a just-in-time (JIT) basis to minimise the size of the empty bottle inventory at the packaging site. Embossing is very common on spirit bottles. This allows brand identification and often increases the attractiveness of the bottle and hence the finished product. Spirit bottles also come in a wide variety of hues. Clear ('flint'), brown, and green are most common, but other colours may be used depending on the product and the product designer or marketer's specifications. In general, glass has no effect on product quality because it is chemically inert and insoluble and will not taint the product. It is also smooth surfaced and easy to clean (usually fluid cleaning using air or water). This makes it the ideal material of choice for the spirits industry.

All commercial glass is based on silica, which is the principal component of sand. Glass bottle manufacturers change the colour of glass by using colourising and decolourising agents within the glass recipe (sand, colouring agents, limestone, other refining agents). Broken glass is hazardous, especially neck slivers from poorly moulded bottles. Critical glass defects (e.g. birdcages, which are caused by an ineffective blowing process when bottles are formed) will make bottles unusable for filling, either through critical weaknesses that will cause the bottle to break or through the generation of glass fragments within the bottle. These are both potentially harmful and undesirable. Exact and repeatable bottle dimensions are important, as bottles may be filled to a particular level in counter-pressure or vent-tube fillers. Variations can cause bottles to be short or overfilled. Dimensions are also critical in ensuring that the bottles can accept a closure, be it cork, plastic, or metal.

Closures

Cork has many advantageous characteristics that make it a suitable material for closure manufacture. It is 100% natural, biodegradable, and recyclable (CQC, 2014). In addition, it is light, elastic, and compressible. Cork is harvested by removing the thick outer bark of the cork oak (*Quercus suber*). As cork is a natural material, it suffers from batch-to-batch variability year-on-year. This variability leads to issues with the surface of the cork that can cause debris and/or taint. The debris is not an issue with wine drinkers who accept a little cork debris in the bottle; however, with premium spirits, this debris can lead to consumer complaints. The debris tends to come from seed holes in the surface of the cork, which can open up further once the cork is under compression in the neck of the bottle. So-called 'cork taint' is a much more serious issue. Taint is caused by the presence of 2,4,6-trichloroanisole. This compound gives a musty odour that is detectable at very low concentrations (typically, parts per billion).

Metal closures are now the most common type of closure used for global spirits bottle packaging, particularly roll-on-pilfer-proof (ROPP) closures. Aluminum is the metal of choice. It is shaped and stretched to attain the closure dimensions before being decorated with inks and varnishes to provide the desired branded finish. The dimensions of these closures are tightly specified, and they do not suffer from the same natural irregularities that cork closures can be prone to. These closures will have a wad (made of PET or cardboard lined with PET) with PET either moulded or glued inside them, which acts as the sealing surface between the closure and the sealing ring of the glass bottle. When these closures are opened, thin metal bridges at the bottom of the closure (near the reinforcing ring of the bottle) snap off leaving a portion of the closure behind on the bottle neck and the top cap in the consumer's hand (called 'pilfer proof'). It shows if the bottle has been opened after packaging before it reached the consumer. Other forms of tamper evident closure packaging include shrink film bands, wax seals, and/or tax labels over the closure.

Plastic closures are also now more commonly used than previously, both as primary closures, in the form of an external screw thread (EST) and as decorative over caps, where the over cap is a marketing tool identifying the brand, which sits atop the primary closure. They are produced by plastic injection moulding, and as with metal closures, this allows for extremely tight dimensional specifications to be met on a highly repeatable basis. Primary plastic closures will also usually have a wad inserted within them, acting as the sealing surface between the closure and the sealing ring on the bottle. The nature of the injection moulding process means that these closures can be more complex in design and may include anti-theft or anti-counterfeit devices as a part of their design.

Labels and applications

Labels are one of the primary decorative parts of packaging; they reinforce brand identity and provide necessary information regarding the contents and type of spirit within the bottle. If the brand has an age statement, it will also be stated on the label. Historically, paper labels applied with wet glue have been the main type of label used to decorate spirits bottles; however, as technology has improved, self-adhesive or pressure-sensitive labels are increasingly being used. An advantage of pressure sensitive labels over cut and stack labels is in the application equipment. Cut and stack require unique parts for each label design and shape, whereas, pressure sensitive labels are roll fed and independent of shape. They also have the advantage of not requiring wet glue, which demands some careful housekeeping on labelling machines to avoid labelling faults such as glue smears on the bottle, which visually spoil the image of pristine quality that whisky producers want to maintain for their products. Labels require careful storage and relative humidity control on the packaging site, especially paper labels, as they are prone to moisture pickup, which affects their application on the label machine, leading to misaligned, poorly applied labels, and machine stoppages. Silk screening, applied ceramic labelling, and foil stamping are forms of traditional printing enhancement applications used in limited production spirits. Today, economical digital printing methods are being used to meet customisation and personalisation attributes through unique designs and elements such as serial numbers for limited and collector editions.

Secondary materials

These are materials that collate primary packages into some new form; that is, they are the second layer of packaging. Secondary materials convert the primary package into a saleable or marketable unit. Examples include cardboard (e.g. cartons, boxes, trays, division inserts, and multipacks), which can be constructed of either corrugated cardboard or paperboard. Corrugated cardboard is usually used for transit cases or shipping containers such as boxes and trays, whereas individual cartons and multipacks are more likely to be made of paperboard. These items are likely to be branded and highly graphic, although some transit cases that contain cartons within them may be plain in appearance, as it is the cartons that will be displayed by the retailer and not the shipping container. Premium and super-premium brands may have their own premium secondary packaging involving more exotic materials. For example, display and presentation boxes may be metal, felt or silk-lined, with fold-out panels or recesses, designed as marketing tools to communicate the premium nature of the spirit within. These materials are usually unsuitable for automatic packaging and require hand finishing. A separate production area may be provided for such packaging operations, with specialist teams assigned to the production process. In-depth knowledge of the pack, coupled with attention to detail and painstaking care in the production process, is very important for such operations as these materials can be extremely expensive.

Tertiary packaging

Tertiary packaging refers to all other packaging materials used to protect the finished product and assist in its handling throughout the logistics chain to the eventual point of sale. Such materials include protective stretch-wrap (around pallets of finished goods), shrink-wrap around individual boxes, and even the pallets themselves. Although their function is purely technical in terms of protection and transport, they are no less important to the product than the primary and secondary materials.

As already stated, packaging plants must pay careful attention to material handling and storage, as the costs involved (related to material losses, production line performance, and potential quality issues in the market) can be substantial if they are not well-managed. Spirits packagers agree on material specifications with their suppliers as a part of the procurement process, and the onus is on the supplier to meet the specification, known as a CofA or certificate of analysis that warrants adherence. Materials that are incorrectly manufactured in the first instance may well be the cause of packaging issues such as those discussed earlier. A positive supplier relationship is essential to dealing with any such issues when they arise. The spirits manufacturer will typically liaison with material suppliers through a working party with representation from production, technical, and procurement departments.

The spirits packaging line

The philosophy of packaging and the equipment used will vary from company to company in the spirits industry. Due to the large and diverse range of products available, production volume, primary and secondary packaging sizes, and packaging formats available, it is unusual to find a 'one size fits all' approach regarding packaging line design. The diversity and expansion of product offerings, mixer-packs, and variant extensions require flexibility, accuracy, and agility to be at the forefront of the decision-making process in equipment and technologies at the distillery or brewstillery packaging line. Layout, equipment, staffing levels, and line capability in terms of stock-keeping units (SKUs) being produced will all influence line design; however, one can generalise on the various stages of the packaging process as summarised below.

The depalletiser

The purpose of a depalletiser is to convert pallets of bulk empty primary vessels into a loose single file stream ready to be handled by the packaging line. The stretch wrap on pallets of empty containers is removed automatically by a stripping device or manually by an operator. The unwrapped pallet is then presented to the depalletiser. The layer pads on top of each layer of containers are removed by vacuum suction cups and collected, ready for redistribution to the material supplier. Each uncovered layer is then swept onto a bulk-flow conveyor using pneumatic air cylinders as a motion control source. The conveyor transports the empty containers away from the depalletiser. This process continues layer-by-layer until the pallet of primary vessels is depleted. The pallet is then ejected from the depalletiser into a pallet collection point, from where the pallets are again taken and returned to the material supplier. Additional caution and safe handling are considered when depalletising glass vessels versus aluminum or PET due to the risk of small glass shards inherent in glass manufacturing and distribution.

The bulk handling involved in the depalletiser operation means that complete safety guarding is necessary, with the loading area and pallet discharge area behind interlocked guarding. The interlock works so that if any access is attempted to the machine or to the heavy loading areas, the machine is emergency stopped and cannot function again until the guarding is closed and reset. For lightweight plastic containers, a bottle unscrambler may be required to orient and single file these vessels. The bulk feed of empty containers from the depalletiser is single filed using a conveyor called a *combiner*. The single-file containers are then conveyed to the next operation on the packaging line—an orientator.

The orientator

Not all packaging lines require an orientator. An orientator is required where primary containers have a specific shape or form design that means they must be presented in a particular orientation to enable labels to be applied to the vessel's specific sides and positions. An example would be the square bottle made famous by the Johnnie Walker brands. Embossing on the bottle dictates that the front and rear body labels must be applied to specific panels of the bottle. To ensure that this happens, the bottles are orientated such that they are all facing the same way when they exit the orientator. The orientator detects a registered mark on the base of the bottle and uses the position of this register mark to determine how much the bottle must be turned before it is correctly aligned. Once a bottle is correctly aligned, it is discharged from the orientator and makes its way to the next operation.

The rinser

The container rinser aims to remove all extraneous material and foreign bodies from the containers before they are filled. This is achieved by blowing the containers with compressed air or by rinsing them with a sterilised liquid. The application of air is by far the most common method currently employed. Bottle rinsers are usually rotary machines, where the vessels are fed in a single file to a timing scroll, which transfers them to an infeed starwheel. From this starwheel, the bottles are picked up by plastic gripper heads, inverted, and blown thoroughly with compressed air. Any debris contained within the bottles drops out into a collection bed. The bottles are then inverted again back to the normal standing position and released into a discharge starwheel, from where they are discharged from the rinser and conveyed to the filler.

The filler

The filler is a critical area on the packaging line. The correct liquid must be filled into vessels at the proper volume. It is also frequently the point at which packaging line performance is measured, as it is often the slowest machine on the line and, therefore, the line's rate-determining step. Spirits are supplied to the filler from finishing tanks, such as the bright whisky vat. Most modern fillers are rotary machines, carousel-like in design, where a tank containing the spirit (the filler bowl) sits above several filling heads. Empty vessels are again timed into infeed starwheels by a rotating scroll, and from the starwheel, the containers are transferred into the individual filling heads on the carousel. Each filling head consists of a number of valves and a filling tube (Fig. 26.3).

For still product spirits, low vacuum-gravity filling systems are common. The filling process involves a series of cams mounted around the filler bowl, which actuate gas evacuation and filling valves as the bowl spins. Spirits flow down outside the hollow filling tubes, while the air within the vessel is evacuated through the filling tube into the bowl above. The level of spirit within the bottle continues to rise until it reaches the bottom of the filling tube, at which point no more air can be evacuated from the bottle, and filling stops. The filling head is lifted from the full bottle, which is then transferred away from the filler bowl by means of a discharge starwheel and exits the machine. Once filled, the conveyor then transfers the product to the capper.

FIG. 26.3 A filler, showing the infeed starwheel, the carousel with filling heads, and the discharge starwheel for a bottle filler.

With the global advancement of mixed spirit product offerings, distilleries and brewstilleries are adopting counter-pressure filling systems that carry out the additional phases of pressurisation and CO_2 injection for carbonation. As technologies advance, operations can increase their automation functionality, expand their manufacturing versatility, improve their data analytics to support quality control, and reduce contamination risks.

The capper—Bottle format

This machine is often closely coupled to the filler, and the operations may be linked by the same starwheels, conveyors, sensors, and guarding. The capper applies the closure to the top of the filled bottle, seals it down, and tightens to the closure at a specific torque level (torque being the amount of force required to break the seal on the closure and open the bottle). A great variety of closure types are available (as well as the various material types employed), meaning that several designs of cappers are available from equipment manufacturers. Correct application and torquing are essential to ensure that the bottle's integrity is not compromised and that the quality of the spirit inside is maintained. Damaged closures may allow the product to leak from the bottle and will not meet quality standard specifications.

The seamer—Can format

For spirits, seltzers, and ready-to-drink (RTD) alcohol beverage offerings packaged in metal, paper, or plastic cans, the equipment used to apply the lid to the can body is a can seamer. Its role is to create a hermetic tamper-evident double seal that forms a leak-proof closure made up of a seamer head, turntable, and rollers through compression and rotation. The seamer head (or chuck) secures the can in place, with compression, against the turntable to allow the seaming rollers to produce the interlocked concave seal between the lid and body flange.

The labeller

The labeller is responsible for bottle decoration and applies the label set to the filled and capped bottle. The label set for each brand will vary widely in terms of design and area of application. Front and back body labels are the usual minimum decoration for a spirits bottle; neck labels and foot labels are also used on many brands. Paper labels require the application of wet gum adhesive before being placed on the bottle. The entire process of separating individual labels, applying adhesive, placing the label onto the bottle, and brushing down the label is achieved by individual 'label stations'.

Fig. 26.4 shows three different design layouts of a labeller. Bottles enter via the infeed worm or scroll before being rotated around the label stations. Each label station is responsible for one label. The first label station might apply the front labels before the bottles are turned on the carousel to the next position, and the second label station might apply the back labels. After each label station, a set of brushes is positioned such that as the bottles pass, the labels that have been applied are gently brushed down onto the container surface into their finished position. Wet gum label application is a delicate process with several variables that can easily cause machine malfunctions or quality problems if not correctly maintained. The temperature

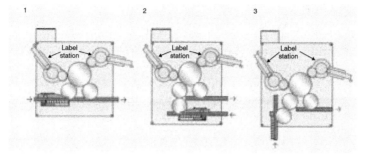

FIG. 26.4 Label design layouts: (1) linear arrangement, (2) parallel arrangement, and (3) angular arrangement.

of the glue must be maintained (to ensure a consistent viscosity and application), and the amount of glue applied to each label must be carefully controlled.

Too much glue will be smeared across the surface of the bottle and the labeller's brushes. Glue smears are unsightly on spirits bottles. Too little glue and the labels will either not stick to the container properly (or 'wing'), with paper surfaces sticking up from the glass or plastic. In extreme cases, the label may even fall off the bottle.

Pressure-sensitive or self-adhesive labels come with their adhesive already applied and are far less messy to deal with, although the operation is just as demanding in terms of machine accuracy. Different designs of label stations are required for this type of application because the labels are printed on rolls as opposed to bulk stacks, as is the case for paper labels.

For can production processes, the cans are shipped to the distillery or brewstillery already pre-printed. Dry-offset printing is the traditional printing process for cans. For lower volume product manufacturing, digital print or pre-printed shrink sleeves are alternative labelling application processes for the spirits industry.

Secondary packaging equipment

These operations are designed to collate filled, labelled, and secured primary containers into multiple units ready for shipping. A variety of pack types can be accommodated, including cartons, multipacks, sleeve packs, individual boxes, and shipping boxes. Finished containers make their way along a conveyor to the staging area, where they will be separated and patterned for their secondary packaging unit configuration (1, 4, 6, 8, 12, 15, 18, or 24 units per carton). For glass containers, where these units involve several bottles together (e.g. a 12-bottle top-loaded corrugated box), dividers may be inserted into the box to keep the glass bottles from rubbing against each other during distribution, thereby lowering the risk of damage or breakage. The packaging orientation for loading can either be top-loaded or end-loaded. A top-loaded carton is automatically erected and sealed (glued) on its base. Containers are placed into the cartons from above and then closed. End-loaded cartons are automatically erected, one end is sealed, and then the containers are placed into the carton from the side. In the case of multipacks, each variant must be separated and organised into a specific configuration before being packaged along the conveyor. A variety-pack system can be used to unload, divide, and separate up to eight different flavours per package using automation, vision detection, and coding.

Decoration considerations

What if the secondary packages are not meant to be customer-facing and are simply used for shipping purposes. In that case, their graphic decoration and material choice can be simplified, such as a kraft corrugated material with a one-colour graphic printed post-print flexography. If, however, the intention is for this secondary package to be a retail selling unit and to find a presence on the shelf, a high graphic full-colour offset or flexography printed paperboard carton would amplify the shelf appeal.

The palletiser/stretch wrapper

The palletiser/stretch wrapper is usually at the end of the production line. Secondary packages are collated, built into individual layers, and stacked on wooden or plastic pallets. The finished pallets can be wrapped in protective stretch wrap, which fulfils two functions. It protects the cardboard cases from water damage and adds stability to the complete pallet, ensuring that cases do not fall during transportation and distribution to market.

Packaging quality management systems

All production stations on a packaging line involve some form of quality checking. Each machine may have a paper or electronic check sheet wherein operators are required to complete periodically, established by standard operating procedures and qualitative measurement specifications. The frequency of such checks will depend on the risk to product quality should a fault occur. Quality assurance depends on the frequency and accuracy of those checks. Operators are often assisted by automatic sensors that stop the line if a critical quality defect is noted (e.g. missing caps, repeated broken bottles, missing or misaligned labels). All products produced have a production code printed on the primary and secondary packaging format for traceability purposes. This code may be applied by laser etching or inkjet printing and will indicate the line and date on which the package was produced.

New investments in intelligent packaging technologies such as QR (quick response) coding or RFID (radio-frequency identification) tagging allow a distillery or brewstillery to track and trace quality and inventory data analytics. This information can also be used to drive further consumer engagement. It can be used for geo-targeting of local businesses and for authenticity verification where counterfeiting is a concern. These advancements provide the spirits industry with end-to-end data analytics information for operational excellence, and as noted by TrendHunter (2021) 'satisfy and anticipate consumer needs, as well as supercharge product experience'. As the spirits production lines continue to diversify, the need for advanced interactive quality management systems is imperative for efficient operations, cost management, and competitive advantage.

Assessment of packaging line performance

Overall equipment efficiency and line efficiency reporting

Packaging operations represent a substantial cost to the business. Putting aside the costs of spirits production and packaging materials up to this stage, operating a packaging line also requires a significant financial undertaking. Capital expenditure is necessary in the first place to buy the production machinery, so procuring the correct assets with a proven track record of reliability and agility goes a long way to ensuring a long-term efficient production process and a return on investment (ROI). Once the equipment is in place, there are substantial day-to-day operating costs, including labour, to consider. The packaging operations team will want to establish an appropriate quality management system to effectively define key performance metrics to monitor the operation's efficiency and effectiveness in order to keep tight control of costs and outputs. The goal in the next section is to give the reader the following:

- An appreciation of the measures that are used to understand the performance of a packaging production line or process.
- An understanding of why overall equipment efficiency (OEE) is important.
- An account of the standard calculations for aggregate OEE, line/process OEE, and line efficiency.

What is OEE?

Overall equipment effectiveness (OEE) is a measure to evaluate a machine's productiveness or that of a production line. It is *not* a measure of people's performance. The higher the OEE measure, the more saleable finished product per shift a machine or line produces, resulting in lower costs per unit delivered and helps the operation be more competitive. OEE analysis is a tool used to analyse equipment performance (to identify process failures versus mechanical failures), accounting for losses due to availability, performance, and quality.

The OEE waterfall

The OEE waterfall is a tool used to represent the performance of a packaging line. It graphically depicts how a line has performed and shows where issues have occurred on the line that may require further attention through, for example, root-cause analysis of equipment or quality defects. An example of an OEE waterfall is shown in Fig. 26.5. Before going into detail regarding the OEE waterfall, it is essential to understand the baseline factors of OEE.

There are several critical factors required to calculate OEE.

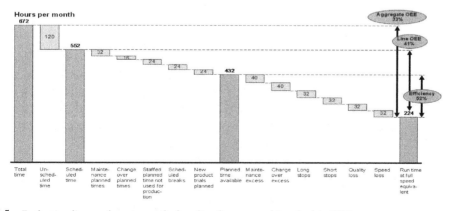

FIG. 26.5 Packaging line performance—the baseline figures used to calculate OEE.

Total time—The total time the line is available to produce is essentially 24h per day, 7days per week (even if it is not scheduled to run this amount of time, it is still available). Total time in a 4-week month is 168 h per week × 4 weeks = 672 h.

Scheduled time—Scheduled time is the total time planned for the line to produce the number of cases required in the period. This time includes changeovers, planned maintenance, breaks, and production trials, and allows for unexpected issues on the line (e.g. breakdowns).

Planned time available—Planned time available is when the line is planning to run without interruption and is typically determined using the expected line speed.

Balanced speed—Packaging lines always have a speed rating, such as units of bottles per minute or cases per hour, known as the balanced speed. This speed provides essential information about the maximum possible output from the line over a specified period of time. Balance speed is the higher of (1) the maximum speed the line is currently capable of running at, or (2) the manufacturer's design speed (from original installation). On the OEE waterfall, this number is used as part of the calculation that determines the *run time at full speed equivalent*:

$$\text{Run time at full speed equivalent} = \frac{\text{Number of bottles or cases to be produced in the period}}{\text{Balanced speed}\left(\text{bottles / cases per hour}\right)}$$

Example calculation

Problem: Over the course of a week, Line 1 must produce 120,000 (12-pack) cases. The agreed balanced speed of the line is 400 bottles per minute. What is the length of time that the line would be required to run at full speed to produce the necessary cases; in other words, what is the run time at full speed equivalent?

Solution:

$$\text{Run time at full speed equivalent} = \frac{120,000\,\text{cases} \times 12\,\text{bottles per case}}{400\,\text{bpm}}$$

$$= \frac{1,440,000\,\text{bottles}}{400\,\text{bpm}}$$

$$= 3600\,\text{min, or } 60\,\text{h running at the full possible speed}$$

Aggregate OEE

Aggregate OEE measures the time that the line is required to run at full speed divided by the overall time available to run the line. Businesses typically use this to determine the utilisation of a packaging production line when deciding where to produce new products. The line's performance can very rarely influence aggregate OEE as it does not account for the production process variables, such as breaks, changeovers, or minor stops. It is a straight division of the run time at full speed equivalent by the total time available as shown below:

$$\text{Aggregate OEE} = \frac{\text{Runtime at full speed equivalent}}{\text{Total time}}$$

Using the example from the previous calculation, where Line 1 is required to produce at full speed for 60h in the week, the aggregate OEE for that week would be as follows:

$$\text{Aggregate OEE} = \frac{60\,h}{168\,h\,available\,in\,a\,week} = 36\%$$

This means that if the line were to run at full speed for 120,000 cases, it would be utilised for 36% of the time available for production without any stoppages or changeovers.

Line/process OEE

Line/process OEE, from a packaging line performance point of view, is the most critical direct production measure on a site. All of the factors that impact line performance can be measured and, more importantly, influenced by the Line/Process OEE. Therefore, aspects can be continuously improved through analysis and corrective action. The OEE waterfall in Fig. 26.6 shows all of the production elements that can impact line/process OEE.

As shown, there are two distinct aspects used in the calculation. First are the planned interventions that occur, such as planned maintenance, scheduled breaks, and team meetings (staff planned time not used for production). Second are the unexpected things that take place on a production line, such as long stops (breakdowns of greater than 5min) or when a changeover takes longer than was planned (changeover excess). Line OEE is calculated by taking the time required to produce all of the cases needed while running the line at full speed divided by the scheduled time, as shown on the waterfall diagram.

Measuring everything that happens on the line while scheduled to operate allows businesses to fully understand how effective the line's machinery is and what people are being compensated for doing. However, the information must be accurately recorded to sustain and continuously improve packaging line performance. Accurate data are the key to understanding where the main issues lie and to provide leadership teams with the opportunity to undertake line improvement exercises using such tools as value stream mapping, root-cause analysis, and lean manufacturing principles. Continuous improvement can be achieved by

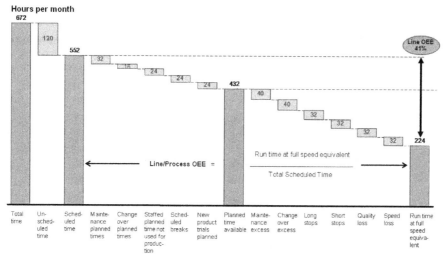

FIG. 26.6   OEE waterfall showing all of the aspects of production that can impact line/process OEE.

methodically collecting data, identifying gaps, setting targets to measure against, and evaluating their outcome in a priority sequence.

Line efficiency

Line efficiency is the measure used to gauge how effectively a line performs at speed without planned or unplanned stoppages such as changeovers, maintenance, errors, etc.

Line efficiency is calculated as follows:

$$\text{Efficiency} = \frac{\text{Runtime at full speed equivalent}}{\text{Planned time available}}$$

Essentially, line efficiency measures the length of time required to run at full speed to produce the required number of cases as a percentage of how long the line took to produce them (when it was planned to). This means that if line efficiency is calculated at 50%, then the line only ran at full speed for half of the time that it could have run, meaning that such things as long stops (breakdowns), minor stops, or quality issues used up the other half of the time.

The efficiency measure should never be used in isolation as, on its own, it can be misleading. A high-efficiency figure does not necessarily mean that the line is performing well in all areas. It does mean that the line is running well when it can; however, this may hide inadequate changeover procedures, long break times, or excessive planned maintenance, all of which impact line OEE. A line that is performing well will have significant performance figures for both line OEE and line efficiency.

The 'V' curve

In this section, the concept of relative machine speeds, as defined by the so-called 'V' curve, is discussed. It is vital to appreciate how a 'V' curve can optimise a packaging production line's running speed and determine the targeted 'balanced speed'.

Line layouts and speeds are of essence to obtain good line performance. There are many layout alternatives. The result may depend on existing arrangements, but modern objectives plan to achieve the best efficiencies and low operator requirements. As a way of determining the optimum running speed for a line and understanding the line's balanced rate, a 'V' curve can be produced using the possible speeds of each piece of machinery. This allows packaging leaders to set the optimum running speeds for each machine's line to ensure that they are provided with enough material so that the product is moved quickly and effectively to the next stage of the packaging process. The operational configuration depends on machine cycle times, and packaging lines are designed around the critical equipment supported up and downstream.

The limiting machine on a line that will determine the balanced speed is usually the filler, although depending on pack size (e.g. 500mL vs. 1L) or the number of bottles in a case (e.g. 6 vs. 24), the fastest speed of the slowest machine on the line may change.

Generally, this means that palletisers, rinsers, labellers, etc. will be able to run at a higher speed than, for example, the bottle filler, as illustrated in Fig. 26.7. As shown, machines before and after the filler are planned to run faster by increments of 5% to 8%. In this way, the line stands the best chance of producing a good OEE. The machine at the bottom of the graph provides the balanced speed for the line. The faster the line, the less robust the efficiency, and stoppages will give a more significant output loss. An example, key performance indicator

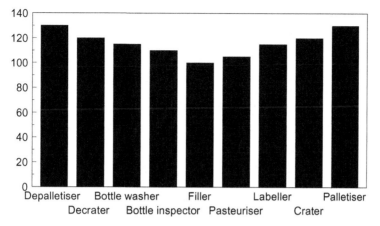

FIG. 26.7   Bottling line speed distribution graph, or 'V' curve.

(KPI) measures the mean time between stops (MTBS) on the packaging line. Also, the more machines there are on the line, the greater the risk is for operational issues and, therefore, a reduced OEE.

It is essential to run the line equipment at the specified rate; otherwise, quality problems could well be experienced, including the following:

- A palletiser that runs over speed could damage pallets or glass.
- A bottle rinser that runs over speed could reduce the time when bottles are blown, leading to dirty bottles and potentially critical food safety failures.
- A filler that runs over speed could result in a higher number of rejects due to under-filled bottles or excessive broken bottles.
- A labeller that runs over speed could result in misaligned or 'winging' labels.
- A packer that runs over speed could result in label damage on the bottles.

Accumulation

Accumulation helps to smooth out the periods when a machine could potentially be starved of product due to a slower running machine either up or downstream on the packaging production line. It should be noted that accumulation can be expensive, both in terms of accommodation to build the line itself and if mechanical issues occur on the line (as there will be faulty product to rework). For slower lines, such as 300 bottles per minute (bpm), accumulation is not as crucial, and the 'V' graph can be flatter. However, for higher-speed production lines, accumulation helps support an effective line balance.

Spirit and packaging material losses

Spirit losses

Typically, if the amount of spirit prepared at the start of the production process is compared with the amount of spirit in bottles or cases at the end, it will be found that some spirit will have been lost during the process.

Losses can occur in volume or strength (alcohol %) or a combination of the two. Losses incur in several ways, such as the following:

- When tanks are filled and emptied, there is inevitably some loss because of the liquid left on the vessel's sides (sometimes called the *wetting loss*).
- Spirit remains in the pipework after it transfers from vat to vat or vat to line. This is often called the *transfer loss*.
- Filtration losses occur because of the spirit that remains in the filter.
- Packaging spirit losses are made when packs are overfilled or when spillages or breakages occur.

Loss control is only possible when the losses are measured. The way to measure losses is to compare the volume before and after a process. Examples of loss measurement are shown in Figs. 26.8 and 26.9.

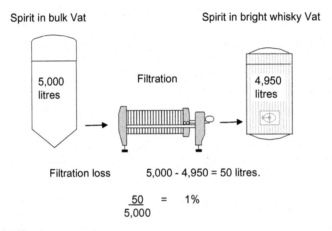

FIG. 26.8   Loss in the filtration process.

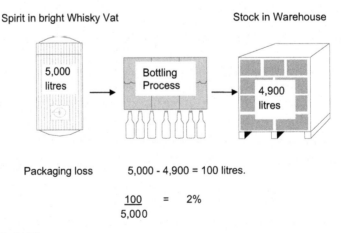

FIG. 26.9   Loss in the bottling process.

The impact of spirit losses

There are three key areas where spirit losses can impact the operation:

1. *Financial*—Spirit waste is highly costly to the alcohol drinks packaging business.
2. *Service*—When spirit losses are excessive, the amount of liquid required for a customer order can be compromised, and as such, the number of cases needed may not be met. This can lead to customer order misses, reputation impact, and loss of revenue.
3. *Legislative*—Spirit is tightly controlled by businesses and governing regulatory guidelines that include taxation, duty, and levies reconciliation.

Spirit losses can be prevented in the following ways:

- Maintain good manufacturing principles.
- Define and adhere to quality checks and procedures on the line.
- Control fill heights.
- Reduce packaging breakages.
- Ensure that all products are appropriately flushed through after transfer.

Measurement and control of packaging materials wastage

Packaging materials include bottles, cans, closures and corks, cases, labels, and cartons. Packaging material costs are a main raw material cost in many manufacturing plants, especially those that run bottling lines. Therefore, the measurement and control of these losses are fundamentally important. Packaging material losses are measured by calculating a theoretical use and comparing that to actual, as illustrated in the example in Table 26.1.

There are three key areas where material losses can impact the operation:

1. *Financial*—Material waste is highly costly in the alcohol packaging businesses.
2. *Service*—When material losses are excessive or unplanned, the number of materials required for a customer order can be compromised. As such, the number of cases needed may not be met. This can lead to customer order misses and loss of revenue. It is important to build your economic order quantities (EOQ's) of all packaging materials to include an average waste figure to ensure standard order fulfilment.
3. *Environmental*—As most businesses are pushing towards a more carbon-neutral footprint, the more damage there is to dry materials that lead to disposal, the more impact there is on the environment. Even when damaged, dry materials should be appropriately recycled. The target of a production facility is zero waste.

The following actions can prevent material losses:

TABLE 26.1   Impact of material losses (packaging 5000L of spirit into 6 packs of 750mL bottles).

Packaging material	Theoretical amount used	Actual amount used	Waste
Bottles	5000÷0.75=6667 bottles	6827 bottles	2.3%
Cartons	6667 cartons	6748 cartons	1.2%
Cases	6667÷6=1112 cases	1184 cases	6.5%

- Implementation of good manufacturing principles (GMP), material change over protocols, quality management systems (QMS), and finished product specifications—standard operating procedures (SOPs).
- Regularly scheduled maintenance of the packaging machinery.
- Taking actions to reduce packaging defects and breakages.
- Adhering to quality checks and procedures on the line to ensure that the product is produced right the first time to prevent rework or holds.
- Accurately setting up of production machines, such as label height settings and change parts.
- Applying effective stock control and aged stock management, such as first-in/first-out, where the oldest materials are used first, and shrinkage is tracked.
- Accurately recording material losses and applying root-cause problem solving where excessive wastage is taking place.
- Expertly managing brand change activity when one design of the material is replaced with another. The older materials are ideally fully consumed before a brand change is implemented on the packaging line.

Global packaging sustainability strategies

The spirits packaging industry, in common with other industries, impacts the environment in many different ways, including the following:

- As a user of energy.
- As a consumer of water and other natural resources.
- As a source of both direct and indirect carbon emissions.
- As a producer of waste.

A damaged environment impairs the quality of life and threatens long-term economic growth. The challenge of sustainable development is to achieve all of the objectives of the triple bottom line; the economic, social, and environmental pillars at the same time. Governing bodies and International organisations establish and guide industries on environmental directions to meet these objectives. These may include the following:

Operational guidelines

- Restrictions on emission levels.
- Restrictions on water use.
- Renewable energy sourcing.

Enforcement of end-producer responsibilities, including fees and deposits

- Packaging materials and composition.
- Energy and natural resources pricing.
- End of life—recyclability, reusability, recovery.
- Chain of custody—supplier management.

Corporate Social Responsibility

- Marketing language—greenwashing.
- Changes in consumer habits.

In turn, businesses implement the enforced legislation and guidelines and recommend additional internal corporate and social progress objectives that integrate sustainability initiatives into the company culture and the business strategy.

These may include the following:

- Comply with all relevant national and local legislation and regulations.
- Design, operate, and maintain processes and plants to optimize the use of all resources (materials, water, energy, etc.) while ensuring that unavoidable wastes are recovered, reused, or disposed of in an economically sustainable and environmentally responsible manner.
- Minimise the potential impact on the environment from site emissions to air, water, and land.
- Regularly assess through life cycle analysis (LCA) methodologies the environmental impacts of processes, materials, and operations. Based on the assessments, set annual objectives and targets for the continual improvement of environmental performance.
- Use and develop packaging development considerations, savings initiatives, and distribution technology systems in which packaging/product combinations will make fewer demands on non-renewable and renewable natural resources. Examples include lightweighting materials, bio or compostable packaging materials, right-sizing packaging formats, and transportation methods.
- Minimise the use of substances that may cause potential harm to the environment and ensure that they are used and disposed of safely—for example, low VOC inks in package printing.
- Encourage corporate social responsibility through a culture of awareness of sustainability issues amongst employees through management commitment, appropriate communications, training, and other initiatives.
- Establish and maintain appropriate procedures and management systems to implement these principles through policy commitment.
- Work with suppliers and other business partners in the supply chain to maintain high environmental standards and chain of custody measures.
- Work with international organisations and foundations dedicated to sustainable development advocacy for the beverage and alcohol packaging industry.

Carbon footprint

Carbon dioxide emission is increasingly seen as a key measure of environmental damage. It is a greenhouse gas which, as it increases in concentration in the upper atmosphere, leads to a greater amount of radiation from the sun's rays being retained as heat within the atmosphere. Distillation, in particular, due to its large usage of heat energy, has the potential to be a major net producer of carbon dioxide, much more so than beverage producers that simply ferment and condition their products. The packaging operation contributes to emissions throughout its operational carbon footprint and distribution, from its suppliers to its retailers. With the combustion of fossil fuels, either during the extraction and conversion of raw packaging materials, at the plant itself for heating purposes or in the generation of the electricity supplied, there is a need for continuous improvement and innovation in methods that support renewable energy sources to minimise the carbon footprint of the spirits packaging process.

Packaging waste

Global alcohol deposit systems that enable a closed-loop returnable system for glass containers and for packaging material choices that are recyclable can create an environmentally efficient industry. However, the cost of spirits packaging as a proportion of sales tends to be higher than is the case with many other consumer goods. This creates compelling financial and environmental arguments for altering spirits packaging choices to align with global sustainability efforts. However, this argument is somewhat in conflict with the need to make premium spirits brands stand out on the shelf. High-end products frequently feature expensive and complicated packaging to support premiumisation, which may not lend itself to easy recycling or conscious environmental considerations. Packaging waste directives are legislated around the world, and with consumer habits changing, corporations are evaluating all areas of the packaging value chain (including its embellishments options) to drive waste out of the process. In turn, they are using their conscious objectives to stake public messaging of their intentions to rally consumer, employee, and industry buy-in of their sustainability commitments.

Acknowledgements

Tony Brewerton of Diageo is thanked for providing the material for the section on the assessment of packaging line performance for the second edition and which is included again in the third edition. Claudia Sobiecki of Molson Coors Beverage Company is thanked for her assistance regarding global packaging developments for this updated chapter.

Note from the Editors: This Chapter has been updated from Chapter 18 'Whisky Global Packaging Developments' originally written by Mike Mitchel for the second edition of this book. Large portions of his original work have been reused in this updated Chapter 26, 'Spirits—Global Packaging Development'. Mike Mitchel is thanked for his contribution to the 2nd edition, and his name has been kept as a co-author of this updated Chapter to acknowledge his original work.

References

CBN, 2021. Molson Coors and J.P. Wiser's Whisky Release Common Bond Lager. https://www.canadianbeernews. com/2021/03/05/molson-coors-and-j-p-wisers-whisky-releas(e-common-bond-lager/.

CQC, 2014. Renewable Harvesting: Harvesting Natural Cork. Cork Quality Council, Forestville, CA. https://www. corkqc.com/harvesting/harvest.html.

IWSR, 2020. Premiumisation. https://www.theiwsr.com/5-key-trends-that-will-shape-the-global-beverage-alcohol-market-in-2021/.

Nielsen, 2020. Insights. https://www.nielsen.com/ca/en/insights/article/2020/covid-19-tracking-the-impact-on-fmcg-and-retail/.

Pine, J., Gilmore, J., 1999. The Experience Economy. Harvard Business School Press, Boston, USA.

Smith, L., 2020. See the world's first whiskey bottle made out of paper. https://www.fastcompany.com/90527727/ see-the-worlds-first-whiskey-bottle-made-out-of-paper.

TrendHunter, 2021. Smart AI Packaging. https://www.trendhunter.com/trends/expand-the-packaging-vision.

Packaging alcoholic beverages

Donna Abdelrazik, Jonghun Park, and Natalia Lumby

School of Graphic Communications Management, Ryerson University, Toronto, ON, Canada

Alcoholic beverage packaging formats

Current spirits packaging

While the overall landscape of alcohol packaging has seen some changes and innovative brands exist in the category, a majority of spirit packaging remains fairly traditional. This is not to say that the package is not critical. With the higher cost of spirits, consumers are often making decisions that reflect their lifestyle. The traditional package must continue to stand out on-shelf to impact the purchase decision and remain an emblem of this reflection inside the home. What is changing, however, is the coming generations' drink and lifestyle preferences. Young people are drinking less, motivated by living healthy sustainable lifestyles, and brands must respond accordingly (Euromonitor International, 2018). While marketing campaigns and branding, in general, play an important role, it is the package that greets the consumer at the store, welcomes them at on-premise establishments as they interact with the liquid, and continues to stay inside the home as a reminder to the customer of their purchasing decisions.

What follows is a review of the current landscape in spirits packaging. Following this baseline analysis, we look at established packaging formats that are being newly introduced in the spirits market, as well as looking at concept packaging that may evolve in the future. Choosing appropriate package format and material is a complex process of managing brand identity, customer desires, product functionality, and corporate social responsibility. Let us review the current and emerging packaging landscape.

Bottles

In Table 27.1, we examine the currently available containers, addressing their functional benefits and drawbacks. We divide this space into three categories: plastic bottles, standard glass bottles, and highly decorative glass bottles.

Brand owners often have a variety of these bottle types to accommodate the brand depth and to allow for a variety of occasions for consumption. Smaller sizes (375mL) are often in

TABLE 27.1   Advantages and disadvantages of currently dominant packaging materials in spirits.

Package type	Advantages	Disadvantages
PET bottle	• Lighter to transport than glass • Recyclable • Cheaper to produce than glass • Cheaper and faster to customise than glass • Less waste in the production process due to reduced breakage • Good use case for outdoor consumption (camping, beach, vacations, etc.)	• Not returnable/reusable • The recycling process is not as effective as glass and under pressure due to changes in the global plastics recycling market • Perceived as lower quality by consumers • Less effective as an oxygen barrier than glass • Perceived as bad for the environment by consumers (Crawford, 2019)
Standard glass bottle	• Recyclable and returnable • Excellent barrier properties • Recognised as high quality by consumers (European Container Glass Federation, 2021)	• High energy use during production • High cost • More difficult to ship [than PET] due to higher weight and breakage (Falkman and Institute of Packaging Professionals, 2014)
Decorative glass bottle	• Same as standard bottle • More recognisable on-shelf • Perceived as very high quality by consumers	• Same as standard bottle • Increased setup costs • Difficult to know how much brand recognition comes directly from the bottle characteristics alone

PET bottles, as plastic bottles are more portable (lighter and little to no risk of breakage). If a brand has several quality levels, it may offer the value tier in a PET bottle (or both PET and glass). Sustainability plays a role in packaging material and format decisions. Life cycle assessments are performed on proposed packaging formats to review their environmental impacts throughout different life cycle phases of the packaged product. For example, the lighter weight of plastic packaging supports a reduced carbon footprint in distribution and material economic savings; however, other areas of the life cycle such as its end of life impacts and current societal focus on plastic reduction and elimination may outweigh a brand's decision to proceed with this material choice. In Canada, between ¼ and ⅓ of all liquor bottles are now PET (Stephens and Beetham, 2019). Sustainability considerations and impacts will be discussed later in the chapter.

Closures

The basic functions of the closure are to ensure that the contents are properly preserved and to create a safety seal that alerts the customer to any product tampering. In addition to these tasks, in some cases, closures also serve more complex functions as part of an active and intelligent packaging system. For example, closures can authenticate the originality of the product as well as extend shelf life by using an oxygen barrier. Anti-counterfeiting technology will be discussed in detail later in the chapter.

One-way closures in spirits can be categorised broadly into twist caps, corks, and synthetic corks. In spirits, corks are commonly called T-stoppers. The top of the 'T' can be made from a variety of materials, including wood, metal, glass, and plastic. Table 27.2 examines each of these options. When selecting a closure, brands must first ensure that the functional benefits are appropriate. A closure that does not adequately preserve the product or that does not perform well during use is not an option. For example, unlike a wine bottle cork, the whisky bottle is consumed over a longer period of time. Thus, the closure has to perform its seal and reseal function multiple times. In addition to function, the closure does have an impact on brand perception. Not only is it a part of the package, but it is also the most handled component, and contributes to the experience of pouring a dram. The feel and sound of the cork are romanticised in several alcohol categories including whisky. Table 27.2 provides an overview of the standard closures seen in whisky spirits (Sackier, 2020).

In addition to the closure itself, brands also use a variety of materials to tamper-seal the bottle as well as help keep the closure in place. The packaging components often used for this purpose are capsules. Traditionally made of tin, capsules could be engraved for branding purposes and add a layer of safety and authenticity to the product. With the increase in the cost of tin, new materials are being used and explored (Schuemann, 2015). This includes polylaminate materials that use laminated layers of plastic and aluminium and PVC capsules and sleeves. All of these components can be printed/decorated to improve brand performance.

Closures also offer some interesting innovations. One example is an all-glass closure released on only 114 bottles of Bruichladdich 12Year Old 2002 Bourbon Cask by Master of Malt. Unlike a decanter, this closure is airtight due to a thin rubberised seal. In addition to being aesthetically pleasing, it offers no risk of the closure impacting the taste profile of the spirit (Hopkins, 2015). Sustainability innovations include T-stoppers created from the residue of the distillation process. Tapi, a luxury closure brand, is able to make a weighted top by pressing the residue, thus diverting it from landfill into a useful purpose (Tapì Group, 2021). Closures

TABLE 27.2   Advantages and disadvantages of different types of bottle closures in whisky.

Closure	Advantages	Disadvantages
Natural Cork	• Perceived as premium and authentic • Good oxygen transmission rate (OTR) • Recyclable	• Can cause quality issues like cork breakdown or cork taint • Expensive • Poor sustainable profile due to harvesting
Synthetic Cork	• Less expensive than cork • Can be recyclable • Tight seal • Offers no risk of mould or breakdown	• Some discussion about the presence of a plastic closure changing the flavour • Perceived as less authentic
Screw Cap	• Less expensive than synthetic and natural cork • Recyclable • Tight seal • Offers no risk of mould or breakdown • Offers more flexibility for creating a solution that has additional functions (for anti-counterfeiting)	• Perceived as low quality • The lightweight and twist nature of the opening is not a typical part of the luxury liquor experience • Poor sustainable profile from the viewpoint of manufacturing

can also be used to prevent the bottle from being refilled by sealing it with a tamper-evident valve system (Fig. 27.1). Refilling is a common method of counterfeiting that will be discussed later in the chapter. Lastly, closures can offer additional functional benefits such as pouring accuracy with no product waste.

Decanters

It is not unusual for whisky consumers to pour whisky into a smaller decanter when the bottle is less full. These vessels are typically made of all-glass with a glass stopper. However, in 2013, Canadian entrepreneur, Joel Paglione, developed the Oak Bottle—a completely new take on the premium alcoholic beverage. This bottle 'adds all of the flavour of traditional barrels in less than 48 hours' (Dillon, 2015). Consisting of American white oak, it is 'charred inside using traditional processes to a medium toast level and dressed with stainless steel rings' (Oak Bottle, n.d.). Avid whisky drinkers now have the freedom to mature their whisky at home in a fast and affordable way.

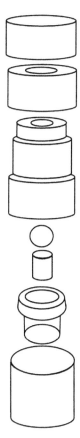

FIG. 27.1   Example of an anti-counterfeiting closure system based on the likeness of a system by Guala Closures Group.

Existing packaging formats—New to the spirits market

While the bottle format dominates the alcoholic beverage market, there are a few package types arriving in spirits that already exist in other beverage categories. Some of these formats are driven by the changing habits in alcohol consumption. Convenience, sustainability, wellness, and facilitating a variety of drinking occasions will continue to gain importance (Euromonitor International, 2018).

Brands have responded to consumer needs with ready-to-drink options, which are growing by double digits and emerging with low-sugar, low-calorie, and low-alcohol products (Carruthers, 2019). In addition to bottles in ready-to-drink packaging, other packaging products include aluminium cans, stand-up pouches, and cartons. What follows is a discussion of each of these formats.

Aluminium cans

Most common in beer and cider, cans are now also being used for ready-to-drink cocktails in support of the single-serve drinking occasion. An example of a popular brand releasing a product in this format is the Jack Daniel's & Cola can. Some advantages of cans are they are easy to transport (little to no breakage and lightweight), have excellent barrier properties to maintain product freshness, and are infinitely recyclable. In fact, cans use on average 70% recycled aluminium during manufacturing, as compared to glass which uses 20%–30% (Hines, 2017). Cans are also appropriate for a variety of drink occasions where cocktails would normally be more difficult to include, such as when camping or to be enjoyed at parties. Akin to aluminium cans are aluminium bottles, which also offer the added advantage of being resealable.

There are two main methods of decorating cans: printing directly on the can or using a plastic printed sleeve. Traditionally printed cans are decorated at very high speeds (which lowers the cost per can) and offer the look and feel of aluminium, as well as being easier to recycle. However, they require high-order volumes and suffer from long lead times and low print quality when compared to sleeves because of their wet-on-wet ink trap printing limitation. Plastic shrink sleeves are printed offline using the flexographic process. They offer high print quality and can be produced at low volumes, with much quicker lead times. Unfortunately, they do introduce a secondary material into the recycling system and do have a higher cost per unit. They can also be more complex to design as the sleeve is shrink-wrapped onto the can and can cause distortions. Regardless of the printing process, cans are a popular option in the RTD category.

Akin to the aluminium can is also another metal option worth noting from Stillhouse Spirits. Released in 2016, Stillhouse launched its whisky line in a new packaging format—a stainless steel can. This package is a nostalgic throwback to moonshine and flasks, with an updated vibrant rectangular red package (Mohan, 2016). In addition to the unique aesthetic, the can is resealable and durable. Unlike vulnerable glass material, the can is unbreakable and can be used comfortably in outdoor gatherings. It also has a strip stamp that covers the closure and a clear shrink seal for tamper evidence (Connolly, 2016). The can itself is 'lacquered, and the font label and strip (both pressure-sensitive) are hand applied" (Connolly, 2016) (Fig. 27.2).

Cartons

Much less commonly available are mixed beverages offered in cartons. Examples using this package include Japanese sake, Onigoroshi, and American alcoholic beverages by Beatbox (which got its head start by appearing on a popular business start-up TV show called Shark Tank). These types of boxes have been most commonly used in wine and are packaged aseptically. Aseptic packaging is a food processing technique that allows liquid products to have a non-refrigerated shelf life of up to approximately one year. The advantages of aseptic processing over conventional package processing include longer shelf life, a greater variety of packaging sizes and materials, and better product preservation (Chen and Rosenthal, 2015). Cartons are also highly portable and can be resealable and fairly lightweight. Unfortunately, due to the multilayer nature of the materials used in the carton, they are not accepted everywhere for recycling.

Pouches

Popular in the frozen beverage category, stand-up pouches offer high performance at low temperatures (making alcoholic slush drinks easy). They are lightweight and portable,

FIG. 27.2 Unique stainless steel can container by Stillhouse.

as well as offering resealable options in a variety of sizes. Much like beverage cartons, they are made from multilayered materials making them difficult to recycle. While they are able to stand, they are quite different from other formats making them more challenging to merchandise. Currently, most pouches on the market are for wine-based cocktails. The format has performed particularly well with Frozé (frozen Rosé) and are well-liked by millennials (Newhart, 2019).

The innovations in this package category have largely been around the dispensing function. Traditional pouches have a tear tab, which requires a glass, or a straw hole which needs to be punctured. Both options decrease the attractiveness of the product. Ampac now offers a pull tab that reveals an opening for a straw more seamlessly (WingDing Cocktails Get Packaged in Stand-up Pouch, 2017). It is used for the WingDing line of drinks. In larger formats, these pouches are competing with the traditional box wine style containers which feature a box with a plastic bladder and pour spout. Ty Ku Soju offers a stand-up pouch with a pour spout in the 1.80L format. The carbon footprint of these large pouches as compared to bottles is promising (because they can ship empty and flat and are much lighter).

Packaging innovation concepts in spirits

With the arrival of new formats of packaging and changes in consumption, concept packaging in spirits have also been evolving. What follows are examples of packaging innovation, either new to the market or still in concept.

The ready-to-drink (RTD) spirit market represents $4.9 billion in the United States (Industry Market Research, Reports, and Statistics, 2021). Several creative designs are responding to this area of opportunity with packaging that creates a new cocktail experience facilitated by the package. One of the first examples of this was portable shots. Here, a small shot glass is sealed and ready to be consumed on-the-go. Companies such as Twisted Shotz create two compartments in the small glass, which are combined in the mouth as the shot is consumed. Another great example is Yumix cocktails, released in 2018 in a twin-chambered bottle. Here, the top of the bottle holds 192mL of drink mix and the snap-off bottom holds the 50mL of alcohol (Bellwood, 2018). The ability to separate the two provides for a fresher flavour profile without additional preservatives or sweeteners. While the resulting package looks simple and not unlike a standard PET juice bottle, designing and manufacturing it is quite complex. For starters, creating a short and wide bottom bottle required a custom preform (Yumix n.d). The top bottle body had to have extra headroom for the alcohol to be added later, making it difficult to hot-fill without warping (Spinner, 2016).

Many design firms have also created innovative packaging concepts for RTD. For example, PB Creative designed a package that features a glass tube, with a small soda can and a sealed alcohol shot. You simply open the tube, take out the two components, and make your on-the-go mixed drink in the tube.

Another example of an innovative packaging concept is the Glenlivet 'capsule collection'. Here, an edible capsule made from seaweed is used to hold 23mL of cocktail made with Glenlivet. The experience is a cocktail without the glass, as you place the capsule in your mouth and bite down. The idea was launched in October 2019 and is not currently commercially available.

Innovative enhancements in alcoholic beverage packaging

With the global distilled spirits volume market share growth up to 3.3% in 2019 (distilled-spirits.org), visual and technical packaging design strategies contribute to a brand's competitive advantage of making an impression on the shelf and winning a piece of this growing segment. According to the 2020 economic briefing for distilled spirits, contributing factors to consumer purchase include 'consumer preference for premium products, prestige bottles, and unique experiences' (distilledspirits.org). Packaging serves a broad variety of roles that connect to those factors. For example, the bottle shape and on-pack graphics help a consumer connect with the brand and find/select it in stores (Silayoi and Speece, 2004). In a category as traditional as whisky, it is important to pay close attention to package design.

Visual packaging design strategies

The distilled alcohol sector has evolved its drinking demographic and occasion over the past decade. What was once retail-lead by 50 and older males is now beginning to shape a new on-premise young men and women consumer segment. In March 2018, the renowned Johnnie Walker brand created a limited edition 'Jane Walker label design targeting their female audience featuring a woman dressed in similar 19th century clothing who also tips her hat mid-stride' (WARC, 2018). The demand for innovation and connection to the story behind the brand is increasing as millennials are experimenting (Dunn, 2019). With this shift in drinker and the continued focus on a product's functional benefits, design thinking that is more universally appealing and has sustained relevance are necessary considerations for category sales and marketing strategies.

Art and personalisation

Premium and super-premium spirits categories look for iconic ways to connect their packaging with fashion trends, artist collaborations, and limited-edition collectibles to create meaningful relationships and enhanced storytelling of a brand. This strategy of packaging design extends the touchpoint of the product vessel into display keepsakes at home or behind the bar.

Variable digital printing allows brands to create unique labels for promotions. Similar to a limited amount of prints made in an art collection, bottle labels can be individually numbered to signify small batch quantities and enhance interest. Dulwich Gin premiumised their brand labels using a combination of smart design and digital printing technology. The brand commemorated South London's oldest gallery by randomising various famous portraits from the Georgian era, with unique colour combinations on their bottle labels. An advanced logic database was used to vary the bold colours and illustrations to ensure that no two labels were alike (McCool, 2020) (Fig. 27.3).

Further to variable printing, one-to-one personalisation is another means to connecting the consumers to the brand. As an example, Jameson partnered with the Irish artist, Hephee,

FIG. 27.3 Digitally printed one-of-a-kind labels for Dulwich Gin (www.dulwichgin.co.uk). Design agency: Derek&Eric. Creative Strategy: Silas Amos.

to design a bottle label for St. Patrick's Day. 'This bottle featured a QR (quick response) code and NFC (near-field communication) chip on the neck label that when interacted with took viewers to a web app that allows them to create a custom label based on the artist's icons and artwork' (Sanchez, 2020). Combining promotions with technology, art, and social media aligns with the rising spirits drinking demographic who appreciate digital engagement.

Packaging embellishments

Beyond using the shape of the bottle and colour to create brand recognition, packaging embellishments amplify the information area of the packaging and differentiate products on the shelf and behind the bar. Tactile (haptic) interactions with packaging are increasingly becoming more common in alcohol packaging design strategy as sensory applications create emotional connections to the products (Fig. 27.4). Embossing or debossing the physical primary packaging such as glass or plastic with a raised or inset element adds a touch factor that consumers can interact with. These features are also possible within the label design, but not as pronounced at first glance. The more intricate the embellishment, the more timely, complex, and costly the application process. Overall glass etchings or matte varnishes that change the surface feel of the primary package create unique craftsmanship identifiers for brand recognition, especially when generic or standard bottle formats are used. Accessibility for universal design is an additional consideration that will be discussed in the coming section on Packaging Technologies.

When it comes to label design, this is where shape, placement, material, and ink or stamping choice require market research, experience, and design thinking to meet the target market objectives. It is those elements complimented by the illustration, photography, typography, and use of colour that finesse the final brand signature.

Technologies for alcoholic beverage packaging

Similar to packages used for other product categories, the primary functions of alcoholic beverage packaging are to contain the product without any leakage and to preserve and protect the product from biological and physical damage during distribution. Additionally,

FIG. 27.4 Example of a metal label on limited-edition Lot 40 *(left)* and a wooden label on a limited-edition Pike Creek *(right)*. Design agency: Davis.

packaging needs to provide the correct product information to users and also attract consumers based on its structural, graphic, and tactile components. Glass has been traditionally used by the alcoholic beverage industry as a primary packaging material for many alcoholic beverages, such as whisky, gin, vodka, and rum, because it is inert, provides good long-term storage, and is aesthetically pleasing.

Today, consumers want product packaging to provide a better user experience and improved convenience when purchasing, storing, using, and disposing products. As a response to the diverse demands of consumers, many brand owners within the alcoholic beverage industry have applied innovative, smart, and intelligent packaging technologies to their products. Robertson (2005) defined intelligent packaging as 'packaging that contains an external or internal indicator to provide information about aspects of the history of the packaging and/or the quality of the food'. Intelligent packaging enables internal or external changes to packaged products to be detected and recorded, provides evidence of tampering and anti-counterfeiting functions, and enhances user conveniences. Also, intelligent packaging is often used to promote sales of the packaged products.

Anti-counterfeit packaging

Since the 1990s, replica whiskies have appeared in auction markets due to their monotone labels and standard shape, making counterfeiting the primary packaging challenge facing the whisky industry. Globally, hundreds of millions of spirit bottles are counterfeited every year. These counterfeited spirits not only result in increased costs to both the brands and to consumers but also cause safety issues to users as they are often refilled with unacceptable or dangerous types of liquids.

Therefore, to prevent counterfeit spirits, several packaging technologies have been introduced. One of the most accurate techniques used today for identifying fake whisky is confirming the authenticity of whisky with archivists whose roles include checking the labels for errors, examining the glass, and testing the liquid (Broom, 2016). Many have also taken the advantage of the increased number of technological advancements to enhance the design of their packaging.

Two types of packaging technology concepts are currently used to prevent counterfeiting of products: tamper-evident packaging and tamper-resistant packaging. While the names of these two technologies sound similar, they have different aims and functions. According to the regulations described in 21 CFR § 211.132 of the Food and Drug Administration (U.S. Food and Drug Administration, 2019), tamper-evident packaging is defined as a package that has 'one or more indicators or barriers to entry which, if breached or missing, can reasonably be expected to provide visible evidence to consumers that tampering has occurred'. Thus, the primary purpose of the tamper-evident packaging is to visually inform users as to whether their package has been accessed by an unauthorised party prior to being accessed by the rightful user. The examples typically include tamper-evident labels, seals, markings, and tags. On the other hand, tamper-resistant packaging is designed to resist access to the product contained in the package. Tamper-resistant packaging deters product tampering and makes it difficult to access the package for reasons other than the intended use. Therefore, tamper-resistant packaging requires special structural designs or materials that can ultimately increase packaging costs. However, tamper-resistant packaging does not provide visual evidence to the users regarding package tampering.

Closure and sealing techniques are simple to perform and are the most direct method for combatting the process of counterfeiting genuine whisky and identifying forged whisky. A variety of closure and sealing formats such as bottle sealing wax, tamper-evident seals, and non-refillable caps have been implemented in the whisky packaging process.

The use of bottle sealing wax is a traditional tamper-resistant method for sealing whisky and wine bottles. Typically, sealing wax is applied by dipping the bottle top into the melted wax. The melting temperature of sealing wax is between 135°C and 150°C and its working temperature is approximately 21°C (Blended Waxes, 2014). Wax seals for bottles provide a tamper-proof function. Wax seals are often embossed with a brand or product logo. Additionally, bottle sealing wax has been used as an iconic symbol of some brands. For example, Maker's Mark bourbon whisky is widely recognised by its trademarked red sealing wax.

Tamper-evident seals are commonly placed on whisky bottles and are used to indicate to users whether a whisky bottle has been forcibly opened by allowing a user to visualise, hear, and/or feel the action of breaking. Several tamper-evident seal types are available in the spirit market, such as tamper-evident with bridges, automatic external/internal tamper-evident bands, and cap cover with a tear-off tamper-evident band (Guala Closures Group, 2020). There are three types of capping systems that differ in their capping mechanism and the area of capping, which are called snap-on, roll-on, and screw-on. The snap-on capping system joins the closure onto the bottle using vertical pressure, and the roll-on system secures the closure onto the bottle using rollers that bend a component of the closure under a bead of the bottle. The screw-on capping system seals the closure on the bottle by screwing. The roll-on-pilfer-proof (RoPP) cap, which is a well-known term in the packaging industry, typically describes an aluminium tamper-evident (screwed or non-screwed) seal secured by the roll-on capping system. The capping process for the RoPP caps is very fast. Tamper-evident seals are often used together with a neck decoration consisting of a hanging charm and tag. For example, Johnnie Walker and Blade & Bow have often used a combination of a tamper-evident seal wrapping and a charm for their packaging.

Some whisky products have adapted non-refillable caps, which prevent refilling of the bottle using one-way valves and internal devices such as plastic or glass balls that complicate refilling of the packaging. However, non-refillable caps may not be a perfect method for preventing forging issues, as there have been some reported counterfeit cases in which bottles having non-refillable caps were refilled using needles and syringes. Therefore, non-refillable caps are often used in whisky packaging in combination with other anti-counterfeit methods such as tamper-evident seals.

Closures that have special structural functions have been introduced in the market as a new anti-counterfeit technology. These special closures offer an advanced tamper-evident sealing function that has irreversible and internal capabilities. Several patented special closures have been developed. For example, indicator, introduced by Guala Closures Group, is 'a device located inside the closure and unhooks at first opening, remaining hung and visible inside the bottle'. Prestige, another closure that uses T-bar corks, works in a way that 'the outer wall of the T-bar lifts on opening, leaving an irreversible gap after replacing the cork inside the bottle' (Guala Closures Group, 2020). This helps users identify at a glance whether a bottle has been tampered with.

Holograms have been applied to many whisky products to prevent counterfeiting. A hologram is made using a photographic technique that records the light scattered from an object and presents it as a three-dimensional image (Benton, 2001). However, traditional holograms such as embossed holographic film can be easily counterfeited and copied using several simple methods. Many counterfeiting cases related to holograms have been reported and are not a newly known fact to the whisky and holographic industry.

Several new holographic technologies have been introduced as countermeasures against forging issues associated with traditional holograms. Some holograms have adapted a UV security technique, which uses invisible photochromic ink that can be visualised under UV light. Since holograms based on invisible photochromic ink have limitations and cannot serve as a perfect countermeasure, new holograms have been created for proving whisky authenticity that are activated either through a smartphone or a manufacturer-specific device. An example of advanced holographic technology is the DuPont Izon WS, which was developed by DuPont Advanced Printing. A 3D image appears when the hologram is rotated using a security film that can be applied directly over printed text, barcodes, or other images. When the security film is viewed off-angle, the holographic image is invisible, unlike traditional embossed foil holograms. It also includes a 2D verification code on the packaging to deter other acts of forgery (McCormick, 2018). Today's technology has made it possible to track and trace such actions, including its implementation closures (Ramos et al., 2015). Not only will today's technology be able to detect duped products through its protective abilities, but users will also benefit from its informative capacities.

QR codes are commonly used to protect spirit products from fraud. A QR code is applied to each bottle in the manufacturing process, mainly as a printed label or tag, and then can be read by users using various smartphone apps to verify that the product is genuine (Emler, 2015). QR codes represent a relatively cost-effective smart packaging technology as compared to other technologies.

Radiofrequency identification (RFID) technology has been applied to spirit products to confirm their authenticity by allowing the tracking and tracing of products through supply chains (Przyswa, 2014). RFID tags can provide much more traceability data than barcodes. Not only for anti-counterfeiting purposes, RFID tags are used to enhance supply chain efficiency and reduce operational costs by increasing inventory visibility (Weinstein, 2005). For example, in South Korea, whisky suppliers or importers have been required since 2012 to attach RFID tags provided by National Tax Service–authorised vendors to their whisky products before distributing them to the supply chain (Oh, 2020). However, the biggest hurdle to applying RFID tags to individual spirit products is the cost of the tags and the devices and servers required to read the information they contain (Covey, 2017). Because of these challenges, brands tend to adopt other interactive tags, such as near-field communication (NFC) tags, rather than RFID tags.

Near-field communication (NFC) technology is being frequently applied to whisky and gin as an anti-counterfeiting measure. NFC tags, often called smart tags, enable short-range communication with compatible devices, such as smartphones (Baek and Youm, 2015). Users can easily use their smartphones to trace and determine whether a bottle had been previously opened (Soon and Manning, 2019). Each NFC tag has a unique identification (UID) code that prevents copying or modification. Packaging technologies in conjunction with NFC technology can be considered as one of the most active and smart technologies that spirit brands can practically apply to their packaging (Chandler, 2012). Many whisky brands, such as Diageo's Johnnie Walker, Mackmyra, and Kilchoman, upgraded their premium Scotch bottles with the NFC technology supported by Thin Film Electronics ASA (Stark, 2019). NFC technology also plays an effective role in adding marketing capabilities to whisky and gin products. The marketing aspects of NFC technology will be discussed later in this chapter.

Another unique anti-counterfeit technology in the premium alcoholic drink industry is glow-in-the-dark whisky, which was first introduced by Professor Uwe Bunz at Heidelberg University (McCormick, 2018). In order to detect a certain whisky or blend, glow in dark whisky uses a synthetic tongue that reacts with fluorescent dyes that glow under a black light and produce different fluorescent patterns (Blanding, 2018). Evaluating the unique colour compounds of each whisky drink reveals its legitimacy. This method is still being tested and possesses a huge potential to be successful within this market.

Spirit packaging technologies for promotion, user experience, convenience, and accessibility

In order to create unique marketing and promotional opportunities, new packaging technologies have been introduced and applied to many spirit products that differentiate the products from the competitors. These packaging innovations include unique inks and substrates, printed electronics, and components that are used in conjunction with near-field communication (NFC) and augmented reality, which enhances user experience, convenience, and accessibility.

NFC technology incorporated into product packaging is widely used to support marketing strategies for spirit products (Hemphill and County, 2018). NFC is activated by simply facing against another compatible NFC device that can either be consumers' smartphones or NFC-specific manufactured devices (Baek and Youm, 2015). This is commonly seen in various forms such as sticker, label, tag, or integrated film and can be applied either inside or outside of spirit packaging. The NFC component on spirit product packaging enables marketers to tell the brands' stories and guides whisky tasting through its various connected media forms such as audio and video. For example, in 2019, the Balvenie introduced The Balvenie series together with Here Design in order to feature their line composed of three unique bottles, all infused with NFC tagging. The use of an NFC neck tag linked consumers to a podcast that revealed the personal stories of the craftspeople (Corbin, 2019).

The use of augmented reality (AR) and virtual reality (VR) has become widespread in various sectors ranging from electronics to fashion to real estate. Several spirit brands have also implemented an AR component to their product packaging to build a more interactive user experience that is incorporated into labels attached to bottles (Brohm et al., 2017). Although both VR and AR provide innovative ways to engage consumers and personal user experiences, they work in different ways and require different devices. VR is a computer-generated simulation of reality and works in a way that is completely separated from the real world. It typically requires special equipment such as VR headsets and replaces and covers the vision of users while they are using the headsets. Although some spirit brands introduced VR-based storytelling content to their users, the need for special equipment limits the opportunity for brands to create a VR-based marketing strategy as well as for consumers to access the content. However, AR combines computer-generated information with what people already see in the physical and real world. AR can be accessed using more common devices such as smartphones, which allows users to see the real world with superimposed computer-generated elements at the same time (Stapleton, 2019). For example, Jack Daniel's Tennessee whisky brand incorporated NFC technology into its products by providing users with an AR experience. Consumers can discover a range of artwork, stories, and other media once they point their smartphone cameras onto the whisky bottles. By showing consumers the production press of their drinks in the distillery, consumers are better informed and construct their own personal connection with the brand (Lydekaityte and Tambo, 2018).

The spirit brands have also used printed electronic technologies for many years as a way to offer more shelf appeal and diverse user experience (Manly, 2014). One of the most popular examples is the electroluminescent ink on the folding box of Bacardi's Bombay Sapphire gin. Bacardi uses a protected technology called HiLight. When a hidden switch at the packaging's base is activated, an electric current is conducted from a battery below the package up to the printed inks, triggering illuminated animation. At first physical contact with the packaging, potential consumers see a five-stage light sequence. Traditional printed electronics are processed by printing conductive circuits on packaging substrates using either traditional high-volume print technology or inkjet and high-speed roll-to-roll printing with advanced conductive inks and substrates. The Thinfilm NFC tag, introduced as a smart anti-counterfeiting technology (Stark, 2019), is also manufactured through the roll-to-roll printed electronics production process (Thinfilm Technology, 2020). There have been many technological advancements in the printed electronics field in the last few decades in conjunction with the Internet of Things in addition to NFC and RFID technologies. Therefore, several spirit brands have used printed electronics in applications, such as monitoring product quality and preventing counterfeiting, not just for promotional purposes (Smyth, 2016).

Because accessibility is considered a crucial element of social sustainability, there has been increasing attention paid to accessible packaging design. Therefore, certain packaging components have been implemented to improve accessibility. The simplest packaging component used to enhance accessibility is embossed braille on the label. Embossed braille application allows visually challenged consumers to gain a sense of independence by providing additional guidance in the purchase and use of spirit products. This benefits not only the company but improves the shopping experience of consumers. A few spirit products have adopted braille on their labels. For example, the Arran Distillery rebranding in 2019 allowed for an accessible design that incorporated embossed braille into its label. NFC tags can also assist visually impaired consumers by offering voice instructions regarding the products (Bhattacharya, 2014).

Sustainability in alcoholic beverage packaging

Environmental consciousness continues to be prevalent in society, as humans look for solutions to reverse, sustain, and prevent ecological damage. Packaging sustainability is at the forefront of this initiative as it is a publicly visible contributor. Governing bodies and corporate business initiatives are taking shape with public impact statements in packaging sustainability goals, legislative amendments in single-use plastics, and internal company culture shifts in manufacturing processes. The triple bottom line (TBL) framework is a three-part model consisting of social, environmental, and economic platforms used to evaluate the performance of an organisation's sustainability efforts. This framework is then laddered against three pillars of focus: people, planet, and profit. Packaging sustainability objectives are established and assessed against its life cycle. Five phases are considered, which include raw material sourcing, manufacturing processes, distribution, use of the packaged product, and end-of-life considerations. This combined process is iterative and supports continuous improvement efforts towards a positive environmental future. What follows is a review of TBL initiatives specific to the spirits industry separated by material, process, and corporate social responsibility efforts.

Materials

Increased public concern over packaging waste and material choices, especially in plastics and straws for the beverage sector, is highlighting the need for corporate environmental platforms. Consumers are voicing their opinions through their purchase decisions, social claims, and overall brand loyalties.

Glass is a leader in sustainable alcoholic beverage packaging materials for its recyclability and reusability. With government-imposed deposit refund systems in place throughout many municipalities around the world and return logistics infrastructures to support the returnable packaging model, it has long been considered the eco-trend starter. Beyond plastics, sustainable material development is being innovated to solve for similar properties to glass without the weight, energy, or resource consumptions. Bio-based materials, such as flax fibre composite, are emerging as a glass alternative (Shaker et al., 2020) and The Absolut Company is testing a paper bottle for their vodka in support of a renewable source alternative (Danigelis, 2019). Will the category take it one step further and begin to initiate a refillable program with any of its brands in the same way that many Kombucha and artisanal beer brands have done? Time will tell if this is an appropriate strategy.

The package is the vessel, the first moment of truth, to hold the liquid and impact the purchase decision; however, the experience vessel is the second moment of truth, where the engagement occurs with the liquid. Glassware, straws, lids, garnishes, and coasters are the accessories that complement the packaged product to create the overall drinking occasion experience. This is where sustainable considerations can amplify the consumer's interaction and emotional connection to a brand. As municipalities are lobbying and passing legislation banning single-use plastics, alternative innovations are in development to support this significant industry change. Edible straws made from plant-based materials such as agave or repurposed pasta are entering the alcohol market for on-premise consumption. Recyclable options such as paper straws or bamboo are also a consideration, with the costliest option of reusable metal or silicone straws (Chirco, 2019). The concern with the reusable solution is with our most recent world pandemic of COVID-19. Bar and at-home sanitation worries may push this idea for other single-use options. For those traditional straight-up whisky drinkers who experience their drinking occasion neat, with a dash of water, or on the rocks, the complete removal of straws is no loss. However, there is an experiential aspect of the mixed-whisky drinking occasion that commonly partners this drinking apparatus which will now adjust the consumer's connection to the experience. The landscape of where whisky drinkers will be consuming products may also look a little different along with the marketing strategies that will correspond. In light of COVID-19, Bills are being brought forward and passed across the United States in support of the sale of bars and restaurants to offer cocktails to go in sealed containers (distilledspirits.org). What will these packaging formats and strategies look like to support the drinking experience and how will they maintain consciously sustainable considerations? With the RTD (ready to drink) market already evolving, there may now be even more growth in this area.

Processes

Whisky and its alcohol counterparts are beginning to focus their packaging environmental missions on a circular economy approach to sustainability. This means that instead of a single-use linear philosophy, the design is systemically infinite like a continuous loop. Processes are developed to either regenerate, recycle, or reuse. Managing natural resources such as water and grains, waste reduction, renewable energy investments such as carbon-neutral goals, and distribution efficiencies such as a package of light-weighting are all processes that contribute to the sustainable triple bottom line of distilled spirits. Bruichladdich Scottish whisky publicly announced their program 'to make the entire distillery "carbon-neutral" by 2025' (WhiskyCast, 2020). This initiative intends to potentially generate power for the distillery and most of the islands' electrical demand, using the waves from the Islay Coast.

Material light-weighting is a packaging sustainability initiative that companies use to support both cost savings, environmental consciousness, and carbon reduction wins. Light-weighting is when a commercialised package 'becomes a lighter version of oneself' (Sterling, 2015). A fine balance must be achieved between package performance throughout the packaging value chain and customer approval. The user experience must not be compromised in terms of perceived quality loss or conveniences, such as grip, dispensing, or handling. It is rewarding when companies like Allied Glass can make a claim that a new 340mL generic spirits bottle 'is 24% lighter than the previous bottle…saving over 1900 tonnes of glass' (WRAP, 2021).

Another area of sustainable packaging consideration is right sizing whereby the structural packaging is developed to properly form the size of the product. This is not as apparent in the primary bottle packaging of the liquid; however, for those brands that use a secondary carton, added packing materials, and for all who use tertiary packaging for shipping and distribution, the relationship of the housing to its product can be considered for review. Again, there is a fine balance between the experiential aspect of a package, like in the unboxing rituals that creates customer engagement with a brand and packaging waste and inefficient use of materials during product development.

Corporate social responsibility

There are competitive business advantages to those spirits companies that use sustainability to message their corporate social responsibility and environmental objectives. It would be additionally positive if the spirits category banded together. This leg of the TBL framework measures 'your business' profits in human capital, including your position within your local society' (Chamberlain, 2017).

B Corp Certified is a global sustainability community organisation that balances profit and purpose for change. There are only six distilleries globally that have achieved the status and the Scottish whisky brand Bruichladdich is one of them (McFarlane, 2020). Other brands are turning their profits over for a sustainable cause. Cooper King Distillery was the first in Europe to join an environmental initiative by giving 1% of their sales to a non-profit organisation (Carruthers, 2019). Bacardi Limited engrains its commitment to sustainability and corporate responsibility with 'good-spirited global goals' (Bacardi Limited, 2019).

Sustainability comes at a cost to implement efficient processes, sustainable materials, etc. Research has highlighted that consumers are willing to pay more for a brand that is consciously supporting the environment with their packaging choices and organisational affiliates. Toluna, an ITWP company, 'confirmed that 37% of (US surveyed) consumers are seeking out and willing to pay up to 5% more for environmentally friendly products and are actively changing their shopping behaviour to do so' (Mohan, 2019).

As the distilled spirits industry continues to take hold of the alcoholic beverage industry, packaging strategies will play a strong role in how consumers continue to interact with the offerings. Strategic visual and format choices, intelligent technological innovations, and sustainability efforts will lead the category forward.

References

Bacardi Limited, 2019. Good Spirited, 2019 Corporate Responsibility Report. Retrieved from https://d3bbd6e-s2y3ctk.cloudfront.net/wpcontent/uploads/2020/02/12184711/CR_Report_2019.pdf.

Baek, J., Youm, H.Y., 2015. Secure and lightweight authentication protocol for NFC tag based services. In: 2015 10th Asia joint conference on information security, May. IEEE, pp. 63–68.

Bellwood, O., 2018. Yumix launches mix-it-yourself cocktail range. 19 March. Retrieved from https://www.thespiritsbusiness.com/2018/03/yumix-launches-mix-it-yourself-cocktail-range/. (Accessed 17 June 2020).

Benton, S.A., 2001. Selected papers on three-dimensional displays. In: SPIE Milestone Series, p. 162.

Bhattacharya, R., 2014. Near field communications: how NFC can assist the visually impaired. User Experience Mag. 14 (1).

Blanding, M., 2018. Is your rare whisky a fake? Robb report. 20 September. Retrieved from https://robbreport.com/food-drink/spirits/rare-whisky-fake-2818361/. (Accessed 11 May 2020).

Blended Waxes, 2014. Bottle Sealing Wax Questions & Answers. Retrieved from: https://blendedwaxes.com/bottle-sealing-wax-questions-answers/.

Brohm, D., Domurath, N., Glanz-Chanos, V., Grunert, K.G., 2017. Future trends of augmented reality. In: Augmented Reality for Food Marketers and Consumers, pp. 1681–1685.

Broom, D., 2016. The long history of fake whisky. Scotchwhisky.com. 22 December. Retrieved from: https://scotchwhisky.com/magazine/from-the-editors/12184/the-long-history-of-fake-whisky/. (Accessed 15 June 2020).

Carruthers, N., 2019. Top five best-selling ready-to-drink brands. 16 July. Retrieved from https://www.thespiritsbusiness.com/2019/07/top-five-best-selling-ready-to-drink-brands/. (Accessed 17 June 2020).

Chamberlain, A., 2017. Sustainability management. In: Triple Bottom Line. Retrieved from https://www.era-environmental.com/blog/sustainability-management-triple-bottom-line. (Accessed 17 June 2020).

Chandler, N., 2012. What's an NFC tag? 14 March. Retrieved from https://electronics.howstuffworks.com/nfc-tag.htm. (Accessed 22 June 2020).

Chen, J., Rosenthal, A. (Eds.), 2015. Modifying Food Texture: Volume 2: Sensory Analysis, Consumer Requirements and Preferences. Woodhead Publishing, Amsterdam.

Chirco, A., 2019. 6 Great Alternatives to Plastic Straws. 19 September. Retrieved from: https://www.pgeveryday.ca/family/activities/alternatives-to-plastic-straws.

Connolly, K., 2016. First Stainless Steel Container for Whisky Reflects Stillhouse brand's Independence. 11 March. Retrieved from https://www.packagingdigest.com/packaging design/first-stainless-steel-container-for-whisky-reflects-Stillhouse-brands-independence-2016-03-11. (Accessed 18 June 2020).

Corbin, T., 2019. The Balvenie unveils bold design for new Stories range. Packaging News. 21 August. Retrieved from https://www.packagingnews.co.uk/design/new-packs/balvenie-unveils-bold-design-new-stories-range-21-08-2019.

Covey, M., 2017. Taking Anti-Counterfeiting Action with Item-Level RFID. RIS. 5 September. Retrieved from https://risnews.com/taking-anti-counterfeiting-action-item-level-rfid.

Crawford, E., 2019. Consumer push-back against plastic will spur manufacturers to seek new packaging in 2019. FoodNavigator, USA. 22 January. Retrieved from https://www.foodnavigator-usa.com/Article/2019/01/22/Consumer-push-back-against-plastic-will-spur-manufacturers-to-seek-new-packaging-in-2019.

Danigelis, A., 2019. The Absolute Company Pursues Paper Bottles. 18 December. Retrieved from: https://www.environmentalleader.com/2019/12/absolut-company-paper-bottles/.

Dillon, G., 2015. Advances in Packaging Technology by GreatDrams. Retrieved from https://greatdrams.com/packaging-technology/. (Accessed 22 June 2020).

Dunn, D., 2019. Robb Report, Food & Drink, Spirits. The New Face of the Average Whisky Drinker: Young, Pierced, and Female. 7 February. Retrieved from https://robbreport.com/food-drink/spirits/who-are-the-new-whisky-drinkers-2840639/.

Emler, R., 2015. Technology used to fight back against fakes. The Drinks Business. 5 October. Retrieved from https://www.thedrinksbusiness.com/2015/10/technology-used-to-fight-back-against-fakes/.

Euromonitor International, 2018. Innovation and Disruption in Alcoholic Drinks: Raising a Glass to the New Consumer. December. Retrieved from https://www.euromonitor.com/innovation-and-disruption-in-alcoholic-drinks-raising-a-glass-to-the-new-consumer/report. (Accessed 17 June 2020).

European Container Glass Federation, 2021. Why Choose Glass? Retrieved from https://feve.org/about-glass/.

Falkman, M.A., Institute of Packaging Professionals, 2014. Fundamentals of Packaging Technology, fifth ed. Institute of Packaging Professionals.

Guala Closures Group, 2020. Retrieved from https://www.savethespirits.com/en/category/save-the-spirits-2/. (Accessed 15 June 2020).

Hemphill, A., County, W., 2018. WVA: New factory near West Virginia school could impact Northern Virginia's wine industry, wineries worry. Control State News. Retrieved from https://www.nabca.org/sites/default/files/assets/files/2018DNU/DNU_AUG_14_2018.pdf.

Hines, N., 2017. Cans or Bottles: What's Worse for the Environment? 12 February. Retrieved from https://vinepair.com/articles/can-versus-bottle-environment/.

Hopkins, A., 2015. 'World's first' single malt glass closures created. 22 April. Retrieved from https://www.thespirits-business.com/2015/04/worlds-first-single-malt-glass-closures-created/. (Accessed 22 June 2020).

Industry Market Research, Reports, and Statistics, 2021. Retrieved from https://www.ibisworld.com/industry-statistics/market-size/rtd-mixed-spirit-production-united-states/. (Accessed 17 June 2020).

Lydekaityte, J., Tambo, T., 2018. Business perspectives of smart interactive packaging: digital transformation of brand's consumer engagement. In: Proceedings of the 8th International Conference on the Internet of Things, October, pp. 1–4.

Manly, A., 2014. Putting the Power of Printed Electronics in Packaging. 8 December. Retrieved from https://www.packworld.com/home/article/13366723/putting-the-power-of-printed-electronics-in-packaging.

McCool, B., 2020. No Two Bottles of Dulwich Gin Are the Same. 3 March. Retrieved from https://thedieline.com/blog/2020/3/3/no two bottles-of-dulwich-gin-are-the-same. (Accessed 18 May 2020).

McCormick, J., 2018. How the Whisky Industry Combats Fakes. Packaging World. 16 January. Retrieved from https://www.whiskyadvocate.com/whisky-industry-combats-fakes/. (Accessed 11 May 2020).

McFarlane, C., 2020. Bruichladdich is 'B Corp' Certified. 18 May. Retrieved from: https://www.bruichladdich.com/bruichladdich-whisky-news/the-distillery/bruichladdich-is-b-corp-certified/.

Mohan, A., 2016. Stainless Steel Can Capture Nostalgia of Moonshine Whisky. 9 March. Retrieved from https://www.packworld.com/design/materials-containers/article/13369858/stainless-steel-can-captures-nostalgia-of-moonshine-whisky#next-slide. (Accessed 22 June 2020).

Mohan, A., 2019. Report: 37% of Consumers Willing to Pay More for Eco-Friendly Products. 6 November. Retrieved from: https://www.packworld.com/issues/sustainability/article/21096310/toluna-report-37-of-consumers-willing-to-pay-more-for-ecofriendly-products. (Accessed 23 June 2020).

Newhart, B., 2019. Chanmé on growing the frosé category: 'We want to change the perception of frozen drinks!'. 7 March. Retrieved from https://www.beveragedaily.com/Article/2019/03/07/Entrepreneurs-launch-RTD-frose-pouches-to-tap-into-frozen-cocktail-craze#. (Accessed 17 June 2020).

Oak Bottle, n.d. Craft Your Own Barrel Finished Spirits & Wine. Retrieved 5 August 2021, from https://oakbottle.com/.

Oh, S., 2020. Distilled Spirits Market Report: KS2020-0023. United States Department of Agriculture Foreign Agricultural Service.

Przyswa, E., 2014. Counterfeiting in the wines and spirits market. In: Key Issues and Presentation of Anti-Counterfeiting Technologies.

Ramos, M., Valdés, A., Mellinas, A.C., Garrigós, M.C., 2015. New trends in beverage packaging systems: a review. Beverages 1 (4), 248–272.

Robertson, G.L., 2005. Food Packaging: Principles and Practice, second ed. CRC Press, Boca Raton, pp. 399–425, https://doi.org/10.1201/9781420056150.

Sackier, S., 2020. The Great Debate: whisky Closures. 4 March. Retrieved from https://moonshineuniversity.com/the-great-debate-whisky-closures/. (Accessed 17 June 2020).

Sanchez, R., 2020. Jameson Debuts Customizable Saint Patrick's Day Bottle. 11 February. Retrieved from: https://thedieline.com/blog/2020/2/11/jameson-debuts-customizable-saint-patricks-day-bottle. (Accessed 18 May 2020).

Schuemann, D., 2015. Alcohol Beverage Branding: Will Capsules Become a Thing of the Past? 2 April. Retrieved from https://drinkwithyoureyes.com/alcohol-beverage-branding-will-capsules-become-a-thing-of-the-past/. (Accessed 17 June 2020).

Shaker, K., Nawab, Y., Jabbar, M., 2020. Bio-composites: eco-friendly substitute of glass fiber composites. In: Kharissova, O.V., Martínez, L.M.T., Kharisov, B.I. (Eds.), Handbook of Nanomaterials and Nanocomposites for Energy and Environmental Applications. Springer International Publishing, pp. 1–25. https://doi.org/10.1007/978-3-030-11155-7_108-1.

Silayoi, P., Speece, M., 2004. Packaging and purchase decisions: an exploratory study on the impact of involvement level and time pressure. Br. Food J. 106 (8), 607–628.

Smyth, S., 2016. Packaging is getting smarter. 20 July. Retrieved from https://www.dlpmag.com/key-articles/20143/packaging-is-getting-smarter/.

Soon, J.M., Manning, L., 2019. Developing anti-counterfeiting measures: the role of smart packaging. Food Res. Int. 123, 135–143.

Spinner, J., 2016. Two-part container disrupts ready-to-drink cocktail category. 3 October. Retrieved from https://www.packagingdigest.com/packaging-design/two-part-container-disrupts-ready-to-drink-cocktail-category-2016-09-26. (Accessed 17 June 2020).

Stapleton, L., 2019. AR vs. VR: What's the difference? Plus, how marketers use augmented reality and virtual reality. 14 November. Retrieved from https://blog.treasuredata.com/blog/2019/11/14/augmented-vs-virtual-reality-difference/.

Stark, K., 2019. Johnnie Walker Blue Label Bottles to Use NFC Technology Giving a More Premium Experience. 22 January. Retrieved from https://rfidworld.ca/johnnie-walker-blue-label-bottles-to-use-nfc-technology-giving-a-more-premium-experience/2287. (Accessed 11 May 2020).

Stephens, R., Beetham, H., 2019. Liquor in Plastic Bottles – The New Consumer Trend. 29 January. Retrieved from https://www.newswire.ca/news-releases/liquor-in-plastic-bottles- - -the-new-consumer-trend-544816002.html. (Accessed 6 June 2020).

Sterling, A., 2015. Lightweighting can be a heavy lift. 5 October. Retrieved from: https://www.packworld.com/home/article/13368732/lightweighting-can-be-a-heavy-lift.

Tapì Group, 2021. Abor – A new concept of Green Design by Tapì Group. Retrieved from http://abor.tapigroup.com/en/. (Accessed 17 June 2020).

Thinfilm Technology, 2020. Retrieved from https://thinfilmsystems.com/technology/.

U.S. Food and Drug Administration, 2019. Tamper-evident packaging requirements for over-the-counter (OTC) human drug products, 21 CFR § 211.132. Retrieved from https://www.govinfo.gov/content/pkg/CFR-2014-title21-vol4/pdf/CFR-2014-title21-vol4-sec211-132.pdf.

WARC, 2018. Johnnie Walker Reaches out to Female Whisky Fans. 28 February. Retrieved June 30, 2020, from https://www.warc.com/newsandopinion/news/johnnie-walker-reaches-out-to-female-whisky-fans/40104.

Weinstein, R., 2005. RFID: a technical overview and its application to the enterprise. IT Professional 7 (3), 27–33.

WhiskyCast, 2020. Creating a Sustainable Future for Whisky. 10 May. Retrieved from https://whiskycast.com/creating-a-sustainable-future-for-whisky-episode-818-may-10-2020/.

WingDing Cocktails Get Packaged in Stand-up Pouch, n.d. Retrieved 5 August 2021, from https://www.packaging-strategies.com/articles/89448-wingding-cocktails-gets-packaged-in-stand-up-pouch?v=preview.

WRAP, 2021. Case Study, Lightweight spirits bottles: a business case for change. Retrieved from http://www.wrap.org.uk/sites/files/wrap/14325-08_Spirits_LoRes.pdf.

Yumix Brand Uses Two Compartment Snap-fit PET Package to Launch Disruptive Cocktail Brand—Brand Launch—Plastic Technologies Inc., n.d. Retrieved 5 August 2021, from https://www.webpackaging.com/en/portals/plastictechnologiesinc/assets/11330814/yumix-brand-uses-two-compartment-snap-fit-pet-package-to-launch-disruptive-cocktail-brand/?alltags=true.

Further reading

Curruthers, N., 2020. Sustainable Spirits Changing the World. 12 February. Retrieved from https://www.thespirits-business.com/2020/02/sustainable-spirits-changing-the-world/11/.

Distilled Spirits Council of the United States, 2020. 2019 Economic Briefing. 12 February. Retrieved from: https://www.distilledspirits.org/wp-content/uploads/2020/02/FINAL-Economic-Briefing-2020-compressed.pdf. (Accessed 3 June 2020).

Euromonitor International, 2019. Alcoholic drinks global industry overview. August. Retrieved from https://www.euromonitor.com/alcoholic-drinks-global-industry-overview/report. (Accessed 8 July 2020).

Lyons Hardcastle, J., 2017. Lightweight Food & Beverage Packaging Saves Money, Reduces Carbon Emissions. 23 February. Retrieved from: https://www.environmentalleader.com/2019/12/absolut-company-paper-bottles/.

McGrew, S.P., 1990. Hologram counterfeiting: problems and solutions. In: Optical Security and Anticounterfeiting Systems. Vol. 1210. International Society for Optics and Photonics, pp. 66–76.

University of Wisconsin Sustainable Management, 2021. The Triple Bottom Line. Retrieved from https://sustain.wisconsin.edu/sustainability/triple-bottom-line/. (Accessed 13 June 2020).

Marketing Scotch whisky

Grant E. Gordon[*]

Introduction

Before exploring the theme of the marketing of Scotch whisky, it would be helpful to define what marketing is about and why it is an important discipline for the successful conduct of any business today.

Philip Kotler, of the Kellogg School of Management at Northwestern University, and one of the leading academics in the field, defines marketing as 'the set of human activities directed at facilitating and consummating exchanges' (Kotler, 1999). This definition is concerned with decisions by people, either individually or in groups, conducting voluntary activities. Marketing covers exchanges that are either one-time transaction or those that concern building lasting relationships between buyers and sellers. The marketing process is a two-sided discipline, where both buyers and sellers can be actively involved. In terms of what is being exchanged, the key point is that it concerns something that has value to one or both parties.

A narrower definition of marketing that is relevant to the practice of the discipline is 'the management of all the functions required to satisfy consumers and maximise profit in the long run for the company'. This definition puts the spotlight on the consumer, being the person who counts most in transactions and is the focus of attention. It also highlights the motivation of companies, which is the pursuit of profits over the long run, which in turn depends on the revenues derived from recurring customer and consumer demand. Returning to Kotler, marketing management is 'the analysis, planning, implementation, and control of programmes designed to bring about desired exchanges with target audiences for the purpose of personal or mutual gain'. Reliance is placed upon coordinating and adapting the key variables for achieving an effective response from the consumer. McCarthy popularised a four-factor classification of the marketing decision variables, the 'four Ps':

Product, Promotion Place, and Price (Fig. 28.1). This chapter explores these four variables and shows how firms practising marketing use these tools to influence sales and pursue

[*] Note from the editors: This chapter is being republished from the 2014 second edition of this book, as the basic marketing principles, which are so clearly described in this chapter, have not diminished in value over time.

The four P's

FIG. 28.1   The four Ps. *Acknowledgements: E. Jerome McCarthy and Philip Kotler.*

long-term profitability. It will also become clear that each of the 'Ps' is in reality a collection of instruments.

Emphasis in these definitions is put on the management process to achieve a desired end result. Without a proactive effort in terms of time, effort, skill, and dedication, the marketing process will just not happen, and achieving any set of objectives will be hit and miss. Marketing demands engagement, and is best carried out with passion, imagination, and flair tempered with discipline. Marketers need a sense of what levers to pull in order to capture the custom of the consumers they are targeting. Successful entrepreneurs usually have an ingrained understanding and appreciation of marketing, but even if marketing is not a science, it can be taught and most certainly learnt through practice. Those who practise marketing should not be afraid of error, because very often the best lessons are gleaned from failed experiments.

Marketing books have become popular, and can be found in many bookstores. They sport catchy headlines such as *The Customer is King, How Brand X Conquered the Market for Widgets, Understanding the Psyche of the Consumer*, etc. Such books, along with many others, have a whole raft of topical messages to deliver. The subjects covered are varied, and offer good guides for those who want to grasp the essentials of best practice. They often include useful lessons in marketing from leading practitioners in the field. However, for all marketers, there is nothing that can replace learning through the experience gained from real business situations.

The origins of demand

Historical development

It is quite likely that distilled spirits in Scotland were an import from Ireland during the Middle Ages, brought over by missionary monks (see Chapter 1). The Scottish Exchequer Rolls for 1494 hold the first written reference to the product, noting that Friar Cor had used 'eight bolls of malt wherewith to make *aqua vitae*'. The product that became known as whisky, from the Gaelic *uisge baugh* ('water of life'), has helped to make Scotland famous around the world. In the early days, the distilling of Scotch whisky was a cottage industry and a natural complement to the traditional farming that supported life in the Highlands. The government soon woke up to the opportunity of using the popularity of the product as a means to raise taxes, and the Excise Act of 1664 fixed duty at 2s and 8d per pint of *aqua vitae*. As a complement to this decision by the Scottish Parliament, the Excise Officer made his appearance. From the eighteenth century onwards, there was an emergence of demand on a significant scale for the water of life, but it took a very long time for the industry to settle down and achieve the transition from the freewheeling days of illicit distilling to the modern, regulated industry, with its strict licensing of producers.

One of the key forerunners in the development of the modern-day industry was The Distillers Company, which was formed on 1 May 1877 as an amalgamation of six Lowland distilleries, all of which produced grain whisky. This was a key milestone in the development of the industry, as it moved it beyond the craft industry stage into a business run much more along commercial lines. Indeed, the advent of grain whisky distilling provided the technology that allowed producers to achieve two objectives: first, they managed to produce a whisky very economically; and secondly, producers now had a quality of spirit offering a lighter and usually more consistent taste than traditional malt whiskies. A new profession emerged, known as blending, along with a new type of whisky that could be mixed to good effect with the traditional malt distillates. Leading blenders would soon launch their own brands, offering consistency of product with a distinct cost advantage. Gradually, the blending houses were able to build consumer franchises for their brands. The names that emerged at that time, such as Johnnie Walker, are today amongst the leading brands in the industry.

Demand in the United Kingdom for Scotch whisky gradually grew throughout the nineteenth century. The companies that were best able to meet this demand were the houses that controlled the supply of the grain whisky. This had become the main ingredient in the end product that consumers now demanded, with malt whisky already playing a subsidiary role. Brands that were unscrupulous with their consumers would include as much as 90% grain whisky, offering doubtful quality. Selling a matured product was yet to become the norm, with consumers still accustomed to drinking whiskies that were bottled straight after distillation—just as vodka is today. In these early days of the modern industry, it can be supposed that product quality varied, but demand grew nonetheless.

With a growing industry, more and more players were being attracted into the market to meet the rising demand. For historians, it is probably not a great surprise that before the end of the Victorian era, the industry had entered a real crisis. Indeed, the growth in popularity with consumers had led to a boom in the production of all types of whisky, which caused a glut in the market at the approach of the turn of the century. In 1898, with the bankruptcy of Pattisons, there was a collapse in confidence in the market. However, from the ashes of this crisis, there emerged a number of players who managed to survive. These companies, of which five are the most famous—Haig, Dewar, Buchanan, Walker, and Mackie—were leaders in the development of the industry at the dawn of the twentieth century. Scotch whisky was about to spread its wings and build demand, starting mainly in the English-speaking nations of the world.

In the run up to the First World War, pressure was increasing to introduce measures to define quality standards in the industry. A benefit of establishing tighter rules of the game would eventually be to improve the product's reputation with consumers, who had grown accustomed to finding products of varying quality. The Immature Spirits (Restrictions) Act was passed on 19 May 1915, making it compulsory to keep spirit in bond for 2 years, with the period extended to 3 years in May 1916. This new legislation was one of the key building blocks that would pave the way to better quality products and, in turn, greater demand. This new legislation led to the demise of the cheaper, immature whiskies that at that stage still accounted for the bulk of consumer demand. After the First World War, demand was reduced, as the more expensive whiskies now being marketed were beyond the reach of the mass market.

There was another major barrier to the development of demand during the first part of the twentieth century. In the home market, as in other countries, a movement emerged against the consumption of alcohol as a significant number of people blamed alcohol for creating social disorder in certain sections of the community. This was aside from any impact excessive drinking was having on health standards overall. The temperance movement, as it was called, began to take hold on both sides of the Atlantic, resulting in a period where the consumption of all forms of alcohol was banned in the United States. This 'Prohibition' era did not last very long, as the benefits that its supporters were looking for did not materialise and popular opinion moved firmly in favour of restoring consumer freedom of choice. However, before the period of restrictions had drawn to a close, the trend against alcohol had spread to other corners of the globe and continued to cast a cloud over the industry's efforts to market its products right up to the advent of the Second World War.

As markets reopened in the 1930s, the industry's fortunes were slowly able to start to recover. This recovery was driven largely by exports. In 1937, for the first time in the industry's history, the export market was larger than the home trade. By this stage, there was also a clear pattern emerging in terms of leading brands that commanded the loyalty of consumers at home and abroad. Names such as Haig, Dewar's and Johnnie Walker were now well established as the most popular choices, and were leaders in the market. Apart from a strong consumer following, these brands shared another matter in common; they were all owned by the same organisation—the Distillers Company Limited (DCL). The DCL was all-powerful in the industry, and enjoyed a quasimonopoly status. Indeed the company, through a series of amalgamations, had collected a portfolio of leading brands. With such a portfolio of products, the DCL had a tight grip on the market, and was able to set prices and thereby define the returns available to all participating companies. The DCL set the agenda in terms of the way the product was marketed, and most of the credit for establishing the foundations of Scotch whisky marketing goes to the efforts of the people around the world who worked in the company, or as its agents and distributors, building its brands. Other companies were generally followers, and their chance of making a mark would have to wait until after the war years.

Winning and losing consumers

The Second World War also had a particular impact on the fortunes of the industry. American GIs streamed through the continent of Europe in the defence of the Allies, and these soldiers brought with them their consumer habits—one of which was the fashion for drinking whisky. Films of the time showed the heroes relaxing with a glass of whisky, and helped to spread the image of a fashionable alcoholic beverage. An immediate consequence of the war, however, was to curtail production of whisky in Scotland, and this reduction in supply was to have beneficial consequences in the years after the war. Indeed as the world emerged from the horror of the fighting to turn to the task of rebuilding peace and prosperity, the demand for Scotch was about to grow against a backdrop of tight availability. A real product shortage was created, and this kept prices firm and helped underwrite the sense of exclusivity of the beverage. Scotch was to become in peacetime the aspirational alcoholic drink sought after by many.

During the postwar era, the industry started to see the emergence of the United States as the main market for Scotch whisky as it took on fashion-drink status throughout the United States. With a flourishing market across the Atlantic, the industry entered its real take-off phase—a phase that was to propel it to the ranks of the leading international spirit of choice for drinkers around the world. Demand for Scotch whisky grew annually at close to double-digit percentages in the three decades after the end of the war up to 1970. These were the glory days for an industry that grew from a volume of 10 million cases sold in 1950 to over 51 million cases in 1970, driven by strong export growth. However, the growth levels the industry had grown accustomed to could not continue indefinitely, and more challenging times lay ahead. Indeed, history was about to repeat itself, and the industry was on the verge of creating the largest hangover it could possibly imagine. Faced with a plateauing world demand, producers continued to maintain output as if growth was still in the offing and, by failing to curtail production, huge excesses of stock were accumulated. The problem gradually built up during the 1970s, and by the mid-1980s, the dam finally burst when prices in the industry collapsed severely as a result of the high volumes of cheap whisky that were flooding the market. Damage to the reputation and image of Scotch whisky was a consequence of the glut that allowed low-priced generic brands to take hold in the market.

In fact, the failure to keep demand and supply in equilibrium was to have a long-lasting impact on the industry as Scotch whisky moved down the path of other consumer products and became commoditised. The worst of the effects were felt in mature markets, where the low-priced products were able to make large inroads in market share terms at the expense of the leading brands. As cheaper substitutes became available, brands started to lose their preeminence in consumers' minds, and they gradually placed less value on higher-priced products. In the United States, the demand boom of the after-war era had taken the industry to new heights, but the consumer following lost momentum as fashion turned the consumer towards other substitute drinks. The cycle had turned, and marketers were unable to keep the product's leading image at the top of the consumers' set of preferred choices whilst other products were able to make inroads and grow in popularity. However, efforts to build demand in other parts of the world started to pay dividends, and the industry's dependence on the United States declined along with sales in its largest market.

The latter part of the twentieth century saw a new phase of the industry's development where a series of markets started to emerge closer to home, and especially in Europe. Here, drinkers were attracted by the image and taste of the product that had previously been the trend in the United States. In Asia, the popularity of Scotch whisky also grew as drinkers found, in the product's quality and image, values that are hallmarks of many luxury western-branded products. For many of the newly acquired customers in these countries, Scotch whisky was a natural beverage accompaniment to social occasions with friends.

The other new element of change that came over the horizon (or more precisely re-emerged from the long past history of the industry) was the emergence of single malt whiskies. To those who could afford to experience the best, the availability of relatively unknown and sought-after brands of single malt heralded the start of a new trend. The result has been the emergence of a new premium category within the industry, identified by products that have their origins in the Highlands and Islands of Scotland. The folklore and mystery surrounding these brands and their deep historic roots give these products a strong traditional background. This imagery, allied to the variety of tastes, has won the loyalty of many new Scotch whisky lovers around the globe. By taking the consumer into new territories in terms of taste and general satisfaction, marketers have managed to create renewed appeal for the category overall, whilst at the same time moving the image onto a higher platform in terms of perceived value.

The industry that has emerged from all these changes and cycles of demand is thriving, and has firmly established a place at the top table in terms of beverages of choice for drinkers around the world. Scotch whisky is a leading category of choice amongst consumers of spirits. Scotch whisky sales totalled 98 million cases in 2012 compared with 63 million cases in 1985, showing the increasing demand for the category. One feature of the pattern of sales during recent history is the rising importance of premium and malt whiskies as subcategories together representing 22% of total volume demand in 2012. There has been a steady switch of consumption towards the higher value added brands, and premium whiskies, dominated by 12-year-old Scotches, have led this trend. Single malts have also played an important part in this trend of trading up by consumers, doubling its share to 6% over the same period. The rise in popularity of these two premium categories has been at the expense of standard brands, with low-priced products maintaining their share overall.

The leading brands that today dominate the category have developed strong loyalty from their consumers; the result of years of building their reputations. Alongside the major brands, there are many niche products that cater for the demands of the connoisseurs. Today, consumers enjoy a diverse choice of brands catering for all tastes and budgets. However, making choices may not be easy for consumers who cannot always differentiate between brands. The power of persuasion through promotion can, therefore, exert a strong influence on choices made, and a complex and sophisticated consumer marketing industry has built up. The modern trade that is the Scotch whisky industry is forever tuning to the consumer and trying to profitably satisfy their every demand.

Current industry dynamics

This section discusses some of the tools that are used to build strong consumer franchises, explores the main building blocks in making a marketing plan, and describes the various approaches of companies in carving a niche for their particular brands. Marketing theory continues to evolve, but at the heart of any campaign to conquer the loyalty of consumers, there lie some basic elements that will never change; an attempt is made here to pinpoint some of these elements and focus on successful strategies to increase consumer demand.

It is vital when looking at any market to be able to view the marketplace through the eyes of the consumer. Consumers, considering their choices in alcoholic beverages, must first narrow the selection to three broad categories: beer, wines, and spirits. In more recent times, the choice has blurred, as there are products that successfully transcend these traditional product categories. Some brands have re-segmented the market and proposed new expressions of their products that are trying to cater for new modes of consumption—for example, there is now a new class of ready-to-drink (RTD) or new-age beverages that can be beer-, wine-, or spirit-based, and sometimes are mixes of all three basic forms of alcoholic drink. RTD products have been fuelling a large part of the growth in consumption of alcoholic drinks during the last few years, especially in the western world. Their popularity with younger drinkers is high—the consumer can chose a trusted brand in a format that fits well with popular drinking modes. The success stories here have managed to offer an alternative to popular traditional offerings, such as draft beer or chilled wine. The consumer is usually comforted by a well-known brand that brings a guarantee of quality, with an image to boot.

The consumer framework in choosing a drink for a particular occasion usually starts with the category of choice, and then narrows to a brand. In terms of category choice, the decision will be partly related to the setting and type of occasion. On the other hand, for some consumers, category choice may be of secondary importance—particularly where there is a social drinking occasion, when choosing the right 'badge' or brand may be more important. When it comes to brand selection, there is no end of choice available to consumers today. Indeed, the choice is bewilderingly large, and consumers can select from scores of different products. A visit to any bar, restaurant, or retail outlet will confirm the plethora of brands available. Probably, the most promiscuous of the three categories of alcoholic beverages is wine. Here, the choice is highly fragmented, with wineries remaining generally small in size, notwithstanding moves by major multinational drinks corporations to consolidate the industry and build a portfolio of wine brands. Compared to many other fast-moving consumer good categories, the drinks industry arguably offers more consumer choice than most—as can be seen from a visit to any major supermarket, whose shelves are abundant with alcoholic drinks of all tastes and variety.

A framework for consumer choice when selecting drinks is set out in Fig. 28.2. The three driving factors are need, occasion, and personal state. 'Need' relates to the person's mood, and 'occasion' to when, where, and with whom the consumer is going to enjoy a drink. 'Personal state', on the other hand, answers the question: 'Who am I at that moment?'. The way in which these three factors intercept help determines what choice of product and brand the consumer makes.

FIG. 28.2   A framework for consumer choice when selecting a beverage.

The consumer selecting a brand of spirits is inclined to make a popular choice and will naturally often chose one of the most popular brands. At present, the five most demanded international brands are Bacardi, Smirnoff, Johnnie Walker, J&B, and Absolut. These five names cover three distinct categories: vodka, rum, and whisky. When confronted with such an array of products, the existence of well-known brands acts as a powerful means of assisting consumers in making choices. The brand is a point of reference for consumers, allowing them easily to identify their brand of choice by its values in terms of taste, image, and other attributes. Of course, choices can be made without reference to a particular brand—for example, by choosing simply vodka or whisky. However, today the vast majority of consumers rely on brands when making choices, as this gives them reassurance to provenance and quality, and is ultimately a guarantee of satisfaction.

With the vast array of spirits that are on offer to consumers around the world, Scotch whisky has to compete with some other very popular alternatives. The choices have tended to grow over the years as consumers have developed a taste for many alternative spirits, ranging in the past from gin to more contemporary drinks such as vodka and tequila. However, this is a dynamic process, and new categories emerge as others lose appeal. For example, few people today know that the bitters that are popular in some European markets used to be a desired tipple in Great Britain. Whisky has developed almost universal appeal, and holds a strong position as the category of choice of many of the world's drinkers. The whisky category also includes some brands that are of international repute and have the strength to compete effectively with strong competitors from other categories.

In terms of category share, Scotch whisky at present enjoys approximately 10.2% of the global spirits market. The total share of whiskies increases a fair amount when all other types of whisky from around the world are added, including bourbon, Canadian, Irish, Japanese, Indian, and other locally produced forms of whisky. Indeed, contrary to common perception, 'brown spirits' are today overall a very large and vibrant part of the total world spirits market. In addition to whisky, this category includes such drinks as cognac, other brandies, and dark rums. To put matters in context, the leading category continues to be vodka, with a commanding position of approximately one bottle out of every five bottles of spirit consumed globally, confirming the huge popularity of white spirits with consumers.

Economics of the whisky industry—The value chain

Cost of goods and the impact of maturing spirit charges

A detailed understanding of the industry and its operations is helped by a full understanding of the way in which costs are built up in the supply chain, starting from production and continuing all the way through to marketing. The most important element that differentiates Scotch whisky in terms of cost from other beverages is the consideration of time as a factor. Whereas a normal industrial product will be ready for sale the day it leaves the factory door, in the case of whisky, law stipulates that the product must be matured in casks for no less than 3 years. There is a delay between the moment of production and the time of sale, and during this period, the funds that have been set aside for the production process are effectively dormant, waiting for the product to become ready for market. This theoretically in itself is not a problem providing that there are sufficient funds available to finance the business pending sale of the stock that has been produced. In practice, most players in the industry have a strong and steady cash flow from their activities, and therefore the financial strength and stability to afford the burden of holding some of their products for very long periods of maturation before bottling and shipping to market.

This position can only be reached once a company is producing healthy profits. Profit margins simply have to be calculated to take the time/cost factor into consideration. However, achieving profitability is much more complex, and results from the marketing action that producers take. How they manage to achieve this is discussed later in this chapter.

Looking at the cost of goods in more detail, the following aspects need to be considered:

- Spirit: the cost of cereals, fuel and energy, direct labour, and production overheads.
- Financing: the cost of funds required to finance the inventory from the time of production until a sale takes place.
- Wood: the depreciation of the cask (over its useful life) and the cost of repairs to the stock of wood.
- Warehousing: the cost of storing casks during the maturation period.
- Transport: the cost of moving whisky from the point of distillation to bottling.
- Bottling: the cost of packaging a case of whisky, including labour, bottling overheads, and depreciation.
- Packaging materials: the variable cost of materials, including an allowance for wastage.

Bringing all these costs together gives a total product cost, and provides a key input variable in the value-chain equation. The gross margin on the product can be calculated by deducting the net revenue derived from the product cost. As in all consumer product industries, emphasis must be placed on minimising the cost of goods in order to help maximise gross margin. The cost of goods is generally being reduced through production efficiencies whilst maintaining quality standards, and this process helps allow room for an adequate marketing effort to take place and for a profit to be earned in the final count. The nature of competitive markets is such that marketing costs have, if anything, grown over the years, as producers compete harder for the consumer's loyalty. Therefore, driving gross margin generation helps to free-up cash for marketing expenditures that will have to be committed in order for brands to flourish.

Marketing costs

The main areas to be considered under the broad heading of marketing costs can be broken down into two key components. First, there are the brand development costs, which include product marketing (advertising, promotion, public relations, and any other paid-for activities designed to influence demand in whatever form). Secondly, there are costs related to channel management, or selling and distribution. This category refers to the entire infrastructure that is required to drive product through the distribution chain to reach consumers. Another way to view cost is that there are the costs of getting the product to market (channel costs) and, once there, of pulling it through to the consumer (advertising and promotion; A&P). The sum of these two major categories is often the largest single cost line on the P & L (Profit and Loss statement). The careful management of these resources is arguably the most difficult area of the management decision-making process in a consumer branded goods company.

The direct output can be measured as profit after marketing (PAM), representing the net revenues minus the cost of goods and the selling/A&P expenses. PAM measures the economic contribution of the product to the company's finances after allowing for all costs related to its production and marketing. Adding value at this level is a key variable in terms of ultimately driving corporate profitability. Success in building long-term value, measured in PAM terms, will drive shareholder value over the long run, and is the goal of modern consumer marketing companies.

Marketing costs vary significantly by brand, as there is generally a distinction between brands that are being built, maintained, or milked. In the build phase, the brand may have a level of spend that is high in relation to revenues, incurring losses. Maintenance applies to brands that are in strong existing market share positions that must be defended against competitive attack. The brand should be making a 'normal' profit at this stage in its lifecycle. Finally, there are brands that are being milked, where the expenditure in marketing is kept low and profitability is maximised in the short-term. This strategy is applied to brands that are deemed to have weak prospects. Common to many consumer goods industries, Scotch whisky producers have somewhat narrowed their portfolios and allowed some brands that were historically of importance gradually to die away, focusing their attention on strengthening their leading brands.

Branding

Companies sell products; companies market brands. *(Bobby Calder and Steven Reagan)*

Making products meaningful

Branding is at the centre of building any business towards sustainable profitability over time. The power of a brand with the consumer will ultimately translate into profits for the producer if this process is managed skilfully. The alcohol beverage industry is no exception, and includes some of the most successful examples of profitable global branded consumer goods companies. Large global players enjoy advantages in the development of portfolios of brands covering key segments of the market, even if the recent trend is for concentration by the key players on a selected number of reputable brands. Valuable marketing resources are channelled into these brands, which in turn are able to reinforce their leadership, gaining consumer trust and loyalty.

The role that brands play in the alcohol beverage market mirrors many other categories. By creating unique imagery and perceived benefits around a product, marketers are able to attract consumers again and again to their leading brands. The brand itself often becomes the main basis today for consumers choosing one product over another. This is the process by which sustained economic value is created, and the product becomes the profit-driver for the company. Without strong brand credentials, the status of the product risks being ill-defined and consumers will struggle to fit the product into their pattern of consumption. On the other hand, a well-defined positioning for a brand, which is well-articulated to the consumer, has the chance of yielding the best overall result. Scotch whisky marketers have to address these very issues, and the strategies they implement determine the long-term success of both the category and their respective brands.

In simple terms, branding is the discipline whereby marketers tell people what it is about their products that make them meaningful. Charles Revlon, of the eponymous cosmetics company, used to say 'I sell hope'. Giving meaning to the product will let the consumer know how it is to be understood and what makes it unique as a brand. The brand becomes a symbol or mark that is associated with a product, and to which buyers attach psychological meaning. Brands can be a form of currency for consumers that can enhance their experience by offering increased certainty and performance. The question is, what are the means by which companies can systematically approach this issue in order to give their products unique personalities that will make them into brands? For the brand to have value for consumers, the associations it offers should ideally become a part of their lives. The brand's equity is created through product design, advertising, distribution, and all the other ways that the company contacts the consumer. The total result of all this effort in some way must in the end reside in the buyer's mind. Unless this is the case, a brand is merely a product with a meaningless name attached to it. Stephen King, of the WPP group in London, said: 'A product can be copied by a competitor whereas a brand is unique. A product can be quickly outdated whereas a successful brand is timeless'.

Successful positioning involves first affiliating a brand with some category that consumers can readily grasp, and then differentiating the brand from other products in that same category. This is referred to as competition-based positioning. For sustained success, it is also helpful to link a brand to the consumer's needs and objectives. This is called goal-based positioning. Brand positioning is an essential discipline to any marketer because it forms a basis for the setting of product attributes and pricing strategy, and selecting the promotional and channel mix to best fit the demands of a particular targeted segment of users. In other words, once the positioning is clear the producer can structure the marketing mix in the context of one overall strategic plan for the brand.

In trying to understand positioning further, it is useful to analyse how people represent information in their memory. This can provide a starting point for developing a competition-based positioning strategy. One way information about brands stored in memory is in terms of natural categories. Taking beer as an example, Heineken is represented in memory as part of the subcategory of lager beer. In turn, lager is part of the category beer, which is part of the category of alcoholic beverages. A level could be added to the hierarchy under Heineken that subdivides the brand form into draft or bottled. For most analytical purposes in consumer behaviour, only two elements of the hierarchy—the brand and the category in which it has membership, or its frame of reference—will suffice.

At each level in the hierarchy, objects can have three types of association: attributes, people, and occasions. Attributes are physical characteristics of a product, such as its colour, size, and taste. People and occasions are together regarded as the image they associate with the brand. Most positions involve some combination of attributes and image. For alcoholic beverages in general, image is normally the dominating attribute, even though many people will try and rationalise their brand choice in terms of the product's physical attributes. It is not surprising, however, to note that consumers do not make their choices based on attributes and image alone. Rather, they use attributes and image to infer some benefit.

The benefit is usually an abstract concept based on appeal to the consumer's emotions, and it answers a need.

The preferred approach to positioning is to inform consumers of the brand's category membership before starting to put forward points of difference in relation to other category members. The rationale is that first consumers need to know what type of needs the brand can fulfil. To this end, the marketer generally first informs consumers of category membership, and then follows by developing awareness of the points of difference. Campaigns that try to conduct both of these tasks at once often fail because the message insufficiently develops either claim. Naturally, the more resources available to the marketer, the easier it is for the brand to deliver these messages home. Whilst it is important to establish the brand's point of difference, this is not usually sufficient for effective brand positioning. At best, this may simply help to grow the category. If many firms engage in category building, the effect may be consumer's confusion. In alcoholic beverages, this is arguably the case in the table wines category, and may also have occurred in single malts, where in both cases, consumers are confronted with a plethora of products making similar claims and thus it is 'harder' to choose between brands. Developing compelling points of difference is normally critical to effective brand positioning, and the message should generally be consistent over time. A sound positioning strategy, therefore, requires specification of the category in which the brand holds membership, allied to a clear reason as to why the brand stands out from others in the category.

Points of difference

Establishing points of difference, even if only perceived, requires an understanding of the consumer's beliefs. Which consumer beliefs about the category can be used to promote a benefit? The strongest positions are usually ones where a brand enjoys a clear point of difference on a benefit that is a key driver of category use. Category leaders supported by large budgets with which to outspend the competition may simply claim the benefit that motivates category consumption for themselves, using this as their unique selling point (USP). This is referred to as trying to own the category generic benefit.

Smaller brands, on the other hand, typically attempt to establish a niche benefit as their point of difference. Niches are achieved by using the primary category benefits to establish category membership, and then by selecting some benefit other than the focal one for the category to establish brand dominance. In normal practice, limiting the number of benefits that are made focal pays off by creating clarity. In fact, the processing of brand benefits requires substantial attention from consumers, and they are likely to become confused or switch off if too much information is put forward. Also, one benefit can easily undermine another—for example, if a brand claims to offer high quality and at the same time be cheap, the consumer may question the real benefit of consuming this brand, as low price and quality do not always go together.

Brand positioning

With competition-based positioning, membership of the category is developed by highlighting the points of similarity with the category, or relating the brand to a category exemplar.

Once the brand has acquired category membership status, its advantage is presented over other category members by representing the point of difference. The consumer rationale to trust the brand is thereby built step by step. So, for example, Laphroaig whisky sets out clearly that it belongs to the single malt category, and differentiates itself from other brands by emphasising unique product attributes in terms of its special taste and flavour.

Once the consumer has established a basic understanding of the brand's category attachment and any point(s) of difference, there exists an opportunity for growth by deepening the meanings associated with the brand's position. This entails demonstrating how the brand relates to the consumer's goals, and having insight into what motivates the consumer to use a brand. The process employed will start with defining key attributes and image, moving through functional benefits and eventually touching on emotional and assumed benefits. By the end of the process, a point can be reached where the benefit defines the essence of the brand. Brand essence thus becomes shorthand for what is at the heart of what makes a particular brand unique in the eyes of the consumer.

The same process that can be undertaken at brand level can be undertaken at category level. The assumption with category essence is that if consumers perceive a brand to be positioned in a manner that is sensitive to their problems, the brand is viewed as a solution to these problems. Appropriating category essence can be a useful way to compete for a brand that does not have a meaningful point of difference.

The challenge that marketers face is to find a viable basis for differentiating their brands. One frequent trap that brands fall into is where they are marketed based on a point of difference with their competition, but one that has no relevance to consumers' reasons for choosing the brand. Claiming multiple benefits is the other frequent error; trying to imply that there is something in the brand for everyone. This approach very often confuses the consumer, leaving it unclear what the brand is about. Once a position is developed, most of the activity behind the brand should be focused on, and aimed at, sustaining the point of difference. Sustaining benefit over time provides a barrier to entry by competitors.

After the key choices have been made regarding the targeting and positioning of a product, it is useful to summarise these decisions in a positioning statement. This statement is a summary of the key aspects of the marketing strategy, and will serve as the foundation for decisions about all elements of the marketing mix. The positioning statement should attempt to answer the following questions:

1. Who should be targeted to use the brand?
2. What is the key reason for buying? (the proposition)
3. What are the emotional and functional benefits?
4. What are the reasons to believe the proposition? (substantiators).

A good positioning statement will definitely set the brand apart from its competitors. It should also provide a clear understanding of when and why the consumer should use the brand, and what would motivate the purchase. Succinct reasons should be available, justifying why the brand is a compelling proposition. Avoiding vagueness is important, as well as making sure there is a clear linkage between the brand's point of difference and the target consumer's needs and requirements. The distillation of all the above into the brand-positioning statement should produce a clear and unequivocal view of 'brand essence'. This statement is a form of genetic code defining the personality and values of the brand.

Creating brands

To be successful, all brands must have product, price, and channel strategies to support the image communicated through advertising or other promotional means. Advertising, whilst a central part of the marketing mix for any brand, is by no means the only vehicle that should be considered for communicating with consumers. Whatever the means chosen, the brand-building process should be guided in all its detail by the desired positioning of the brand. This is particularly true with brands of international repute sold across many different markets. The overall positioning of the brand must be rigorously upheld and implemented at the local level in a manner consistent with the agreed positioning. Implementation is by way of all the decisions related to the marketing mix. Depending on the type of brand being developed, the elements in the marketing mix will be varied. Further discussion on each of the variables in the marketing mix follows.

When it comes to alcoholic beverages, brands tend to be valued primarily based on imagery. Scotch whiskies are no exception, and the vast majority of the best-known brands are reliant on their imagery for their position in the market and their consumer following. Image brands distinguish themselves by offering unique sets of associations, and these associations are usually based on an emotional (nonnatural) appeal to the consumer. For example, Johnnie Walker wanted to associate its brand with 'celebrating male success', and its 'Keep Walking' campaign featured male celebrities who had succeeded on their own terms and often against the odds. The principle behind an image brand is that consumers are more likely to engage with the brand if it can tap into these types of powerful associations, rather than if the product is portrayed as tasting better or smoother. This can be a powerful route to success, but can require large financial resources to develop and to sustain it.

Managing image brands

Image brands will succeed, if and when, they make an emotional connection with the consumer. Advertising and the other forms of promotion play a prominent role in developing the image of any brand. The process of building the brand takes a lot of time and effort and considerable resources, because the image has to be created in the mind of the consumer. Caution should also be exercised, and the limitations of the brand-building process understood—for example, products that rely heavily on user imagery may have limited appeal across generations. Young people in many countries may see Scotch whisky as the drink of the older generation. Where this situation is faced, it provides a challenge for Scotch marketers who are seeking to discover ways to regain support for their brands with a lost generation of drinkers. There is no ready and easy solution to this challenge, although much effort continues to be made to conquer new and younger groups of consumers.

Determining the optimal branding strategy will, at the early stage, mean making choices between stand-alone branding and family branding. In the former case, each of a company's brands is unrelated, whereas with family branding multiple brands are linked under a common house name. The familial association can provide an assurance of quality up to a certain level. On the other hand, product branding can enable the firm to address distinct market segments of consumers who may not want to be associated with each other. Also, buyers seeking relatively exclusive products may disdain brands that are ubiquitous, seeking more obscure and esoteric choices in place of the bestsellers. An issue with multiple branding is the cost of sustaining a broad portfolio of unrelated products when the brand-building process is by definition a costly and long haul process. In the end, success rides upon consumer acceptance, with the winning brand resonating strongest in the consumer's mind. A brand that manages, through a carefully chosen and properly executed positioning strategy, to carve out a unique and relevant position in the eye of the consumer, will be in the best position to succeed.

Routes to market

Channels of distribution—Introduction

In terms of the four key marketing variables ('four Ps'), 'place' refers to all choices related to the design, selection, and management of the channels of distribution. Coughlan and Stern define a marketing channel structure as a set of pathways a product or service follows after production, culminating in purchase and use by the final end-user. Some of the key questions that arise are

- What role do consumer characteristics and demands play in the appropriate channel design?
- Why do marketing channels change in structure over time?
- How should a manufacturer decide what types of intermediaries to use in the channel?
- What problems can arise in the ongoing management of complex marketing channels?

A marketing channel is normally a set of interdependent organisations involved in the process of making a product available. A key point is that the company is not going about its business in isolation. Indeed, channels rely on assembling sometimes complex networks of independent parties—manufacturers, wholesalers, retailers etc.—to serve the needs of the end user. Each one of the parties involved depends on the performance of all the other parties in the chain, and orchestrating this complexity requires discipline from all parties. The interaction of the key players and their ability to work together is a key success factor.

In the chain, the focus should be on the end user irrespective or where any party stands in the distribution chain. Satisfying the ultimate consumer is the goal that all parties in the chain need to address. Indeed, simply passing the product down the line does not guarantee success. The marketing channel structure is a strategic asset for the producer. Managed optimally, it not only allows the product to be physically brought to market but also it helps implement and support the brand's positioning strategy in a consistent manner right through to the point of consumption. Many acquisitions are helped by the ability of manufacturers to obtain distribution synergies and increase their muscle with the trade. In the drinks industry, many recent mergers have been predicated on obtaining large synergistic benefits by channelling previously competing products through a common distribution system. Successful international drinks marketers have channel capabilities that are designed to bring to market a broad portfolio of brands, each with a distinct positioning, and effectively execute the brand message at the point of sale.

Reaching the drinks consumer

In the case of alcoholic drinks, it helps to understand both where the product is purchased and where the actual consumption takes place. These two places may or may not be the same. There are broadly two locations of end usage: on licensed premises (also known as on-premise), or at home. Commonly in the drinks trade, the two channels are referred to as the on and off trades. The on-premise trade relates to consumption in bars, hotels, and restaurants, whereas the off-trade is defined by channels (including supermarkets and other specialist outlets) that supply the market for retail purchases to satisfy personal consumption at home and with friends, as well as gifting. Internet retail is a new form of specialist retailing often tied with established bricks and mortar retailers. It is also an area where some producers are offering their brands directly to consumers by way of their own websites. The travel market is an important category that includes elements of both on and off consumption. For example, airlines serve passengers beverages in flight as well as run their own retail shops on board.

Channel design

The orthodox approach to channel design is to start by examining the target audience in terms of end users, and to work back from there. The market needs to be segmented into groups of end users that have similarities in terms of their needs and lifestyles. Defining groups based on their demands should be related not just to the product itself, but also to the service needs of the group. The same product can be in demand from channels with widely varying needs in terms of service level. We should recognise the value-added element that must be provided to meet the needs of specific groups of customers.

With a clear idea of who the marketer is targeting, an important step is to understand what are the main occasions on which consumption will take place. A decision can follow regarding the most likely type of outlet to focus on. Early in the decision-making process, there is the question of whether consumption will be mainly on or off premises, and the relevant importance of each sector. The answer to this basic, but sometimes difficult, question will help determine which channels need to be included in order to get the brand in front of the potential user at the right time and place. The alcoholic drinks trade has always relied heavily on the on-trade for the purpose of brand building.

The marketing channel challenge involves two major steps: first, selecting the right channels; and secondly, ensuring successful implementation of that design. As discussed previously, the design step involves segmenting the market and understanding the demands of end users. The channel design will focus on utilising the routes to market that are the most appropriate to reach the main points of usage for the brand. Whoever is the ultimate consumer being targeted, the marketer should be able to tailor the channel strategy accordingly. Until such time as the channel strategy is established, other marketing efforts (for example, in advertising and promotion) will likely be wasted resource and energy as long as the consumer cannot effectively connect with the product.

The optimum channel is usually selected based on the channel that is the most efficient at meeting the needs and servicing the demand of the specific consumer segment being targeted. There are two main building blocks in channel design: first an understanding of the types of intermediaries that are going to be involved in the particular channel, and then of who they are specifically. The choice has implications not only on efficiency but also on the image of the product, depending on the final selection of retail outlets. It is also necessary to determine the level of intensity with which the channel will be used. This could range from a high level through to only selective or exclusive arrangements. The company must distinguish between the channels that the company considers necessary and those that should not be used. Some channels will just not be viable because the support required cannot be met by the company's current resources. Knowing what segments to ignore is very important, because it keeps the channel effort focused on the key segments from which it plans to reap profitable sales.

Invariably the planning will involve the use of internal and external resources in the company. Most channel activities rely on the interplay between a number of entities—for example, the brand owner may or may not have their own sales force or distribution division to conduct the selling of the product to the relevant intermediaries, and the function may be outsourced to a wholesaler of some description. Wholesalers include companies that can either be generalists or are specialised in servicing a limited segment of the market. The critical point to have in mind at all times is the ultimate end-user, and the relative abilities of the chosen intermediaries to execute effective merchandising strategies and bring the product in front of the target audience in the most appropriate way. In the case of markets that are managed by exclusive distributors, the choice can have significant consequences on the success or failure of the brands. The level and quality of the relationships of the distributor in their particular country or area of operation can be a key success factor. In the drinks industry, as elsewhere in commerce, fortunes of brands have risen and fallen based on the performance of local market distributors and their abilities to meet their consumer's needs. The trend of global retail consolidation is now confining the role of the exclusive distributor to supporting the overall marketing effort in different ways. The distributor needs to complement the efforts of the producer, who is seeking to capitalise upon relationships with global retail giants to expose their own brands more widely to consumers regionally and around the globe. The distributor's task is to have the knowledge to be able to implement international positioning strategies for brands in a way that is relevant and effective in their marketplace. To summarise, marketers have to think global and act local!

The costs and efficiencies of the channel system also need constant monitoring. The challenge that the marketer faces is reducing costs without endangering the service needs of intermediaries. This exercise is constantly being revisited by producers, as such a large amount of cost is tied up in channel management and there is a danger of wasting these resources. Failed channel design and implementation can condemn brands to poor performance and cause profit targets not to be met.

Channel power and conflict

Once intermediaries have been identified and the terms of reference have been agreed with respect to the objectives to be achieved, the discussion on strategy can commence. It is important that a good dialogue exists between all key players in the channel system, especially at the planning stage. At this time, it should be clear to all participants what their roles are and the part they will play in moving the product into the ultimate consumer's hands. Certain players fulfil specialised roles in the channel process, and they need, with all other channel players, to be coordinated so that the entire chain works in harmony. Any weakness in the system will cause the overall performance of the system to suffer.

It may be that a choice is made to opt for a parallel approach, with a number of parties competing within the same channel. If this is the case, rivalry will be the norm, and this needs to be managed in a way that it is constructive, not destructive. The manager needs to be able to identify the source of the conflict and differentiate between poor channel design and poor distributor performance. Once the problem is identified, swift action must be taken to reduce the conflict and restore the normal operations of the distribution system. Controlling rivalry is often achieved by use of channel power. Power in this context is the ability of one player to control the decisions taken by another member at a different level in the distribution chain. When the disparate members of the channel system are brought together to advance overall performance in line with the overall brand goals, then the channel system is well-co-ordinated. Coordination is an overriding objective of the entire channel management process.

Multiple marketing channels

More and more attention has been focused on multichannel approaches to marketing. The advent of the Internet has been a catalyst in this debate, because consumers can now be reached through a variety of brick and mortar retail outlets, direct mail catalogues, and on-line web stores. The producer needs to recognise the needs of consumers and to adapt the channel strategy to give them the choice and convenience of accessing the product in the manner that suits them. New channels do not replace old ones but tend to be complementary, and may simply allow buyers more flexibility in how they make their choices. With multiple channels, the potential for channel conflict increases. In this context, marketers must also manage the issue of unplanned movement of goods between markets referred to as 'parallel' or 'grey'. Intermediaries and dealers are always ready to gain profits from price arbitrage, given that the product can easily be moved from one market to another, and prices will naturally be driven down to the level established by the most efficient channel. Even with increasing channel choices, the objective of channel management is still to maintain viable channels to reach the chosen target market segments the company has identified.

Conclusions

Constantly educating the consumer can be very costly, although the payback can be rewarding in terms of market share, sustained uniqueness, price premium, and, ultimately, greater profits. To support this, creating, maintaining, and managing the channel system are important strategic undertaking for the company. The high fixed-cost nature of channel systems necessitates careful consideration of how channels should be structured and managed. Accessing performance information, keeping constant vigilance, and having the willingness and ability to respond are some of the keys to successful ongoing channel management.

The product

Historical development

The development of the marketing strategy for any brand requires a clear frame-work for understanding what the consumer wants in order to know what the product has to offer.

However, this has never stopped entrepreneurs from using their instinctive feel for what consumers want in coming up with new products.

Those who pioneered the development of the large-scale whisky business a century or so ago were confronted with these very choices. Whilst historically the type of product on offer was traditional malt whisky made by the old-fashioned methods pioneered in Scotland and Ireland, a real breakthrough took place in the 1830s with the advent of the Coffey still. The invention of the patent still by the Irish Engineer Aeneas Coffey allowed producers to offer drinkers a product with a lighter aroma than the traditional malt whisky. From this time forward, whisky drinking grew as the taste became popularised and it was developed into a product with mass appeal. In these early days of the industry, there was no minimum maturation period, and a large part of the market was represented by consumption of 'on-tap' whisky in pubs and bars around the country.

A campaign ensued at the beginning of the twentieth century to raise the quality standards of the industry. The means chosen was the imposition of minimum ageing rules. Those who proposed this new legislation understood that through the imposition of tighter rules on the manufacturing of the product, they would be able to improve the appeal of the category overall and their own brands. This move proved extremely effective, and led to an overall improvement in the Scotch whisky consumer franchise. Producers working together in such cross-industry initiatives are able to help improve quality for consumers across the board, and thereby the category's competitiveness overall.

Premium whiskies

Marketers of Scotch whisky over the years have made many efforts to promote their higher quality special whiskies. The best example of successfully seizing this opportunity to appeal to the consumer's need for more aspirational brands was the development of the aged 12-year-old blend category. The market was led from an early stage by two leading brands, Chivas Regal and Johnnie Walker Black. Consumers flocked to these products because they offered prestigious imagery combined with product quality of a high order. The brands became accessories of drinkers whose lifestyle associated positively with the upscale imagery of these brands. The age of the product also became an important criterion for product differentiation, especially with less knowledgeable consumers.

Single malts

Companies had detected a growing demand for products that would cater for a more particular clientele. The scene was set for the development of a new category, single malts, that emerged as a product with appeal to more discerning users. These brands had been closeted for years as hidden treasures and kept away from the consumer limelight except in the North of Scotland, where they historically enjoyed a good following. However, noticing that consumers were looking for a greater choice, some marketers seized the chance and started to make serious efforts to promote these whiskies. One brand, Glenfiddich, acted the locomotive role for the category, and has retained its leadership ever since spotting early on the first shoots of demand for these products. What this category has done collectively is to take consumers back to the original roots of the industry and the malt distilleries of the Highlands and Islands of Scotland, where they can explore the pleasures that come from the complexity and variety that are the hallmark of the single malt whisky. In recent times, there has been a trend towards producing new and interesting product variations, principally by using maturation methods as a variable. Special wood finishes in particular have added new interest to the whisky category, appealing especially to connoisseurs.

Product differentiation

Seeing scope for product differentiation within the Scotch category itself opened up opportunity for growth, and allowed those brands that were able to see the trend to emerge and gain market share. More product differentiation ensued, and today the market for single malts is fragmented as producers offer consumers more choices to satisfy their curiosity and a desire to experiment in seeking new taste experiences. This evolution has helped to alter overall perceptions of whisky, heightening consumer interest, and creating new appeal for the category overall.

Taking a broader view, there have been other successful attempts at achieving consumer success through clever product differentiation. A good example is the creation of whisky liqueurs. By marrying the finest whiskies with fine herbs, spices, and sugar, whisky liqueurs were able to address demand for a product that would have an appeal to a different user. One brand, Drambuie, stands out above all others. This company managed to create a niche to satisfy drinkers who preferred a sweeter-tasting product. The success of this brand and others that followed was product-driven marketing in action.

Deluxe blends, single malts, whisky liqueurs: these are three manifestations in product terms of whisky made to meet the differing demands of consumers. Each of these categories has played a role in adding complexity and value to Scotch whisky in the mind of the consumer. The search for new and better products will continue unabated, and marketers will always be trying to find new niches to exploit. One area that marketers of Scotch whisky are wrestling with is the RTD product category, where they are yet to find a formula for success in the way that Smirnoff and Bacardi have with their 'Ice' and 'Breezer' products. The search for innovative ideas will go on as brand owners relentlessly pursue the creation of new streams of income to leverage the equity of their existing brands. Experimentation with new forms of distillation, maturation, and other aspects of the product will also continue as part of a concerted effort to find new ways to bring greater satisfaction to consumers.

Packaging—Being best dressed

As with human beings, whose clothes send out strong signals of their individual style and character, the same applies to any packaged consumer product. The design and form of the container in which the product is sold send out messages that the consumer will assimilate, and plays an important role in determining the purchasing decision.

The greatest whisky in the world incorrectly packaged will lack perceived value for the consumer by being wrongly dressed! Having decided on the liquid, the process of preparing all aspects of the product in terms of packaging forms one of the cornerstones of successfully grounding the marketing strategy for the brand. No product strategy is complete without bringing together the contents and packaging strategy into a unified plan. Consumers are not attracted by either liquid or packaging alone; it is achieving the right overall mix of both liquid and pack that is the starting point for any serious effort to market the brand. Putting in place the appropriate packaging strategy usually involves a number of key steps, as described below.

Forming the packaging strategy starts with defining some parameters that will help frame the choices to be made by all involved in the process. The strategy needs to cover all elements, including the design of the bottle, the labels, and any gift and outer packaging. There are a number of participants in the process, ranging from engineers to design specialists. At its most simple, packaging must carry out a physical role in acting as a vessel to deliver the liquid to consumers. The container needs to meet criteria, including size, and the ability to be transported and to be leak-proof. The container will normally also be filled using machines of various types, and consideration must be given to the practicalities and cost of differing forms of bottle.

More important, for the marketer, is a clear understanding of the intended positioning of the brand. Once consumers know a brand's category membership, the focus is usually on establishing a point of difference. Finding a product differentiator is important to consumers, and distinguishes the brand from alternative offerings. The overriding goal is to establish an understanding of how the product is intended to be targeted, and what the unique messages are that the marketer wants delivered. By translating this into a succinct brief, the marketer can start to involve other parties in the process of developing suitable packaging. The end result should be in harmony, with the intended essence of the brand, its position within the category, and, most importantly of all, the consumer's perception of what the product should look like.

Another vital consideration is the trademark. If the brand has already existed for some time, most likely it will have acquired a reputation of its own. This reputation is embodied both in the taste that the consumer is used to and in the element of imagery. This imagery will have been formed through contact with the brand through various means. Users of the brand will have experienced the brand first-hand, and will recognise the trademark and the design and packaging of the product. All these elements come together to create an image for the brand. Nonusers may also be aware of the brand through advertising and other means of communication, and may equally hold an image of the brand, albeit possibly without ever having consumed a drop. Brands with history have historical baggage, and these elements, including the brand roots, can be factored into the brand positioning and, ultimately, the design brief. The designer will create an end result in terms of logos, typeface, symbols, and overall design, keeping all elements in harmony whilst at the same time giving the brand some differentiating features that help it stand out of the pack. Conformity is a double-edged sword in the world of packaging; it can send out strong category clues, but may be a poor means of achieving differentiation. Creating highly individual presentations can be successful (if kept within the boundaries of what consumers will accept), or can fail completely because they do not conform to category norms. To illustrate, one company seeking to differentiate its main brand from the pack but wishing to remain firmly within category norms decided to adopt a triangular bottle as its trademark design. From that day forward, the brand in question, Grant's, developed a personality of its own, whilst continuing to conform overall to category requirements.

The importance of making correct packaging decisions is a reason why there are advantages in conducting exhaustive technical and consumer research in advance of putting products to market. The objective of most research of this type is very often to counter the risk of taking incorrect decisions that could jeopardise any efforts and result in wastage of scarce resources. However, care must be taken not to put one's entire faith in such studies, and to trust one's natural instincts to take the right decisions. Many a marketing decision in the field of branding generally, and packaging specifically, has been left to the instinct of the marketer. Some people develop a strong instinct for what works and what will appeal to consumers. Left to their own devices, these people are responsible for many of the best-dressed and presented brands.

Another key issue with packaging is the need to regularly adapt the presentation of the brand. This adaptation will have two goals: keeping the positioning in line with the desired strategy for the brand and ensuring that the brand is appropriately dressed for the times. Just as in the car industry, yesterday's model with time looks dated, and drivers want to be seen behind the wheel of the car that reflects their image. Likewise, drinkers are making a statement about themselves every time they order a particular brand. So the brand needs to outwardly reflect that image, and the look and feel of the bottle and label play major roles in sending our visual signals about the brand's personality. Many marketers follow the Unilever adage of using the 'bacon slicer' approach, and make regular and gradual adjustments to the packaging of their brands. Only if a brand is encountering a serious problem in terms of market share, volume sales, or other key performance indicators, should radical changes to the packaging be contemplated, remembering that product packaging is only one element of the overall marketing mix. Radical change risks confusing current users and losing their loyalty and custom. If such change is not heavily outweighed by new customer acquisition, to replace lapsed users, the game can be lost completely.

Managing price

The importance of price

Price is one of the 'four Ps' in the marketing mix, and in terms of the decision-making process, it is important that finance and accounting work closely with marketing to reach the right policy. Decisions relating to price management are key drivers for any company in achieving profitability and long-term shareholder value. Today's consumers are even more determined to search for the very best value for their hard-earned cash, and are increasingly willing to change their buying habits in their search for better value propositions. Value is an aggregate of the functional and emotional benefits that the product creates—for example, whilst relatively expensive, Dom Perignon offers good value for money to many buyers of champagne. Effective price management is, however, key to developing and building our brands in both the short- and long-term. Of all the components of the marketing mix, price is the only one that directly generates income. It has an immediate and direct impact on a brand's ability to generate revenue. All the other components (promotion, product, packaging, and distribution) generate costs and represent investments.

If the strategic objective is profitability more than market share, the marketer will price less aggressively and seek those target segments that are willing to pay the set price, rather than cut prices to appeal to a larger segment. It is not very realistic to expect to maximise both profits and market share in highly competitive markets.

In overall terms, the price a marketer asks consumers to pay for a brand:

- Drives the perceived value of the total proposition offered by the brand.
- Provides a key signal of product quality and competitive positioning of the brand.
- Dictates the level of funds available for brand building investments.

Price directly influences consumer brand choice, and has a critical role to play in creating the equity of the brand. If the overall influence is positive, the brand can develop successfully to achieve its maximum profit potential. Pricing, in combination with the benefits associated with the product, its packaging, and such promotion as is carried out, is central to the development of a brand's perceived value and its ultimate success in the marketplace. Financial contribution generated through pricing also enables further investment in marketing activities of all forms, and is the foundation for maximising long-term profitability. Thus, achieving the optimal price/volume/ equity equation will improve the company's chances of maximising long-term shareholder value.

There are a number of formulae that can be used for setting price, but very few can be used in total isolation. Some of the major determinants of the pricing decision for marketers are set out below.

Cost of goods-related pricing

The product's variable cost is generally the pricing floor. This section examines each of the major components that are relevant to building the cost structure of a Scotch whisky brand.

One of the main cost factors that plays a major role in pricing within Scotch whisky (and alcoholic beverages in general) is the taxation element. It is commonplace in most countries to impose some forms of excise duty on alcoholic beverages, and a relatively high rate has tended to be imposed on spirits compared to other forms of alcohol, including beer and wines. Whatever the specific level of tax, all marketers must factor duties and other taxes into their cost equations in order to calculate the cost of goods. In some countries, the tax component may represent well over 50% of the final consumer price.

Given the heavy tax burden, marketers are obliged, to some degree to use cost-plus calculations when determining the price at which they intend to sell. Another element that complicates the equation when considering cost-plus pricing is the level of overheads allocated to a particular product. Generally, allocations are made on an arbitrary basis, based on the expected operating level of the company and factors or keys that are determined by the accounting department. Depending on how these keys are set, each product ends up picking up a share of the company's overhead and this is in turn factored into the calculation of cost. The third component of cost is, of course, the direct costs of the product, and there should be little doubt as to the accuracy of these numbers.

The three main cost elements—tax and duties, direct costs, and overhead allocation—can then be added together and the marketer can consider what mark-up or margin the product can bear. Obviously, this decision cannot be taken in isolation of other factors, such as the price of competing brands, but the benefit of this type of analysis is to allow companies to determine, out of their portfolio of products, which are those that can generate the highest economic profit.

A variant on cost-plus pricing is using pricing to fill the profit gap to meet the plan, or where a price increase is dictated by the planned profit needs of the business.

This approach could be seen as short-term thinking, and may have the potential to damage the long-term health of the brand. It is also based on the premise that price increases always result in an increase in profit, which may or may not be the case depending on the price elasticity factor.

Competitive set pricing

Companies can also set pricing based on what the competition is charging. This is referred to as competition-oriented pricing. In this scenario, the marketer keeps a watch on competitive pricing and sets the company's prices relative to one or more competing brands. In this situation, the company is not maintaining a strict relationship between demand for the product and cost.

The most popular type of competition-oriented pricing is where a firm tries to keep its price at the average level charged by the industry, also called going-rate or imitative pricing. Product differences, whether packaging, quality, or other product features, serve to desensitise the buyer to existing price differentials.

Matching competitor price moves is often the chosen strategy on the grounds that this represents a least risk approach. This assumes that competitors know what they are doing, and ignores potential competitive advantage from alternative pricing strategies. In other words, pricing a brand solely in relation to a competitive set is usually not advisable, although for brands that are followers in the market, this form of benchmarking is unavoidable, and failure to heed the impact of changes in competitor brand prices could have serious negative consequences for the marketer. In this context, price elasticity (which is discussed below) is important because it determines the way in which the volume of the brand fluctuates when price adjustments are made both in absolute and in relative terms.

Market-oriented pricing

This refers to pricing a product according to what the market will bear. It tries to exploit perceived brand strength and competitive weakness. Pricing to what the market will bear risks pushing the price too far, thus damaging its perceived value and the long-term potential of the brand. Pricing a brand too highly may create opportunities for lower-priced competition to develop, which may change the category dynamics altogether, gradually cannibalising market share.

The product, including its features, brand name, mode of distribution, and mode of communication, creates the value. The price captures the value. Pricing is a complicated decision, because value can be manipulated. It is, therefore, hard to establish objectively the market price that any product can bear in isolation of other considerations, such as the competitive set of products.

It is sensible to consider that the value of a particular brand to a targeted group of consumers, but equally to be able to ascribe pricing levels in a vacuum or without reference to other products is almost impossible. This stresses the complication of the pricing decision-making process, and the degree of skill and understanding that is required by marketers to take the correct decisions.

Pricing management principles

International Distillers and Vintners (IDV), now part of Diageo, identified for their brands three basic principles that need to be carefully considered before taking any important pricing decisions for a brand:

1. Price elasticity.
2. Price positioning.
3. Perceived value.

Price elasticity

Price elasticity refers to the ratio of percentage change in demand (quantity sold per period) caused by a percentage change in price. It is defined as the percentage increase/decrease in sales volume that would be produced by a 1% decrease/increase in its relative price within the competitive set of the category. Price elasticity is always negative. The higher the price elasticity value, the larger is the response of sales to each 1% change in price. If a brand has a price elasticity between 0 and −1, then demand for the product is said to be 'inelastic' because the percentage change in sales is less than the percentage change in price. If the price elasticity is above −1, then the demand for product is said to be 'elastic' because the percentage change in sales is greater than the percentage change in price.

Determining a product's price elasticity helps to answer the question regarding what happens to category share if the company changes the retail price relative to its competitors. In practice, price elasticity is extremely difficult to measure, and is dependent on a high level of data and relatively sophisticated analysis techniques. However, price elasticity of a brand is a good indicator of its relative strength within the category. Brands with low price elasticity will, by definition, have a high proportion of loyalists amongst consumers and/or have fewer competitors. Brand strength can be monitored by comparing category share and price elasticity. The effectiveness of cumulative marketing effort can be measured by its effect on reducing price elasticity. Price elasticity can also vary according to how the competitive set is defined. This makes it important to define the most appropriate category and competitive set for the brand in any particular market, to ensure that meaningful price elasticity value can be derived.

In general, higher-priced brands and brands that have been developed through consumer marketing-led strategies tend to have lower price elasticity than lower-priced brands and brands that have been developed through sales-led strategies. Successful brands usually have more powerful 'value added' reasons to buy other than just price—image, taste, and convenience, for example.

For most mature brands, elasticity changes slowly over time because brand equity takes time to build and relative prices tend to move slowly within a narrow range. Generally, in mature categories, the price elasticity factor is only disturbed by major external factors such as large duty increases or severe competitive price-cutting. It should also be borne in mind that price elasticity operates at many levels: markets, categories, brands, different product lines, and sizes within a brand all have price elasticities, which may vary considerably.

Price elasticity helps determine whether a price promotion will generate a sufficient increase in sales and profit for the promotion to pay back. If a marketer does not know the price elasticity of the brand, it is unwise to run a price promotion or change its price. It is also important to consider competitors' sensitivity to changes in price differentials, and their likely action, before making price change decisions. Price elasticity analysis can help identify when a brand moves outside its 'acceptable price' band—when it changes its price position to such a degree that consumers re-evaluate their brand choice. The edges of these price bands are sometimes referred to as price thresholds. Even unique brands that may dominate a category have thresholds beyond which consumers stop buying, switch to a competitor, or move outside the category.

Price positioning

This principle relates to pricing a brand at the optimal long-term level in relation to its competitors. Consumer markets tend to segment into distinct price bands, often some distance apart, and each price band is associated with a level of product performance or quality. Depending on the market structure and the price gap between segments, there is scope for positioning products at a different price point within a segment to try and create a competitive advantage.

Taxation, referred to above as a key cost driver, can also create significant distortion on both the definition and spread of price segments. The low price segment often accounts for a larger proportion of the category in higher taxation markets. To retain market share, brands in the standard and premium price segments must manage price relative not only to other brands within their own price segment but also to the cheaper alternatives on offer.

The emergence of own-label brands and low-priced commodity labels has meant that the floor level pricing in the spirits market generally, and Scotch whisky in particular, has dropped significantly in real terms. This phenomenon, sometimes referred to as commoditisation, has an adverse effect on category dynamics, puts prices of standard and premium brands under severe pressure, and destroys economic value within the categories overall. No simple remedy exists for this situation, which any single brand owner is generally powerless to stop. In the long run, the trend towards commoditisation will polarise markets in two directions. Many consumers will seek the reassurance brands with trusted values and continue to support those brands that are perceived to offer sufficient added value. On the other hand, large groups of consumers will move in the direction of cheaper brands and will not display the same level of loyalty. They will be more inclined to switch preference based on small reductions in price.

Once a brand has a well-established positioning image, it becomes difficult fundamentally to change this positioning and adjust relative pricing over time. A consumer's predisposition towards a standard category brand generally creates barriers to trading up to a higher price, even after improving brand benefits. It is also difficult for a brand to trade down into a lower price segment, as short-term volume gains may not generate sufficient incremental profit for this to pay back. There is also the danger that any short-term gain will be offset by long-term erosion in brand image, which would reduce the brand's capacity to sustain its volume in the long-term.

However, a brand may have sufficient scope to move price within its price segment, subject of course to its price elasticity. In general, leading brands within the segment may have the ability to move their relative price up or down within the segment; weaker secondary brands more usually only move price down. Leading brands tend to have lower price elasticities, whereas secondary brands usually do not and the only option that is available to them is to price-down relative to brands in the same price segment. Another common strategy for brands wanting to appeal to a wide target audience is to offer a range of products covering a variety of price points.

Perceived value

It is useful to understand how consumers link the perceived benefits of a brand to its price. Perceived value is a trade-off between perceived benefits and the price the consumer must pay to acquire these benefits. This is a qualitative measure in the sense that it refers to the assessment the consumer makes in the appraisal of the total benefits that are being offered by a brand in relation to the price being asked for. The higher level of price must be balanced in the consumer's mind by a higher level of benefits. Products that offer more benefits will be perceived to deliver more value, and thus improving the benefits that can support a higher price, or a higher share at a lower price. Moving to a higher price without offering more benefits (whether functional or emotional) to compensate will very likely lead to share erosion.

Advertising and promotion

The brand-building process

The marketing mix is rounded out by the fourth 'P', which plays a key role in developing demand. The marketing of alcoholic beverages is no different in this regard to that of other consumable products. Manufacturing a good product and bringing it to market at a fair price is usually not sufficient to maximise the potential of any brand. Normally, every time a brand or category is being developed, the product is being promoted in some way or another. In the case of a Scotch whisky, it may range from a barman recommending the product to customers to a broad-reaching advertising campaign designed to build awareness and product trial. Whatever the means of promotion chosen, the ultimate goal of all brand builders is to maximise the value of their products in the eyes of consumers in order to achieve a share of the market and, ultimately, make a profit.

Marketers use a variety of tools to promote their brands. The overall aim of promotion is to build bonds with consumers by means of communicating messages to them. With consumers, for relationships to be lasting and of value, they generally need to be emotional in nature. The message also needs to convey real values and be communicated in an effective manner. Being sure that the message is targeted at the right audience is part of ensuring the wise use of resources and gaining a positive effect. In summary, with promotion, the key issues that need to be answered are how much to spend and in what way.

So how can promotion be defined? Kotler's definition is that promotion encompasses all the tools in the marketing mix whose major role is persuasive communication (Kotler, 1999). There is a whole raft of tools that fit this definition, but in overall terms, they can be broken down into four broad categories:

1. Advertising.
2. Personal selling.
3. Public relations and publicity.
4. Sales promotion and merchandising.

Each of these basic four components offers a wide panoply of alternative means to communicate with the consumer. 'Above and below the line' is another way to refer to marketing activity; the former refers to advertising, and the latter to all other activities. This section gives only a general overview of each of the elements of the promotional mix.

Promotional activity can cover a very wide range of actions, but the common denominator is that it is normally targeted in nature and is objective-led—i.e., it is aimed at achieving a set of consumer goals. In other words, the purpose is to achieve a behavioural outcome on the part of the targeted audience. Planning activity is an important discipline that all marketers attempt to adhere to, although ad hoc activity can, with luck, yield good results. Having clear consumer goals is vital in order to know what aspect of consumer behaviour the marketer is trying to influence—for example, is it to try and encourage trial, to persuade consumers to adapt the product as one of their main brands, or simply to make the consumer aware of the brand? A structured approach to advertising and promotion that has measurable targets and is properly tracked for effectiveness will help ensure that the expenditure impacts on the consumer. However, it is wise to heed the comment by Lord Lever, the soaps and detergent 'king', who mused that only half the money spent on advertising was effective, and then went on to complain that no one could tell him which half it was!

Advertising

Advertising consists of paid-for messages by an identified party using a wide array of media to reach the target audience. The purposes of advertising are manifold, but can usually be broken down into two underlying objectives: to convey awareness of the product, and to establish a clear image of it in the minds of the target audience. In other words, advertising plays a dual role: informing consumers and getting the brand more 'top of mind', and developing an image of the product. Advertising, very often, plays a critical role in establishing a brand's equity. It can be particularly effective when a brand is differentiated from its competitors on dimensions that are important to consumers. Consistently, advertising a brand's position also serves as a barrier to competitive entry.

The company's approach to planning any communications will be guided by the brand positioning statement and a definition of who is being targeted. Developing this statement will, of course, depend on a discussion on consumer insights, these being the foundation for effective advertising. What consumers believe about a brand and the category is one important element of the consumer's beliefs; the other is how consumers use advertising information to make brand decisions. With effective consumer insights leading to a clear statement of the brand positioning strategy, the marketer can commence developing communication objectives. Leading up to this phase, there will usually be some market research carried out to reveal consumer insights—usually using focus groups, in-depth interviews, and surveys. These lay the foundations for designing creative and media strategies. The company also has commercial goals for the brand in terms of targets for recruitment of new users, on top of the important job of retaining the brand's existing consumer franchise. With a clear idea of what needs to be communicated, to whom, and with what intensity, an agency can be given a detailed brief and can start the process of recommending a communications strategy to fit the overall brand objectives. At this stage, close cooperation is often needed between all the 'actors' involved in the nascent marketing campaign. The company marketing team, advertising agency, and research company team all need to work in a coordinated manner to plan and execute the development of the communications campaign.

The alcoholic drinks industry spends relatively large amounts on advertising. This is not necessarily the case for Scotch whisky, where absolute levels of expenditure have not generally been high and relative amounts have not been great when compared with expenditure on other drinks categories. With few exceptions, drinks advertisers use the media in order to provide messages to consumers about their brands in order to sustain and grow their share of the market. The motive can often be defensive, particularly for brands that are in leadership positions and where consumers need constant reminding of their benefits. Advertising may also be playing a role in the drinks industry in promoting the much talked about trend of consuming 'less but better'. Whatever the case, much emphasis is placed on developing strong brands and on creating the foundations for a product's unique image.

Advertising based on accepted consumer beliefs is generally more effective than advertising requiring consumers to change beliefs. Most often, therefore, advertisers will develop messages that conform to accepted beliefs. If this is not possible because the brand equity is out of line with the consumer's beliefs, the knowledge of the beliefs can be used to develop arguments to change the consumer's disposition. It is also useful to understand how consumers process the advertising messages received. It is commonly realised that they process the information of what is said about the brand in relation to their current repertoire of information about the advertised brand and its competitors.

Brand linkage

The elaboration of a brand's benefits is obviously effective only if it is linked to the brand name. It is clearly useful to introduce information about a brand that is consistent with what people already know about it. An example of this approach was a campaign for Cutty Sark whisky, a brand recognised by drinkers for its tall ship sailing logo. The brand was losing share at one point, and launched a campaign: 'don't give up the ship'. Subway advertisements in New York said: 'When you have had it with graffiti, don't give up the ship'. The slogan was memorable because it was the cry during the Revolutionary War, and also reminded consumers of the brand's symbol. Sales of the brand were seen to recover at that time.

However, brand linkage is difficult to achieve. The reason is that a persuasive campaign has more associations with a person's life than with a brand name. The small number of associations that people generally have in relation to brands makes them hard to remember. On the other hand, brands that are rich with associations in people's minds are easier to recall—for example, Coca Cola for soft drinks and Bacardi for spirits. It is less difficult with these well-known brands for consumers to link the message to the brand, than for products that are less prominent in their daily lives. Brand linkage is even less likely to work when the brand does not have a strong point of differentiation. When several producers use the same messages, the linkage between the brand and message is weakened further.

Creative strategies

The aim is obviously to find the message that will induce the desired action by the consumer. There are a number of creative strategies integral to the delivery of the message that can help consumers evaluate an advertising message. For example, using message discrepancy between the position advocated in a message and people's present beliefs is one potentially strong motivational device. Persuasion is usually only minimal when the communication argues a position people currently already hold and may be only reinforcing people's current beliefs. On the other hand, a message that is very discrepant with the consumer's current views may just prompt disbelief. An approach somewhere in the middle is, therefore, usually preferable.

Threat appeals are another strategy employed, although not all threats will result in a persuasive response. Focusing too hard on dire consequences may just bring about an emotional response. If this approach is to be used, it is important to focus on helping people to recognise danger and be better able to cope with it. More common is to use humour as the means to persuade the consumer. An advantage is that this usually succeeds in getting the consumers' attention and generally also motivates them to process the message. However, there are serious drawbacks, and humour is not always effectively used in advertising. To work to good effect, the humour should generally be related to conveying the brand benefit and ideally be focused on the product itself. If the focus cannot be on the product, it may be better to focus on nonusers than on the actual consumers.

Media strategies

The resources available to run any campaign are always limited, and therefore there must be wise use of media in planning any advertising. Media strategy starts with gaining an understanding of who is the target consumer. In order to establish how to target the consumer, a view also needs to be taken regarding the target's media habits. In overall terms, media selection is a trade-off between reach and frequency. Reach relates to the number of people in the target who have been exposed to a message during the period of the campaign; frequency is the average number of times the target sees the message during the campaign period. For a given advertising budget, increasing one means reducing the other, and this trade-off is a key decision that advertisers must take. There is no evidence pointing to any one correct way to manage this trade-off. More relevant, perhaps, than the amount of times an audience sees one message is the type of exposure.

Krugman (1977) has advocated three types or stages of exposure to advertising:

1. Establishing that a brand belongs to a category.
2. Focusing on the brand's point of difference.
3. Prompting a decision by the consumer.

The overall focus is not so much on the number of exposures that an advertisement is given, but rather on the fact that consumers respond to mounting exposures in different ways. In overall terms, persuasive impact increases with every impact, but substantial repetition prompts a decline in message impact. This is commonly referred to as campaign or message wear-out. To overcome this phenomenon, advertisers commonly use the technique of changing the context but maintaining the central theme. The problem with this approach is that unless the change in context offers consumers new insights, it is hard to sustain their interest, and the message is simply not processed. Wear-out can be forestalled by presenting new information to stimulate the audience to process the message.

Another dimension of the media strategy is when to advertise. Continuity throughout the year is an expensive proposition and is affordable only to the largest brands. On the other hand, brands can use a concentration strategy, focusing the campaign on periods where consumption is historically highest for the category. The danger with this approach is that it can become self-fulfilling in terms of concentrating brand sales even further and exacerbating the seasonality in the brand's sales. Hybrid approaches try and seek the middle ground. Another media strategy is 'flighting', where a campaign is run in short bursts interspersed with periods of silence. The idea with this approach and other variations on the same theme is to develop the impact of concentration along with the sustaining value of continuity.

When dealing with the issue of tight budgets, the advertiser has the choice of narrowing the target or reducing the media used to target the audience. Reducing the geographical scope of the campaign and the times during the year when there is an advertising presence are other simple means of restricting the reach and frequency of the campaign. The normal objective is to have a strong presence amongst as large a proportion of the target audience as is affordable.

Integrated marketing communication

The principle of integrated marketing communications is not a novelty for drinks marketers. This refers to the broad range of communication tools that are available in the armoury of the company to spread their messages to consumers. Apart from advertising, these include public relations, direct marketing, event marketing, and Internet marketing, and all are explored in a little detail below. Of all these tools, whilst advertising is the largest in terms of the total amount spent, it does not always play the leading role. Its role depends on the brand objectives and the audience being targeted. Indeed, there are some well-known and not unsuccessful brands that choose to ignore paid-for media and concentrate on other communications means to get their message over. In spirits, the majority of brands tend to attach importance to advertising because it allows the brand to be brought to the attention of a wide audience, but there are many niche brands, particularly at the premium end of the spectrum in terms of quality, that eschew mass media and concentrate on word of mouth recommendation, perhaps supported by some public relations and event marketing. The overriding principle, normally applied, is that the message about the brand needs to have a consistency across all channels of communication. If an integrated campaign is properly planned and executed, the synergy of the same message appearing in different manners can be highly effective and lead to building intimacy with the customer.

Measuring effectiveness

The last step in the communications process is to measure the effectiveness of the campaign. There is a range of standard measures to be considered, but all ultimately point at the key question of whether the target audience has awareness of the brand resulting from the campaign. Then, more importantly, what does the consumer know and believe about the brand and its personality? This is usually captured by measuring the attitudes of consumers towards the brand, including their likes and dislikes. This is usually in the form of quantitative market research, and through repeated tracking of such variables the marketer starts building a picture of how the attitudes of consumers are evolving through time. The research findings can be useful in picking up warning signals and allowing the messages and appeal to be modified to attempt to improve the position of the brand in the consumer's eye.

Direct marketing

One of the key tools employed by marketers is direct marketing, which has many similar attributes to advertising. It allows the marketer to enter into a direct dialogue with the consumer and to transmit a message. Direct marketing has become a very widely used medium for communication with consumers, and one of the main reasons is that the brand owner can generally target messages with a fair degree of accuracy. Its impact is also often easier to measure. Indeed, the principle of much of direct marketing is that companies identify their target audience and then seek to establish a mailing list to match. Another group that marketers often try to target by way of direct marketing is their existing customer base. Names of a company's customers can be a valuable source of continuing revenues, and customer loyalty programmes can be conducted to maintain interest in the brand high by way of an ongoing dialogue with the consumer.

Direct marketing does have limitations. Its effectiveness has been undermined by the constant barrage of unsolicited materials that consumers receive through their letterboxes and that fall out of their newspapers and magazines. However, the growing use of the medium is testimony to the fact that response rates in the majority of cases more than cover the direct costs of the campaigns, thus yielding profits. In more recent times, a new dimension to direct marketing has emerged in the form of telemarketing. In this case, consumers receive unsolicited calls from brands offering their goods and services.

In the context of drinks, direct marketing can be a useful tool, particularly in the form of targeted direct mail. Premium brands can derive good value from this using this medium, sending promotional literature, perhaps augmented by a special promotional trial offer, on the brand.

Public relations

In the view of some, this is one of the most potent means for transmitting messages about brands. In this case, the objective is to transmit the message to the audience in an indirect manner. Once again, starting with careful consumer targeting allied to achieving clarity with the message, the strategy is to have the message relayed through other organs. A key point of difference with advertising is that the featuring or transmission of the message is not paid for directly. The conduit for relaying the message in general tends to be the media, where the journalists, writers, and other contributors to the media write and/or talk about the product. Of course, a major challenge with all PR is to ensure that the message stays intact and close to the intended line. Getting the message through to the intended audience can be a challenge, given that the transmission of the message is not under the company's direct control. Public relations is a must for most marketers of drinks. One reason is the interest that the sector's products attract from the media. In this context, relationships with the writers need to be well maintained so they are kept abreast of the developments of the company and its brands. Consumers of drinks, and particularly connoisseurs, are avid readers of the newspaper columns, and a brand should aim to obtain a good share of voice in this area and to have the right things said. Negative PR is generally avoidable, provided the brand owner is careful to foster good relationships with the key writers and experts in the field.

Event marketing and sponsorship

This tool can be one of those with the highest impact. The principle involved is to relate the brand to an event that is appropriate in terms of the audience and message that the marketer wishes to convey. For example, to promote new-age ready-to-drink spirits, brands may look for events that are attended by a younger audience and that have a fun and social aspect. At the heart of a good event and a sponsorship campaign is the skill of carefully matching the brand personality to the event. Failure to do this can lead to sending misleading messages about the brand to the consumer, whereas if the association is strong, an element of synergy can be built and such promotions can run effectively over long periods.

The sporting world has always been an area targeted by drinks brands. Obviously, the brands are attracted to the audiences, particularly those who offer a good match with the consumer profile of the brand. An example of this is the long-standing sponsorship of Scottish rugby by Famous Grouse. In this case, the relationship was prosperous for the brand by association. The effect transmitted a clear message in terms of the Scottish credentials of the brand, as if it had been chosen on its merits. The message reached audiences outside the home country and assisted in building the awareness and credentials of the brand more widely. The leading Scotch brand worldwide, Johnnie Walker, has in a similar fashion stamped its mark on the sport of golf in a consistent manner over the years, thereby using the sport to transmit a constant message about the prestige and quality of the brand to a global audience.

Digital marketing

Digital marketing continues to evolve and brands are still experimenting with finding the optimal uses afforded by the online medium. The digital revolution is providing opportunities to interact with consumers in a different way. It is a rapidly changing medium, reflecting the increasing use and evolving behaviour of consumers across multiple and connected platforms. We have transitioned from website and email marketing to social media and digital consumer engagement at a very rapid rate. A simple banner ad is no longer sufficient to engage consumers. Social media websites such as LinkedIn (founded 2002), Facebook (founded in 2004), YouTube (founded in 2005), Twitter (founded in 2006), and Weibo (launched in 2007), etc., have helped us change in terms of how we interact with consumers.

The pace with which digital media options come and go requires that marketers focus on developing strategies to manage their brands within a rapidly changing landscape in order to deliver ever more efficient and effective campaigns. They must keep themselves up-to-date with the current tools and maintain a flexible mindset on how they will approach digital marketing.

For example, consumers today use their mobile phones more frequently than any other devices with the majority connecting to the mobile internet. This trend has led to marketers increasing investment across not only mobile apps, but also ensuring that brand websites are responsive and adaptive to mobile screens. With increased mobile advertising, the frequency of smartphone use, along with that of other devices, provides vast amounts of data available to marketers to identify and deliver tailored and more engaging consumer experiences.

Never before has so much content been created and shared across the digital landscape; it is estimated to have grown ninefold globally in the last 5 years. In this environment, marketing professionals are creating and optimising high quality digital content. In addition, they must seek out ways to distribute this. Social Media also continues to evolve at pace. The traditional big players such as Facebook continue to grow, with 1.2 billion active monthly users (January 2014). However, the mobile messenger and chat apps are also expanding rapidly, impacting consumer behaviour across traditional social media channels. Against this background, marketing teams continue to increase social media advertising, as well as monitor growing mobile and local social media and messaging channels, in order to reach and engage consumers. New and innovative technology also continues to be introduced quickly and is then refined. This revolution constantly presents marketers with opportunities to add value in different environments, whilst being aware of the risks of engaging consumers through new technology.

These examples illustrate the extent to which modern marketers need to adapt to keep up with consumers changing use of technology and online platforms. It highlights the need to remain up-to-date and maintain a flexible and innovative mindset towards digital marketing. This is crucial to find new ways and reasons to influence time-poor, savvy consumers to engage with brands, over competitors.

Conclusion

Advertising is an inexact science, as is marketing overall. Intuition plays a large role in finding the messages that build consumer appeal and create brand equity over the long run. Planning tools are also there to help make rational choices and approach decision-making in a disciplined and structured manner. In the final analysis, the impact of advertising depends to a large extent on whether it resonates with the consumer. Indeed, much of the impact of advertising works through self-persuasion. Highly effective advertising will tend to build a strong link between the brand and the perceived benefit to the consumer, and this is often achieved by clearly distinguishing the brand from its competitors using a unique style of creative execution. With alcoholic beverages, there are limitations on the mediums for advertising, and care needs to be exercised to respect both the legal framework and any voluntary codes of conduct. Common sense is required to ensure that the messages are responsible and do not mislead in any manner. The question of how to publicise the message, or of the choice of media, is another important step in planning the communications strategy for the brand. There are a myriad of ways and means for conveying the message, and this represents a vast specialised discipline in itself.

Market research

In today's world, the combination of strong competition and scarce resources means that advertising and promotion need to be approached with a high degree of professionalism. This has led to the emergence of the market research industry, where the positioning of products is exhaustively explored to find optimum strategies. All the elements of the promotional mix are also put through tests to ascertain whether they are likely to produce the intended results. This might, for example, include establishing what image an advertisement for a particular brand creates in the minds of all targeted consumers. Communications are often pretested in 'rough' before large budgets are spent on producing the final advertisement. The answer to these questions and others will indicate to marketers if the message is in line with the positioning statement, and give them the confidence to commit expenditure of scarce financial resources. On the other hand, a poor diagnosis sends out warning signals and allows the marketer to go back to the drawing board and amend any aspect of the marketing mix. In all of this, it is worth stressing that market research should not be a substitute for common sense. Sometimes, the solution is obvious and it is not necessary to incur valuable resources in terms of time and money to conduct long pieces of research. It is particularly important to interpret any data coming from market research studies very carefully, to avoid falling into misinterpretation traps.

Used properly, market research can be a very powerful tool with which all aspects of the brand development process are explored. The usefulness of research can range from the basic diagnosis of consumer needs to mapping out potential new product development strategies, and more specific testing of campaign material for a well-established brand. It is usual to distinguish between two general types of research: exploratory and evaluative. Assembling the right team of experts to conduct any form of research and interpret the results is a good starting point. Also, it is vital to give the team a carefully prepared and thorough written brief. Without a clear understanding of the background and the objectives of the research, money can easily be wasted and, in extreme cases, misleading findings can emerge that give spurious information leading to wrong decisions. If research indicates that consumers prefer brand X over brand Y, it does not always lead to the former being more successful, because perhaps there may be other deeper factors motivating brand selection that the research has overlooked and that may not have been volunteered by the research sample. Another classical trap in market research is to have an ill-defined target that results in sampling consumer groups who are not representative of the actual target audience.

General conclusions

The Scotch marketer of today must, at all times, remain focused on the needs, desires, and preferences of the consumer. It also helps to be aware of the broader alcoholic beverage market in which producers are competing, and to recognise that Scotch whisky is just one of the many choices from which the consumer can select. With the eye firmly on the consumer during all stages of the marketing process, the chances of success and continuing prosperity are greatly increased. The customer is said to be 'king, queen and master', and is on the lookout for consumer trends and meeting consumers' needs is the task of every marketer. Therein lies the route to long-term profitability and success.

Acknowledgements

Thanks and grateful appreciation for all their efforts are due to Jonathan Driver who gave me full access to his library, and to Tim Dewey and Laurence Miklichansky-Maddocks from William Grant & Sons, who read my script and brought many valuable insights that otherwise would have been missing. Last but not the least, acknowledgement is due to The Worshipful Company of Distillers, a guild that is dedicated to supporting education and the development of knowledge within the distilled spirits profession. By virtue of The Distillers links with the International Centre of Brewing & Distilling at Heriot-Watt University, the author was given the opportunity to write this chapter.

These words are dedicated to tomorrow's professionals who will choose to make their careers within our industry, and as a reminder to them that 'customers are scarce, not products' (Kotler, 1999).

References

Kotler, P., 1999. Marketing Management: Millennium Edition. Prentice Hall College Division, Upper Saddle River, NJ.

Krugman, H.E., 1977. Memory without recall, exposure without perception. J. Advert. Res. 17 (4), 7–12.

Recommended reading

Calkins, T., Tybout, A., (Eds.), Kotler, P. (Foreword), 2005. Kellogg on Branding, first ed. Wiley, USA.

Feldwick, P., 2002. What Is Brand Equity, Anyway? World Advertising Research Centre, (WARC), Henley-on-Thames p. 160.

Iacobucci, D. (Ed.), 2000. Kellogg on Marketing: The Kellogg Marketing Faculty Northwestern University, first ed. John Wiley & Sons, New York, NY, p. 427.

Kotler, P., Armstrong, G., 2013. Principles of Marketing, fifteenth ed. Prentice Hall.

Tybout, A.M., Sternthal, B., 2010. Developing a compelling brand positioning (Chapter 4). In: Tybout, A.M., Calder, B.J. (Eds.), Kellogg on Marketing, second ed. Wiley, New York, NY.

Further reading

Coughlan, A.T., Stern, L.W., 2000. Market channel design and management. In: Iacobucci, D. (Ed.), Kellogg on Marketing. Wiley, UK (Chapter 11).

Fortenberry, J.L., 2010. Calder & Reagan's brand design model. In: Health Care Marketing: Tools and Techniques, third ed. Jones & Bartlett Learning, USA, pp. 62–68 (Chapter 9).

Marketing whisky and white spirits in 2021

Julie Kellershohn

Ted Rogers School of Hospitality and Tourism Management, Ryerson University, Toronto, ON, Canada

Introduction

In this chapter, the goal is to provide an overview of the marketing of whisky and some white spirits, with a focus on the changes that have occurred in the marketplace over the past few years. Chapter 28, 'Marketing Scotch Whisky' by Grant Gordon, from the second edition of this book has been included in this third edition without revision. It captures the classic basic marketing principles in a fashion that has withstood the passage of time. This chapter builds on the marketing information from Chapter 28 by addressing the rise in the use of social media for marketing, mobile and other technologies, e-commerce, and the changes in consumer purchase habits arising from the COVID-19 pandemic.

Marketing in 2021 has proven to be an evolving challenge for the potable alcohol industry. The COVID-19 pandemic has brought with it changes in both consumer perceptions and purchase venues. Social media and influencers, the demand for greater transparency in the product production chain, and the desire for innovative new products must all be considered when planning how to market spirit products. In addition, consumers are demanding more from companies, not only due to the consumers' changing views on the social responsibility of companies as it pertains to the sustainability of the planet, but also due to the expectation that companies will use technology to provide better consumer experiences, such as easy e-commerce options.

E-commerce has been growing exponentially due to the COVID-19 pandemic, as consumers are now routinely online and discovering convenience, speed, and the option of doorstep delivery. The concept of 'direct-to-consumer' is increasingly becoming the new normal, as the retail industry has pivoted in response to industry challenges. Home drink delivery for alcoholic beverages during the time of COVID-19 restrictions has become a trend that is not likely to change in the future, with almost 70% of consumers surveyed indicating that they will continue to consider this option post-pandemic. Environmentally friendly packaging is also on consumers' minds, resulting in innovations in certain packaging formats. Packaging formats such as wine in a pouch, or wine in a can, are gaining increasing consumer acceptance when accessed through the direct-to-consumer approach. This suggests that growth in more delivery friendly package formats will likely continue.

This chapter uses examples of products on the current market to illustrate the changing environment in the promotion, purchase, and consumption of alcoholic beverages, with the caveat that in a short period of time, these examples will be superseded by new innovations and technology changes.

Global market size and demographics

The global alcoholic spirit market in 2019 was valued at 524 billion USD, and it is forecasted to reach 709 billion USD by 2028. In 2019, 71% of spirit consumption was at-home and 29% was consumed out-of-home (Statista, 2020).

The spirit segment of the overall alcoholic beverage market revenue in 2019 was 30% of the worldwide total, with the largest components being beer at 40%, wine at 23%, and a segment of 7% that includes cider, perry, and rice wine.

In 2020, due to the COVID-19 pandemic, there was a major shift in online food and beverage delivery, including spirits. A 2020 survey of respondents, who indicated that they had used food and drink delivery between March and May of 2020, showed an increase in food and drink delivery, with 49% of respondents in China reporting having used food and drink delivery and the United States and the United Kingdom with 26% and 25% delivery, respectively (Statista, 2020). Alcoholic drinks accounted for total revenues of over 1.6 trillion USD in 2019, but a sharp decline is expected in the out-of-home segment in 2020/21 due to the pandemic-induced closing of restaurants, hotels, pubs, etc.

Consumers post-2021

What will the 2022 post-pandemic consumer be looking for? Emerging markets in China and India are projected to be the key growth drivers due to growing disposable incomes and premiumisation, which is a trend expected to continue with traditional spirit brands.

Changing demographics will be an important driver in the global alcohol market. One of the fastest-growing demographics is the population over the age of 55. In many countries, alcohol consumption often declines in this demographic after peaking in the younger years. Cutting back on consumption is not limited to an ageing population. There is also a change in terms of more consumers looking for a healthier lifestyle as well as generational shifts in attitudes to alcohol. Millennials, in particular, have a heightened awareness of the importance of a healthy lifestyle.

Social media platforms

Today, over 4.5 billion people use the Internet, and active social media users number over 3.8 billion. In 2020, Facebook was the largest social media platform globally, with 2.6 billion monthly active users worldwide, 1.7 billion daily active users, and with over 98% of users accessing the site via mobile phone. This was followed by YouTube at 2 billion users, and WhatsApp, Facebook Messenger, WeChat, and Instagram, all having over a billion users. However, the newcomer site of TikTok, launched in 2016, by early 2021, already had over one billion users and is one of the fastest-growing social media platforms, starting out with a teen audience but now with a growing number of adult users (around 30% of users are in the age range of 20–29) (Wallaroomedia, 2021).

Social media can connect with key target consumers due to the increase in mobile device use, in a way that traditional advertising could not be targetted as easily in the past. In Q4 of 2019, over 96% of advertisements were viewed on mobile devices, with Facebook receiving 58% of the relative ad spend, followed by the Instagram feed/stories that received 28% of the ad spend. Women (age 25–34) were the largest group engaging with brand pages on Facebook. By Q4 of 2019, Instagram's total audience surpassed that of Facebook, and more importantly, there was higher engagement with posts on Instagram (Socialbakers, 2020a).

In Q4 of 2020, the 50 biggest brand portfolios' total audience size was 39% larger on Instagram than on Facebook. Engagement was significantly stronger on Instagram, which had 21 times more interactions than on Facebook. While alcohol interactions on Facebook overall decreased by 33.7%, on Instagram, little change was noted (Socialbakers, 2020b).

Social media is a very powerful way to connect to key consumers, and Instagram stories are an especially important way to communicate. Tap back (where consumers rewatch an Instagram story) was highest in the categories of e-commerce and all beverages (a median rate of 2.72%), with alcohol at 2.66%. Good storytelling combined with social media marketing, when engaging enough, has consumers choosing to rewatch the advertising (Socialbakers, 2020a,b).

Social media platforms are where Millennials like to document their lifestyles, thereby reinforcing current trends. They embrace third-party recommendations, value brand authenticity, and dislike traditional advertising. Social media marketing is one of the most effective ways to reach Millennials. Millennials check their smartphones on average over 43 times a day, making this a key route to capture their attention through advertising in various ways, such as peer reviews and social media influencer recommendations (SocialToaster, 2020). Millennials use social media both for its entertaining content and for keeping up with friends' and families' lives, while Generation Z, the future generation of alcohol consumers, have a greater focus on social media as a way to access entertaining content and following online celebrities and content creators (YPulse, 2021). Millennials are more likely than Gen Z to be following small business and big brands, and 57% of Gen Z say they are using social media while second screening at the same time (for example, watching TV while using social media) (YPulse, 2021). It will be harder for brands to connect with Gen Z on social media and convince them to 'follow', as there is more competition to access Gen Z's divided screen time.

Influencer marketing

Influencer marketing uses endorsements and product mentions from influencer individuals. These are individuals with dedicated social media followers who have built a high amount of trust and interest from their followers. When they recommend a product, it acts as social proof for the brand's potential customer. For example, the Tequila brand Casamigos, founded in 2013 by the actor George Clooney, nightlife entrepreneur Randy Gerber, and real estate mogul Mike Meldman, is an excellent example of the power of influencer marketing. The emphasis on the product was that it was not just another spirit brand but a lifestyle label as well. This image is bolstered by the original ownership team, who today still play an active role in the brand's development and marketing, even though Casamigos was purchased by Diageo in 2017. The original owners developed their product with a focus on authenticity, quality, smoothness, and a specific flavour profile. They positioned it as a product created by friends for friends. With over 7000 organic and sponsored mentions, their sponsored posts have the potential to reach as high as 935,000 users (Upfluence, 2020).

Instagram is the most popular platform used by alcohol influencers; however, influencers are often active on more than one platform (Sproutsocial, 2020). The Instagram audience is mostly in the age group of 25–34 and is 50.8% female to 49.2% male. Micro-influencers (with under 15,000 followers) have the highest engagement rate with brands. However on YouTube, the more heavily male-oriented influencers are the macro- and mega-alcohol influencers (500,000 to over a million) that generate the most views. TikTok currently has fewer alcohol influencers, but they have the largest number of followers, and on TikTok, there is a much higher engagement rate than on the other platforms.

The cancel culture

Today, one of the marketing concerns on social media is the rise of, and harm that can occur, to a product or brand due to 'cancel culture'. This is the practice of withdrawing support after a person or company has said or done something, often on social media, considered to be objectional or offensive (Forbes, 2020). With folks spending more time online due to pandemic lockdowns, the impact of cancel culture, especially where influencers have offended, can have catastrophic effects on a brand.

When brands do not deliver on their promises, they can be called out by informed consumers, whose voices can be quickly magnified through social media. A global survey by Edelman (2018) reported that 64% of consumers would buy or boycott a brand solely based on its position on a social or political issue. This indicates that it is now more important than ever for a brand to build trust with consumers through authentic interactions in order to build long-term sustained growth (Forbes, 2021).

Watchdogs on the brand

In 2021, due to COVID-19, there has been a change in how companies are perceived, and the growing consumer expectation is that companies should protect the health and interest of both society and the planet. Environmental damage is top of mind for many consumers, and they want to know if companies are committed to environmental protection and the details of what they are doing in terms of reducing emissions.

Bulleit Frontier Whiskey, which already had a long legacy of supporting sustainability, recently teamed up with the US non-profit organisation American Forests to plant one million white oak trees over the next 5 years. Planting 2,000 acres of forest across the Eastern United States will help to restore the white oak population (white oak is used in bourbon barrels) and illustrates the company's commitment to sustainability and to fighting climate change with reforestation (Bulleit, 2020).

Packaging materials are undergoing extensive changes in order to become more light-weight. Innovative bottle designs are reducing the weight of glass used in the bottle, and there is reduced plastic use in new closure types. Recycled glass, aluminium caps (rather than plastic caps), and paper labels (rather than PVC) are all now much more important to the consumer and are becoming the mainstream approach. Consumers are paying attention to brands that act in transparent and authentic ways that assist consumers to achieve more sustainable lifestyles.

An example of using sustainability and packaging as a marketing tool is the award-winning Bullards Spirits' packaging innovation. Although the invention of a bottle that is of size and material that will go through the mail slot of a UK postbox was launched several years ago, this concept has been taken further by the UK company, Bullard's Spirits, with the brand's eco-refill pouch, which holds 700 mL of gin. The consumer only needs to buy the glass bottle of gin once, and then with the eco-pouch, they can decant the gin into the original bottle, and more pouches can be ordered online to refill the bottle. What is innovative about the gin pouch is that the empty pouch can be put into any UK postbox without the need of a stamp or envelope, and it is returned to the distillery for careful recycling. By returning it to the distillery, the consumer is assured that the plastic will be appropriately recycled (since recycling plastic can be problematic), thus giving the consumer an extra level of trust, along with the convenience of ordering the eco-pouches for home delivery and the ease of recycling (Bullards, 2020).

From traditional marketing to virtual and augmented reality marketing

Visitor centres and tasting tours

Visitor centres, tasting tours, and road trips are effective vehicles to tell a brand's product story to a large number of interested consumers. It helps them form a connection to the brand through the experience of tours, which offer insights into how the product is made and what is unique about the particular product.

In the United States, the famous Kentucky Bourbon Trail Tour was founded in 1999 as a road trip-style experience, and later, the Kentucky Bourbon Trail Craft Tour was added. Combined, the two tours have now attracted over 2.5 million visitors in the past 5 years. Visitors have the option of visiting just a few distilleries or all 38 on the various tours (Kentucky Bourbon Trail, 2020). This provides an experiential journey that creates a deeper emotional connection and leads to brand loyalty.

In Europe, the Guinness Storehouse visitor centre in Dublin, which opened in 2000, has hosted over 20 million visitors, with 1.7 million in 2019. The tour tells visitors the story of Guinness and gives them a taste of the product. To evolve and enhance the experience for visitors, in 2018, they introduced an innovative experience for Instagrammers called a 'Shout*ier*'. As a part of the 'Shout*ier*', the visitor's selfie photo is imprinted in amazing detail on the foam head of a pint of Guinness (a natural malt extract is used to create the design with printing technology from Ripples), for the price of only six Euros. This personalised technology has now been rolled out further to other Guinness locations with over 50,000 unique 'Shout*ier*' pints poured to date. This provides the customer with a unique premium experience that many love to share on social media (Ripples, 2020).

Scotland, with its five distinct whisky regions and whisky trails, each producing unique products to that region, has a total of 133 malt and grain distilleries, making the region the highest concentration of whisky production in the world (SWA, 2020). There are 2.2 million yearly visits to Scotch whisky distilleries, making these visits the third most popular tourist attraction in Scotland. There is no lack of choice with the famous malt whisky trail (through the Speyside region, which is home to the largest concentration of malt whisky distilleries) and the Hebridean whisky trail (to the wild Atlantic islands of Skye, Raasay, and Harris). In 2019, Scotch whisky accounted for 75% of Scottish food and drink exports at a value of £4.9 billion (~5.8 billion USD) (SWA, 2020). With the consumer faced with a plethora of choices, it can be challenging as a marketer to help an individual distillery make itself stand out and be memorable!

With the lockdown due to the pandemic, creativity and speed of implementation have been vital to survival. An example of this creativity is the whisky distiller WhistlePig and Flaviar, who held a 'blend your whisky at home' event in the United States, which later won them a 2020 Spirits Business Award (The Spirits Business, 2020a). The event allowed thousands of whisky fans to create a new whisky expression together (a crowd blending event) in the midst of the pandemic with a four-step process: a blend at home step with a prior ordered blending kit, a favourite recipe feedback, an online live blending event, and then a final vote to choose a winner. The winning blend (rye 45%, wheat 30%, and barley 25%) was named 'Homeblend', and it was released with the tagline, 'Blended Together, While Apart'. The event also included a charitable component tied to it that benefited a COVID-19 emergency assistance programme.

During the pandemic, consumer surveys indicated that brands that donated a portion of their revenue to charity were highly regarded, and 62% of consumers stated that they were interested in 'drinking for a cause' (Datassential, 2020). Finding new ways to engage with the customer in a way that is entertaining and educational while also addressing a social need during the time of pandemic restrictions is clearly valuable.

Selling with storytelling

Selling with storytelling is a very effective and popular marketing technique. A recent survey shows that over 56% of global consumers indicate that stories about a brand influence their purchase decisions (Innova, 2020). With the strong interest in a company's position and work on sustainability and social responsibility, a well-crafted story can reassure the consumer while conveying this information, weaving it together with the product's history and ingredients, thereby building trust in the brand. Awards for green practices are an opportunity for marketing to consumers who are concerned with the environment.

An excellent example of this type of storytelling is the Nicaraguan premium rum company Flora de Caña, a global spirit brand. This product is the world's first spirit to be carbon neutral and Fairtrade certified. They hold two sustainability certifications—for being carbon neutral and for fair trade. In the production of its rum, all carbon emissions are offset during the rum's lifecycle (field to market), and it is produced in compliance with over 300 labour, social, and environmental standards (Flora de Caña, 2021). Their marketing includes: the touchstones of 0 g sugar, gluten-free, and no artificial ingredients—these labels are sought out by many consumers. They have won numerous prizes in spirit competitions, lending credibility to their product's quality. They have a 130-year-old family story that they tell about their company, which makes it clear that being a good corporate citizen is not something that is new for them, but rather a long-term commitment and a part of the company's heritage.

In storytelling, customers will seek out authenticity, and small can be an advantage. An excellent example of the use of storytelling is by a boutique distillery in Dublin, located in the historical district called 'The Liberties', where once there were 40 distilleries operating. While most visitor centres now use technology to enhance visits, this centre has a very different focus and in 2019 won the award for the best Dublin tourist experience in their size category, after only being open for tours since 2017. The distillery was built within a heritage site—the former St. James Church at a refurbishment cost of over 20 million USD and features custom-designed stained-glass windows and a beautiful lit glass spire dome. For the owners, Dr. Deirdre Lyons and the late Dr. Pearse Lyons, renovating and designing the site and turning it into an operating distillery, which is also a visitor centre, was a labour of love!

What makes this distillery visit unique? A visit to the historic St. James Church site is all about telling a personalised and oral story to the visitors (Pearse Lyons Distillery, 2021). Distillery tours are kept very small, never rushed, and include a visit to the adjacent graveyard (where the owner's grandfather is buried). The tour tells the story of the whisky, the story of the Liberties, the story of the church, and the story of the graveyard—and it becomes a very intimate and enjoyable experience, which then ends with a tasting of the highly rated products of the distillery. The tour touches the brain, the palate, and the heart, with a unique and memorable story (Fig. 29.1).

Millennials value experiences, and as a result, visits to distilleries, with tours and tastings, are very popular. In 2020, virtual online food tours, with pre-ordered samples (e.g., cheese, chocolate, olive oils, and wines), filled that consumer void when travel was not an option. Distilleries also started to offer virtual tasting events where consumers could pre-buy the drinks, have them shipped to their home, and then participate in an online event. This allowed the distillery host to tell the story of their product and guide consumers through the tasting along with the other participants—all online and with a global reach, limited only by where the shipping rules allowed for delivery of the product. Although these events were started to address the issue of lockdowns, events such as these will likely continue into the future as experiential events, since participation is only limited by where the product can be shipped in order to join the event from your own home, attracting a much larger potential market.

Wild Turkey, a Kentucky Bourbon brand, has introduced a new way to connect with its customers using storytelling and offering virtual bourbon tastings leveraging the technology of Alexa and Google Assistant (Fig. 29.2). Already over 25% of US households currently have at least one smart speaker in their homes (that figure is projected to grow to 75% by 2025), so it is an easy step for a customer to obtain the information just by saying, 'Alexa, enable Wild

FIG. 29.1   (A) The Pearse Lyons Distillery built in the historic St. James Church site in Dublin. (B) One of the custom-designed stained-glass windows in the distillery. Photographs by Donal Murphy reproduced with permission.

Turkey Tasting' or to Google Assistant, 'Hey Google, talk to Wild Turkey Tasting'. The audio tour uses the voices of the legendary distilling family—the Russells, to share their heritage and knowledge, along with tall tales about the brand. They share humorous anecdotes while teaching about different styles of Wild Turkey whisky, the distilling process, the palate, and a guided tasting (Wild Turkey, 2021).

FIG. 29.2 Wild Turkey bourbon offered virtual tastings, using technology consumers already had at home. For example, a consumer's Google Assistant could be activated with the phrase 'Hey Google, talk to Wild Turkey Tasting'.

Augmented reality (AR) and virtual reality (VR) tools

Augmented and virtual reality tools were already growing in popularity pre-COVID-19, but during the pandemic, with folks unable to travel, they have become an even more popular way to explore and highlight spirit products.

Augmented reality (AR) technology should not be confused with virtual reality technology (VR), as virtual reality technology requires immersive headgear. Virtual reality technology immerses the consumer in a digital world, with 360-degree views and with little or no input from the physical world they are in. Augmented reality technology overlays virtual 3D graphics onto the consumer's real world (augmenting) and has a much lower entry barrier, typically requiring only the download of an app onto the consumer's smartphone.

Patron Tequila has been a marketing leader in virtual reality marketing, with the use of Oculus VR headsets, which allow the viewer to take the perspective of a bee flying through the site of tequila production in Mexico. The blending of storytelling with the latest technology of virtual reality devices allows the viewer to feel as though they are an active participant instead of just a passive viewer, making the experience more compelling (Patron Tequila, 2017).

Virtual reality (VR) and augmented reality (AR) are immersive technologies that bring a new dimension and perspective to storytelling, virtual tours, local promotions, etc. They add increased depth to the customer experience. An example of the innovative use of web-based augmented reality is the BON V!V Spiked Seltzer advertising campaign, which featured out-of-home augmented reality advertising. From a printed mural in the street, a passer-by could scan the QR code to activate the experience. Once activated, using web-based augmented reality, the person could see a fully functioning virtual vending machine in front of them, from which they could see a selection of flavoured seltzers, pick their favourite, and have it dispensed in 3D animation. Once the product was virtually dispensed, the consumer was then offered a custom map to navigate them to the nearest store that offers the product or a link to purchase it online (Aircards, 2020).

Augmented reality is not a new concept. It became mainstream with the global mobile game phenomenon, Pokémon Go, whose app has now been downloaded over one billion times worldwide. Augmented reality experiences can engage massive audiences. By 2022, the number of smartphone augmented reality users is expected to exceed 3.5 billion users (~44% of the world's population) (Digi-capital, 2020). Augmented reality technology has surpassed virtual reality technology as it requires no special set-up or hardware such as the headgear associated with virtual reality. With the increasing computing power of smartphones and the widespread use of filters and animations, the ability to overlay the geo-specific content on live streams offers many new opportunities to create a unique content as real worlds merge with digital worlds.

What are consumers looking for?

Omnichannel shopping

Although the terms multichannel and omnichannel are often used interchangeably, there are significant differences. Multichannel is the historical interaction with customers on multiple channels such as email, social media, and websites, but they are not tied together for a consistent experience. A true omnichannel experience uses multiple marketing channels but creates a single-user experience whether they are shopping in-store, through the website, or using a mobile device. Amazon's omnichannel approach allows the customer to access its site through various devices. Amazon understands that today's consumer changes devices throughout the day, whether at work on a desktop or commuting and using their smartphones. Amazon's goal is to make it simple to place, track, and arrange the delivery of orders. The goal is a consistent approach, provided in a flexible and convenient manner to the customer (CM Commerce, 2020).

Omnichannel shopping has now become entrenched in the behaviour of consumers due to extended stay-at-home periods. Before COVID-19, only 9% of global consumers were regularly shopping online. By May of 2020, 44% of global consumers indicated that they were shopping online each week, and many of these consumers were new to online shopping (Nielsen, 2020). Many retailers have removed or lowered the delivery cost, which was an early hurdle to widespread e-commerce adoption. Due to COVID-19, there has been an unprecedented adoption of online shopping, but the consumer's expectations of the experience focuses more and more on ease of use, product availability, and delivery times. The consumer expects a seamless experience across online and offline shopping.

Although shopping for spirits during the pandemic has accelerated digital adoption, care must still be paid to the needs of customers of different ages and what the customer considers to be optimal. While a younger consumer may prefer a total digital interaction, an older consumer may still want to talk to a human service representative. Chatbots had already been rising in use prior to the pandemic, but the adoption of chatbots has since accelerated, especially in terms of conversational artificial intelligence (AI) abilities. This was one way for industry to handle the sudden high increase in customer questions during the pandemic. It is predicted that by 2022, 70% of white-collar workers will interact with a virtual agent or chatbot on a daily basis, in particular the Millennial cohort (Gartner, 2019). It is easy to see how the chatbot experience can significantly impact a customer's shopping experience.

Consumers are looking for facts and expect marketing materials to dispel misinformation and offer truths. Transparency will continue to be crucial, and technology that better informs consumers needs to be embraced. To receive personalised offers and deals, 37% of consumers already willingly share some of their personal data. Consumers will continue to look for convenience, increased online options, and fast delivery, requiring companies to be much more flexible and agile with their marketing strategies (Euromonitor, 2021).

Novel tastes and new combinations

Consumers are looking for innovation, novel experiences, and unique tastes. Some products on the market address this by infusing classic products with new flavours, such as whisky blended or aged in casks used previously for other alcoholic products. This has led to the launch of new products such as Pernod Ricard's beer-infused Irish whisky. Other examples of new product blends include Absolut Juice by Pernod Ricard USA, a blended vodka with juice and natural flavours, positioned as a beverage for earlier in the day occasions such as brunch. This product taps into the move to lower alcohol products (at 35%) and low calories at 99 calories per serving, and 5 % juice (described as perfect for mixing with soda) (Fig. 29.3).

Jameson Cold Brew, a combination of Jameson Irish whisky and cold brew coffee, is another example of a blending of popular beverages for different occasions. With a lower alcohol content of 30% ABV (vs regular 40% for Jameson whisky) and caffeine at 17 mg per shot of whisky, they have entered the flavoured spirit market, addressing the segment of consumers who are seeking new taste experiences, while remaining with familiar brands that they trust (Fig. 29.4).

Buying local

Consumers are increasingly looking to support products made locally. With the 'buy local, drink local' movement, food explorers need more local ingredients, and they also need a story that helps them connect with the ingredients. A Canadian example is Northern Landing's Cranberry Ginberry, which uses cranberries (not a typical gin ingredient) sourced locally in

FIG. 29.3   New product blends combine classic products with new flavours, such as the Absolut Juice (Apple Ed.)—a vodka blended with juice.

Muskoka, Canada (Toronto's cottage country). This red gin product stands out due to its colour. It claims all-natural ingredients, no added sugar, no artificial colouring, or flavours. Most importantly, it is marketed as 'local' for the highly populated area where the distillery is located. Distinguishing a product in a crowded marketplace requires creativity and close attention to what today's consumers value, many of whom are seeking new tastes and unique experiences.

Premiumisation

There has been increasing premiumisation of spirit brands in recent years as consumers place increased value on certain spirits' production processes and origins. However, during the pandemic lockdowns of bars and restaurants, as consumers were no longer able to access traditional luxuries, many opted to splurge on premium and ultra-premium spirits. At-home indulgences became important, and consumers looked for superior quality or exclusivity. With the 'healthier for you mentality', many consumers are choosing a 'drink less, but better' approach. Premium spirits, often considered an affordable luxury during times of recession, provide an upscale feeling. In China and India, the popularity of the cocktail culture is expected to rise, and this will drive the purchase of premium spirits as consumers continue to search for authenticity and status.

FIG. 29.4   Familiar brands extend their product offers through the introduction of new blended flavours, such as the Jameson Cold Brew, a combination of Jameson Irish whisky and cold brew coffee.

Ready-to-drink (RTD) cocktails

The ready-to-drink (RTD) cocktail consumer is looking for drinks that offer convenience. There has been a steady growth in the RTD category, but the consumer's desire for RTD cocktails accelerated with the closing of bars and restaurants during the pandemic. RTD cocktails presented a convenient in-home solution.

Much of the growth in RTD cocktail drinks has been captured by the large traditional distillers. Simple RTD cocktail combinations, such as Brown-Forman's Jack Daniel's & Cola (sold in 330 mL cans at 5 % alcohol content), boasted sales of 13 million 9-L cases in 2019—a 51.5% increase over the previous year, making it the second bestselling RTD brand in the world (The Spirits Business, 2020b). The Kahlúa Espresso Style Martini product includes a packaging innovation in that it produces a nitro foam when the can of the triple ingredient 'vodka-Kahlúa-coffee' drink is opened (sold in 200 mL cans at 4.5% alcohol). NOVO FOGO's Sparkling Caipirinha cocktail contains Brazilian cachaça, mango, and lime (sold in 200 mL cans at 8.5% alcohol). From old local favourites to new international flavour experiences—the selection of RTD options is enormous. Drinks that contain interesting ingredients in single-serve cans are ideal for food and drink explorers. However, in the RTD market, it is hard to differentiate, and creative marketing of the product becomes key.

The rise of White Claw—A case study

Millennials in 2019/2020 were looking for 'better-for-you' options in alcoholic beverages. Lower calories, lower alcohol content, an emphasis on high-quality ingredients, and the convenience of canned beverages all offered extra appeal.

The rise of White Claw as the largest selling new beverage introduction is worthy of study. The hard seltzer category showed phenomenal growth in 2020. It was the fastest-growing menu item—topping all other beverages. For hard seltzers at the end of 2019, three brands dominated (White Claw, Truly, and Bon&Viv) with over 80% of category sales. White Claw held 50% of the share, and its viral appeal on social media platforms led to high growth rates driven by Millennials and Gen X (Fig. 29.5).

Mark Anthony Brands' White Claw hard seltzer was first launched during 2016 in the United States, and by 2019, it was the top-selling RTD product on the market. In one year, it grew its volume from 6.9 million cases in 2018 to 27.5 million cases in 2019, a 298% increase, making it the fastest-growing RTD brand. White Claw is classed as a 'hard seltzer', usually defined as a highball drink of carbonated water, alcohol, and often fruit flavouring. It attracted both genders equally, especially in the age group of 21- to 44-year-olds (Nielson, 2020). The drink's popularity is driven by its appeal to casual drinkers who do not enjoy strong alcohol drinks, consumers who are looking for a lower alcohol option, and those

FIG. 29.5   White Claw—a sparkling hard seltzer, a market leader with strong social media appeal.

who are spirit enthusiasts but looking for something lighter. Even beer drinkers, looking for something different, have joined the trend of drinking hard seltzers. Beer companies, such as Boston Beer and Anheuser Busch, responded to this trend by quickly bringing their own seltzers to market. In 2019, White Claw was the nation's number one seltzer, generating about $1.5 billion in sales with an over 40% seltzer market share, and it is now the top-selling hard seltzer brand globally.

The marketing of White Claw was aimed at Millennials, who were embracing the wellness culture and 'better-for-me' products. White Claw offered a product that was low in calories (~ 70–100 calories), 3.7% alcohol, and gluten-free, was packaged in convenient cans, and offered flavour variety, making it a good fit for this market segment.

It was positioned as a gender-neutral product that one would enjoy with friends. The gender positioning, combined with its aspirational brand marketing and an affordable price point, made for compelling advertising on Instagram, Facebook, and Twitter. However, much of its social media success was due to free social media marketing in the summer of 2019, with events and memes popularising the drink. All of this attention reinforced the drink's popularity, leading to a nationwide shortage. The scarcity of the product then made it even more desirable. Its trend-based viral marketing campaign successfully captured the minds and hearts of the social media-driven, 'better-for-you' culture.

The future

There is little doubt that the pandemic will leave a changed consumer world for spirit marketing. With the growth of online sales, increasing use of mobile phones for obtaining instant information (on everything from a product's history to a company's values), and the expectation of convenience both in packaging and rapid delivery, this will continue to direct how spirits are marketed in the future. Rapid advances in technology, especially in terms of what is possible with advances in augmented reality technology and other new technologies on the horizon, will forever change the mode of shopping for spirits, bringing the option of purchasing products instantly to consumers browsing on mobile phones and other electronic search devices.

Links to websites

Aircards, 2020. https://www.aircards.co/blog/unforgettable-ar-out-of-home-advertisements.

Bullards, 2020. https://bullardsspirits.co.uk/ecoproject/.

Bulleit, 2020. https://www.multivu.com/players/English/8793351-bulleit-frontier-whiskey-american-forests-one-million-trees/.

CM Commerce, 2020. https://cm-commerce.com/academy/ultimate-list-of-omnichannel-marketing-examples-statistics/.

Datassential, 2020. https://datassential.com/keynote-reports/.

Digi-capital, 2020. https://www.digi-capital.com/news/2020/08/the-ar-vr-ecosystem-are-we-there-yet/.

Edelman, 2018. https://www.edelman.com/news-awards/two-thirds-consumers-worldwide-now-buy-beliefs.

Euromonitor, 2021. https://go.euromonitor.com/white-paper-EC-2021-Top-10-Global-Consumer-Trends.html.

Flora de Caña, 2021. www.flordecana.com.

Forbes, 2020. https://www.forbes.com/sites/kianbakhtiari/2020/09/29/why-brands-need-to-pay-attention-to-cancel-culture/?sh=452ac53645e8.

Forbes, 2021. https://www.forbes.com/sites/augustinefou/2021/02/03/the-disconnect-between-brand-advertisers-and-consumers-has-never-been-wider/?sh=f951e622ab9c.

Gartner, 2019. https://www.gartner.com/smarterwithgartner/chatbots-will-appeal-to-modern-workers/.

Innova, 2020. https://www.innovamarketinsights.com/.

Kentucky Bourbon Trail, 2020. https://kybourbontrail.com/.

Nielsen,   2020.   https://www.nielsen.com/ca/en/insights/article/2020/covid-19-has-flipped-the-value-proposition-of-omnichannel-shopping-for-constrained-consumers/.

Nielson,   2020.   https://nielseniq.com/global/en/insights/analysis/2020/hard-seltzer-defies-categorization-and-limits-as-the-most-resilient-alcohol-segment-in-u-s/.

Patron Tequila, 2017. https://www.prnewswire.com/news-releases/patron-tequila-unveils-the-patron-experience-one-of-the-first-brands-to-create-a-hand-held-augmented-reality-innovation-300522885.html.

Pearse Lyons Distillery, 2021. https://www.pearselyonsdistillery.com/.

Ripples, 2020. https://www.drinkripples.com/success_stories/guinness-bar-pilot. https://www.drinkripples.com/blog/new-drinkx-podcasts-reveal-the-digital-transformation-of-the-beverage-industry.

Socialbakers, 2020a. https://www.socialbakers.com/web-api/wp/study/webinar-q42020-trends?studyId=32713.

Socialbakers, 2020b. https://www.socialbakers.com/web-api/wp/study/social-media-trends-report-q4-2020?studyId=32489.

SocialToaster, 2020. https://www.socialtoaster.com/tips-marketing-to-millennials/.

Sproutsocial, 2020. https://sproutsocial.com/insights/influencer-marketing/.

Statista, 2020. https://www.statista.com.

SWA, 2020. https://www.scotch-whisky.org.uk.

TheSpiritsBusiness, 2020a. https://www.thespiritsbusiness.com/2020/12/the-spirits-business-awards-2020-winners/.

TheSpiritsBusiness, 2020b. https://www.thespiritsbusiness.com/2020/06/top-10-fastest-growing-spirits-brands-5/7/.

Upfluence, 2020. https://www.upfluence.com/industry-reports/alcoholic-beverages.

Wallaroomedia, 2021. https://wallaroomedia.com/blog/social-media/tiktok-statistics/.

Wild Turkey, 2021. https://www.prnewswire.com/news-releases/plan-a-virtual-escape-as-wild-turkey-transports-travel-seekers-and-bourbon-enthusiasts-to-kentucky-with-its-new-guided-tasting-for-amazon-alexa-and-google-assistant-301219531.html.

YPulse, 2021. https://www.ypulse.com/article/2021/02/22/gen-z-millennials-use-social-media-differently-heres-x-charts-that-show-how/.

Index

Note: Page numbers followed by *f* indicate figures, *t* indicate tables, and *b* indicate boxes.

Printed in the United States
by Baker & Taylor Publisher Services